Smart Innovation, Systems and Technologies

Volume 273

The Smart Innovation, Systems and Technologies book series encompasses the topics of knowledge, intelligence, innovation and sustainability. The aim of the series is to make available a platform for the publication of books on all aspects of single and multi-disciplinary research on these themes in order to make the latest results available in a readily-accessible form. Volumes on interdisciplinary research combining two or more of these areas is particularly sought.

The series covers systems and paradigms that employ knowledge and intelligence in a broad sense. Its scope is systems having embedded knowledge and intelligence, which may be applied to the solution of world problems in industry, the environment and the community. It also focusses on the knowledge-transfer methodologies and innovation strategies employed to make this happen effectively. The combination of intelligent systems tools and a broad range of applications introduces a need for a synergy of disciplines from science, technology, business and the humanities. The series will include conference proceedings, edited collections, monographs, handbooks, reference books, and other relevant types of book in areas of science and technology where smart systems and technologies can offer innovative solutions.

High quality content is an essential feature for all book proposals accepted for the series. It is expected that editors of all accepted volumes will ensure that contributions are subjected to an appropriate level of reviewing process and adhere to KES quality principles.

Indexed by SCOPUS, EI Compendex, INSPEC, WTI Frankfurt eG, zbMATH, Japanese Science and Technology Agency (JST), SCImago, DBLP.

All books published in the series are submitted for consideration in Web of Science.

More information about this series at https://link.springer.com/bookseries/8767

Fausto Pedro García Márquez
Editor

International Conference on Intelligent Emerging Methods of Artificial Intelligence & Cloud Computing

Proceedings of IEMAICLOUD 2021

Springer

Editor
Fausto Pedro García Márquez
Ingenium Researh Group
University of Castilla-La Mancha (UCLM)
Ciudad Real, Spain

ISSN 2190-3018 ISSN 2190-3026 (electronic)
Smart Innovation, Systems and Technologies
ISBN 978-3-030-92904-6 ISBN 978-3-030-92905-3 (eBook)
https://doi.org/10.1007/978-3-030-92905-3

This Springer imprint is published by the registered company Springer Nature Switzerland AG
The registered company address is: Gewerbestrasse 11, 6330 Cham, Switzerland

Preface

The International Conference on Intelligent Emerging Methods of Artificial Intelligence and Cloud Computing 2021, IEMAICLOUD conference, has been designed to help research organizations and business leaders from across industries transform their organizations into AI-driven disruptors. The utility of the technology in the face of massive globally interconnected complexity will be explored. The significant characteristics of IEMAICLOUD is the promotion of inevitable dialogue between scientists, researchers, engineers, corporates and scholar students to mitigate the gap between academia, industry and governmental ethics which will be fostered through keynote speeches, workshops, panel discussions and oral presentations by eminent researchers in relevant fields. The industry personnel depict cutting-edge researches in Artificial Intelligence and Cloud Computing to convey academia regarding real-time scenarios and practical findings. Conference will be well equipped with talks by industry experts on the state of the art in computer science, lectures by eminent scientists designed to inspire and inform, and presentations by innovative researchers coming from 20+ countries, from Europe and abroad. Moreover, IEMAICLOUD facilitates better understanding of the technological developments and scientific advancements across the world by showcasing the pace of science, technology and business areas in the field of Artificial Intelligence and Cloud Computing.

Ciudad Real, Spain

Fausto Pedro García Márquez
FaustoPedro.Garcia@uclm.es

Introduction

The International Conference on Intelligent Emerging Methods of Artificial Intelligence and Cloud Computing 2021, IEMAICLOUD conference, has been designed to help research organizations and business leaders from across industries transform their organizations into AI-driven disruptors. The utility of the technology in the face of massive globally interconnected complexity will be explored. The significant characteristics of IEMAICLOUD is the promotion of inevitable dialogue between scientists, researchers, engineers, corporates and scholar students to mitigate the gap between academia, industry and governmental ethics which will be fostered through keynote speeches, workshops, panel discussions and oral presentations by eminent researchers in relevant fields. The industry personnel depict cutting-edge researches in Artificial Intelligence and Cloud Computing to convey academia regarding real-time scenarios and practical findings. Conferences has been considered talks by industry experts on the state of the art in computer science, lectures by eminent scientists designed to inspire and inform, and presentations by innovative researchers coming from 20+ countries from Europe and abroad.

Moreover, IEMAICLOUD facilitates better understanding of the technological developments and scientific advancements across the world by showcasing the pace of science, technology and business areas in the field of Artificial Intelligence and Cloud Computing.

IEMAICLOUD 21 is a four dais International Conference specially designed with a cluster of scientific and technological sessions, providing a common platform for researchers, academicians, industry delegates, scientists, professionals, industry and government policy-makers, engineers and students across the globe to share and exchange their knowledge and contribution. This conference attends to emerging areas of research and development related to the new challenges presented by the COVID-19 pandemic, focusing specifically upon aspects and applications of Cloud Computing, Artificial Intelligence, Big Data and the Internet of Things.

The conference is equipped with well-organized scientific sessions, keynote and plenary lectures, research papers, poster presentations and world-class exhibitions.

This conference also has some special sessions related to innovative applications with Artificial Intelligence, Internet of Things and Cloud Computing, and a demo-session for innovations achieved through prototyping and simulation.

Advisory Committee

- Fausto Pedro Garcia; University of Castilla-La Mancha, Spain
- Christopher Nugent; Ulster University, UK
- KokkonenTero; JAMK University, Finland
- Harry Agius; Brunel University, UK
- FerranteNeri; Nottingham University, UK
- ShahramLatifi; University of Nevada, Las Vegas, USA
- Liz Browne; Oxford Brookes University, UK
- Francesco Colace; Universitàdegli Studi di Salerno, Italy
- Salah Al- Majeed; University of Lincoln, UK
- PeriklisChatzimisios; International Hellenic University (IHU), Greece
- Shuo Wang; University of Birmingham, UK
- Shiyan Hu; University of Southampton, UK
- Kirk Martinez; University of Southampton, UK
- Tony Prescott; The University of Sheffield, UK
- Larisa Soldatova; Goldsmith University of London, UK
- Juan M. Corchado; University of Salamanca, Spain
- Ranga Rao Venkatesha Prasad; Delft University of Technology (TU Delft), Netherlands
- MonomitaNandy; Brunel University, UK
- PietroOliveto; University of Sheffield, UK

Organizing Committee

- Rachel Kent, Kings College London, UK (General Chair)
- SatyajitChakrabarti, Smart Society, USA (General Co-Chair)
- Fausto Pedro Garcia, University of Castilla-La Mancha, Spain (Technical Chair)
- Sunday Ekpo, Manchester Metropolitan University, UK (Publicity Chair)
- Yulei Wu, The University of Exeter, UK (Outreach Chair)

Keynote Speaker

- Cristiano Paggetti, Prof. of Digital Health at the School of Computing at Ulster University, Belfast

- FerranteNeri; Nottingham University, UK
- Harry Agius; Brunel University, UK
- Rachel Kent; Kings College London, UK
- Larisa Soldatova; Goldsmith University of London, UK
- Liz Browne; Oxford Brookes University, UK
- Tony Prescott; The University of Sheffield, UK
- Md. Hossein Zoualfaghari, Ph.D., MEng, MIET, Research Manager, IoT Architect & Technical Lead, Applied Research, BT Technology, UK
- J. Mark Bishop, Scientific Advisor Fact360 and Professor of Cognitive Computing; Goldsmiths, University of London, UK
- PietroOliveto; University of Sheffield, UK
- Dr. Jian-Guo Zhang, Senior Lecturer; London South Bank University, UK
- Dr.-Ing. DetlefStreitferd; Technical University Ilmenau, Germany
- Professor HuanshengNing; University of Science and Technology Beijing
- Ranga Rao Venkatesha Prasad; Delft University of Technology (TU Delft), Netherlands

Fausto Pedro García Márquez

Contents

About the Editor

Fausto Pedro García Márquez Ingneium Research Group, University of Castilla-La Mancha, Spain

Fausto works at UCLM as Full Professor (Accredited as Full Professor from 2013), Spain, Honorary Senior Research Fellow at Birmingham University, UK, Lecturer at the Postgraduate European Institute and he has been Senior Manager in Accenture (2013–2014). He obtained his European Ph.D. with a maximum distinction. He has been distinguished with the following prizes: Runner Prize for Management Science and Engineering Management Nominated Prize (2020), and Advancement Prize (2018), First International Business Ideas Competition 2017 Award (2017); Runner (2015), Advancement (2013) and Silver (2012) by the International Society of Management Science and Engineering Management (ICMSEM); Best Paper Award in the international journal of Renewable Energy (Impact Factor 3.5) (2015). He has published more than 150 papers (65 % ISI, 30% JCR and 92% internationals), some recognized as follows: "Applied Energy" (Q1, as "Best Paper 2020"), "Renewable Energy" (Q1, as "Best Paper 2014"); "ICMSEM" (as "excellent"); "International Journal of Automation and Computing"; and "IMechE Part F: Journal of Rail and Rapid Transit" (most downloaded). He is author and editor of 31 books (Elsevier, Springer, Pearson, Mc-GrawHill, Intech, IGI, Marcombo, AlfaOmega,...), and 5 patents. He is Editor of 5 International Journals, and Committee Member of more than 40 International Conferences. He has been Principal Investigator in 4 European Projects, 6 National Projects and more than 150 projects for Universities,

Companies, etc. His main interests are Artificial Intelligence, Maintenance, Management, Renewable Energy, Transport, Advanced Analytics and Data Science. He is being an expert in the European Union in AI4People (EISMD) and ESF. He is Director of www.ingeniumgroup.eu.

Mathematical Modelling to Predict Fuel Consumption in a Blast Furnace Using Artificial Neural Networks

Wandercleiton Cardoso[1](✉), Renzo di Felice[1], and Raphael Baptista[2]

[1] Genoa University, Via all'Opera Pia, 9, 16145 Genoa, Italy
wandercleiton.cardoso@dicca.unige.it
[2] ArcelorMittal Tubarão, Av. Brig. Eduardo Gomes, 526, Serra/ES 29160904, Brazil

1 Introduction

The iron reduction process is ancient and started in an artisanal and empirical way and today it is consolidated in a strong steel industry, with an important economic role in the global production chain [1].

The main form of processing is the reduction of iron ore by the blast furnace and although the principles of pig iron production are the same as those of a century ago, technology and the understanding of how the blast furnace works have evolved a lot from that time to the present day, ally with technological advances that allowed important evolutions in the monitoring of the process and procedural changes, such as, for example, the injection of fines, the use of top gases and the use of slag as a by-product for different applications [2]. Considering the importance of the pig iron production stage, for the production chain, there is a need for studies and the search for tools that can optimize or help control the process of obtaining it and consequently the production costs [3].

The Blast Furnace uses the counter current principle, where rising gases react and transfer heat to descending solids and liquids. Figure 1 illustrates the operation process of a blast furnace.

In the field of simulation of complex processes, the application of solutions based on neural networks has gained space due to its versatility of application and possibility of development and increased reliability of responses as the neural network receives new data in the training process [4].

In this scenario, the possibility of developing a model based on a set of neural networks presents itself as an interesting alternative, being possible to use it in different blast furnaces, adapting only the database to be adopted for network learning neural network to be used [5, 6].

Currently, applications using neural networks have been gaining ground due to the countless possibilities of use with positive results, above all, in the problems of greater complexity, and it presents as a possibility their use as committee machines, which are constituted by the combination of the result of several neural networks acting separately for modeling the process [7].

Thus, considering the importance and complexity of operation of the blast furnace, combined with the potential reached by neural networks in process optimization, the

F. P. García Márquez (ed.), *International Conference on Intelligent Emerging Methods of Artificial Intelligence & Cloud Computing*, Smart Innovation, Systems and Technologies 273, https://doi.org/10.1007/978-3-030-92905-3_1

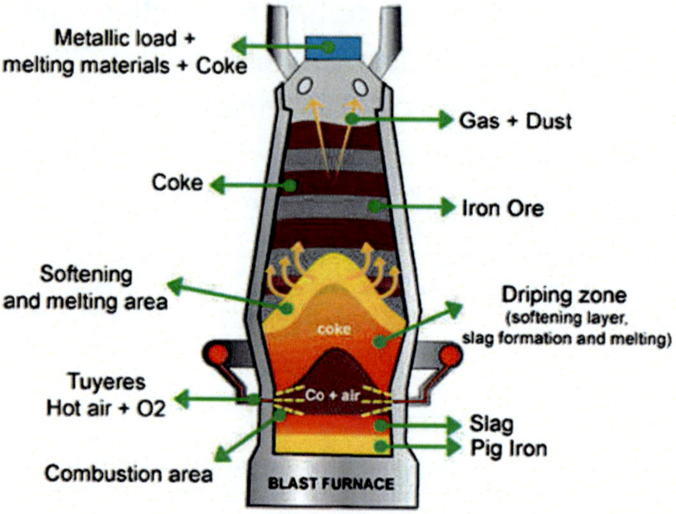

Fig. 1. Blast furnace operation

purpose of this paper is to develop a source code based on artificial neural networks in the format of a committee machine to perform the monitoring the operation of the blast furnace and the prediction of results related to fuel consumption.

2 Materials and Methods

2.1 Artificial Neural Network Configuration

Artificial neural networks are characterized as artificial intelligence techniques inspired by the structure of the human brain, simulating mathematical operations in computer systems in an efficient and simplified way.

Artificial neural networks perform three essential operations: learning and storing knowledge, applying acquired knowledge to solve proposed problems and acquiring new knowledge of constant learning.

The artificial neuron is the basic processing element of an ANN, that is being formed by a set of input connections (xj), synaptic weights (Wkj), where, (k) is the number of input neurons and (j) is the input stimulus, and the bias (bk) which is a weighting parameter that increases or decreases the value of the linear combination of inputs of the neuron activation function (f.).

Figure 2 below illustrates the simplified model of an artificial neuron, that presents a simplified model of an artificial neuron, where (uk) represents the linear combination of the input signal sand (yk) corresponds to the output value of the neuron.

Thus, the input weighting process represents the learning rate attained/achieved by an ANN. The weights are adjusted as an entry data set and presented to the network. The supervised learning process in an ANN is based on the adjustment of synaptic weights so that the output value is as close as possible to the expected value.

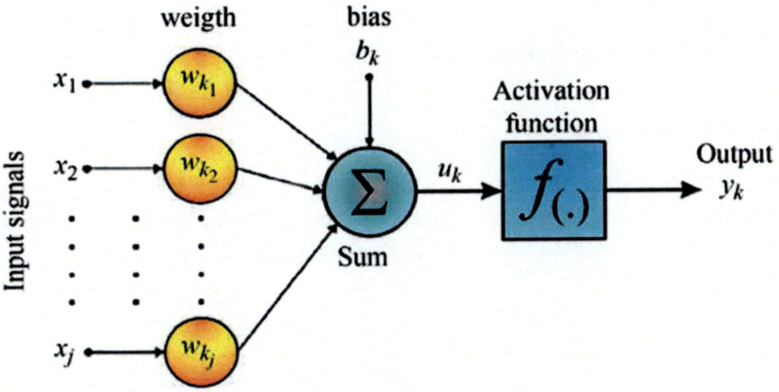

Fig. 2. Architecture of an artificial neural network

The purpose of the activation function ($f_{(.)}$) is to limit the network input signals to a specific range, usually ranging from (0 to 1) or (−1 to 1). It generates an output neuron based on the input values (x_1, x_2, x_j) of the neural network and the adjusted synaptic weights. The most used functions for engineering research are the linear, log-sigmoid and tan-sigmoid functions.

Artificial neural networks have unique and specific arrangements and characteristics that adjust to the type of problem to be solved and may have a single hidden layer or in several layers. In architecture, multilayer perceptron (MLP), the artificial neural network is composed of multilayers with nonlinear activation of the sigmoidal type in hidden layers giving the network a genuinely nonlinear mathematical modelling.

The Levenberg–Marquardt algorithm is an optimization of the ANN application which uses an iterative numerical optimization technique, capable of locating the minimum of a function expressed as the sum of squares of other non-linear function. The Levenberg–Marquardt backup is an adaptive network that uses the Jacobian matrix for calculations that assume This research, estimated the quality of the chemistry composition of the blast furnace slag used in cement production by developing different ANN's models.

The number of neurons in the mid-layer is usually empirically defined and depends on several factors, such as: (a) number of training examples; (b) amount of noise in the examples; (c) complexity of the function to be learned; (d) statistical distribution of training data.

In most metallurgical problems, the architecture of an ANN is generally obtained by trial and error, however, a single hidden layer is sufficient to approach any continuous function. The number of hidden neurons (h) in a single-layer network depends on the number of input variables (I), as shown in Eq. 1:

$$H < (2I + 1) \tag{1}$$

The simulations were performed in MATLAB R2020b environment, using the "nftool" toolbox and the Levenberg–Marquardt training algorithm, so, neural networks were trained with a single layer, with the number of neurons varying (10, 25, 50, 75 and 100) and the use of 18 input variables with a log sigmoid activation function in the hidden layer and the linear function in the output layer.

2.2 Data Normalization

Data normalization is used to model the database, organizing how information will be stored in order to eliminate or minimize redundancies during mathematical modelling. The normalization of the data input was performed in MINITAB software according to Eqs. 2 and 3:

$$delta_i = \frac{(\max - \min)}{\max\_input_i - \min\_input_i} \tag{2}$$

$$norm\_input_i = \min - (delta \times \min\_input_i) + (delta \times input_i) \tag{3}$$

The initial pre-processing step consists of normalizing the variables used, avoiding numerical problems during the training phase and to improve the performance of the backpropagation algorithm. Considering the variety of the data, the data normalization was performed aiming at the optimization of the results and the reduction of the convergence time of the model.

The pre-processing step consists of normalizing the data input variables that was used to match the order of magnitude of the input variables between 0 and 1, to avoid numerical problems during the training phase and to improve the performance of the backpropagation training algorithm.

2.3 Data Collect

The data set used in this research comes from the operation of blast furnace 1 of the company ArcelorMittal in Brazil, which has an average daily production of 7200 tons. The operational data correspond to 150 records (150 days of operation) related to the average daily values of operation of 19 input variables and 2 output variables.

During the period selected for data collection, the metallurgical reactor operated without major operational variations and practically the same reactivity and resistance value as the coke loaded in the furnace. The input variables used in the model are illustrated in Table 1.

The output variables analysed in the mathematical model were blast furnace fuels: Coke Rate, PCI Rate and Fuel Rate. Table 2 illustrates the descriptive statistics of the 150 records in the model.

Pulverized Coal Injection (PCI) through blast furnace nozzles is widely used by steelmakers around the world to reduce the consumption of coke loaded from the top of the furnace, and thus the cost of pig iron. The fuel rate is calculated by Eq. 4 illustrated below:

$$\text{Fuel Rate} = \text{Coke Rate} + \text{PCI Rate} \tag{4}$$

In this research, the "fuel rate" output variable was not calculated by the neural network, being obtained indirectly using the equation above.

Table 1. List of model input variables

Variable	Unit	Minimum	Maximum	Mean
Pellet	kg/t	633.8	876.5	754.4 ± 62.4
Sínter	kg/t	604.4	867.2	754.4 ± 50.2
Iron ore	kg/t	12.4	102.1	37.1 ± 27.4
Dolomite	kg/t	6.9	7.9	7.1 ± 4.6
Slag basicity (B2)	%	1.14	1.26	1.20 ± 0.02
Slag basicity (B4)	%	1.01	1.10	1.06 ± 0.02
Carbon of pig iron	%	4.5	4.9	4.7 ± 0.1
Pig iron temperature	C	1431.3	1539.1	1502.7 ± 18.8
Blowing flow	Nm^3/min	6505.3	7093.7	6884.9 ± 105.5
Coke ash content	%	6.7	9.3	8.9 ± 0.9
Coke moisture	%	2.5	5.7	3.9 ± 0.7
Nitrogen	Nm^3/t	4619.9	4702.5	4658.7 ± 11.1
Oxygen flow	Nm^3/t	1501.0	1606.8	1557.0 ± 53.1
Oxygen enrichment	%	3.2	4.9	4.1 ± 0.9
Flame temperature	C	1200	1206	1203 ± 2.4
Airspeed tuyère	m/s	214.4	227.6	221 ± 21
Permeability	–	3.92	4.42	4.21 ± 0.21
Daily production (slag)	ton	1536.3	2240.8	1840.2 ± 221.7
Daily Production (pig iron)	ton	6231.4	7775.0	7283.1 ± 696.7

Table 2. List of model output variables

Variable	Unit	Minimum	Maximum	Mean	Standard deviation
Coke rate	kg/t	275.2	347.8	299.6	12.1
PCI rate	kg/t	156.9	225.5	199.1	14.2
Fuel rate	kg/t	472.5	540.0	495.7	11.1

2.4 Cross-Validation

Cross-validation is a technique used to assess the generalizability of a model, based on a set of data. This technique is widely used in problems where the purpose of modeling is prediction.

In this research, the sample was partitioned into subsets and later used to estimate the model parameters (training, validation and test data). Table 3 illustrates the division of variables for the construction of the mathematical model:

Table 3. Division of variables

Step	Variables
Training	73
Validation	16
Test	16
Cross-validation	45
Total	150

The neural network was built using 105 variables (training, validation and testing) and 45 variables were used to assess the model's responsiveness (cross-validation).

3 Results and Discussion

The development of a mathematical model that can predict the fuel consumption of a blast furnace is not easy, therefore, the choice of the best estimation method to be used suggests the use of statistical techniques to evaluate the efficiency of the artificial neural network, as well, in this research the efficiency of the mathematical model was evaluated through the values of regression and linear correlation.

The performance is evaluated by using the correlation coefficient for equations (R), calculated as follows in Eq. 5.

$$R = \sqrt{\left[\sum_{i=1}^{n}(C_{neural} - C_{real})^2\right] \Big/ \left[\sum_{i=1}^{n}(C_{real} - \overline{C}_{neural})^2\right]} \tag{5}$$

where (n) is the number of observations, (C_{neural}) is the value calculated by the artificial neural network, and (C_{real}) is the value measured.

In general, it aims to assess the relationship between the number of observations and the value calculated by the artificial neural network, from (n) observations, indicating how much the independent variable can be explained by the fixed variable.

Statistical analysis has been carried out using the Minitab statistical software and as proposed in the methodology, Fig. 3 illustrates the linear correlation rate considering the variation in the number of neurons (10, 25, 50, 75 and 100).

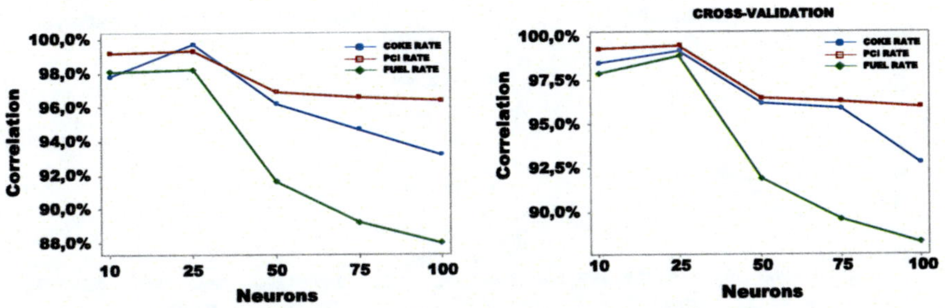

Fig. 3. Performance of the neural network considering the variation of neurons

In Fig. 3 on the left it is possible to evaluate the result of the neural network with 10, 25, 50, 75 and 100 neurons that was modeled using 105 variables (training, validation and test) and we concluded that the neural network that presented the best results was with 10 and 25 neurons.

When we analyzed the results of the neural network with 10 and 25 neurons, we concluded that the performance of the neural network with 10 neurons was 0.2% lower when compared to the neural network with 25 neurons.

As the number of neurons increased, performance decreased, and the neural network with 100 neurons showed the worst result (10% lower) when compared to neural networks with 10 and 25 neurons.

In Fig. 3 on the right, it is possible to evaluate the result of cross validation with 10, 25, 50, 75 and 100 neurons. In this step, we used 45 groups of variables that were not used to model the neural network.

When the 45 groups of variables were tested to calculate the prediction performance of the neural network, we obtained results similar to the initial one with an average standard deviation of 0.2% approximately.

The artificial neural network with 25 neurons showed the best results of linear correlation during the initial mathematical modeling (training, validation and testing) and also during the cross validation.

In this way, Figs. 4, 5 and 6 present the results of the initial mathematical modeling and cross-validation of the artificial neural network with 25 neurons of the output variables: Coke Rate (Fig. 4), PCI Rate (Fig. 5) and Fuel Rate (Fig. 6) and Figs. 7, 8, and 9 show daily behavior of the neural network with 25 neurons.

Analyzing Figs. 4, 5, and 6 on the x-axis, the real variables are presented, that is, the variables of the data collection step as proposed in the methodology while on the y-axis the neural variables are presented, that is, the respective values calculated by the neural network artificial. The figures on the left show the results of the initial modeling (training, validation and testing) using 105 variables (105 days of production) while the figures on the right show the results of cross-validation using 45 variables (45 days of production), totaling so 150 days of operational production.

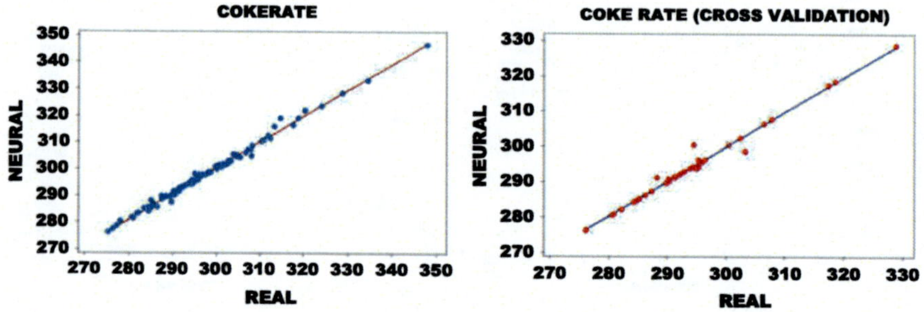

Fig. 4. Behavior of the Coke Rate variable in the artificial neural network with 25 neurons

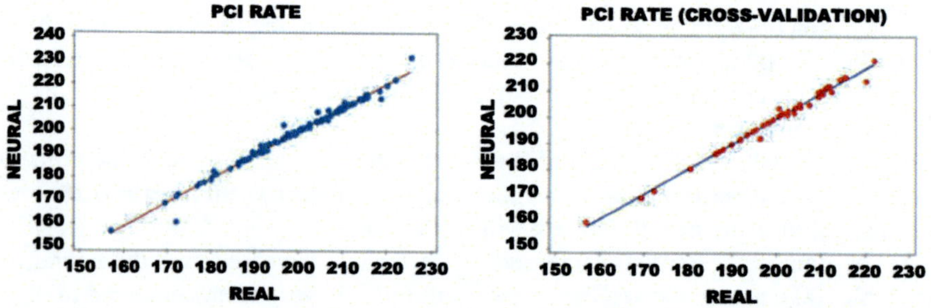

Fig. 5. Behavior of the PCI Rate variable in the artificial neural network with 25 neurons

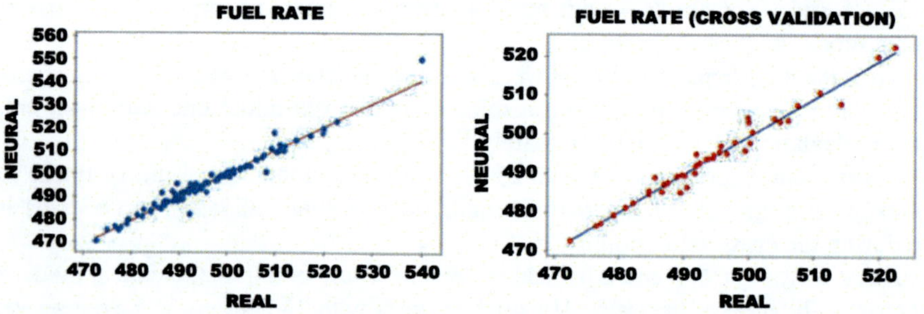

Fig. 6. Behavior of the Fuel Rate variable in the artificial neural network with 25 neurons

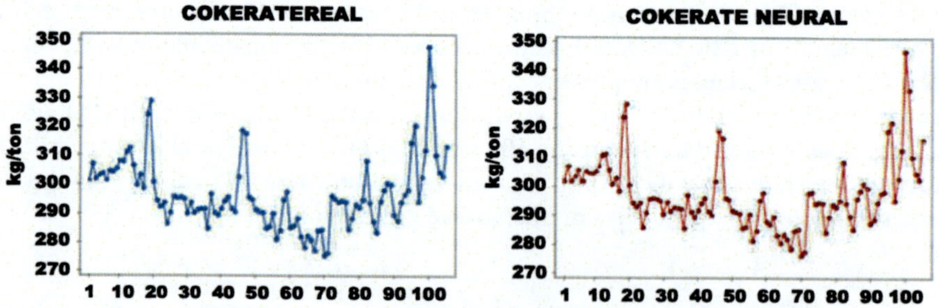

Fig. 7. Daily behavior of the output variable with 25 neurons: Coke Rate

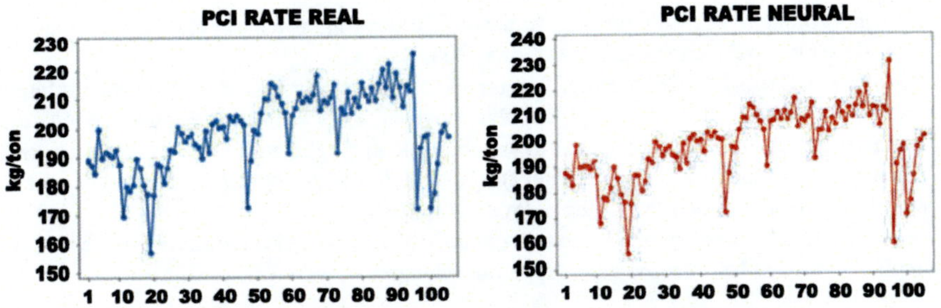

Fig. 8. Daily behavior of the output variable with 25 neurons: PCI Rate

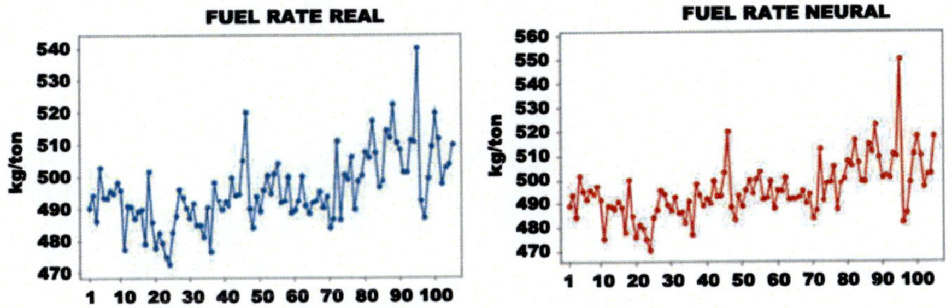

Fig. 9. Daily behavior of the output variable with 25 neurons: Fuel Rate

4 Conclusions

Could be concluded that the neural model is a useful tool for supporting an iron blast furnace operation since some corrections and retraining are carefully carried out by expert human operators in a systematic basis.

Concluding, the high values of correlation mathematical show the good statistical performance of ANN and It shows that the mathematical model is an effective predictor of blast furnace fuel consummation. The results obtained in conjunction with the cross-validation of the data demonstrate the ANN's ability to generalize the acquired knowledge.

References

1. D. Fontes, L. Vasconcelos, R. Brito, Blast furnace hot metal temperature and silicon content prediction using soft sensor based on fuzzy C -means and exogenous nonlinear autoregressive models. Comput. Chem. Eng. **141** (2020)
2. H. Tang, J. Li, B. Yao, et al., Evaluation of scheme design of blast furnace based on artificial neural network. J. Iron Steel Res. Int. **15**, 1–36 (2008)
3. A. Kandiri, E. Golafshani, A. Behnood, Estimation of the compressive strength of concretes containing granulated blast furnace slag using hybridized multi-objective ANN. Constr. Build. Mater. **248** (2020)

4. J. Chen, A predictive system for blast furnaces by integrating a neural network with qualitative analysis. Eng. Appl. Artif. Intell. **14**, 77–85 (2001)
5. I. Matino, S. Dettori, V. Colla, et al., Two innovative modelling approaches in order to forecast consumption of blast furnace gas by hot blast stoves. Energy Procedia **158**, 4043–4048 (2019)
6. J. Zhang, L. Shucai, L. Zhaofeng, Investigation the synergistic effects in quaternary binder containing red mud, blast furnace slag, steel slag and flue gas desulfurization gypsum based on artificial neural networks. J. Clean. Prod. **273** (2020)
7. F. Pettersson, N. Chakraborti, H. Saxén, A genetic algorithms based multi-objective neural net applied to noisy blast furnace data. Appl. Soft Comput. **7**, 387–397 (2007)

A Hybrid Model Based on Behavioural and Situational Context to Detect Best Time to Deliver Notifications on Mobile Devices

Rashid Kamal(✉), Paul McCullagh, Ian Cleland, and Chris Nugent

School of Computing, Ulster University, Jordanstown BT37 0QB, Northern Ireland, UK
kamal-r@ulster.ac.uk

1 Introduction

Notifications delivered through a smartphone have become a crucial part of everyday life. The Ubiquitous nature of this technology (anywhere and anyplace computing) provides an opportunity to keep up to date with various aspects of everyday life, such as current affairs, social media accounts and, incoming emails. For technology providers, the primary method to keep in touch with the user is to push a notification [1, 11]. Industry giants such as Amazon, Facebook, Twitter, and Google have this as a core requirement, their business model depends on attracting and retaining the user's attention [11]. With the number of mobile applications growing and users now having access to multiple devices, the number of notifications from these applications/devices will continue to increase. According to [12], on average, users received around 100 notifications per day.

Many of these notifications are irrelevant or distracting and reduce efficiency when they interrupt the user from an important task. This repeated disturbance can also cause a burden on the mental load of the user, affecting their efficiency and even their overall health and well-being [1, 11].To cope with this issue, many users disable their notifications, which may subsequently lead to them missing important information [1, 11].

To address this overload, researchers have suggested the use of user-aware notifications [1, 5, 11, 12, 17, 18]. Approaches, have included intelligent notification systems, based on physical activity recognition systems and, detecting user context [1, 11, 17]. Increasingly these solutions incorporate aspects of behavioral science and underpinned by machine learning [17, 18].

For an important notification, finding the opportune moment to engage the user increases the chance of successful delivery. The key interval is known as the "Break point". In such a breakpoint, either the user has taken a break from some work and perhaps has some free moments where they can be engaged with the important notification. In this paper we present a new model based on both situational activity recognition and psychological context to assist with fining the best time to deliver notifications. The smartphones is used for sensing, processing and delivery. Activity recognition is

F. P. García Márquez (ed.), *International Conference on Intelligent Emerging Methods of Artificial Intelligence & Cloud Computing*, Smart Innovation, Systems and Technologies 273, https://doi.org/10.1007/978-3-030-92905-3_2

underpinned by pervasive sensors, but the problem with psychological context is that there are no obvious direct sensors that can be used for quantification. It is, however, possible to fuse different sensors and phone event's be yield features from which we can infer psychological context see Table 1. The fusion of this information with situational activity recognition (using native sensors and Google Activity Recognition API provides a hybrid model. The remainder of the paper is structured as follows Sect. 2 explains terminologies in behavioural science and human-computer interaction and how different methods have been used by other researchers to detect situational and psychological context. The key challenges with monitoring context are also presented. Section 3 presents the proposed framework and present the major components of this framework. Extension to this work is addressed in Sect. 4.

2 Related Work

This Section introduces, terminologies used in behavioral science and human-computer interaction and appraises methods and models used by current research to identify the best context to deliver the notification to the user.

2.1 Behavioral Science

Behavioural science deals with the cognitive processes within organisms, and the behavioural interactions between organisms in the natural world [9]. Aspects of behavioural science that are relevant to user engagement are attention, interruption, and processing of external and internal stimulus [11].

Attention may be considered as selective information coming from different sensory inputs [4]. According to [6], attention is referred to as the ability to neglect the irrelevant information. On the other hand, interruption is considered a distraction from the primary task by external or internal stimuli. This distraction from the primary task often results in loss of focus to the current activity [13].

Internal stimuli are considered as motivation, emotion or thoughts, whereas external stimuli are linked with environmental surroundings such as the phone ringing and colleagues interrupting for consultation [1]. Attentional interruptions caused by internal stimuli are defined as internal interruptions, whereas attentional interruptions caused by external stimuli are considered as external interruptions [3, 13]. The proposed framework will be based on external interruptions and how to handle them in Ubiquitous computing.

2.2 Detecting Ideal Time

Researchers have implemented behavioural prediction models using mobileand Internet of Things (IoT) devices, which can harvest data to predict the best time to engage the user with notification. The following sections describe a range of the most commonly used sensors and models in this domain.

Table 1. Interpretation and values of the parameters used in the framework

Data	Data features relevance to context	
Psychological context		
Call app	Indicate whether Phone app can detect user is busy and phone app is used by not a good time to prompt notification the user	
Audio	True if phone is connected to audio jack or Bluetooth speakers	Audio can give insight of user daily routine phone
Music playing	Indicate whether user is playing music on the phone	This feature can be used to classify the user into groups depend on the usage
Charging	True if phone is on charge	This feature can be used to classify user into groups e.g. heavy phone usage versus light phone usage
Day of the week	Indicate the day of the week	User behaviour may vary depend on the day of the week
Hour of the day	Indicate the hour of the day	User behaviour may vary depend on the hour of the day
Proximity	True if the screen is covered	To detect phone is in pocket or near the phone etc.
Ringer mode	Different ringer mode e.g. silent etc.	An important feature to detect the user routine
Semantic location	Semantic location work, home etc.	Indicate user current whereabouts which may effect the mood of the user
Notification	Get info about each notification phone receive	How the notification was removed. Clicked by the user, removed by the user, reaction time and posted time, how long user take to react to notification. Name of the app push the notification
Situational context		
Accelerometer	Accelerometer	Accelerometer data is key sensor to detect phone movement
Gyroscope	Gyroscope	Gyroscope rotation data on all three axis, Gyroscope data can be helpful to detect phone movements

(*continued*)

Table 1. (*continued*)

Data	Data features relevance to context	
Linear acceleration	Linear acceleration along any axis	Can be use to detect phone movement at particular axis in m/s
Google activity recognition	Give the list of physical activities with level of confidence assign with each activity	Google API is very good source to detect user physical activity at any given moment
Data labels		
User state label 1	Prompt notification for ground truth labels	User-perceived the emotion state on scale 0–5 (e.g. 0 Sad, 5 Very Happy)
User state label 2	Prompt notification for ground truth labels	User-perceived the user availability state on scale 0–5 (e.g. 0 very busy, 5 labels Free to engage)

2.2.1 Psychological Context

In this Section we describe different methodologies used by researchers which can be used to detect user psychological context. There are no direct methods or sensors which can be used to detect user psychological context. However, Pielot et al. [17] describes how some of underlying phone usage, event's listeners and other sensors can be used to detect user boredom. These features can also be used to detect user overall psychological context.

In [14, 15] the research gives an exciting aspect of application usage; the authors are the first, to the best of our knowledge, to use the term "Break point". Break point, they argue, is the time between the user's primary task and the time they do some other work or task. The user's reaction to a notification widely depends on the timing of the "Break point,". Martin et al. [16] conducted a diary log study which recorded, the user's response to incoming notifications. One key finding from the study was that the user typically accepts or dismisses the notification in a 10-minute window; otherwise, the user does not respond to it or they assume the user is not interested in that notification.

In [17, 18] Pielot created an android application to obtain the contextual data, using a in the wild study methodology (i.e. study conducted in use natural environment). They also suggest possible features to detect boredom from the mobile phone data and implemented a machine learning model based on these features. Research reported in [7] defines a boredom as being; when the user moves from their routine; this can be aligned to the physiological definition of interruption. This provides a concept/notion of interruptibility.

Fisher [5] created a machine learning reinforcement model based on 'time of the day data' for user interruptibility. In 2017 [15] conducted one of the most extensive studies in user interruptibility and engagement with 680,000 users of Yahoo! Japan. Their research demonstrated that the user delays reading the notification until a breakpoint occurs. In a further study by [19], the researcher provides the user with a new option to postpone the notifications. The study argues that the user has not entirely refused the notifications. Besides the extensive work completed in the time domain of user interruptibility, several researchers have found that time alone is not a good predictor for user engagement [10]. For this reason, time should be used with other features for improved results. For example, [17, 18], demonstrated that time can be an essential feature for the interruptibility, when combined with other sensors from the smartphone. The concept of time can be both used as an explicit feature or as context to the ordering of users tasks. Consider a scenario for time as a feature where user A will be more welcoming to incoming notifications in the evening. Nevertheless, that might not be the case for User B. Time can also be contextual, e.g. duration of breakpoints to individual users [15, 18].

2.2.2 Situational Context

One common way to detect user availability has been by assessing user physical context. Many researchers have implemented models that detect a user's physical activity and send notifications to engage users accordingly. The most common sensors used to detect physical activity, are the accelerometer, and gyroscope [1]. Kern and Schiele [8] used low contextual data from these sensors to distinguish a user's social interruptibility versus personal interruptibility. Urh created an Android application TaskyApp, which used sensors to infer user task engagement [20]. A review of this work has shown that physical activity alone is not a sufficient candidate for detecting an improved opportunity for engaging the user.

Researchers also used the ambient sound conditions to detect a user interaction with the phone [5, 18]. With the help of a microphone, they can detect if the user is either indoors or outdoors using the surrounding sound. The microphone can be a good feature used by [17, 18], however, alone the microphone cannot be used as a good predictor. Also, the microphone might not be feasible to use in a real-world environment due to privacy issues concerned by the users. Researchers used other sensors along with the aforementioned that include battery [17, 18], GPS [17, 18], Light [17, 18]. Nevertheless, none of these sensors alone has been shown to be a suitable predictor to model user engagement.

More promisingly application usage has been shown as a good candidate for a feature in many studies [17, 18]. In [18], the researcher uses foreground apps as a feature to detect user-device interaction. In recent work by [17, 18] it has been shown that the combination of the smartphone event's and smartphone physical sensors can provide better results to detect the best time to engage with the user.

2.3 Challenges with Monitoring Context and Research Gaps

There are some major challenges to monitoring context in a ubiquitous nature. The biggest challenge is the labeling of the data set which can be used to train the model. The collected data set needs ground truth labels (from the user) for interpretations. In recent work by [2] it was mentioned that one of the reasons why Human Activity Recognition (HAR) models don't perform well is because they are trained in specific groups, which will result in overfitting of the training data. Several other researchers have mentioned the same reason [1, 11]. The paradox, however, is that too much user burden (i.e., a significant requirement for ground truth labeling) may cause the user to abandon the study.

Another constraint with monitoring context is the limitation of data points that can be collected from the user due to privacy and system limitations. For example GPS can be a good sensor to collect vital information about the user's location but some users might not agree to this level of location tracking. One solution might be that a generalized location tracking asked from user such as (e.g. work, home and other) instead of specific coordinates.

3 Framework

In this Section, the proposed framework will be explained in detail.

3.1 Hybrid Architecture

As discussed in the literature Section, situational context alone will not be a good predictor, it needs the input of user psychological context for better prediction. There is, however, no direct sensors which can be used to collect the features for psychological context. Nevertheless, sensor fusion can be used to provide these. Computational models can then be fed by physical sensors (e.g. from the smartphone) along with psychological context. Refer to Fig. 1 for the Hybrid Architecture. To the best of our knowledge this is the first time that a hybrid model has been used to detect user engagement.

3.2 Mobile Usage Data Collection

A data collection App has been developed for Android phones with OS 6.0 or newer. Data usage will be inferred from the user's mobile phone via sensors data and event's listener (i.e. data from different events such as battery usage etc). The data collection is split into four groups: (1) Data collected at each 1 sec time interval. (2) Data collected at each 20 Hz/sec time interval. (3) Data collected only when the notification from all installed apps is prompted. (4) Data collected from ground truth labels prompt. This approach provides the required data with the minimum possible burden on the phone battery.

In the data collection phase, participants will be required to install the App on their phones. They will be asked for their explicit consent for their participation in the study. Therefore, the first screen will explain the study, what type of data is collected, how and when the data will be collected and how it will be stored. The participant will also be informed how they can quit the study at any time. The process can be viewed in Fig. 2.

3.3 Data Fusion

In this Section, we will discuss the data fusion from different sensors and event's listener which will be used in the proposed framework. Android phones provide many sensors

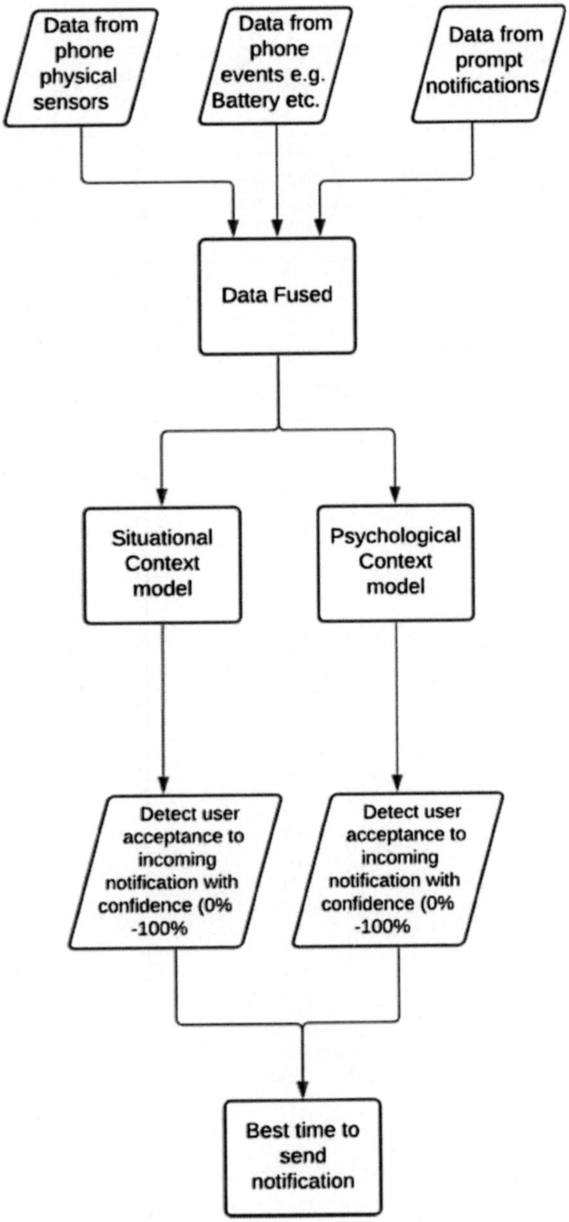

Fig. 1. Overall view of the hybrid architecture which shows how data from different sources fused together for the input of two models

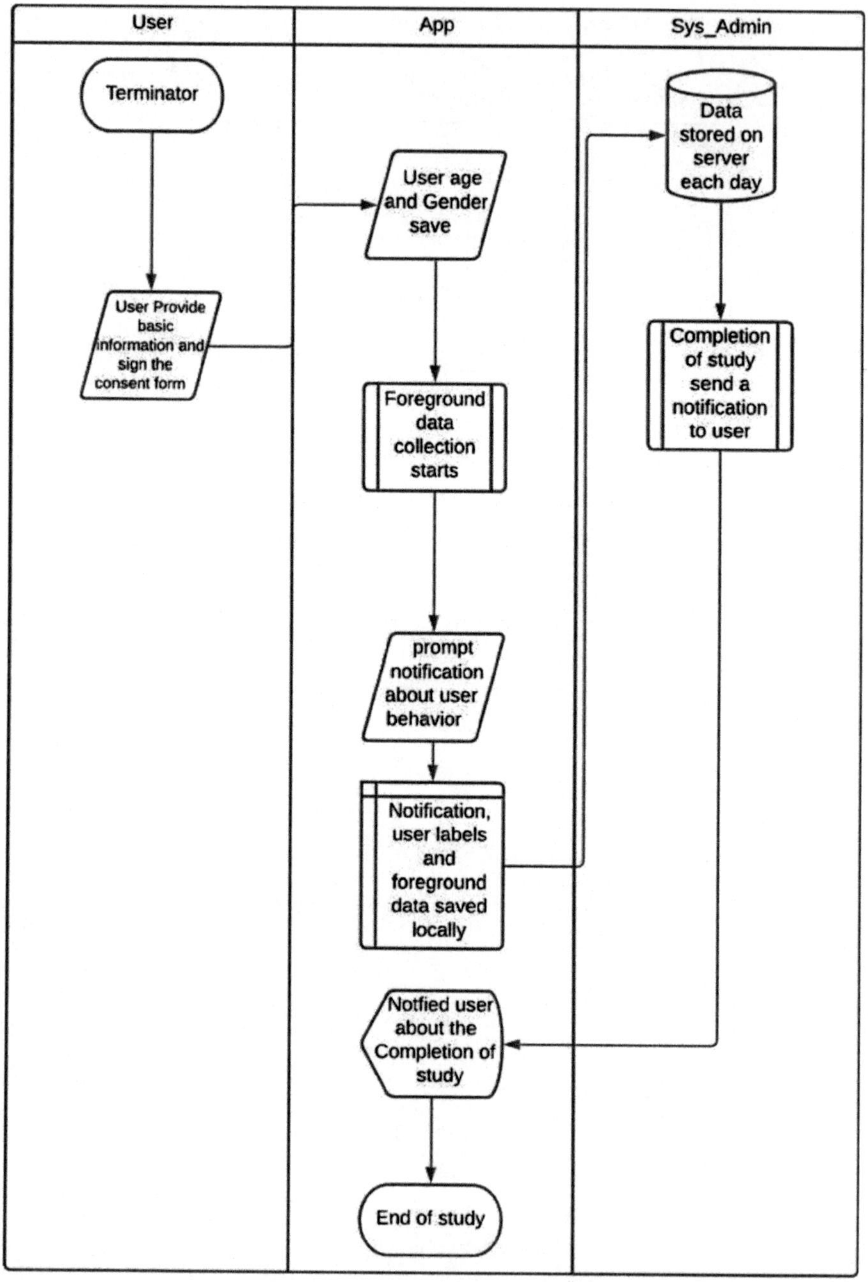

Fig. 2. The complete process of data collection from the participant using data collection app

that can be used to collect data. The accelerometer can be used to detect phone movements in all three x,y, and z-axis. Accelerometer data can be used to detect user physical activities e.g. user is moving, still, etc. A gyroscope is another sensor that will be used in our framework. The gyroscope gives the rotation data of the phone in all three axes. Gyroscope data can also be used to detect user physical activity. Linear acceleration in any axis will also be used which can give the speed of the phone in any particular axis in m/s. We have also proposed to use battery usage, proximity sensor, audio, screen orientation, and either phone is connected to a charger or not. These sensors can be fused to give more detailed data for prediction. E.g. if the accelerometer detects user movement in a particular direction and linear acceleration shows speed in a particular direction it means the user is on the move.

3.4 Data Labelling

As discussed in Sect. 2.3 labeling of data is one of the key challenges. Cruciani et al. [2] used a technique of semi-label data. Which can be used on a larger data set. Our proposed framework will collect data from a small number of participants first with ground truth labels and then in a second phase, it will collect a data set without ground truth labels. The semi-label approach will be used to classify the data set and used for the prediction of the best time to engage the user with notification.

To collect ground truth labels prompt notifications will be delivered by the data collection application. A prompt notification can be sent to the participant using any scheme.

1. Send a prompt notification at any time of the day, however, with the gap of 60 min between the last prompt notification.
2. Send a prompt notification at specific times (morning, afternoon, etc.)
3. Send prompt notification to the participant exactly after they click or ignore the notification.

The above schemes have their pros and cons. The most reasonable option can be option 3 which can give a very exciting insight into the user's thinking by looking at the previous 1-minute sensors data which may give the hint of why the user has opted for the specific decision.

4 Conclusion

We have proposed a new framework that is based on two models situational con-text and psychological context. As mentioned in Section 2 psychological context should be explored to make a more accurate prediction. We also discussed the importance of an unbias data set.

This can be achieved by collecting data from small number of participants in wild natural environment. The semi-labeled approach can then be used to label data from a larger number of participants in their natural environment. Details of the proposed framework are presented in Fig 1. We also discuss each sensor and the different type of data which will be collected for both models in Table 1. Future work will implement and evaluate this data collection application and develop the models for situational and psychological context modelling.

Acknowledgement. Invest Northern Ireland is acknowledged for supporting this project under the Competence Centre Programs Grant RD0513853—Connected Health Innovation Centre.

References

1. C. Anderson, I. Hu¨bener, A.K. Seipp, S. Ohly, K. David, V. Pejovic,A survey of attention management systems in ubiquitous computing environments. Proceedings of the ACM on Interactive, Mobile, Wearable and Ubiquitous Technologies 2(2), 1–27 (2018)
2. F. Cruciani, Personalisation of machine learning models for human activity recognition. PhD thesis, University ofUlster (2020)
3. L. Dabbish, G. Mark, V.M. Gonz´alez, Why do i keep interrupting myself? environment, habit and self-interruption, in *Proceedings of the SIGCHI Conference on Human Factors in Computing Systems*, pp 3127–3130 (2011)
4. J. Driver, A selective review of selective attention research from the past century. Br. J. Psychol. **92**(1), 53–78 (2001)
5. R. Fisher, R. Simmons, Smartphone interruptibility using density-weighted uncertainty sampling with reinforcement learning, in *2011 10th international conference on machine learning and applications and workshops*, IEEE, vol. 1, pp. 436–441 (2011)
6. E.B. Goldstein, J. Brockmole, *Sensation and Perception* (Cengage Learning, 2016)
7. K. Kapoor, K. Subbian, J. Srivastava, P. Schrater, Just in time recommendations:Modeling the dynamics of boredom in activity streams, in *Proceedings of the Eighth ACM International Conference on Web Search and Data Mining*, pp. 233–242 (2015)
8. N. Kern, B. Schiele, Towards personalized mobile interruptibility estimation, in *International Symposium on Location-and Context-Awareness*, Springer, pp 134–150 (2006)
9. E. Klemke, R. Hollinger, A. Kline, *Introduction to the Book in 'Introductory Readings in the Philosophy of Science'* (Prometheus Books, Buffalo, NY, 1980)
10. A. Mashhadi, A. Mathur, F. Kawsar, The myth of subtle notifications, in *Proceedings of the 2014 ACM International Joint Conference on Pervasive and Ubiquitous Computing: Adjunct Publication*, pp. 111–114 (2014)
11. A. Mehrotra, M. Musolesi, Intelligent notification systems. Synth. Lectur. Mobile Pervasive Comput. **11**(1), 1–75 (2020)
12. A. Mehrotra, M. Musolesi, R. Hendley, V. Pejovic, Designing content-driven intelligent notification mechanisms for mobile applications, in *Proceedings of the 2015 ACM International Joint Conference on Pervasive and Ubiquitous Computing*, pp. 813–824 (2015)
13. Y. Miyata, D..A. Norman, Psychological issues in support of multiple activities. User centered system design: New perspectives on human-computer interaction, pp. 265–284 (1986)
14. T. Okoshi, J. Nakazawa, H. Tokuda, Attelia: sensing user's attention status on smart phones, in *Proceedings of the 2014 ACM International Joint Conference on Pervasive and Ubiquitous Computing: Adjunct Publication*, pp. 139–142 (2014)
15. T. Okoshi, K. Tsubouchi, M. Taji, T. Ichikawa, H. Tokuda, Attention and engagement-awareness in the wild: A large-scale study with adaptive notifications, in *2017 IEEE International Conference on Pervasive Computing and Communications (percom)*, IEEE, pp. 100–110 (2017)

16. M. Pielot, K. Church, R. DeOliveira, An in-situ study of mobile phone notifications, in *Proceedings of the 16th International Conference on Human-Computer Interaction with Mobile Devices & Services*, pp. 233–242 (2014)
17. M. Pielot, T. Dingler, J.S. Pedro, N. Oliver, When attention is not scarce-detecting boredom from mobile phone usage, in *Proceedings of the 2015 ACM International Joint Conference on Pervasive and Ubiquitous Computing*, pp. 825–836 (2015)
18. M. Pielot, B. Cardoso, K. Katevas, J. Serr'a, A. Matic, N. Oliver, Beyond interruptibility: Predicting opportune moments to engage mobile phone users. Proceedings of the ACM on Interactive, Mobile, Wearable and Ubiquitous Technologies **1**(3), 1–25 (2017)
19. L.D. Turner, S.M. Allen, R.M. Whitaker, Push or delay? decomposing smartphone notification response behaviour, in *Human Behavior Understanding*, Springer, pp. 69–83 (2015)
20. G. Urh, V. Pejovi´c, Taskyapp: inferring task engagement via smartphone sensing, in *Proceedings of the 2016 ACM International Joint Conference on Pervasive and Ubiquitous Computing: Adjunct*, pp. 1548–1553 (2016)

Smart IoT System for Chili Production Using LoRa Technology

Fatin N. Khairodin(✉), Tharek Abdul Rahman, Olakunle Elijah, and Haziq I. Saharuddin

Wireless Communication Centre, Universiti Teknologi Malaysia, Skudai, Johor, Malaysia
{fatin.nasuha,haziq1995}@graduate.utm.my, tharek@fke.utm.my

1 Introduction

Chili has become a high-demand production since the improvement of agronomical technologies and agribusiness marketing but has a low sustainable way of production [1]. Chili is considered a major cash crop for local farmers in Malaysia. The Malaysian chili is preferred to the imported chili from China, Thailand, and Vietnam due to its better taste, texture and is very hot. However, Malaysian chili is more expensive compared to the imported ones from neighboring countries. In [2], the authors show the cost for the small, medium, and large groups on the production of chili for specific practices. The high cost could be a result of low production due to several factors. The factors that affect the chili plants are inadequate supply of water and fertilizer, inadequate exposure to sunlight, attack from pests and diseases [3]. The health of the chili can be monitored from the color of the leaves. The green color of the leaves shows the chili plant is healthy else it is considered unhealthy or damaged.

Two methods can be adopted to improve the production of chili which is semi-automated or fully automated. A semi-automated system employs a scheduler which uses a timer for irrigation and fertigation purpose. The use of a fertigation system allows the application of irrigation and fertilizer to a crop in some principle for water management solution [4]. On the other hand, a fully automated system employs the use of sensors to monitor the condition of the soil and its environment to control the irrigation and fertigation system. The use of a fully automated system is needed to optimize the use of resources (water and fertilizer) [5] and to reduce the amount of time and labor needed. Thus, the use of IoT for agriculture is considered the most appropriate but critical as it needs real-time monitoring and a controlled environment for better application [6].

2 Related Work

In [7], the authors proposed a smart water sprinkle and monitoring system for chili plants to intending to optimize the water consumption using IoT to manage and monitor the crops. The system uses a pH sensor to monitor the acidity of the soil and a soil moisture

F. P. García Márquez (ed.), *International Conference on Intelligent Emerging Methods of Artificial Intelligence & Cloud Computing*, Smart Innovation, Systems and Technologies 273, https://doi.org/10.1007/978-3-030-92905-3_3

sensor to define the humidity of the soil. The Electrical Conductivity (EC) sensor was used to monitor the nutrient condition in the soil. Furthermore, the system makes use of automatic watering and fertilizing based on the value of the pH and EC sensor. Arduino was used as the microcontroller while the connectivity is based on the ethernet shield. The data from the sensors are displayed using a web mobile application. The parameters monitored by this system are soil moisture, pH value, EC value, and temperature. Overall, the irrigation system is automatically controlled by the pump valve that allows the flow of water to the plants. However, the analysis of the performance was limited to the growth of the chili leaves. Further analysis is needed to demonstrate the advantages of the IoT system in chili production. The use of IoT in precision agriculture using wireless moisture sensor network (WMSN) for chili greenhouse farm is presented in [8]. The project compared the use of automatic irrigation-based WMSN and a scheduled irrigation system. The use of WMSN resulted in less consumption of water and fertilizer compared to the scheduler irrigation system. In [9], the authors said that precise and controlled irrigation not only increases the production but the water and other sources are optimized. The system resulted in optimizing the use of water and stable soil humidity. However, the impact on fruit production, plant growth was not reported. In [10] the author studied the effects of fertigation and water application frequency based on soil amendment capacity for chili production. The use of a fertigation system resulted in better yield production and chili growth compared to solid soil application. The scheduling method employed has several limitations which can be overcome by using IoT for better monitoring, reduction of labor, optimal resource allocation, and improving accuracy. The authors in [11] proposed smart agricultural solutions to farmers for getting better yields using IoT. It combines IoT with cloud computing and the use of devices and sensors to modernize agriculture for farmers' benefit. The project aims at better production of crop yield and automating the agricultural task that was manually done by the farmer. It uses a GPS connection and a system that has a real-time monitor for the crop field. The system allows for monitoring of the efficiency of energy usage, moisture measurement, and leak detection by an electronic automated system. The use of 3G and 4G networks were employed for connectivity. This poses the challenge of short connectivity range and hence requires long-range communication technologies.

In this paper, the use of IoT and LoRa connection is proposed for monitoring of growth parameters which are temperature, soil moisture, EC values and status of the system. This project makes the system fully automated which other than only monitoring the soil and fertilizer environment, this smart chili system controls all the sensors which are temperature, soil moisture, EC value, and water level in the tank that triggered the water pump, fertilizer pumps, water valve, mixer valve, and fertigation valve to be functioning based on the range value of the sensor that has been program in the Arduino microcontroller.

3 System Design

The system employs the use of IoT for monitoring and automating the operations of growing the chili. The use of pumps and valves are employed for the application of fertilizer A and B in water. The amount of fertilizer A and B needed is determined by using the EC sensor while measuring the water level in the tank and the process is automated. This reduces the amount of time required by the farmer to monitor and apply the fertilizers. A set of fertigation systems that is fully automated is prepared. Fig. 1 shows the system design for the fully automated chili fertigation system.

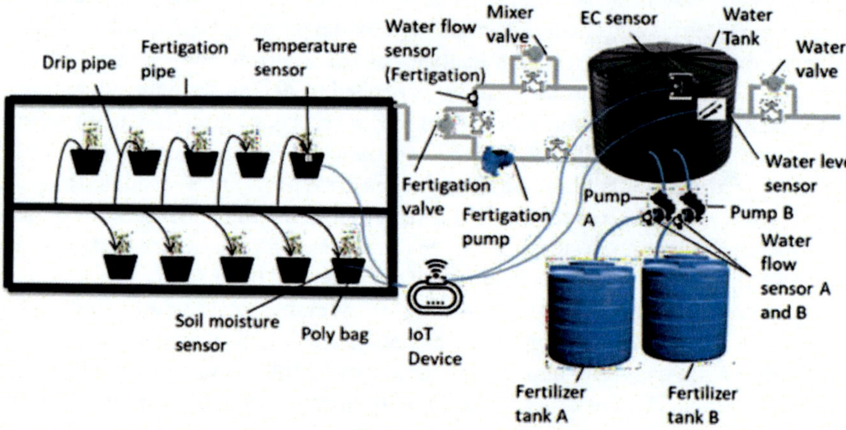

Fig. 1. Smart Chili system design

A low-power wide-area (LPWA) communication technology LoRa will be used to transmit data from sensors to the IoT Gateway. The Gateway will collate and transmit data to the network server using LAN. LoRa offers low-power consumption, long battery life, and low cost, low throughput, and has a deep coverage for a large geographical area.

The most important thing is LPWA is suitable for use cases that do not need a high data rate and power consumption, none crucial tolerant, and low-cost infrastructure [12]. The LoRa will be deployed due to flexibility and support for the flexible size of data that can be transferred from IoT devices and the IoT gateway. Finally, the number of resources used, growth, and the production of the chili plant between the semi- automated systems and fully automated systems will be compared. The automation will reduce more manpower and control the basic scheduler for fertigation while the water flow can be measure by the water flow sensor [13]. In [14], the authors said that the more sensor used required more energy consumption, thus, LoRa or any LPWA network is a better solution. The description of the devices use in the fertigation system is shown in Table 1.

4 Flowchart of the System Operation

The operation of the fully automated chili fertigation system is shown in Fig. 2.

In IoT platforms which is inexpensive and has ubiquitous connectivity in outdoor IoT applications while keeping the management and network simple. The endnotes consist of sensors, LoRa transponder that transmits a signal over LoRa physical layer, and the microcontroller. The LoRa gateway transmits the data from the sensors to the server

Table 1. Description of the devices for the fertigation system

Description	Functions
Pump A	To pump the Fertilizer A into the water tank
Pump B	To pump the Fertilizer B into the water tank
Fertigation pump	To pump the water in the tank into the plants
Water valve	Allow the water to flow from the inlet
Mixer valve	Allow the water to circulate in the tank
Fertigation valve	Allow the water to flow into the plants
Flow sensor A	To measure the amount of fertilizer A flow into the tank
Flow sensor B	To measure the amount of fertilizer B flow into the tank
Water flow sensor	To measure the amount of water flow into the plants

or cloud and it acts as the transparent bridge that connects to the network server via standard IP connections.

5 Experiment Setup

In this section, we describe the LoRa setup which are LoRa node and LoRa gateway, the hardware, and software used, and the fully automated chili fertigation system setup.

A. *LoRa Setup*

This system uses LoRa communication technology connected to IoT devices and the chili plants. A LoRa gateway has been set up at Wireless Communication Centre (WCC) as shown in Fig. 3.

In this IoT project, LoRaWAN has been used as it offers a wide range of connections, long battery life, and low payload internet connection [15]. The LoRa (Long-range communication link) is the physical layer while LoRaWAN is the communication protocol and architecture of the network. Reliable and secure communication is the advantage of using the LoRaWAN protocol. LoRaWAN is the most used LPWA

Cytron LoRa-RFM Shield and Arduino are used to set up and configure the LoRa node which is connected to The Things Network (TTN). First, set up the hardware by prepares the node and Arduino board then install the antenna. Arduino IDE software is needed to install the LoRa node that includes necessary libraries before register an account in TTN. TTN is the network server that can view and monitor the LoRa node that gathers all the sensor's values. The LoRa node is registered at the application in the TTN by enters the application ID and other information. Activation by Personalization (ABP) is used as the device activation by generating the session keys [15].

Cytron 915MHz LoRa Gateway Raspberry Pi Hat (HAT-LRGW-915), SD card, 5V 2.5Apower adapter, and Ethernet cable are needed to set up the gateway. Etcher software is used to burn the Raspbian image and LoRa gateway software pre-installed into the

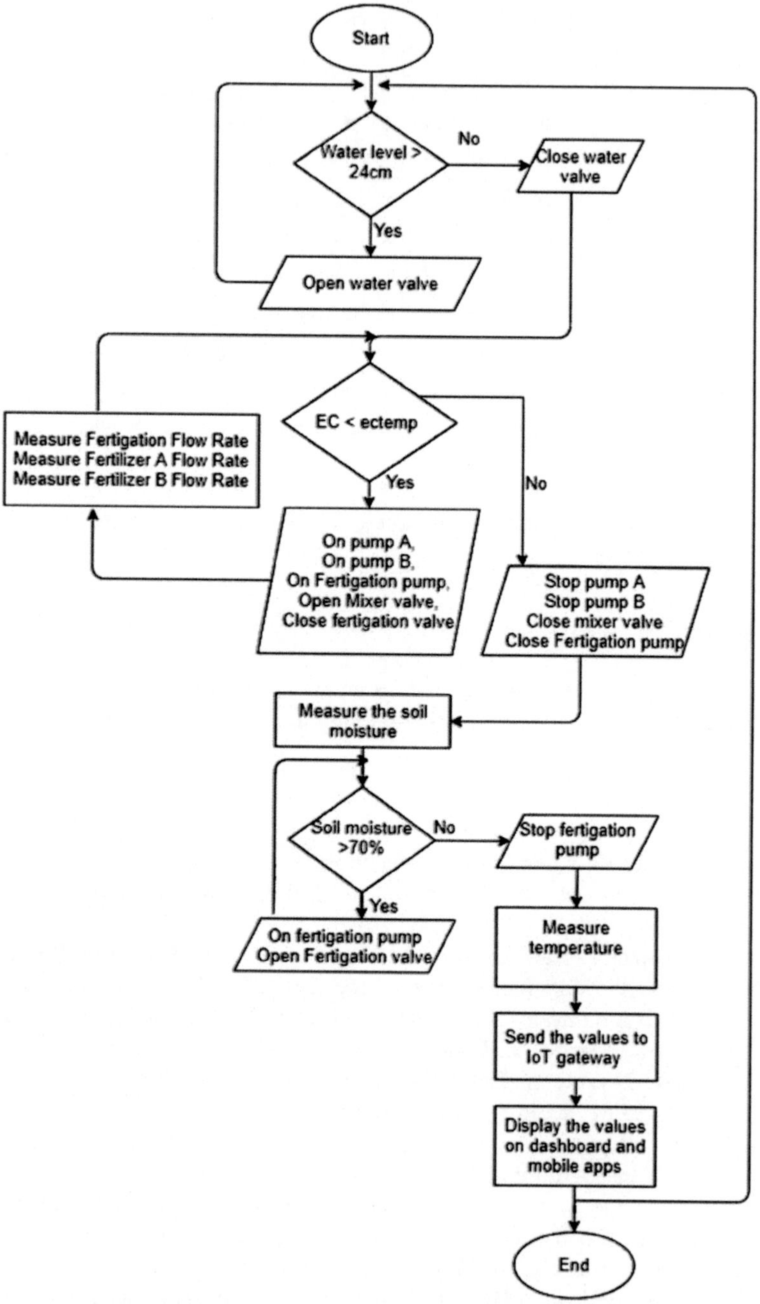

Fig. 2. Flowchart for the fully-automated chili fertigation system

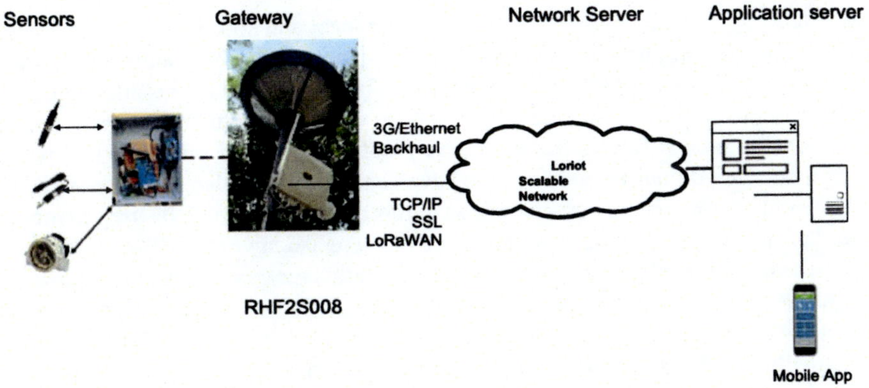

Fig. 3. The LoRa and application setup for the system

SD card and inserted into the Raspberry Pi SD card slot. WCC broadband is used as the internet for the LoRa gateway to transfer data packets to the cloud server. The packets that have been transmitted by LoRa node to LoRa gateway will be displayed on TTN at the data menu on the application page. Several nodes can be connected to LoRa gateway to display the data in TTN [16]. Table 2 shows the LoRa parameters.

Table 2. The LORA settings

Description	Unit
Transmit power LoRa node	14 dBm
Spread factor	10
Code rate	4/5
Transmit power for The LoRa gateway	27dBm
Payload	48 bytes

B. *Fertigation Setup*

The irrigation and fertilizing system are monitored and controlled with sensors. A water level sensor is used to measure the level of water inside the mixer tank as it will trigger the water and mixer valve based on its value. The EC sensor is used to determine the amount of fertilizers A and B required for the fertigation. The fertilizer A and fertilizer B pump will be triggered when the EC value is inside or out of the range. The EC value is not fixed as it needs to changes vary on the growth of the chili. Soil moisture value will trigger the fertigation pump based on the moisture of the soil instead of irrigating the water based on a timer without considering the soil environment. Data analysis will be carried out to determine the correlation between environmental parameters and soil parameters. By using a temperature sensor, water level and soil pH sensor, EC sensor, soil moisture sensor, image collection of the chili plants in the fully automated system, the amount of water and fertilizer will be quantified. The setup of the fully automated system is shown in Fig. 3. The use of greenhouse will also be studied as it is one of the critical environmental factors that affect the production of chili plants [17]. Later, the use of robots and drone also are useful to have better use of fertilizer management [18]. Mobile applications also are crucial things in this era, thus they can be used to monitor instead of only using web pages to monitor anywhere and anytime [19].

The setup of the fully automated is described in this section. A fully automated chili fertigation system was set up on the roof of Wireless Communication Center (WCC), Universiti Teknologi Malaysia (UTM). The system consists of a scheduled fertigation system with 0.5 HP electric water pump used for watering the 101 polybags of planted chili. The chili seeds were planted and transferred after two weeks and the quantity of mixed fertilizer A and B in water was applied in the system (Figs. 4 and 5; Table 3).

Fig. 4. The chili plants setup at the rooftop of WCC

6 Result and Analysis

Monitoring software for remote monitoring for chili management needs to have the features of real-time and can be monitored anytime and anywhere. An IoT dashboard and mobile application have been developed and need to be improving over time. Figrues 6 and 7 show the dashboard and mobile application used to monitor the chili. Table 4 shows the parameters and the status of the system.

A. Ubidots

Ubidots is used to display the data for the chili fertigation from TTN as a dashboard. The current parameters used in this smart chili fertigation project are temperature, soil moisture, EC value of the mixed water and fertilizer, the water level in the main tank, and the water flow rate for fertigation, fertilizer A and B. More parameters could be used to monitor and those available sensors already can trigger the pumps and valves to open or close based on the range set of values for each sensor.

MIT app inventor is used as the mobile application to display the values of each sensor and the status for the pumps and valves. This application is useful as farmers

Fig. 5. The IoT devices and system for the automated chili production setup

Table 3. The fertilizer requirement for chili plant

Age after transplant (weeks)	EC (mS/cm) of mix fertilizer A and B in water	Quantity (ml/day/plant) mixed solution (Fertilizer A and B in water)
1	1.6	500
2	1.6	600
3	1.6	800
4	1.8	1200
5 – 6	1.8	1500
7 – 9	2.0	1800
10	2.5	2000
11	2.8	2000
≥12	>2.0	>1800

Note. The redraw table data for chili plant requirements from Vegetable, Organic & Fertigation Section Agriculture Research Centre Semongok (2018)

Fig. 6. The dashboard of the chili monitoring

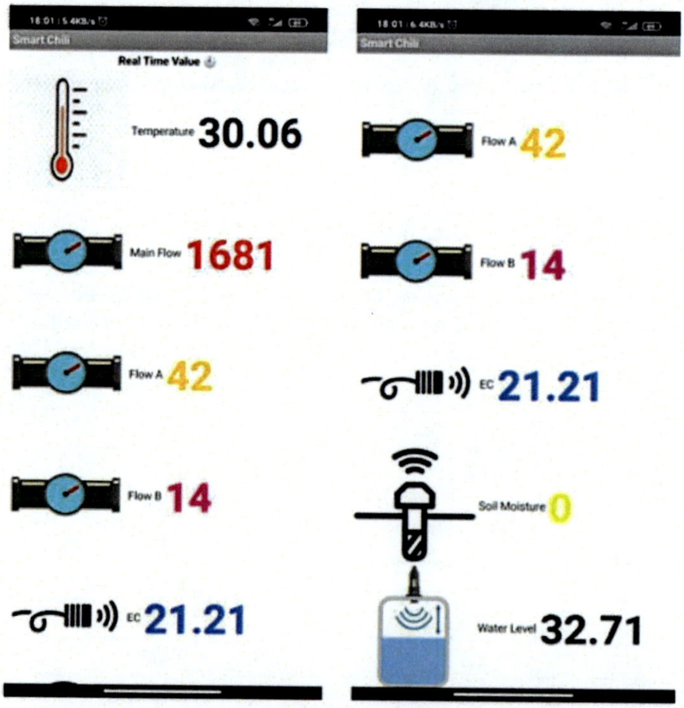

Fig. 7. The mobile apps of the chili monitoring

Table 4. The parameters and status of the chili fertigation system

Parameters	Status
Temperature	Water pump
EC value	Fertilizer A pump
Soil moisture	Fertilizer B pump
Water level	Water valve
Water flow rate A	Fertigation valve
Water flow rate B	Mixer valve
Water flow rate (Fertigation)	

or users are using mobile phones in their daily life thus they can monitor their farms anytime and anywhere.

7 Preliminary Result

Thingspeak tools are used to view and visualize the data from TTN and it shows the numerous data relationship between each sensor in time varies. The data shows the dependencies of the values based on the weather, the amount of fertilizer in the tank, and the level of water left in the tank. Figure 8 shows the graph chart for the sensor's value in Thingspeak.

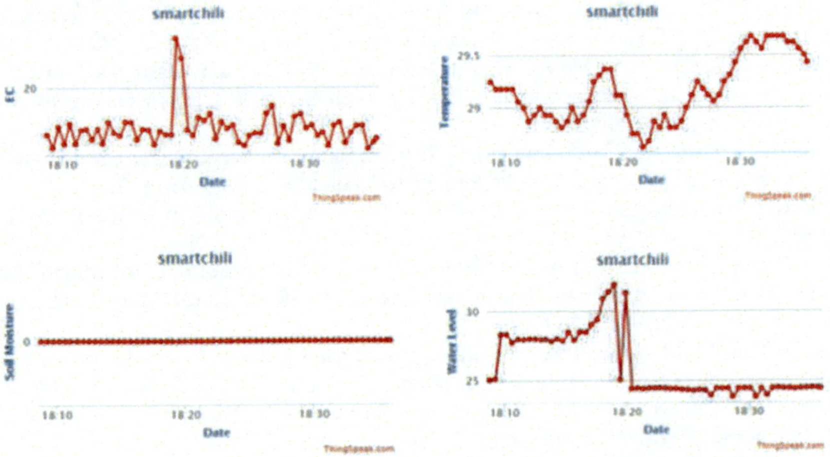

Fig. 8. The graph analysis for the sensors values in Thingspeak

8 Conclusion and Future Work

In conclusion, the use of IoT as an automated system in chili farms is useful as it helps the farmers and reduces the amount of work and needed at the farm to monitor the chili plants. Chili plants need a lot of care and frequent monitoring as it is easy to have plant disease which damages the plants. Therefore, the modernization of agriculture using IoT is expected to overcome those problems. A fully automated chili rooftop production has been presented in this paper that manages to overcome the inefficient of semi-automated fertigation system. The design and the flow of the fully automated chili fertigation system have been presented. In future work, more parameters or sensors can be used to improve the quality and more chili environment details can be managed by only using this Smart Chili IoT system. The performance analysis of the system will be presented in our future work.

Acknowledgements. This research is supported by the Ministry of Education (MoE), Malaysia and Universiti Teknologi Malaysia under the project Vote No. 19H38 and 04G26. The authors wish to acknowledge the assistance and support of the Wireless Communication Center (WCC) UTM.

References

1. E. Science, Impact of starter solution technology on the use of fertilizers in production of chilli (Capsicum frutescens L.) Impact of starter solution technology on the use of fertilizers in production of chilli (Capsicum frutescens L.) (2019). https://doi.org/10.1088/1755-1315/230/1/012063
2. M. Tariq, I. Khan, Q. Ali, M. Ashfaq, M. Waseem, Economic analysis of open field chilli (Capsicum annuum L.) production in Punjab, Pakistan. J. Exp. Biol. Agri. Sci. **5**(2320) (2017)
3. A.J. Rau, J. Sankar, A.R. Mohan, D. Das Krishna, J. Mathew, IoT based smart irrigation system and nutrient detection with disease analysis, pp. 3–6 (2017). https://doi.org/10.1109/TENCONSpring.2017.8070100
4. N. Bafdal, S. Dwiratna, D. R. Kendarto, E. Suryadi, Rainwater harvesting as a technological innovation to supplying crop nutrition through fertigation **7**(5), 1670–1675 (2017)
5. C. Joseph, I. Thirunavuakkarasu, A. Bhaskar, A. Penujuru, Utilization of fertilizer and water (2017)
6. M.S. Farooq, S. Riaz, A. Abid, K. Abid, M.A. Naeem, A survey on the role of IoT in agriculture for the implementation of smart farming. IEEE Access **7**, 156237–156271 (2019). https://doi.org/10.1109/ACCESS.2019.2949703
7. J.H. Gultom, M. Harsono, T.D. Khameswara, H. Santoso, Smart IoT water sprinkle and monitoring system for chili plant, in *ICECOS 2017 - Proceeding 2017 Int. Conf. Electr. Eng. Comput. Sci. Sustain. Cult. Herit. Towar. Smart Environ. Better Futur.*, pp. 212–216 (2017). https://doi.org/10.1109/ICECOS.2017.8167136
8. I. Mat, M. Rawidean, M. Kassim, A.N. Harun, I.M. Yusoff, IoT in precision agriculture applications using wireless moisture sensor network, pp. 24–29 (2016)
9. R. Prabha, E. Sinitambirivoutin, F. Passelaigue, M.V. Ramesh, Design and development of an IoT based smart irrigation and fertilization system for chilli farming, April, 2018
10. S. Chanthai, S. Wonprasaid, Effects of fertigation and water application frequency on yield, water and fertilizer use efficiency of Chili (Capsicum annuum L.). **3**(2), 209–213 (2016)

11. M.K. Gayatri, J. Jayasakthi, G.S.A. Mala, Providing Smart Agricultural solutions to farmers for better yielding using IoT, *Proc. - 2015 IEEE Int. Conf. Technol. Innov. ICT Agric. Rural Dev. TIAR 2015*, no. Tiar, pp. 40–43 (2015). https://doi.org/10.1109/TIAR.2015.7358528
12. U. Raza, P. Kulkarni, M. Sooriyabandara, Low power wide area networks: an overview. IEEE Commun. Surv. Tutorials **19**(2), 855–873 (2017). https://doi.org/10.1109/COMST.2017.2652320
13. M. Karaşahin, Ö. Dündar, A. Samancı, Turkish Journal of Agriculture - Food Science and Technology The Way of Yield Increasing and Cost Reducing in Agriculture : Smart Irrigation and Fertigation **6**(10), 1370–1380 (2018)
14. J. Ruan et al., Accepted from open call a life cycle framework of green IOT-based agriculture and its finance, operation, and management issues. IEEE Commun. Mag. **57**(March), 90–96 (2019). https://doi.org/10.1109/MCOM.2019.1800332
15. Ain, Lesson 1: Build a simple Arduino LoRa Node In 10 Minutes | Tutorials of Cytron Technologies. https://tutorial.cytron.io/2017/09/15/lesson-1-build-simple-arduino-lora-node-10-minutes/. Accessed 10 Apr 2021
16. Ain, Lesson 2: Setting up a Raspberry Pi 3 LoRa Gateway with HAT-LRGW-915 | Tutorials of Cytron Technologies. https://tutorial.cytron.io/. https://tutorial.cytron.io/2017/09/15/lesson-2-setting-raspberry-pi-3-lora-gateway-hat-lrgw-915/. Accessed 10 Apr 2021
17. H. Gong, Z. Wu, Research of the pepper environment monitoring system based on IOT, pp. 567–571 (2014). https://doi.org/10.4028/www.scientific.net/AMR.850-851.567
18. I. Mat, M. Rawidean, M. Kassim, A. N. Harun, I. M. Yusoff, M. Berhad, Smart agriculture using internet of things, in *2018 IEEE Conference Open System*, pp. 54–59 (2018)
19. G. Parameswaran, K. Sivaprasath, Arduino based smart drip irrigation system using internet of things. Int. J. Eng. Sci. **5518**(5), 5518–5521 (2016). https://doi.org/10.4010/2016.1348

Peculiarities of Image Recognition by the Hopfield Neural Network

Dina Latypova and Dmitrii Tumakov(✉)

Institute of Computational Mathematics and Information Technologies, Kazan Federal University, Kazan, Russia
dtumakov@kpfu.ru

1 Introduction

Artificial neural networks are used as a method of deep learning, one of the many subsections of artificial intelligence. Thanks to advances in hardware development, humanity has been able to build networks that can be trained on a huge set of data in order to achieve breakthroughs in machine intelligence. These discoveries allowed machines to match and exceed the capabilities of humans in performing certain tasks. One of these tasks is object recognition. The problem of pattern recognition by neural networks is one of the most popular today. This leads to a wide variety of neural network architectures for solving the problem.

One of the most popular and frequently used neural network architectures is the multilayer perceptron. However, one of the biggest drawbacks of this network is its redundancy. For example, if the input is an image in the form of a 34 × 34 matrix, the neural network will have 1156 inputs. This indicates a large amount of computing power that is expended for this algorithm. One of the most successful neural networks for solving the problem of pattern recognition is the hierarchical neural networks [1, 2]. For example, convolutional neural networks represent the best algorithms for face recognition in terms of accuracy and speed. The brain has a peculiarity, which is receiving any incoming information; the output of the brainwork gives a lot of information that is somehow related to the input. Since traditional neural networks cannot mimic this peculiarity, recurrent neural networks have been developed. One such network is the Hopfield neural network.

In the present work, we consider the problem of recognizing images of handwritten digits contained in the MNIST database. Recognition is performed using the Hopfield recurrent neural network [3, 4]. Due to the fact that the Hopfield neural network has a limit on the number of objects to "remember", the Kohonen neural network is used as the first stage [3, 5, 6]. Thus, at the first stage, the cluster centers are obtained. It is these centers that are the sample for "memorization" for the Hopfield neural network. In the present paper, the Hopfield neural network is modified by us. As a response, the network produces an image that is most similar to an image from the "memory" of the neural network. The paper suggests two methods for comparing two images. The work consists of two stages: the Kohonen neural network, which clusters images of each digit

F. P. García Márquez (ed.), *International Conference on Intelligent Emerging Methods of Artificial Intelligence & Cloud Computing*, Smart Innovation, Systems and Technologies 273, https://doi.org/10.1007/978-3-030-92905-3_4

separately, and the Hopfield neural network, which recognizes images of handwritten digits. This results in a hierarchical neural network.

Today, there exist many ways to cluster data. For example, in [5], we proposed an algorithm for consensus clustering on a set of handwritten digits from the MNIST database. The so-called reliable continuous clustering was considered in [7]. In [8], an algorithm was proposed to improve the classification performance in handwritten digit recognition, and the efficiency of the algorithm was evaluated on the MNIST dataset. C-means [9] and k-means [10, 11] approaches are often used to cluster handwrit- ten digits. For example, in [6], clustering of handwritten letters was performed using the Kohonen neural network. In [12], we reviewed various categories of clustering algorithms: partitioning-based clustering, hierarchical clustering, density-based clus- tering, and grid-based clustering. In that work, the main task was to determine the most suitable algorithm for clustering any big datasets. In [13], one of the most popular clustering algorithms, k-means, was described. When using this method, an accuracy of approximately 90% is usually achieved.

Recurrent neural networks are successfully used for solving a number of different problems. Excellent results were achieved by LSTM (Long short-term memory) networks [14]. The Hopfield neural network is also often used for pattern recognition. For example, in [15], the Hopfield neural network was used to recognize images in gray shades. The paper [16] describes the process of memorizing image objects using the Hopfield neural network. In [17], statistical estimates of the probability of correct image recognition were compared with each other. In that paper, the correlation algorithm was found to be generally worse than the modified Hopfield network algorithm. In the present work, the Kohonen neural network is used for clustering to identify new clusters. There are many other methods for clustering. In [18], the multilayer hierarchical self-organizing map is discussed as an uncontrolled clustering method. The self-organizing map method created by Kohonen is a well-known method of neural networks [19]. Neural networks are a good tool for working with a large amount of data [20]. Traditional methods of computer vision and machine learning cannot compare with human performance in tasks such as recognizing handwritten numbers or road signs. Our biologically plausible, broad and deep architectures of artificial neural networks can [21].

2 Clusterization of Handwritten Digits by the Kohonen Network

The task of clustering is to divide a set of objects into groups called clusters. Objects are grouped in such a way that within one cluster they are more similar to each other than objects in other clusters.

We will cluster the digits contained in the MNIST database. The MNIST database is a database of handwritten digits, which contains 60.000 images of digits in the training sample and 10.000 images in the test sample. Table 1 presents the numbers of images of each digit in the training and test sample.

The number of clusters into which the input images need to be split is unknown. Let us determine the optimal number for each digit. For automatic clustering, we use the following algorithm [16, 17]. Using automatic clustering, we obtain the number of clusters for each digit (Table 2).

Table 1. The number of images of each digit in the training and test samples from the MNIST database.

Digit	Training sample	Test sample
0	5923	980
1	6742	1135
2	5958	1032
3	6131	1010
4	5842	982
5	5421	892
6	5918	958
7	6265	1028
8	5851	974
9	5949	1009

Table 2. The optimal number of clusters for each digit.

Digit	0	1	2	3	4	5	6	7	8	9
Number of clusters	39	49	30	47	28	33	49	29	38	28

After determining the optimal number of clusters for each digit, we use the Kohonen neural network to find the clusters themselves.

The Kohonen neural network is of the "unsupervised" learning type and consists of a single layer of configurable weights. The weights of the neural network change in such a way that vectors that belong to the same cluster activate the same output neuron. The Kohonen neural network architecture is shown in Fig. 1. For example, for the digit 0 (Table 2), the number of neurons in the output is 39 (m = 39).

The outputs Z_j are calculated using the formula:

$$Z_j = \sum\nolimits_{i=1}^{n} w_{ij}x_i, \quad (1)$$

where x_i are the inputs of the neural network. The images in the MNIST database have a size of 28 by 28, so the number of neurons in the Kohonen neural network will be 28 × 28 =784. The weights w_{ij} represent the cluster centers and are found during training. We describe the algorithm for training the Kohonen neural network as follows:

1. Normalization of input vectors x^p.
2. Initialization of weights w_{ij} by random variables. Most often, the input vectors x^p are distributed unevenly, and then the vectors of the weights w_j can be removed from the input, and, therefore, will not participate in the training algorithm, and the remaining vectors are not sufficient to be divided into clusters. To solve this problem,

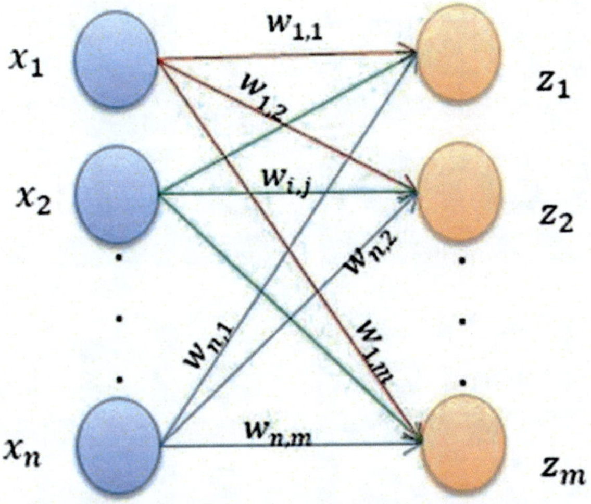

Fig. 1. Kohonen neural network architecture. To cluster handwritten digits from MNIST: n = 784, and the value of m for each digit must be selected from Table 2.

we initialize the weights by placing them in a cluster of input vectors. Therefore, for the initial values of the weights, we choose random vectors from the training sample.

3. Between the input vector x^p with coordinates x_i^p and all vectors w_j, we calculate the distance by the formula:

$$D_j = \left(\sum\nolimits_{i=1}^{784} \left(x_i^p - w_{ij}\right)^2\right)^{1/2}$$

We choose a winner vector, i.e. the weight vector w_j, from which the distance D_j to x^p is the smallest.

4. We change the coordinates of the winner vector selected in the previous step, according to the formula:

$$w_l = w_l + \theta(x^p - w_l) \tag{3}$$

where θ is learning speed.

5. Steps 3 and 4 are repeated until we have passed all the vectors x^p.
6. We reduce the value of θ by the formula $\theta = \alpha\theta$, where $\alpha = 1$. If $\theta > \varepsilon$, then go to step 3 and again, passing through all the input vectors, the weights are adjusted.

To cluster the digits, we choose the following parameter values: $\alpha = 0.96$, $\varepsilon = 0.02$, and set the initial value $\theta = 0.6$.

We present examples of the centers of the obtained clusters in Fig. 2. The next step is to binarize the images of the obtained cluster centers: we proceed from the image in grayscale to the black-and-white image. Note that, in this problem, there are 784 pixels of gray shades. Experimentally, the value of border is selected, if the value of the image pixel is greater than the border's value, then we change the value of the pixel by 1, otherwise, we change it by −1. The value of border in our work is 125. This indicates that we adopted the values less than 125 for the white color.

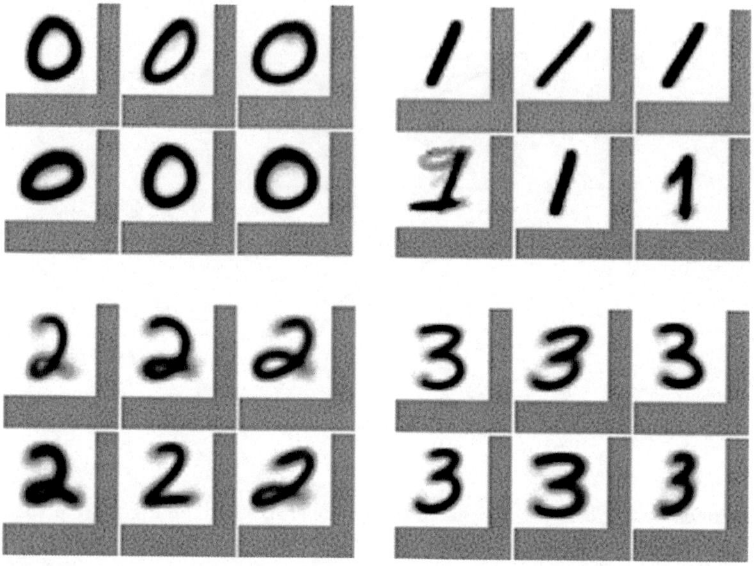

Fig. 2. Examples of cluster centers obtained using the Kohonen neural network.

3 Hopfield Neural Network

The Hopfield neural network (hereinafter, the HNN) consists of a single layer of configurable weights w_{ij}. Each neuron in the network is connected to all neurons: all neurons return their signals back to the inputs. The HNN operates with the values $+1$ and -1.

The problem that is solved with the help of this network is formulated as follows. Suppose that we are given a set of binary signals: images, audio digitization and the like. The HNN must single out ("recall") the corresponding sample from the non-ideal signal supplied to the input, if the sample is available in memory, or give a conclusion that the input signal does not match the available samples.

The input signals to the HNN are fed once. Each signal from the training sample is described by the vector X= *{x1, x2, ..., xm}*, where m is dimension of the vector X. This is the set that the neural network "memorizes". The input signal that needs to be recognized, or "recalled", is indicated as Y = *{y1, y2, ..., ym}*, where m is dimension of the vector Y. Let X_k be a vector that describes the k-th signal. If the Hopfield network recognizes any sample, then $X_k = Y$, otherwise the output vector will not match any vector from memory.

Before the Hopfield neural network starts functioning, the values of the w_{ij} weights are calculated using the formula:

$$w_{ij} = \begin{cases} \sum_{k=0}^{m-1} x_i^k x_j^k, i \neq j, \\ 0, i = j. \end{cases} \tag{4}$$

Then the HNN starts functioning.

1. The signal Y, which needs to be recognized, is supplied to the input of the HNN.
2. Calculation of the new state of neurons:

$$S_j(p+1) = \sum_{i=0}^{n-1} w_{ij} y_j, \tag{5}$$

where p is a number of operations.

3. Calculation of the new state of axons:

$$y_j(p+1) = f\left[s_j(p+1)\right] \tag{6}$$

where f is threshold function:

$$f(S) = \begin{cases} 1, S_j > T_j, \\ -1, S_j < T_j, \\ does\ not\ change, S_j = T_j, \end{cases} \tag{7}$$

where T_j is neuron's threshold.

4. Checking for changes in the outputs of neurons for the last iteration. If the outputs have changed, then go to step 2, otherwise the outputs have stabilized. If there is no vector that exactly matches the vectors in the memory of the neural network, then the answer is the sample that best matches the input.

The next question arises, what is the capacity of the network, that is, how many images can the Hopfield neural network memorize. Experimentally, it was shown that the NSC, which consists of N neurons, can only remember fifteen percent of N [3]. Thus, "memory", which is implemented using an artificial Hopfield neural network, is associative. In the present work, the input sample size for the HNN does not exceed 50 images for each digit.

4 Methods for Comparing Two Images

The result of the HNN operation is an image vector that matches the vector from the network memory after a certain number of iterations; otherwise, the most similar vector is selected. In the present work, as already mentioned, we propose two methods for determining the similarity of two image vectors. A separate HNN is constructed for each digit. The result of each network is given with a percentage of how much the original image matches the image from the memory of the neural network. After all the networks (for ten digits, there will be 10 networks) have completed their work, as the final result, the three images that best match the original image are selected, that is, they have the highest percentages.

The first method is based on a simple comparison of the image under study and the image stored in the network memory. In the last step in the Hopfield algorithm, one needs to check how much the sample to be recognized is similar to the k-th sample in memory. The check is performed using the following formula:

$$C = C + 1,\ if\,(y_j = x_j^k = 1) \tag{8}$$

where C is number of pixels. As can be seen from the formula, the check is carried out only for matching the "1", since it is the value making the greatest contribution to the samples. Thus, for example, if there is an image recognition task, then −1 is for white pixels, and 1 is for black ones. Thus, black pixels are of more importance.

After the number of matched values equal to 1 is calculated, this value is recalculated as a percentage, that is, divided by the dimension of the vector. In the case of the MNIST base, the dimension of the vector is 784 (28*28). If the percentage is less than a certain number of percent, then the algorithm continues its work (moving to step 2 of the Hopfield algorithm). The percentage number is selected experimentally. The choice of the number depends on how many units (black) are contained in the samples. If the number of units is small, then the percentage number will also be small.

The method is evaluated on all images of the MNIST test sample. Each image from the test sample becomes an input of the HNN for each digit, and it is evaluated whether the network correctly recognized the digit. Table 3 shows the results of the HNN recognition using the first method.

Table 3. Results of recognition using first method.

0	1	2	3	4	5	6	7	8	9
75%	91%	74%	66%	79%	71%	74%	85%	47%	70%

Based on the results obtained, the following conclusions can be made:

1. The digits 0, 1, 2, 4, 5, 6, 7 are recognized well: more than 70%.
2. The digit 8 is recognized very poorly: a little more than 45%, that is, more than half of the digits 8 from the test sample are recognized incorrectly.

Table 4 shows a summary table showing where the HNN made errors using the first method.

Based on the data obtained in Table 4, it can be concluded that most of the digits as an erroneous result produce the digit 1. This is because the image for the digit 1 is written as a straight line, which is contained in images for almost all digits. We can also conclude that the digit 8, which is recognized the worst, is most often recognized as the digit 1.

The total recognition error of the neural network is approximately 26%. We present the F-measure for the HNN when using the first method (Table 5).

The method was evaluated on all images of the MNIST test sample. Each image from the test sample is fed to the input of the HNN for each digit, and it is evaluated whether the network correctly recognizes the digit.

Here are examples of images in which the neural network made errors and recognized the image incorrectly. The red indicates the intersection of the image that needs to be recognized and the image from the network memory that the network gave out as the correct result. This is clearly demonstrates the common elements of digits (Fig. 3).

We also give an example when the HNN recognized the image correctly when using the first method (Fig. 4).

Table 4. Summary results of the HNN recognition using the first method.

	0	1	2	3	4	5	6	7	8	9
0	732	21	41	15	28	50	24	26	32	11
1	8	1035	2	3	22	0	0	48	14	3
2	7	153	768	13	15	7	3	61	2	3
3	0	102	14	661	9	60	1	28	16	119
4	1	100	1	1	777	2	4	12	0	84
5	3	66	1	41	41	636	8	36	22	38
6	7	117	3	2	67	34	709	15	3	1
7	1	52	21	4	23	4	0	870	1	52
8	1	166	23	83	48	56	5	46	454	92
9	2	56	3	9	148	4	2	76	5	704

Table 5. F-means of first method.

Dig	0	1	2	3	4	5	6	7	8	9
Rec	0.9	0.5	0.8	0.7	0.6	0.7	0.9	0.7	0.8	0.6
Pre	0.7	0.9	0.7	0.6	0.7	0.7	0.7	0.8	0.4	0.6
F-m	0.8	0.6	0.8	0.7	0.7	0.7	0.8	0.7	0.5	0.6

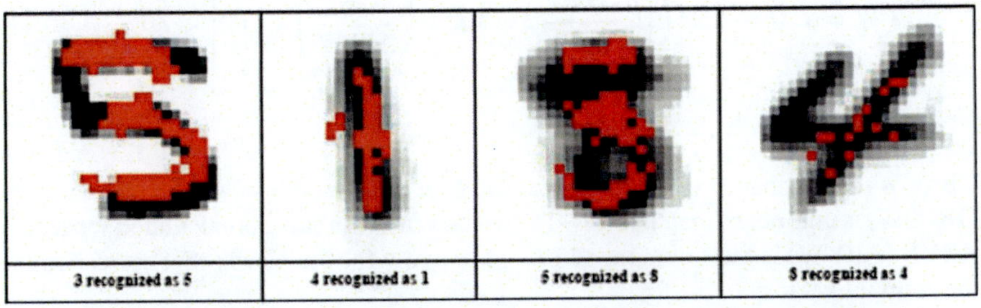

Fig. 3. Examples of incorrectly recognized images when using the first method.

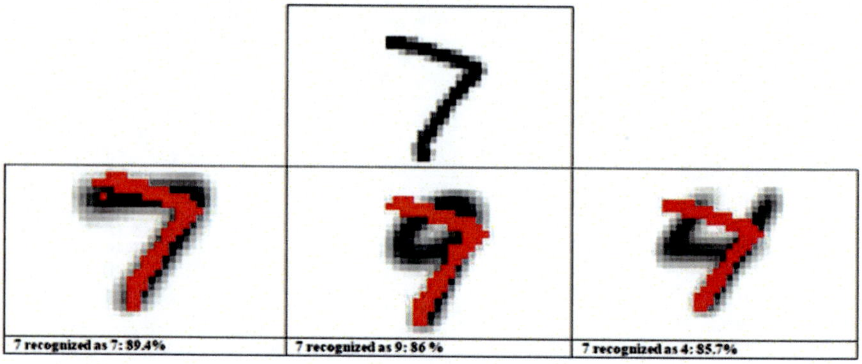

Fig. 4. Examples of correctly recognized images when using the first method.

In this case, the neural network recognizes the digit 7 as the digit 7 with a probability of 89.4%, as the digit 9 with a probability of 86% and as the digit 4 with a probability of 88.7%. We choose the digit 7 as having the maximum probability.

In the second method, to find the image vector from the network memory that is most similar to the original image vector, the vectors must be represented as matrices. In this problem, each vector is a 28 by 28 matrix. Next, we need to compare the matrices by row as well as by column. Let us describe the comparison algorithm for rows (for columns, the comparison is performed using the same algorithm).

Each row counts the number of matched values equal to 1 and the number of matched values equal to -1. It also counts the number of values equal to 1 in the matrix that corresponds to the desired vector, and the number of values equal to 1 in the matrix that corresponds to the k-th vector from the memory. Similarly, the number of values equal to -1 is calculated. Next, the result of how much the row of the matrix corresponding to the original vector is similar to the matrix corresponding to the vector from memory is calculated using the following formulas:

$$CB = CB + 1, if \left(\left| \left(\frac{OB}{AllYB} \right) * 100 - \left(\frac{OB}{AllYB} \right) * 100 \right| < ovB \right) \tag{9}$$

$$CW = CW + 1, if \left(\left| \left(\frac{OW}{AllYW} \right) * 100 - \left(\frac{OW}{AllYW} \right) * 100 \right| < ovW \right) \tag{10}$$

Where

- CB (CW) is number of rows with great matches of values $1(-1)$;
- OB (OW) is number of matched 1 (-1); for each row this value is calculated separately;
- AllYB (AllYW) is total number of 1 (-1) in a row for the matrix that corresponds to the original vector;
- AllMB (AllMW) is total number of 1 (-1) in a row for a matrix that corresponds to a vector from memory;
- ovB (ovW) is number that describes how similar two strings are to each other.

The difference in the formulas limits to zero—the more similar the rows are to each other, that is, the more matches occur, the smaller the difference becomes. In this paper, ovB = 15, ovW = 3. Like in the first method, these parameters were selected experimentally. How to choose such values depends on which images are in the database: what shade prevails. The number of ovW is significantly less than the number of ovB, because in the images of digits, the number of white pixels significantly outweighs the number of black pixels. Therefore, there will be a lot of white pixel matches. We will perform similar calculations for the columns.

After calculating the CB and CW for rows and columns, we need to calculate the similarity of the source vector with the vector from memory using the formula:

$$Res = \left(\frac{CBS}{NBS} + \frac{CWS}{NWS} + \frac{CBC}{NBC} + \frac{CWC}{NWC}\right) * 100 \tag{11}$$

Where

- CBS is number of rows where matches of 1 are sufficient;
- CWS is number of rows where there are enough matches of −1;
- CBC is number of columns where matches of 1 are sufficient;
- CWC is number of columns where there are enough matches of −1;
- NBS is number of rows containing at least one 1;
- NWS is number of rows containing at least one −1;
- NBC is number of columns containing at least one 1;
- NWC is number of columns containing at least one −1;

If the Res value is less than a certain percentage value, the algorithm continues its work (moving to step 2 of the Hopfield algorithm). The percentage value, as in the case of the first method, is selected experimentally. The results of the work are shown in Table 6.

Table 6. Results of recognition using second method.

0	1	2	3	4	5	6	7	8	9
89%	95%	54%	56%	64%	53%	85%	65%	56%	66%

Based on the results obtained, the following conclusions can be made:

1. The digit 1 is recognized more accurately than all the others.
2. The worst recognized digits are 2, 3, 5, 8: a little more than 50%. Summary Table 7 shows the results of the second method of digit recognition.

Based on the data obtained from Table 7, it can be concluded that when the Hopfield neural network recognizes using the second method, the following errors are often made:

Table 7. Summary results of HNN recognition using the second method.

	0	1	2	3	4	5	6	7	8	9
0	874	25	5	3	1	11	15	5	37	4
1	10	1074	1	3	5	2	7	10	21	2
2	124	100	553	58	5	64	79	23	23	3
3	50	46	64	569	14	109	26	40	75	17
4	31	112	10	9	626	6	11	6	24	147
5	95	47	23	94	15	471	45	23	66	13
6	50	69	7	1	1	5	814	0	11	0
7	10	117	9	32	52	4	1	671	21	111
8	149	118	6	39	8	49	9	34	544	18
9	29	53	0	16	140	8	3	74	19	667

- Digit 0 is recognized as digit 8;
- Digit 8 is recognized as digit 0;
- Digit 4 is recognized as digit 9;
- Digit 9 is recognized as digit 4;
- Digit 1 is recognized as digit 7;
- Digit 7 is recognized as digits 1 or 9.

Such results can be explained by the fact that these digits contain parts of each other, that is, the images of the digits have common features.

The recognition error of the neural network is 32%. We give the F-measure for the HNN when using the second method (Table 8).

Table 8. F-means of second method.

Dig	0	1	2	3	4	5	6	7	8	9
Rec	0.6	0.6	0.8	0.7	0.7	0.7	0.8	0.8	0.7	0.7
Pre	0.9	0.9	0.5	0.6	0.6	0.5	0.9	0.7	0.6	0.7
F-m	0.7	0.7	0.7	0.6	0.7	0.6	0.8	0.7	0.6	0.7

The method was evaluated on all images of the MNIST test sample. Each image from the test sample is fed to the input of the HNN for each digit, and it is evaluated whether the network correctly recognizes the digit.

Here are examples of images in which the neural network made errors and recognized the image incorrectly (Fig. 5).

We also give an example when the HNN recognized the image correctly when using the second method (Fig. 6).

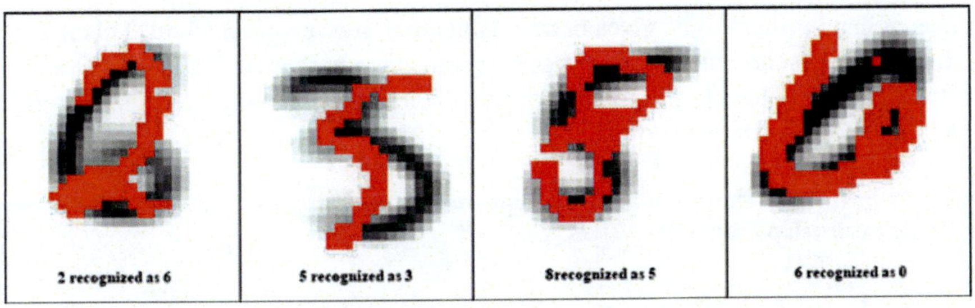

Fig. 5. Examples of incorrectly recognized images when using the second method.

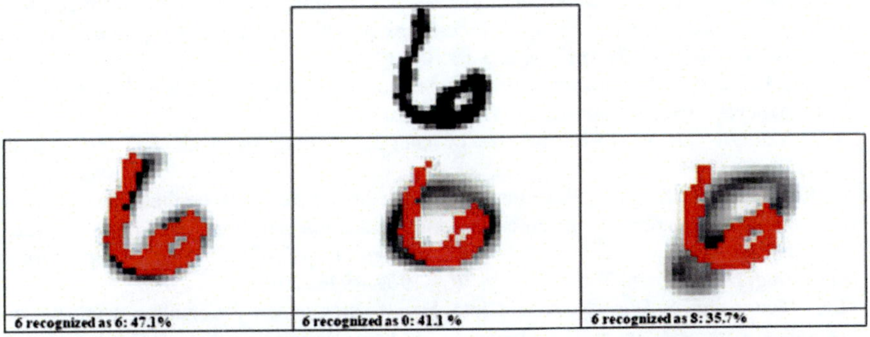

Fig. 6. Examples of correctly recognized images when using the second method.

In this case, the neural network recognizes the digit 6 as the digit 6 with a probability of 47.1%, as the digit 0 with a proba- bility of 41.3% and as the digit 8 with a probability of 35.7%.

Thus, two methods were analyzed to compare the two images. As a conclusion, we can state that the first method is more suitable for recognizing handwritten digits than the second method. Errors in the recognition of handwritten digits are caused by the fact that when using the F-measure, it is not taken into account that the images are shifted, that is, the images have a different slope and can be located above, below, to the right or to the left.

Based on the results obtained, we can conclude that these two methods are suitable for different digits separately. The second method recognizes the digits 0, 1, 6 and 8 better than the first method. Both methods recognize the digit 1 quite well, more than 90%. Both methods recognize the digit 8 very poorly, but the second method recognizes this digit slightly better.

5 Conclusion

The Hopfield neural network is used to recognize handwritten digits. The cluster centers constructed using the Kohonen neural network served as objects for memorizing.

For the Hopfield neural network, two methods for comparing two images were studied and analyzed. As a conclusion, we can state that the first method for recognizing images of handwritten digits gives better results than the second method. This is due to the fact that the second method does not take into account that the image can be shifted: located above or below, to the right or to the left. In addition, when recognizing digits, digits often contain elements of each other.

Acknowledgements. This paper has been supported by the Kazan Federal University Strategic Academic Leadership Program ("PRIORITY-2030").

References

1. Z. Kayumov, D. Tumakov, S. Mosin, Combined convolutional and perceptron neural networks for handwritten digits recognition, in *Proceedings of 22th International Conference on Digital Signal Processing and its Applications*, pp. 1–5 (2020)
2. Y. Xu, W. Zhang,On a clustering method for handwritten digit recognition. 3rd Int. Conf. on Intelligent Net. and Intelligent Syst. **112**, 115 (2010)
3. S.V. Aksenov, Organization and use of neural networks (methods and technologies). Tomsk:NTL (2006)
4. C. Ramya, G. Kavitna, K. Shreedhara,Recalling of images using hopfield neural network model. national conference on computers, communication and controls -11 (N4C-11) (2011)
5. M. Rexy, K. Lavanya, Handwritten digit recognition of MNIST data using consensus clustering. Int. J. Recent Technol. Eng. **7**(6), 1969–1973 (2019)
6. L.C. Munggaran, S. Widodo, A.M. Cipta, Handwritten pattern recognition using Kohonen neural network based on pixel chatacter. Int. J. Adv. Comput. Sci. App. **5**(11), 1–6 (2014)
7. S. Nhery, R. Ksantini,M.B. Kaaniche, A. Bouhoula, A novel handwritten digits recognition method based on subclass low variances guided support vector machine. 13th Int. Joint Conf. on Comput. Vision, Imaging and Computer Graphics Theory and App. (4), 28–36 (2018)
8. S.A. Shal, V. Koltun, Robust continuous clustering. Proc. Natl. Acad. Sci. USA **114**(37), 9814–9817 (2017)
9. E. Miri,S.M. Razavi, J. Sadri, Performance optimization of neural networks in handwritten digit recognition using intelligent fuzzy c- means clustering, in *1st International Conference on Computer and Knowledge Engineering*, pp. 150–155 (2011)
10. S. Pourmohammad, R. Soosahabi, A.S. Maida, An efficient character recognition scheme based on k-means clustering, in *5th International Conference on Modeling, Simulation and Applied Optimazation*, pp. 1–6 (2013)
11. B.Y. Li, An experiment of k-means initialization strategies on handwritten digits dataset. Intell. Inf. Manag. **10**, 43–48 (2018)
12. A. Fahad, N. Alshatri, Z. Tari, A. Alamari, A. Zomaya, I. Khalil, F. Sebti, A. Bouras, A Survey of Clustering Algorithms for Big Data: Taxonomy & Empirical Analysis. IEEE transactions on emerging topics in computing, (2014)
13. A. Ullah, J. Ahmad, K. Muhammad, M. Sajjad, S.W. Baik, Action recognition in video sequences using deep bi-directional LSTM with CNN features. Special Section on Visual Surveillance and Biometrics: Practices, Challenges, and Possibilities **6**, 1155–1166 (2018)
14. K.N. Mutter, I.I. Abdul Kaream, H.A. Moussa, Gray image recognition using hopfield neural network with multi-bitplane and multi- connect architecture, in *International Conference on Computer Graphics, Imaging and Visualisation (CGIV'06)* (2006). https://doi.org/10.1109/CGIV.2006.49
15. A. Basistov, G. Yanovskii, Comparison of image recognition efficiency of bayes, correlation, and modified hopfield network algo- rithms. Pattern Recognit. Image Anal. **26**, 697–704 (2016)

16. I.S. Senkovskaya, P.V. Saraev, Automatic clustering in data analysis based on Kohonen self-organizing maps. Bulletin of MSTU. Nosova I., G.: No. 2, pp.78–79 (2011)
17. Z. Kayumov, D. Tumakov, S. Mosin, Hierarchical convolutional neural network for handwritten digits recognition. Procedia Comput. Sci. **171**, 1927–1934 (2020)
18. J. Lampinen, E. Oja, Clustering properties of hierarchical self-organizing maps. J Math Imaging Vis **2**, 261–272 (1992)
19. F. Murtagh, M. Hernández-Pajares, The Kohonen self-organizing map method. an assessment. J. Classif. **12**, 165–190 (2012)
20. P.Y. Simard, D. Steinkraus, J. Platt, Best practices for convolutional neural networks applied to visual document analysis, in *Seventh International Conference on Document Analysis and Recognition*, vol. 1, pp. 958–963 (2003)
21. D. Ciresan, U. Meier, J. Schmidhuber, Multi-column Deep Neural Networks for Image Classification. in *IEEE Conference on Computer Vision and Pattern Recognition*, CVPR, pp. 3642–3649 (2012)

Automatic Sentiment Analysis on Hotel Reviews in Bulgarian—Basic Approaches and Results

Daniela Petrova(✉)

Technical University, Varna, Bulgaria
daniela.petrova@tu-varna.bg

1 Introduction

Human life is filled with emotions and opinions and they define the way people act and interact with each other. Emotions influence the way people think and react. But these feelings and beliefs are foreign for computer programs and they cannot define them alone. Due to the accumulation of a huge amount of user reviews, impressions, inquiries and questions in social media, blogs and web applications, opinion mining and sentiment analysis have become extremely popular. The analysis of such texts and documents is very important in marketing, advertising, as well as assessment of people's moods and attitudes on politics and even of security point of view. In the present opinion mining and sentiment analysis are scientific fields in the crossroad between information retrieval (IR) and natural language processing (NLP), as well as some other disciplines as text mining and information extraction [1].

The term "opinion mining" is first introduced by Dave, Lawrence and Pennock. They define it as "a process of processing the search results for a certain element and generating opinions about its product characteristics" [2]. Sadegh, Ibrahin and Othman evolve the idea that those are methods for finding and extracting subjective information from texts. Opinion mining is done through natural language processing by defining the perceptions, views and understandings on a given topic, while the automated approach aims to extract the features of the object and thus to determine whether the text is positive, negative or neutral [3].

Dash, Chen and Tong are the ones to use the term "sentiment analysis" in automatic text evaluation for the first time. In the majority of researches opinion mining and sentiment analysis are used as synonyms and very often they overlap [3].

F. P. García Márquez (ed.), *International Conference on Intelligent Emerging Methods of Artificial Intelligence & Cloud Computing*, Smart Innovation, Systems and Technologies 273, https://doi.org/10.1007/978-3-030-92905-3_5

2 Overview of the Previous Work

The different methods that can be used for ASA could be classified according to the way the system is trained: machine learning and lexicon-based. Most of the approaches could be summarized in the following Fig. 1.

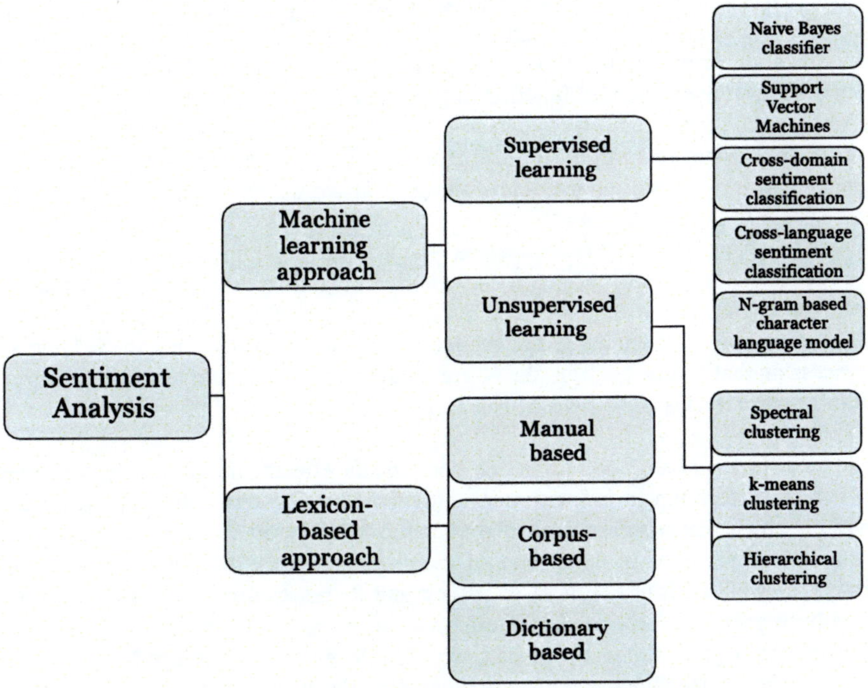

Fig. 1. Classification of sentiment analysis approaches [4]

2.1 Supervised Learning

Supervised learning algorithms build a mathematical model of a set of data that contains both the inputs and the desired outputs. The data is known as training data, and consists of a collection of coaching examples [5].

The most commonly used supervised learning algorithms in sentiment analysis are:

- Naïve Bayes classifier—despite its simplicity, the Naïve Bayes classifier is a popular method for classification of texts and is very effective in numerous spheres. During working, Naïve Bayes takes a stochastic model for generating a document. Using the rule of Naïve Bayes this model is inverted to predict the most probable class of the new document [6]. The Naïve Bayes classifier assumes that the attributes, describing the entries are independent from the context. This assumption is false in most of the real life tasks, but in practice the Naïve Bayes classifier defines the right class with a high accuracy. Another advantage of the classifier is its speed, it calculates fast even with huge datasets [4].

- Support vector machines (SVM)—it is a statistical classification method, first introduced by Vapnik (1995). This model could be used for binary and multi categories classification. SVM seeks for the hyperplane, represented by vector w that divides the positive from the negative vectors with an optimal margin [6].
- Logistic regression—Logistic regression is a regression model but can be used for classification problems when thresholds are used on the probabilities predicted for every class. It uses Sigmoid function to get the probabilities of the classes [7].
- Cross-domain sentiment classification—it is applying a sentiment classifier trained using labeled data on a particular domain to classify sentiment of reviews on a different domain. This often ends up in poor results because words within the train domain may not appear in the test domain.
- Cross-language sentiment classification—aims to use predefined sentiment resources in one language (usually English) for sentiment classification of text documents in another language.
- N-gram based character language model—this is a new model in natural language processing that originates from the N-gram language models. For the basic feature of its algorithm it uses signs (letters, spaces and symbols) instead of words.

In their work Bo Pang and Lillian Lee, compare the effectiveness of a couple of supervised learning techniques for sentiment classification of documents. They use Naïve Bayes classifier, maximum entropy classification and support vector machines, to categorize movie reviews. In the course of their work they try out using unigrams and bigrams, as well as different parts of speech and its position in the text to determine which base gives the best results. From their research can be concluded that Naïve Bayes classifier gives the worst results, while SVM gives the best results—over 81% accuracy. They determine that for more precise results should be explored the focus of every sentence in the document [4, 8].

Rudy Prabowo and Mike Thelwall offer a hybrid approach for sentiment classification in a couple types of documents—movie reviews, product reviews and comments in MySpace. Before using SVM they apply different classifiers, based on rules and study the effectiveness of the different combinations. There are different results for the different types of data, but the optimal turns out to be RBC (rules based classifier) - >SBC (statistics based classifier) followed by SVM [4, 9].

2.2 Previous Work on Bulgarian Texts

To the knowledge of the author, there are not many researchers that have done sentiment analysis on Bulgarian texts. The complexity of Bulgarian language makes it difficult but interesting object for research and there is still necessity for work in this direction. This section of the overview makes an attempt to summarize the previous studies.

Maybe the first to automatically classify Bulgarian adjectives are Boris Kraychev and Ivan Kochev. In their work they are presenting a method for automatic classification using pre-selected emotional axes like love-hate, generosity-greed, goodness-evil and others. They use an unsupervised learning method with the primary data source—the frequency of occurrence of words in documents from the index of the popular search engine Bing [4, 10].

There are several works on Bulgarian language texts that are not exactly connected with opinion mining but could be used as a basis for a sentiment analysis. Such is the semantic classification of adjectives in the Bulgarian WordNet by Tsvetana Dimitrova and Valentina Stefanova. Their classification is based on the information that is already available in WordNet from other synsets (noun, verb and others) that are linked to the adjective synsets via lexicon semantic relations [11]. Another work is the dissertation of Ivelina Stoyanova about automatic detection and tagging of phrases in Bulgarian language. According to her, multiword expressions are a major part of the lexical system of the language and in the same time they are difficult to detect by an automatic system. Dealing with the problems connected with multiword expressions will enhance significantly the results of different natural language processing applications, including sentiment analysis [4, 12].

All this leads to the need for more research in the field of sentiment analysis and opinion mining regarding texts in Bulgarian language and the motivation for the current project of the author.

3 Creating a Database and Applying Supervised Approaches

Every text mining project consists of four major steps of implementation:

- Information retrieval—collecting the needed data for the analysis.
- Data preprocessing—this stage has several sub steps, which can all be made or some could be skipped depending on the results. They are as follows: tokenization; removing stop words, punctuation and empty spaces; stemming or lemming; word tagging.
- Applying the automatic classification methods.
- Analysis of the results.

Figure 2 shows the proposed high level algorithm for ASA of Bulgarian texts, based on the research and the performed calculations, which are described below.

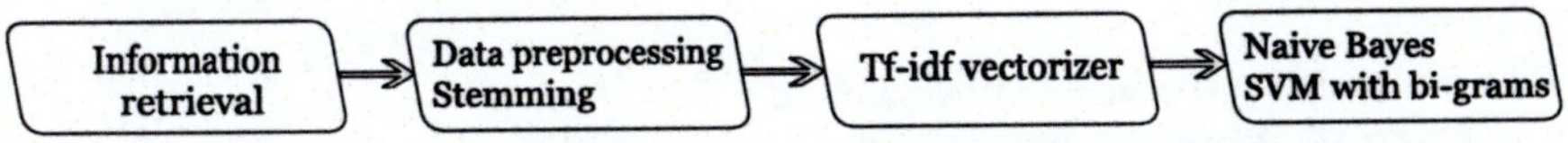

Fig. 2. High level algorithm for automated sentiment analysis of Bulgarian texts

3.1 Information Retrieval

During the search for the needed data for the current research project was discovered that there are almost no ready-made databases in Bulgarian language that could be used. This led to the creation of a database by the authors. It was chosen to apply sentiment analysis on hotel reviews, as they are the most common topic people (in particular Bulgarians) leave comments. For the source of the information at first was chosen the international site www.booking.com, as it provides numerous and easy to extract data in Bulgarian language. The hotel reviews in booking.com are separated in two fields—positive and negative, which allowed the separation of the reviews in two files consisting of respectively positive and negative reviews (from one word to a few sentences). Very often the clients of the hotels leave only positive or only negative review, which resulted in a significantly higher count of the positive reviews.

Positive and negative reviews were extracted for the major big cities and resorts in Bulgaria.

Additional manual correction of the data was made, since there were a lot of comments left in the Negative section that said "Nothing", "Everything was perfect", "Nothing to complain for" and similar reviews. As they do not have any negative sentiment and should not be among the negative reviews, the Negative data base was searched manually for such comments and they were deleted.

After the machined parsing and the manual correction was created a data base of hotel reviews from booking.com with 31 720 positive comments and 17 230 negative comments in Bulgarian language, separated in two files.

For additional expansion of the database was found another source of information—a site for vouchers—www.grabo.bg, which has data for over 2400 hotels or houses in Bulgaria and reviews for them. The reviews from grabo.bg are not divided into positive and negative fields, but have a score from 1 to 5 in stars.

After all the reviews were scraped they had to be modified so that they can fit the current model of the data base. All reviews ranked 4 and 5 stars were tagged positive and the rest as negative. Another way of tagging the reviews could be used in the future work and planned research in the field—the data could be split in three parts—positive, negative and neutral.

The final data base for the current needs of the project consists of 100 082 reviews, 72 078 of which are positive and 28 004—negative.

3.2 Data Preprocessing

The preprocessing of the data included the following steps:

- Tokenizing;
- Removing empty spaces and punctuation;
- Removing all words and letters that are not in Cyrillic;
- Removing e-mail addresses and web sites links;
- Removing stop words—stop words are a group of words and prepositions that have no meaning in text analysis, but are in fact the most common words in texts. They don't have any impact on sentiment analysis and should be removed. To our knowledge, stop words lists are available and ready to use in the programming language Python in many languages but not in Bulgarian. This necessitated the creation of an own list of Bulgarian stop words. The difference between texts with stop words and texts without stop words can be seen in Table 1, which shows the 10 most common words in the

data base before and after removing the stop words. It is clear that the second column consists only of words that have their own meaning and most of them bear sentiment.

Table 1. Most common words before and after stop words removal

Before stop words removal		After stop words removal	
Word	Word count	Word	Word count
и (and)	91,402	храната (food)	10,196
на (of)	55,385	персонал (staff)	9253
е (is)	50,233	има (there is)	7749
в (in)	35,290	стаята (room)	7612
много (many)	34,583	обслужване (service)	7273
за (for)	34,444	вкусна(delicious)	7179
беше (was)	30,878	любезен (kind)	7084
да (to)	27,100	чисто (clean)	6650
се	25,190	храната (food)	6474
от (from)	24,482	персонала (the staff)	6255

- Stemming –in the whole word list there are words that are with the same lemma, but exist in several word forms. This leads to the need to unify the words using lemming or stemming. The process of lemming is done by using a dictionary with all the words and their lemmas. It is more efficient as a method, but constructing such a dictionary is a time consuming work. There is not such for Bulgarian language yet. This is the reason why for the current research was chosen stemming, which is the removal of prefixes and suffixes from the words. There is only one working module integrated in Python for the Bulgarian language, which is developed by Preslav Nakov from the University of California at Berkeley [13]. It removes the suffixes from the words and was used in the preprocessing of the data.

3.3 Applying Classification Methods

For the current state of the project were chosen only methods using supervised learning. After preprocessing of the hotels review database on the data were applied the following supervised learning methods:

- Naïve Bayes
- Support Vector Machines
- Logistic regression
- Logistic regression with bi-grams
- Support Vector Machines with bi-grams.

Each method has been applied both on not stemmed and stemmed data for evaluation of the impact of this preprocessing step. The results from all calculations are presented in Table 2, where they are divided into two columns for stemmed and not stemmed data.

Table 2. Final results for accuracy of prediction of supervised learning models

Model for classification	Without stemming		After stemming	
	Count Vectorizer	Tfidf Vectorizer	Count Vectorizer	Tfidf Vectorizer
Naïve Bayes	**0.873**		0.864	
Logistic regression	0.836	0.847	0.837	0.849
LR with bi-grams	0.839	0.846	0.841	0.852
SVM	0.843	0.857	0.843	0.857
SVM with bi-grams	0.845	**0.864**	0.846	**0.864**

Table 3. SVM results with Tf-idf and count vectorizer

	Precision	Tf-idf recall	f1-score		Precision	Count recall	Vecorizer f1-score	Data
Positive	0.85	0.96	0.90	Positive	0.84	0.95	0.89	36,039
Negative	0.83	0.57	0.68	Negative	0.82	0.52	0.64	14,001

where
Precision = the ability of the classifier not to mark a positive review as a negative.
Recall = the ability of the classifier to find all positive review.
f1-score = 2. mean harmonic of precision and recall.

Naïve Bayes method. For the purpose of the different methods the data were additionally modified. For the Naïve Bayes method the data are combined in a single data base, type dictionary, where every entry is marked positive or negative. Then they are randomly shuffled and divided into two parts—70% for learning and 30% for testing purposes. After applying the Naïve Bayes method it resulted in accuracy of prediction of 0.873. An experiment was done with separating the data in 80%—20% and the result is slightly lower—0.868. Both of those calculations were performed on data that have not been stemmed.

The same system was used to classify the data after stemming. Although there is a slight decrease of the result, it is not significant—**0.864**.

Support Vector Machines (SVM). Before applying the SVM model the reviews should be vectorized. For the purpose of the project were chosen two different types of vetorizers:

- CountVectorizer—number of frequency with respect to index of vocabulary.
- Term Frequency and Inverse Document Frequency (Tf-idf) —a very common algorithm to transform text into a meaningful representation of numbers, that weights the word counts by a measure of how often they appear in the documents [14].

The SVM model was applied on the two types of vectorized data and the results are shown in the following Table 2.

It can be seen from Table 3 that since the positive reviews are more than twice than the negative, the f1- score of the positive is significantly higher than the negative. It is also obvious that Tf-idf vectorizer is more suitable for this classification as it gives higher results.

Logistic regression. The estimation is made by applying binary classification with Logistic Regression (LR) on the data allocated to training and test data, separated into two data sets each containing equal count positive reviews and negative respectively, the second file being used 75% for testing and 25% for validating the results.

The final result of the precision of the LR is **0.847**. After application of stemming to the data set, the result from the LR is **0.849**, which is insignificantly higher.

Logistic regression and SVM with bi-grams. Additional research was done to find if during vectorization are found connections between the individual words. Both Count Vectorizer and Tf-Idf Vectorizer give the opportunity to choose the number of words it can combine. For the purpose of the project were chosen unigrams and bi-grams. Calculations were done on the same data set (both with stemming and without stemming) by applying LR and SVM and the results show that SVM with bi-grams and Tf-idf gives slightly higher accuracy.

4 Conclusion

From the research done so far could be concluded than even with a very small margin, the results are higher with stemming done in the preprocessing of the data. In addition, from the methods for vectorization the more appropriate one is Tf-idf vectorizer. The final accuracy of precision of all methods is very similar and could not stand out one before the others. These results could lead us to a high level algorithm for automated sentiment analysis on Bulgarian texts that gives a relatively sufficient predictability (Fig. 2).

Although all the trained models show similar results, they are under 0.90, which is not completely satisfactory. This leads to the future research that will be done by the author—additional classification using unsupervised methods, exploration of the possible implementation of lexicon based approaches and optimizing the existing models in order to create a new approach that gives maximized results from ASA of customers' reviews in Bulgarian language.

References

1. M. Sadegh, I. Roliana, Z. Othman, Opinion mining and sentiment analysis: a survey. Int. J. Comput. Technol. 171–175 (June 2012)
2. S. Sulova, An approach for automatic analysis of online store product and services reviews. IZVESTIYA J. Varna Univ. Econ. **60**(455–467), 020 (2016)
3. T. Atanasova, N. Filipova, S. Sulova, Y. Alexandrova, J. Vasilv, Intelligent analysis of students dataset. Publishing house Knowledge and business (2019)
4. P. Nakov, Design and evaluation of inflectional stemmer forBulgarian (2003)
5. https://www.parsons.com/2021/03/qrc-technologies-focus-on-ai-transfer-learning-and-ai-transparency-in-ems-survey/
6. Q. Ye, Z. Zhang, R. Law, Sentiment classification of online reviews to travel destinations by supervised machine learning approaches. Expert Syst. Appl. **36**, 6527–6535 (2009)
7. D. Petrova, Overview of the methods used for opinion mining and sentiment analysis and their application in Bulgarian language so far. Comput. Sci. Technol. **XVIII**(1/2020), 126–132 (2020)
8. P. Bo, L. Lee, Thumbs up? Sentiment Classification using Machine Learning Techniques, Proceedings of EMNLP, pp. 79–86 (2002)
9. R. Prabowo, M. Thelwall, Sentiment analysis: a combined approach. J. Informet. **3**, 143–157 (2009)
10. B. Kraychev, I. Koychev. Automatic classification of Bulgarian adjectives on emotional semantic axes, in *VII national conference "Education and ......"*, pp. 121–130
11. T.S. Dimitrova, V. Stefanova, The semantic classification of adjectives in the Bulgarian WordNet: towards a multiclass approach, Cognitive Studies 18, Warsaw (2018)
12. I. Stoyanova, Automatic recognition and tagging of compound lexical entities in Bulgarian language (in Bulgarian), BAS, Sofia, April 2012
13. https://towardsdatascience.com/sentiment-classification-using-logistic-regression-in-pytorch-e0c43de9eb66
14. https://medium.com/@cmukesh8688/tf-idf-vectorizer-scikit-learn-dbc0244a911a

Voice-Controlled Intelligent Personal Assistant

Mikhail Skorikov(✉), Kazi Noshin Jahin Omar, and Riasat Khan

Department of Electrical and Computer Engineering, North South University Dhaka, Dhaka, Bangladesh
{mikhail.skorikov,kazi.omar,riasat.khan}@northsouth.edu

1 Introduction

The world has advanced quite rapidly with the widespread availability and use of computing technologies. One field of computing that is reaching new heights is the domain of artificial intelligence (AI). But even AI systems are not precisely at the level of being called intelligent. This is a natural consequence of lacking computing power and the difficulty of modeling abstract concepts that humans can easily grasp. The domain of natural language processing (NLP) has advanced by leaps and bounds, but we are still quite far from achieving prolonged natural conversations. To demonstrate and advocate for such advancements in AI technologies, we developed an intelligent personal assistant (IPA) that lives inside the smartphone and assists in general tasks without explicit instructions. The personal assistant application will interact with the smart device's user naturally, follow his communication, and perform actions via voice commands of the user.

The idea of IPAs for phones and other platforms is not new considering almost every smartphone comes equipped with the operating system's IPA inside. Examples of prominent intelligent or virtual personal assistants are Google Assistant, Apple's Siri, Amazon's Alexa, and Microsoft's Cortana. There have been many attempts to develop assistants that have their domain vastly restricted to improve performance.

Iannizzotto et al. [9] have designed an architecture for intelligent assistants in smart home environments using a Raspberry Pi device. Their prototype device consisted of the Raspberry Pi along with a small screen, microphone, camera, and speaker. Using these components, the device can'see' the user while speaking to them. The screen of the device is used to display a virtual red fox that moves its mouth while speaking and can make several expressions. Such functionally unnecessary features are useful in making users have a positive impression of the assistant. The authors have used several existing software tools to make the entire system work, such as the Text-to-Speech (TTS) system, Speech-to-Text (STT), Mycroft smart assistant [13], as well as several other such tools. In the end, they seamlessly integrated various services and independent systems into a full-fledged intelligent visual assistant that received positive test evaluation.

Matsuyama et al. [11] present a social focus on virtual assistants. Their assistant, made to help conference attendees find their seats and meet like-minded people, speaks with the user and builds rapport with them through analysis of visual, vocal, and verbal

F. P. García Márquez (ed.), *International Conference on Intelligent Emerging Methods of Artificial Intelligence & Cloud Computing*, Smart Innovation, Systems and Technologies 273, https://doi.org/10.1007/978-3-030-92905-3_6

cues. The proposed assistant can generate animated behavior that matches the level of rapport with the user to make the user more comfortable with the assistant while also making personalized recommendations. Their work consists of tasks that cover a small domain, and their emphasis is on the social aspect of conversations.

Felix et al. [6] have made an Android application intended to help people with visual impairments. The proposed application uses AI technologies to increase the independent activities of blind people and help them interact with their surroundings by using voice input and output as the interface. They proposed a system that leverages Google Cloud APIs for helping identify objects, perform text recognition, and maintain conversations with the user. It can act as an audio- book while also having the capacity to answer queries such as the weather. Their system maintains focus on helping the visually impaired.

In [3], Chowdhury et al. presented a restricted-domain assistant that uses finite state automaton to perform language processing over a small vocabulary. They implemented and trained their own Automated Speech Recognition (ASR) module for two languages—English and Bengali. The scope of their assistant is limited to opening and closing the Facebook and Google Chrome apps on the phone and so the required data was very small. Their focus was on building the system with the speech recognition and user intent identification aspects as their primary features.

Khattar et al. [10] have created a smart home virtual assistant based on the Raspberry Pi. The device is extended through components such as microphones, speakers, and cameras placed at various locations around the house. Existing modules such as Google Speech Recognition API, Google Text-to-Speech (TTS) API, and the Eigenface algorithm are used to provide speech recognition, text-to-speech, and face recognition features. The assistant can be controlled through voice to control home appliances, and answer basic queries related to the weather, stock market, and word definitions.

This paper is organized such that, in Sect. 2, the proposed system is discussed with appropriate visualizations for clarity. The results of our work are discussed in Sect. 3 in detail. Finally, Sect. 4 concludes the paper with some directions on future work.

2 Proposed System

This paper proposes a system for a voice-controlled AI assistant. The design of the system is entirely software-based, needing only a smartphone and an internet connection to operate. A server is required to host the assistant's processing modules, which will communicate with the phone application to take appropriate actions. The system can be described as a chatbot extended to have voice capabilities.

2.1 Software Components

The software dependencies and components of the proposed assistant are described in Table 1. The most critical two software components are Rasa chatbot development tool [2], which consists of Rasa Natural Language Understanding (NLU) and Rasa Core for dialogue management, and the React Native mobile development framework [5].

Table 1. Software components

Component	Library/Program used	Description
Speech-to-text	Native function of phone	Converts user voice input to text
Text-to-speech	Native function of phone	Converts assistant's text output to voice
Intent classifier	RASA NLU	From a given text, identifies the intention of the text
Dialogue manager	RASA CORE	Based on a predicted or given intent, select what response and action to take
Action executor	Phone app and server	Depending on the selected action, execute the action either on phone or on server
Phone app	React native	The interface for the users connecting all services with the core assistant

2.2 Hardware Dependencies

The base hardware requirements for this system on the user side are only enough memory and storage space to run the smartphone application. The server hardware requirements will depend on the number of users actively using the application. At the bare minimum, it is required to have at least 2 GB of memory and a moderately powerful processor.

2.3 System Architecture

The assistant is designed to be an online one, where natural language processing is done on the server and relevant actions are taken on the phone. Figure 1 illustrates how the system is structured and how the components interact with each other.

2.4 System Workflow

The system works in the following manner: a voice input from the user is converted to its text form, which is then passed to the intent classifier. Next, the intent classifier gives out the intent along with any entities while the dialogue manager chooses what to do. Then, a text response is generated from templates in addition to any further actions in code, which is finally converted to voice. The sequential working procedure of the proposed system is illustrated in Fig. 2. The sequence of events occurs every time the user talks to the personal assistant through the application.

2.5 System Features

The features of the assistant are concisely described in Table 2. The proposed automated personal assistant can perform several tasks for the user based on his voice commands. For instance, it can set alarms and reminders and look for the definition of any specified word. It can notify the location and weather information and read the incoming text messages for the user. The assistant can play videos from YouTube, and it can inform

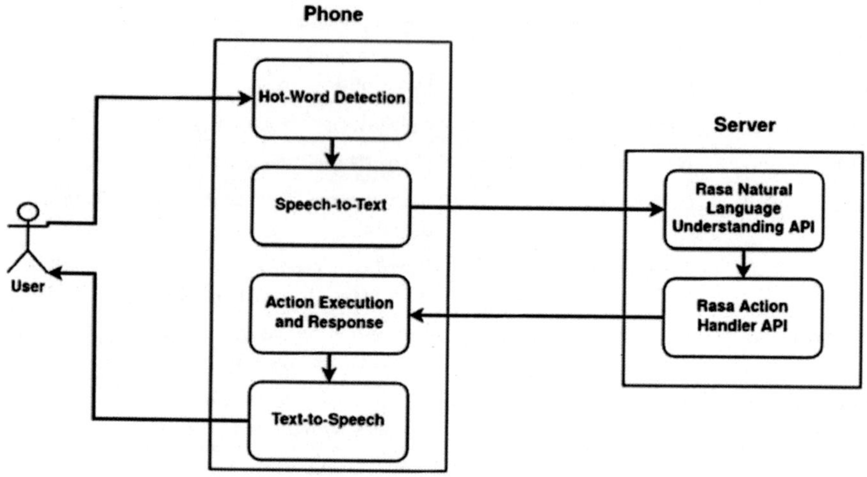

Fig. 1. System architecture.

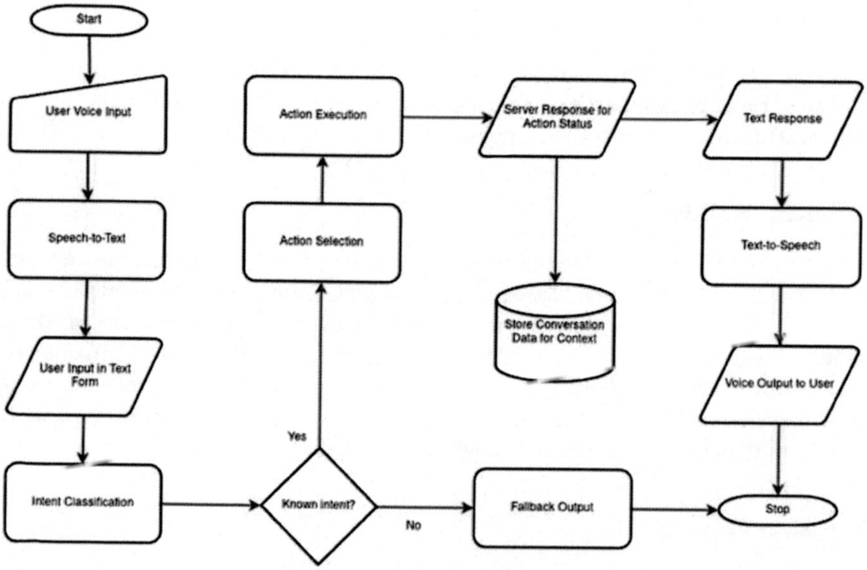

Fig. 2. System flowchart.

the most trending local and international news. It should be noted that, these features are not focused on assisting certain groups of people but are meant to be for anyone's general purpose use.

Table 2. System features

Feature	Description
Weather	Answer queries about weather
Reminder	Set reminders and the assistant will remind with a notification on the phone
Alarm	Set alarms and the assistant will set an alarm on the phone's default alarm app
Read Aloud SMS	Read out loud incoming text messages from the default Messaging app on phone
News	Read aloud and display latest or most trending news either locally or internationally
YouTube	Play YouTube videos on the phone from the search-term given
Definition	Find the definition of the given word and read aloud
Location	Display the current location of the user

3 Results and Discussion

The final application was tested with ten unbiased individuals who rated the performance of the system in terms of various features.

3.1 Server

We have leveraged a Virtual Private Server (VPS) from IBM Cloud [4], which contains our core assistant services. There are two application programming interfaces (APIs) hosted on the server: Rasa chatbot API and Rasa Action Handler API.

The server hosts the intent classification model as well as the dialogue manager. The action execution is performed by the phone, barring the weather and definition features, but the text responses come from the dialogue manager on the server.

Intent Classifier The intent classification model is trained by providing a list of intents with many examples. For example, for an intent of greeting, it may be provided examples of greetings such as "Hello" and many variations of such. The intent classifier comes with a pre-trained language model that is later trained on these provided intents. The final result is a robust model that can handle intent classification quite well. Each of the features listed in Table 2 needs its own intent, along with relevant parameters and supporting intents.

Entity Extractor Entity extraction is the process of retrieving useful information such as dates, numbers, and names from a text. For most of our features, we needed necessary information, e.g., for alarms, we needed alarm time and day; for the weather, there is the option of asking the weather for a specific place or time, and so on. In this work, we have used an advanced natural language processing library called SpaCy [8] and its models to extract the entities in our pipeline.

Dialogue Manager The dialogue management and action selection is handled by a Rasa component called Rasa Core, which uses a transformer-based neural network model to identify the most likely action to take with respect to a given intent. Rasa Core also has several other methods to determine the next action to take, one of which comes in the form of mapping to stories. Stories are sequences of conversations and operations that take place in a chat environment. It is in the form of intents-then-responses and can be chained to be very complex.

3.2 Phone Application

The phone application is built using React Native, a cross-platform mobile development framework. It features taking in voice input, using the native speech-to-text service to convert it to text, and then making an API call to the Rasa NLU service. The text response from the API is then converted to synthetic voice using the native text-to-speech service, while a JSON object is received with action data for the execution.

Hot-Word Detection Hot-word or wake-word detection is the application's constant waiting mechanism where it waits for a specific word to activate voice input for the user to start a conversation. We have used the Porcupine [16] wake-word detection engine, which allows for a limited number of hot-words to be set up, all predefined by the engine. The words that have been selected for our application are blueberry and bumblebee.

Speech-to-Text (STT) and Text-to-Speech (TTS) The application uses the phone's native STT and TTS services made available by Google. Though this reduces the complexity of implementation, there are concerns over the performance of these services. Our tests show that clarity and volume of voice are critical in successful implementation of the executable commands. Although there are better performing alternatives such as Mozilla DeepSpeech [7] and Mozilla TTS [12], the increased storage burden on the user side encourages us to keep the default services.

Action Execution Once an action is identified, we have the option of executing the actions either on the server or on the phone. The phone execution option is more reliable and fast since most of the features are native to smartphones. So once the action is identified, the API returns a JSON object along with a text response for the user. The phone application is responsible for using its native functions to execute the required actions. With the exception of weather and definitions search features, the remaining features are all executed on the phone, although they do require the server to pass the phone the extracted entities that are needed. Figure 3 demonstrates screenshots of some samples of the executed actions.

Constraints Finally, the complete system performs as intended, with certain restrictions and performance reductions. The most significant debilitating factor of our application is the fact that the phone must keep the app open in the app list for it to function.

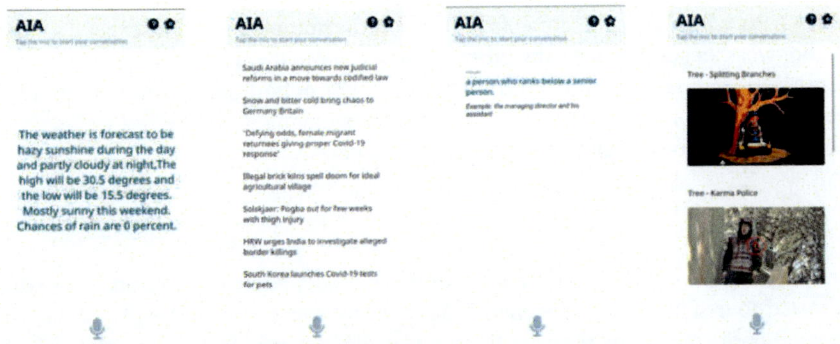

(a) Weather response from asking for the weather 3 days into the fu- ture

(b) Assistant reading and display- ing the latest local news

(c) Assistant displaying the definition of "assistant" after having spo- ken it

(d) Showing the list of videos found on YouTube for "tree" and is about to play the top re- sult automatically

Fig. 3. Screens of the phone application.

The user also has to manually enter the app settings to allow it to display over other applications, although there is no other alternative than to set it manually.

3.3 Communication

The application communicates with the server through API calls. The server internally communicates also through API calls, but it is entirely handled by the framework and thus unnecessary for us to design.

REST API The app needs to use two API calls each time the user speaks to the assistant. First, it sends the text that is retrieved from the speech to the server in the form of a rest request with a JSON body. The server responds with a text response that is intended for speech synthesis. The phone application also has to make a request for the conversation tracker, which is a JSON object that has all the conversation history and data stored. When a task is expected, the tracker has a member called slots that act as the memory for the assistant, and the slot member task-to-do has the value for the action the phone application has to execute. The slots also hold any values for time and date and other entities which have been filled. Using these details, the app can execute actions, after which the app makes a request to reset the slots.

RASA The server internally has two servers running. Both of these are Sanic- based web servers [14] where one hosts the assistant and the other hosts an action execution server. The action server is necessary for running the custom code if a text response is not sufficient. A text response cannot be generated for most of the features without running some specific code initially. For example, to respond with weather details, the weather API needs to be called first. The communication between the action server and the assistant happens within the framework, with minimal configurations required.

Other APIs The implemented features of the personal assistant in this work required API calls to many online services. For location and YouTube, we used Google Cloud

Services APIs which does not come free of cost if the number of usages exceed a certain value. For weather service, we subscribed to the AccuWeather [1] APIs which also come at a cost if exceeding a certain number of usages. For definitions, we use Owlbot [15], a free-to-use API for English word definitions. For showing news, we use web scraping to retrieve news information from a popular local news site.

3.4 Survey

After designing the personal assistant application, finally some of its important features were tested to evaluate its performance. The survey was conducted with the intent of testing the performance of the assistant in real-life settings. It was completed by ten individuals of ages between 20 and 22 and unrelated to the development of the assistant and hence were not biased in the way of their instructions as the developers would be.

The survey contained questions regarding the user's comfort in keeping the phone application open in the background all the time, whether the speech of the user was properly understood, and the performance rating of each feature out of 5. Figure 4 shows the distribution of average scores for the features. According to the survey results, half the surveyors said they are uncomfortable keeping the app always open and all of the users agreed that their speech was properly understood by the assistant.

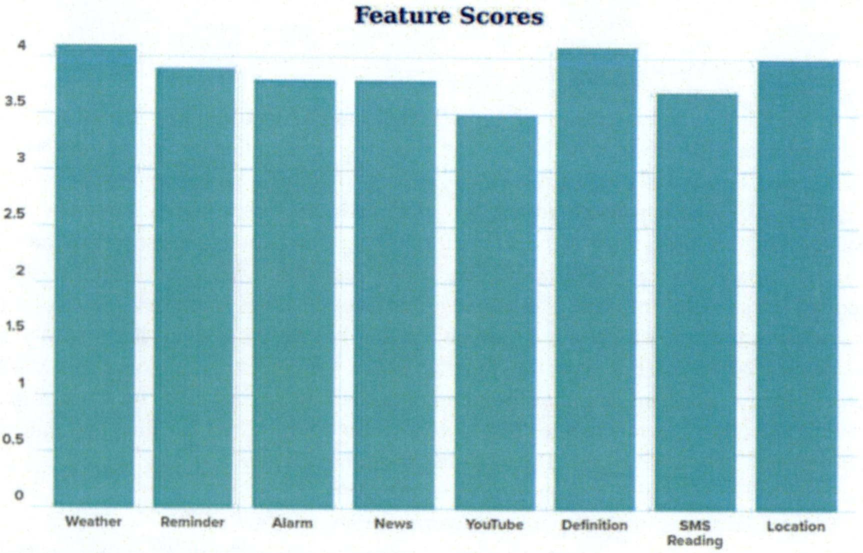

Fig. 4. Performance rating of the features

4 Conclusion and Future Work

This paper proposes the development of an intelligent personal assistant for Android phones. The assistant design is similar to a chatbot with extended abilities and given a voice. The proposed assistant can perform several actions and interpret queries by taking the voice inputs by implementing natural language processing. Although the final app has some minor limitations, there have been positive responses from the test users about the usefulness and performance of the assistant. In the future, the capabilities of the assistant can be extended to include more unique features as well as increase the sophistication of the existing features. The app can be improved by being made to be able to run in the background.

References

1. AccuWeather: Accuweather apis. https://developer.accuweather.com/apis (2021)
2. T. Bocklisch, J. Faulkner, N. Pawlowski, A. Nichol,Rasa: Open source language understanding and dialogue management. arXiv preprint arXiv:1712.05181 (2017)
3. S.S. Chowdhury, A. Talukdar, A. Mahmud, T. Rahman,Domain specific intelligent personal assistant with bilingual voice command processing, in *IEEE Region 10 Conference (TENCON)*. pp. 731–734 (2018)
4. L. Coyne, S. Gopalakrishnan, S.J.R.I., Ibm private, public, and hybrid cloud storage solutions. http://www.redbooks.ibm.com/redpapers/pdfs/redp4873.pdf (2014)
5. Facebook: React native. https://github.com/facebook/react-native (2021)
6. S.M. Felix, S. Kumar, A. Veeramuthu,A smart personal ai assistant for visually impaired people, in *2018 2nd International Conference on Trends in Electronics and Informatics (ICOEI)*, pp. 1245–1250 (2018)
7. A. Hannun, C. Case, J. Casper, B. Catanzaro, G. Diamos, E. Elsen, R. Prenger, S. Satheesh, S. Sengupta, A. Coates, et al., Deep speech: Scaling upend-to-end speech recognition. arXiv preprint arXiv:1412.5567 (2014)
8. M. Honnibal, I. Montani, spaCy 2: Natural language understanding withBloom embeddings, convolutional neural networks and incremental parsing (2017)
9. G. Iannizzotto, L.L. Bello, A. Nucita, G.M. Grasso, Avision and speech enabled, customizable, virtual assistant for smart environments, in *2018 11th International Conference on Human System Interaction (HSI)*, pp. 50–56. IEEE (2018)
10. S. Khattar, A. Sachdeva, R. Kumar, R. Gupta, Smart home with virtual assistantusing raspberry pi, in *2019 9th International Conference on Cloud Computing, Data Science & Engineering (Confluence)*, pp. 576–579. IEEE (2019)
11. Y. Matsuyama, A. Bhardwaj, R. Zhao, O. Romeo, S. Akoju, J. Cassell,Socially aware animated intelligent personal assistant agent, in *Proceedings of the 17th Annual Meeting of the Special Interest Group on Discourse and Dialogue*, pp. 224– 227 (2016)
12. Mozilla: Tts. https://github.com/mozilla/TTS (2021)
13. MycroftAI: Mycroft core. https://github.com/MycroftAI/mycroft-core (2021)
14. Organization, S.C.: Sanic. https://github.com/sanic-org/sanic (2021)
15. Owlbot: Owlbot dictionary api. https://owlbot.info/ (2021)
16. Picovoice: Porcupine. https://github.com/Picovoice/porcupine (2021)

PDCloudEX Software Defined Network

Deepak Mishra(✉), Dinesh Behera, Alekhya Challa, and Chandrahasa Vemu

Prodevans Technologies Pvt. Ltd, Bengaluru, India
deepankar@provedans.com

1 Introduction

PDCloudEx SDN (Software Development Network) is a modular platform that makes intra-datacenter connectivity for virtual and physical workloads feasible. It provides modularity, extensibility of functionality, and REST API for external applications. Along with that, it enables other features like uniform interface, layered system, code on demand, etc. The support for automation of networks and SDN for Network function virtualization while managing the underlay physical fabric makes it industry-ready. It provides domain-wide visibility and analytics to manage physical and virtual domains while enhancing the ability to administer network-wide threat monitoring. PDCloudEX SDN technology enables full network virtualization and allows enterprises, data centers, and service providers to easily deploy, control [1], monitor, and manage secure multi-tenant network infrastructure. In other words, it is a comprehensive solution that makes the network as readily consumable as compute resources across the data center, enterprise WAN (Wide Area Network), and public cloud providers. It does so by providing the missing link to ensure rapid and efficient delivery of highly customizable application services, in and across multi-tenanted data centers. It also provides ease of network configuration for the data centers, which makes the task of configuring the large fabrics much easier and efficient. The product also provides orchestration for tenants, virtual machines in terms of network configuration and network security. The product leverages OpenStack, an open-source cloud computing platform, allowing PDCloudEx SDN to look deep into the aspects for next-generation real-time analytics, security, and remediation.

PDCloudEx SDN has the capability to establish connectivity between bare-metal services and the virtual infrastructure, integration with multiple fabrics, and network devices. It boasts support for layer 2/3 virtual private networks, which is the representation/imitation of a physical network on a cloud, as well as L4–L7 services, which are responsible for data storage, manipulation, and communication-related services. The SDN solution enables the tenants with features like monitoring and troubleshooting while not altering or affecting any mission-critical communications. Equipped with fine-grain management, to enforce related policies based on applications, groups, and services. Provides the application with a number of high-level abstractions, through which the applications can learn about the state of the network and through which they can control the flow of traffic through the network, for optimal performance. It's capable of interfacing with legacy network infrastructure, defining its flexibility. Logically centralized,

F. P. García Márquez (ed.), *International Conference on Intelligent Emerging Methods of Artificial Intelligence & Cloud Computing*, Smart Innovation, Systems and Technologies 273, https://doi.org/10.1007/978-3-030-92905-3_7

physically distributed for resiliency and scalability. The creation of multilayered service makes it possible to create, activate, modify, and restore dynamic multilayer services in a fraction of the time required today. It brings a physically distributed, logically centralized, SDN controller that can interface with other systems. Built on Open Source SDN Software Distribution [2, 3].

High availability of PDCloudEx SDN controllers is ensured without compromising running services and performance. The software architecture handles failures gracefully, providing anytime availability for operating, activating, and managing your applications and network configurations. To sum up all of the above-mentioned features, it can be stated as highly Scalable, Cost Efficient, Secure, Flexible, and User Friendly [4, 7].

2 Features

- *Visualization*: The product provides visualization of Network Topology and Network statistics along with the overlay components in the Datacenter.
- *Multiple Protocol Support*: Enables a range of protocols like OpenFlow, OVSDB (Open vSwitch Database), NETCONF (Network Configuration Protocol), BGP (Border Gateway Protocol), PCEP (Path Computation Element Communication Protocol), LISP (Locator ID Separation Protocol), SNMP (Simple Network Management Protocol).
- *Integration*: The capability to integrate third-party applications that leverage Northbound APIs and Integrate with private cloud management platforms like OpenStack.
- *Support for REST API*: The ability to integrate third-party APIs and features bidirectional Northbound APIs, which are basically the link between the applications and the SDN controller. The applications can communicate to the network regarding their requirements (data, storage, bandwidth, and other related resources) and the network can fulfill those demanded resources, or communicate what it has, enabling a wide range of application support.
- *Cloud Networking and Security*: Responsible for the security and isolation Authentication, Authorization and Accounting (AAA) Services Scalable and stable in performance, boots routing capabilities, and high-performance virtual switching, along with providing multi-tenancy and control over the security policies for the virtual networks.
- *Service Function Chaining*: It is responsible for Network Slicing by providing chaining logic and APIs for provisioning service chains in the network and end-user application.
- *SDN For NFVI (Network functions virtualization infrastructure)*: It is responsible for providing Network Functions Virtualization Infrastructure along with resilience, i.e. flexing its robustness.
- *On-Demand Customization:* Based on business policies, the dynamic policy-driven, software-controlled service chain customization is achieved.
- *High Availability*: In case of crashes and failures, data persistence, i.e. the data will be safe under any unwanted/unavoidable crashes/mishaps.

3 Architecture

- *PDCloudEX SDN inside Openstack*: The PDCloudEX SDN will be installed on the Open Stack by the OOO (TripleO), the API service runs on the Controller role, while the OVS (Open VSwitch) service runs on Controller and Compute roles. It also integrates with the ML2 (Modular Layer 2) core plug-in by providing its own driver called networking-odl. This eliminates the need to use the OVS agent on every node [3, 4].
- *PDCloudEX SDN HA (High Availability) on Openstack*: PDCloudEX SDN High Availability clustering is well-tested architecture for both the neutron and the Controller. It scales the number of API service instances by scaling the number of Controllers to three. This scenario uses network isolation to separate the Management, Provisioning, Internal API, Tenant, Public API, and Floating IPs network traffic [7] (Fig. 1).

4 Key Benefits

- *Scalability*: Ability to ramp-up/ramp-down SDN Controllers; thus efficiently managing and distributing workloads.
- *Cost Efficient*: Automated virtual network resource provisioning, configuration, and operation of networking and security resources minimizes manual intervention while improving operational efficiency and cutting operational costs [5].
- *Secure*: Provides security and isolation by decoupling control and forward planes and enabling the AAA services
- *Flexibility*: Solution for cloud and NFV environments to improve business flexibility and enhances security, availability, and performance
- *User Friendly*: Ease of operations and easy to understand UI which improves operational efficiency.

5 Use Cases

The following are the Use Cases and Compliances

- Centralized management for SDN Controller and enabling the networking policies consistently across any workloads, be it physical or virtual environments.
- Creation and maintenance of Overlay tunnels for widely-deployed and standard VXLAN.
- Enabling integration of third-party network and security solutions through standard REST-APIs and Protocols.
- SDN support for scalability and VM mobility to further increase the flexibility of deployment [6].
- Configuring and Enabling port Mirroring options in Underlay and Overlay to analyze the traffic.
- Performing Micro-segmentation and port mirroring for hybrid workloads.
- Integration with cloud management systems using open interfaces like REST API and plugins.

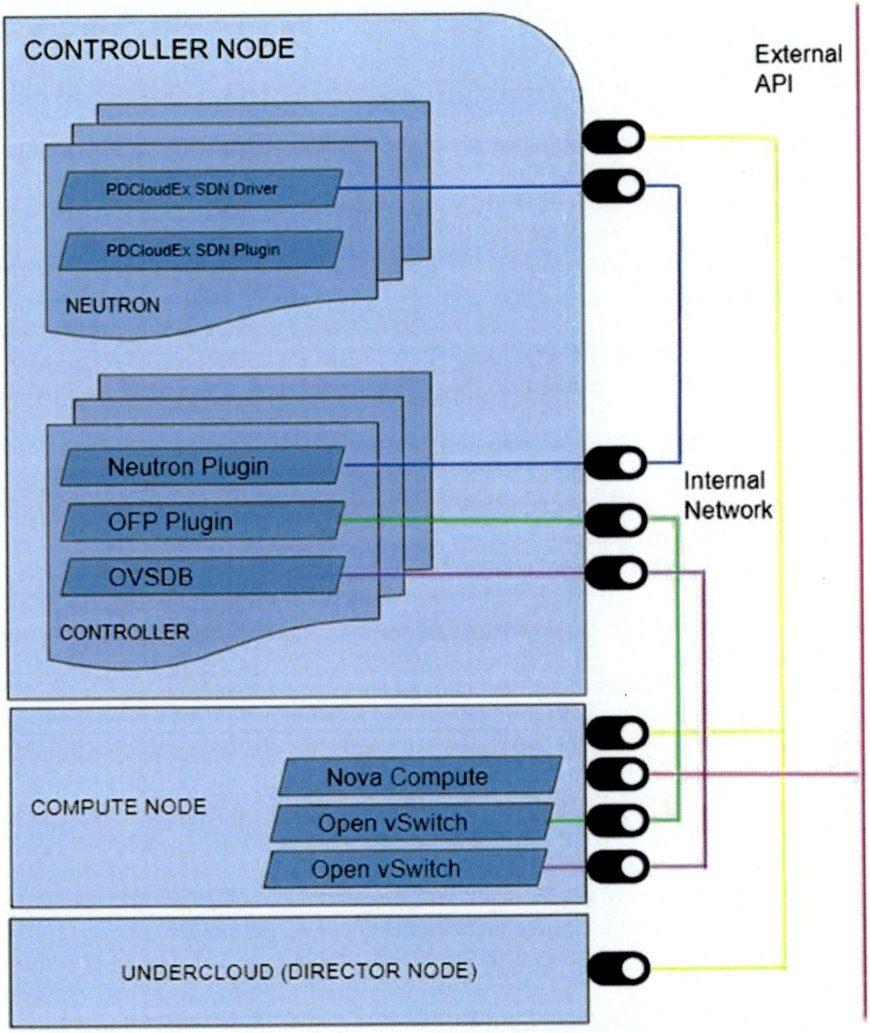

Fig. 1. Architecture of PDCloudEx SDN

- Multi-tenancy and Role-Based Access Control for the tenant management via the centralized management platform.
- Configuring High Availability cluster for data persistence and compliance of performance for running services during failures.
- Centralized management appliance or SDN Controller redundancy to provide availability during a component failure.
- Communication to southbound devices using open standard protocol OpenFlow and communication with Northbound APIs.
- Provision of L4–L7 Services physical or virtual appliance and integration with Virtual Machine Manager.
- Provision of DC Overlay/Underlay Correlation, Overlay Service debugging, and Real-Time Overlay to Underlay Correlation.
- Administer effective network-wide threat monitoring and check Port Reachability State.
- Leverage OVSDB Hardware vTEP plugin for device configuration and connection to HW-VTEP.
- Enable enhanced network operations to support diagnosis, troubleshooting, and analysis of historical failures.
- Configure Real-Time Overlay Failures Root Cause Analysis and Impact Analysis for underlay reachability and problems between the SDN controller and SW/HW-VTEP.
- Visualization of Data Center Overlay Components and Network topology with the Network statistics
- Enable security and isolation by decoupling control and forward planes.
- Enable the Authentication, Authorization, and Accounting (AAA) Services framework for controlling access to resources, enforcing policies, and auditing the resources.
- Enhance the Security by creating and managing the security policies across all Virtual Networks.
- Provision micro-segmentation and ACLs federation between sites for the Virtualized environment
- Creation of zones and policies for hierarchical network optimization.
- Enabling the role-based access control policies and AAA using Local User authentication.
- Support and provision of centrally managed distributed L2–L4 stateful firewalls can be successfully implemented.

References

1. B. Goswami, Implementation of SDN using OpenDaylight Controller (2017)
2. Network Application Testing Platform using Openstack and OpenDayLight
3. The Analysis of OpenStack Cloud Computing Platform: Features and Performance (2015)
4. Advancing Research Infrastructure Using OpenStack
5. R. Van den Bossche, K. Vanmechelen, J. Broeckhove, Cost-optimal scheduling in hybrid iaas clouds for deadline constrained workloads (2010, July)
6. M. Jammal, T. Singh, A. Shami, R. Asal, Y. Li, Software defined networking: State of the art and researchchallenges
7. Release boron at. https://www.opendaylight.org/odlboron

Food Aayush: Identification of Food and Oils Quality

Richard Joseph(✉), Naren Khatwani, Rahul Sohandani, Raghav Potdar, and Adithya Shrivastava

Department of Computer Engineering, Vivekanand Education Society's Institute of Technology, Mumbai, India

1 Introduction

The quality of food consumed by a person plays a significant role in determining a person's health and quality of life. The food consumed must be edible and fresh to avoid the risk of food-borne diseases. The nutritional value of the food is an equally important parameter in determining the quality of the food. Food rich in nutrients is essential to develop a person's immunity. Lack of nutritious food may harm immunity and the person's health in general. The quality of oils used in cooking the food also needs to be taken into consideration. If a particular oil sample is used repeatedly, it may become rancid due to the exposure to high temperatures. Rancid oils not only spoil the taste of food but may also be harmful to human health. Since it is not always possible to know the quality of food and cooking oils, a system must be designed to accurately determine the freshness of food and the rancidity levels of oils. This will allow a person to be sure of the food or oil quality and make an informed decision as to whether the food is suitable for consumption or the oil is ideal for cooking. The system must also predict the nutritional value of food items and determine the required consumption of various nutrients for a particular person based on their daily calorie consumption. This will facilitate the consumption of a healthy, balanced diet by the person and thus be conducive to the right immunity level, thereby preventing susceptibility to diseases.

2 Existing Methodology

Food quality verification is usually done manually or with automated systems that use extensive hardware and complex methods. Manual verification is tedious, time-consuming, and may also be inaccurate. The current automated systems, on the other hand, are tremendously expensive and complicated. Therefore, they are suitable only for industrial or laboratory use and not for everyday use. Also, they are not easily portable. For checking the rancidity of oils, the pH value of oils is not considered as a factor. Usually, the rancidity of oils is checked manually by observing the visual properties and odor. This method may lead to inaccuracy in the determination of rancidity. Currently, calculating the nutritional value of dishes consumed is a cumbersome manual task. Hence, it is challenging to keep track of one's daily nutrient intake.

F. P. García Márquez (ed.), *International Conference on Intelligent Emerging Methods of Artificial Intelligence & Cloud Computing*, Smart Innovation, Systems and Technologies 273, https://doi.org/10.1007/978-3-030-92905-3_8

3 Literature Survey

Many methods have been proposed to classify the quality of food and oils, using machine learning, deep learning, computer vision, and image processing. A classification model using Support Vector Machine (SVM) was developed for the classification of food images. This model made use of the pre-trained AlexNet and VGG16 models for feature extraction. The model was experimentally tested on three different datasets, and the average accuracy was found to be higher than that of CNN models [1]. Among the methods proposed for food quality check, there is one that involves microcontrollers for detecting gases released from food to detect food spoilage. This method also uses machine learning models for predicting the probability and the time required for the spoilage of foods [2]. Other methods use the visual properties of foods, such as color. A plan has been proposed, which first classifies various food items from the images and then, for a particular food item, extracts information about the color using machine learning algorithms and uses the HSV values to detect spoilage in the food item [3]. Classification methods specific to particular food items have been proposed as well. A two-layer method for banana grading has been devised, which classifies bananas based on color and texture using a Support Vector Machine. The second layer uses the YOLOv3 model to check the banana peel's affected areas to prevent the banana's ripening [4].

Similarly, a method for mangoes has been proposed which uses computer vision to determine properties such as mass, volume, density, and defects, enabling the classification into different categories of quality, even considering sweetness as a parameter [5]. For the rancidity check of cooking oils, a system was developed that used color and photodiode sensors to extract visual information about the oils and accordingly detect the oils' frequency of use. The oils are classified into five usability categories with the k-nearest neighbor's algorithm implemented on an embedded system platform. This will help to determine if a particular oil sample is suitable for further use in cooking. The use of pH sensors, temperature sensors, and gas sensors has been made to classify food items [6–8].

4 Proposed Methodology

The model proposed in this paper uses artificial intelligence and image processing to classify food items into various levels of freshness based on the food images captured using a mobile phone camera. In the case of oils, the classification into different rancidity levels is done based on oil sample images captured through the mobile phone camera and the pH value of oils captured through a pH sensor and given as an input to the application. For finding the nutritional value of food items, either the ingredients of the dish or an image of the dish are taken as input. Additionally, the daily dietary requirements for a particular person are calculated from daily calorie consumption. The application thus has the following main features:

- Identification of the freshness of food items
- Identification of repeated frying on cooking oils from the rancidity levels of oils
- Nutritional evaluation of dishes
- Calculation of nutritional requirements for a particular individual

The system is a simple mobile application developed using the Flutter Toolkit, which uses the DART programming language. For the rancidity level check feature, there is an option to integrate the application with a pH sensor using a microcontroller to evaluate the rancidity levels using pH values and the oil sample images.

The nutritional analysis feature and dietary requirements calculation require data analysis and are developed using Streamlit, a Python framework.

In addition to these functionalities, the application also has some customer care features that help guide customers and answer common customer questions. These include a grievance portal, a frequently asked questions page, and an about us page.

The following figure shows a block diagram representation of the application (Fig. 1).

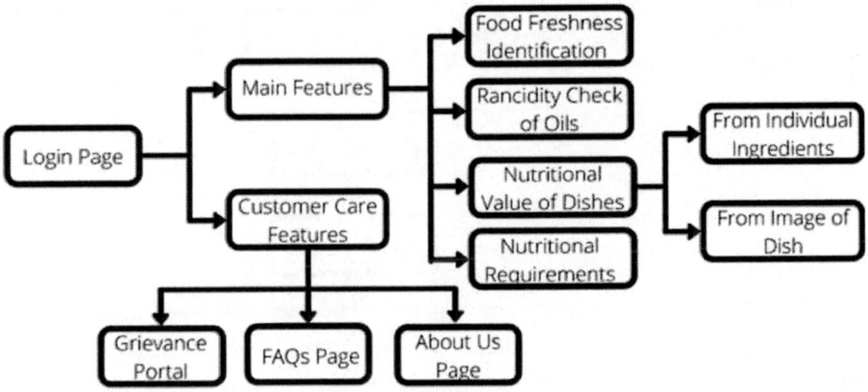

Fig. 1. Block diagram

The figure shows the features of the application. The two categories of features are the main features and customer care features. The main features are the freshness identification of food items, rancidity check of oils, analysis of the nutritional value of dishes from ingredients present, and the calculation of daily dietary requirements for a particular user based on daily calorie intake. Customer care features include a grievance portal where customer-specific issues or questions can be addressed, a frequently asked questions page, and an about us page for contact information.

Each of the main features of the application shall now be described in detail.

A. *Identification of the freshness of food items*

For the development of this feature, firstly, a training dataset of food images is created. Pictures of different food items are captured daily. The images are captured using a simple mobile phone camera. With the progression of days, the food items would become stale. Depending on the day when the food images were captured, each image is labeled into one of three categories:

- Fresh
- Medium
- Stale

This creates the training dataset. The classification model is trained on this dataset. The model is trained first to identify the food item from the image. After identifying the food item, the model must identify the food item's visual properties from the image and accordingly classify it into the appropriate freshness level. Subsequently, when the image of a new food item is captured, the model will identify the food item and its freshness level from the image (Fig. 2).

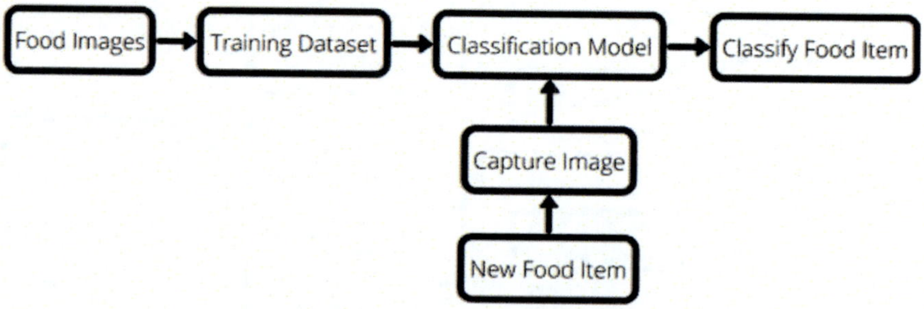

Fig. 2. Diagrammatic representation of food freshness detection feature

The figure shows the working of the food freshness detection feature. Initially, food images are captured, which form the training dataset. The classification model is trained on this dataset. Subsequently, when new food items are to be classified, the new food item's image is captured. The trained model then classifies the new food item into a particular freshness level.

B. Identification of repeated frying on cooking oils from the rancidity levels of oils

Here, first, the training dataset of oil sample images is created. The oil sample images are labeled into different categories of rancidity based on their visual properties. The model is trained using this training dataset. For this feature, the mobile application is also integrated with a pH sensor that will record the oil samples' pH values. An IoT microcontroller will act as an interface between and integrate the mobile application and the pH sensor.

pH value shows the level of acidity or basicity of any solution. It ranges from 0 to 14. pH values below 7.0 indicate that the solution is acidic, while pH values above 7.0 indicate basic solutions. pH 7.0 indicates a neutral solution. As oils become rancid, their acidity levels increase. Oils with lower pH will thus be more rancid.

The pH sensor records the pH value of the oil samples. pH sensor outputs analog value in range of 0.5V to 3V. ESP32 microcontroller processes these values and decides the rancidity of oil.

The model thus considers the visual properties of the oil and the pH values to classify a new oil sample into a particular rancidity level (Fig. 3).

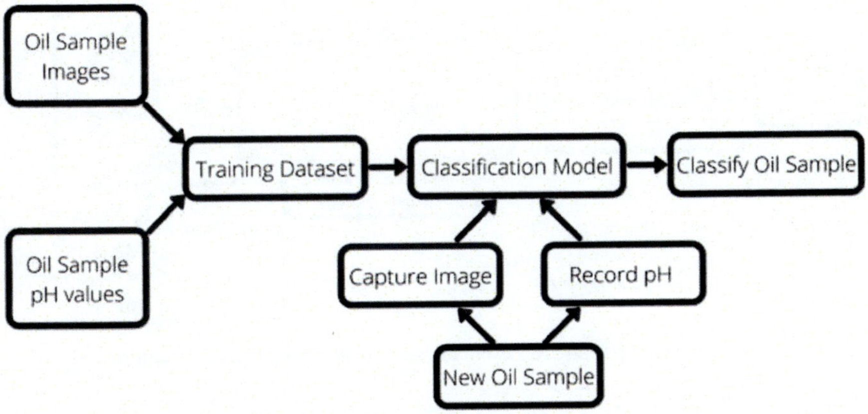

Fig. 3. Diagrammatic Representation of Oil Rancidity Check Feature

The figure shows the working of the oil rancidity check feature. Oil sample images and pH values of oils are recorded to create the training dataset. The classification model is trained. For any new oil sample, the image is captured, the pH value is recorded, and then classification is done using the trained model.

C. Nutritional evaluation of dishes

This feature uses a dataset of dishes and their ingredients. An analysis of the data is performed. When a particular dish is given as input, its nutritional value is calculated from the ingredients present in the dish. Additionally, a matrix is designed, which contains various combinations of food ingredient types. There are certain combinations of ingredients that, when used together, lower each other's nutritional value. In some cases, certain combinations may even be harmful to health. The matrix contains information about each combination of ingredients, regarding whether it is suitable or harmful. Thus, the nutritional analysis feature also helps a user identify if a particular dish contains an unhealthy combination of ingredients (Fig. 4).

The figure shows the working of the nutritional analysis of dishes. The ingredients of the dish are given as an input to the application, either individually or by capturing the image of the dish. Based on the dataset of dishes and their ingredients, the nutritional analysis of the dish is performed. Additionally, a matrix is used that identifies suitable and harmful combinations of ingredients.

D. Calculation of Nutritional Requirements for a Particular Individual

The daily nutritional requirements, i.e., the required daily consumption of various nutrients such as proteins, carbohydrates, and fats for a particular person, depends on the

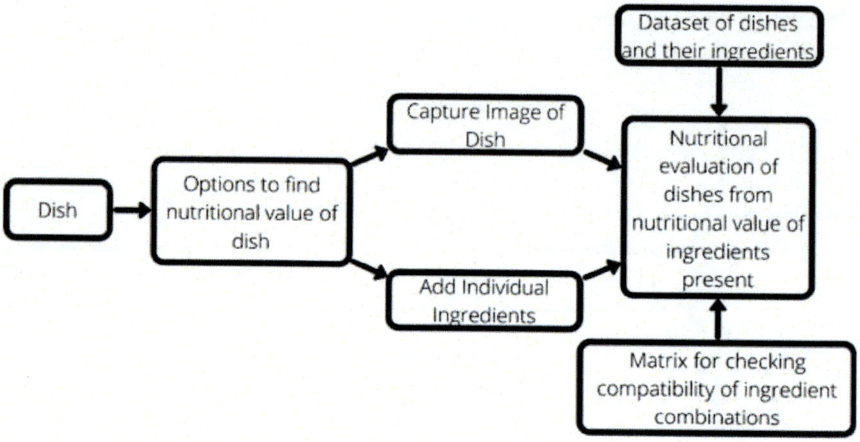

Fig. 4. Diagrammatic representation of nutritional evaluation feature

person's daily calorie consumption. Additionally, the ideal daily calorie consumption also varies from person to person. One of the factors that determine suitable daily calorie consumption is gender. This feature takes as input the number of calories consumed by the user daily. Accordingly, the output is the quantity of proteins, carbohydrates, and fats required daily for the specific user. The feature also asks the user to enter their gender, and according to the gender tells the user the ideal daily calorie consumption for the user. The user can now use the feature again to calculate the daily nutritional requirements as per this ideal calorie consumption (Fig. 5).

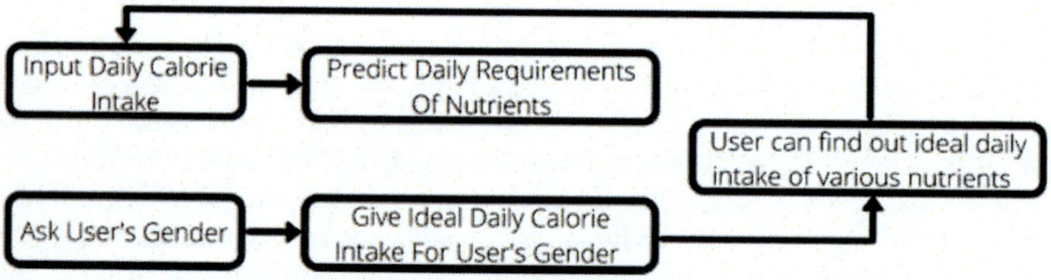

Fig. 5. Diagrammatic representation of dietary requirements feature

The figure shows how the nutritional requirements for a particular individual are determined. Based on the daily calorie intake, the required amounts of various nutrients are calculated. The ideal daily calorie intake is also informed to the user based on their gender.

Classification Model:

The models for the classification of food images and oil sample images use convolutional neural networks (CNNs) for image classification. Convolutional neural networks are suitable since the process of feature extraction is performed conveniently. The CNN consists of four layers: the convolution layer, the nonlinear layer, the pooling layer, and the fully connected layer. The features and visual properties of the images (color and texture for both food and oil images, and the presence of surface defects in case of food images) are identified using the CNN layers, and the images can be classified accordingly.

The following figure shows the modular diagram of the application, i.e., a brief overview of the application's entire functioning and its various features (Fig. 6).

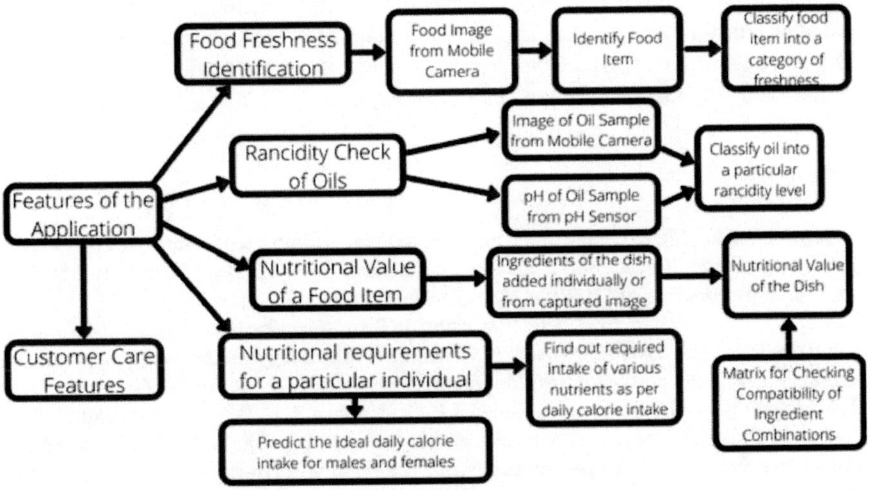

Fig. 6. Modular diagram of the application

The figure shows a summary of the features of the application. Food freshness detection uses images for training the dataset and performs classification for new food items. For rancidity check of oils, images, as well as pH values, are used. The nutritional evaluation of food items considers the ingredients of the food item and performs the analysis based on the food ingredients dataset. The matrix for checking compatibility of ingredient combinations is also used. Nutritional requirements for an individual are calculated from the daily calorie consumption.

5 Results and Discussions

The accuracy of the classification model depends on the size of the training dataset. As the size of the training dataset is increased, the accuracy of the classification model increases. This is because the separation between the different classes becomes more visible. Therefore, sufficient images of each food item are required in each category (fresh, medium, and stale). The same holds for the oil samples, i.e., sufficient images, as well as pH values of oil samples, are required in each rancidity level.

6 Conclusion

The proposed system will prove useful in ensuring a good quality of food and oils before consumption. It would help prevent health-related problems due to the consumption of stale or low-quality food and the consumption of food cooked in rancid oils. The nutritional analysis feature would allow the system users to enter the dishes that they are consuming daily and inform them about these dishes' nutritional value. The users would thus be able to keep track of their daily nutritional intake. Users would also find out their ideal daily calorie consumption and ideal daily nutritional intake according to the calories consumed. Thus, the application would also act as a means of ensuring a healthy and balanced diet, which would ensure adequate immunity levels and resistance to diseases, thereby incorporating a healthy lifestyle.

References

1. A. Şengür, Y. Akbulut, Ü. Budak, Food image classification with deep features, in *2019 International Artificial Intelligence and Data Processing Symposium (IDAP)*, Malatya, Turkey, 2019, pp. 1–6. https://doi.org/10.1109/IDAP.2019.8875946
2. N. Hebbar, Freshness of food detection using IoT and machine learning, in *2020 International Conference on Emerging Trends in Information Technology and Engineering (ic-ETITE)*, Vellore, India, 2020, pp. 1–3, https://doi.org/10.1109/ic-ETITE47903.2020.80
3. R.K. Megalingam, G.S. Sree, G.M. Reddy, I.R. Sri Krishna, L.U. Suriya, Food spoilage detection using convolutional neural networks and K means clustering, in *2019 3rd International Conference on Recent Developments in Control, Automation & Power Engineering (RDCAPE)*, NOIDA, India, 2019, pp. 488–493. https://doi.org/10.1109/RDCAPE47089.2019.8979114
4. L. Zhu, P. Spachos, Food grading system using support vector machine and YOLOv3 methods, 2020 IEEE Symposium on Computers and Communications (ISCC), Rennes, France, 2020, pp. 1–6. https://doi.org/10.1109/ISCC50000.2020.9219589
5. N.T. Thinh, N. Duc Thong, H.T. Cong, N.T. Thanh Phong, Mango classification system based on machine vision and artificial intelligence, in *2019 7th International Conference on Control, Mechatronics and Automation (ICCMA)*, Delft, Netherlands, 2019, pp. 475–482. https://doi.org/10.1109/ICCMA46720.2019.8988603.
6. D. Syauqy, H. Fitriyah, M.N. Marofi, Detecting repeated frying on cooking oils based on its visual properties using embedded system, in *2019 International Conference on Sustainable Information Engineering and Technology (SIET)*, Lombok, Indonesia, 2019, pp. 223–227. https://doi.org/10.1109/SIET48054.2019.8986088
7. J.A. Garcia-Esteban, B. Curto, V. Moreno, I. Gonzalez-Martin, I. Revilla, A. Vivar-Quintana, A cloud platform for food sensory estimations based on artificial intelligence techniques," 2018 13th Iberian Conference on Information Systems and Technologies (CISTI), Caceres, 2018, pp. 1–5, https://doi.org/10.23919/CISTI.2018.8398635
8. A. Prajwal, P. Vaishali, z. payal , D. Sumit, Food quality detection and monitoring system, in *2020 IEEE International Students' Conference on Electrical,Electronics and Computer Science (SCEECS)*, Bhopal, India, 2020, pp. 1–4. https://doi.org/10.1109/SCEECS48394.2020.175

LEAST: The Smart Grocery App

Mohd. Zeeshan(✉), Navneet Singh Negi, and Dhanish Markan

School of Computer Science and Engineering, Galgotias University, New Delhi, India

1 Introduction

The problem that we are trying to solve by building this project is: For example, A person buys bread every third day, so it will automatically recommend bread to the user every third day, furthermore as a smart assistant In suggestion box will also suggest products related to bread, such as jam, peanut butter, etc. on the pattern of what he likes to buy. What can bring more convenience to the user than just scanning his receipts by which the app will learn about his shopping habits, add items using voice commands, or just watch their barcodes [1–3].

- Learning Goals
  - Android/IOS app development using Flutter.
  - Building, Serving and using APIs.
  - Handling Backend with Flask.
  - ML model to make predictions.
  - Firebase for Authentication and Storage.

2 Problem Definition

Remembering everything at the shopping complex or keeping track of everything we need or may have expired is quite hard. We have made a smart assistant solve this problem. Least is a complete package to manage a virtual list of groceries and day to day things to buy from malls or shops without frequent user modification. Also, it has learning capabilities from your shopping habits with just some clicks and scans. Also, adding items is very handy and comfortable in our app either you can type out or only use your voice to add the items to the list or directly push the suggestions to your list.

- Required Knowledge
  - Firebase
  - Flask
  - Dark Flutter
  - Heroku
  - API
  - Machine Learning

F. P. García Márquez (ed.), *International Conference on Intelligent Emerging Methods of Artificial Intelligence & Cloud Computing*, Smart Innovation, Systems and Technologies 273, https://doi.org/10.1007/978-3-030-92905-3_9

3 Feasibility

- Receipt Scanning
  - The app can use the phone camera to take a photo of the receipt
  - The app shows the list of items bought by the user in that transaction on the interface
- QR/Barcode Scanning
  - The app can use the phone camera to scan the QR or barcode and get the product name using an API
  - It can help indirectly adding the user to add the product without typing its name
- Adding items using voice commands
  - The app will be featured with voice recognition to add a product.
- Smart Shopping list
  - Based on products scanned and stored from the user transaction receipts, barcodes/QR, the app creates a list of items for the user
  - Based on how many days there are multiple purchases. The app turns the articles on/ off in the mobile interface in the user's "current shopping list."
  - The user can decide to swipe an item off for the time being. If they choose not to see it on their list,
  - they can add an item not showing in their inventory.
- Product recommendations
  - The app can utilize a machine learning algorithm for finding distances between products based on user-product similarity.
  - Based on the distance between products, a new product may be recommended to a user based on their previous purchase patterns.
  - The user can decide to add the recommended item to their shopping list.
  - The user can decide to delete an item from their recommendations list.

4 Complete Work Plan Layout

- Basic Implementation (Grocery List):
  - This will be the basic implementation of the list where users can modify the plan.
  - Add items from their barcodes and give voice commands to add. All manual features can be accessed.
  - This phase will be focused on the UI of the app and naive parts of the App. Interface designing and User compatibility will be dealt with.

- After this Implementation of Machine Learning starts where the above implementation, with the next phase will include the implementation of Machine learning and connectivity with databases.
- After the successful integration of ML the app will be able to suggest and recommend to the user. The database will help in the same features, such as providing data to monitor user purchases and allows the machine to learn. After the compilation of these two phases, the app will be all ready to perform the same as mentioned.

5 Technology Stack Used

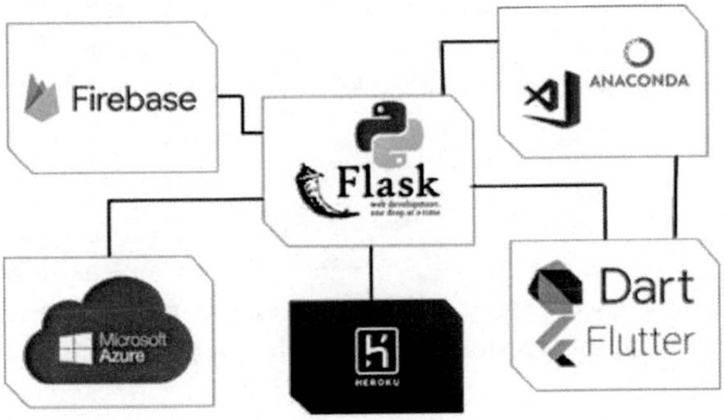

6 Use Case Diagram

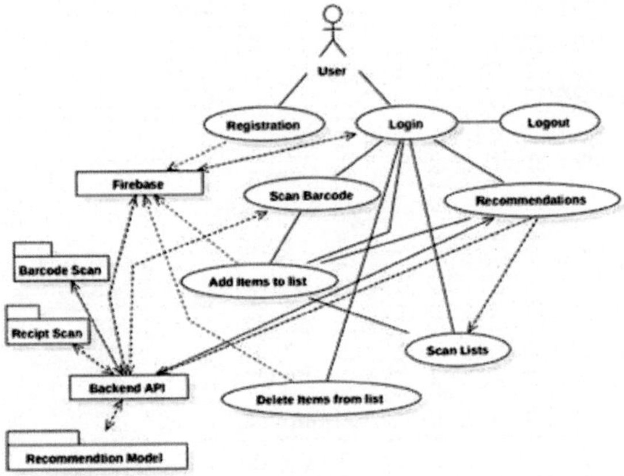

The work will start when the user signup with the mobile application he will be authenticated with firebase. After that, he will have options to either add entities to the shopping list or scan a receipt or scan a barcode. If he chooses to scan the receipt or the barcode, a request will be made to the API hosted on the Heroku platform.

If a request for receipt scanning is made, it will first click an image and send it to the firebase storage from when it will be converted into a PDF for further use using the Python Imaging Library. After that, it will take the help of the Microsoft Azure Cognitive Tools from where their Scanning API will be used to extract all the data of the receipt from which the data is then filtered and formatted and sent back to the flutter app, which will be then rendered there.

If you go for the barcode scan, it will only call the EAN API, giving the EAN result, which can be then formatted and sent to the flutter app to either add to the list.

The K-Nearest Neighbour algorithm will be used for the recommendation system, which will be explained more further in this paper. This will help to give similar suggestions that most of the people like to buy and they can add them to their lists.

7 Datasets

The datasets will be updated as the users use the application. It will extract all the data from the firebase of the users' habits, and when a user wants some recommendation, it will use the machine learning model to compute the predictions (Tables 1 and 2).

Table 1. User info and rating table

UserId	Item	Rating
0I9ZfPXjTud0C2ovAmQKY a41dRa2	Milk	10.0
0I9ZfPXjTud0C2ovAmQKY a41dRa2	Bread	5.0
CUZ7QpBGv7UUNLiWzxV 7	Cream	5.0
CUZ7QpBGv7UUNLiWzxV 7	Apple	5.0
g9MoGb4BWUSqcJnfpeJr	Milk	10.0

Table 2. Items table

ItemId	Item
1	Apple
2	Banana
3	Bread
4	Cream
5	Milk

Above the first data set is the data of a particular user. It contains the user id, the items he buys, and the rating is calculated by the formulae below.

$$\textbf{Rating} = (\text{Total Number of times a particular item bought } * 10) / \text{Total number of unique products}$$

This will help us understand the items the user most often buys.

8 Preliminary

There are three types of recommendation techniques in the market.

- **Collaborative Filtering**: Suppose there are similar users. If a user buys milk from company A and the other person also buys the same milk, then they will be recommended the other products that they buy.
- **Content-Based Filtering**: Suppose person A and B. Person A read an article and gives it a rating, and person B also gives the same rating after reading that article. Then they will be recommended the other article they read.
- **Hybrid Filtering**: In this type of recommendation system, both the other filtering techniques are combined.

So, in this model, we will be using K- Nearest Neighbor using Collaborative Filtering.

So, firstly to work this model, we will require some dummy data so we can use the receipts by checking out a grocery store nearby and get their data. We can then scan them, and the data will be automatically added to the database of the user purchases, which will help the algorithm work.

The datasets will be measured, and the further process will begin. So, we will start with clean merging the dataset and then taking out the counts. After that, we will extract the pivot matrix, which will have the index as item names, column as user id, and the values as the ratings. We will then use the CSR matrix from the scipy sparse library and pass the pivot table to the CSR matrix.

Then we will import the K - Nearest Neighbour, which has a concept of Cosine similarities, which we will discuss here further, and this algorithm is not K Nearest neighbor classifier or regression. It is unsupervised learning. It works on the same basis that we try to find the nearest neighbor and group them. We will record a particular item and then specify the number of suggestions we want. Initially, the dataset is small, so the value should be small, but we can increase that number as it grows. We will get the indices and the distance corresponding to that product and then flatten the distance and give all the nearby values' results.

Cosine Similarities: It is a measure of similarities between two vectors of an inner product space that measure the cosine of the angle between them. For product clustering, different measures are available where cosine similarity is one of the most commonly used.

The similarities between the two items can be calculated using the formulae below:

$$\cos(A, b) = A.B/\|A\|\,\|B\| = \sum_{i=1}^{n} AiBi \sqrt{\sum_{i=1}^{1}}$$

when this measure takes bigger values, i.e., close to 1, then the two items are identical. When it takes values close to 0, this indicates that there is nothing in common between them, i.e., their vectors are orthogonal to each other. Please notice that the attribute vectors A and B are usually the term frequency vectors of the item.

9 Problem Resolution

The app can utilize a machine learning algorithm for finding distances between products based on user-product similarity.

Based on the distance between products, a new product may be recommended to the user based on their previous purchase patterns.

The user can decide to add the recommended item to their shopping list. The user can choose to delete an item from their recommendations list.

To make the app more hassle-free, we have added receipt scanning, which will automatically add his shopping preferences into the database. We have also added Bar Code scanning to make it easier to add the product.

We are also trying to add voice to text features to add items more efficiently by using his voice command.

10 Conclusion

After extensive research and analysis, we have drawn the following conclusions.

Yes, online shopping is pretty popular among the young Indian blood. Many teenagers and bachelors are now using E-Commerce for fulfilling their shopping desires.

Most of them are completely aware of all the pros and cons of online shopping. Graduates are the majority of users of E-retail. Cash on delivery remains the best choice for payment, followed by Debit and credit cards. Females are marginally more interested in shopping through the internet due to its convenience. Most users are shopping once or twice a week through the web, with spending ranging from rupees 100 to 2500 monthly.

With most E- retailers selling branded goods and having flexible return policies, they are well trusted by the users. Apparel, footwear, and Accessories lead as the most demanded goods online, followed by Software and Music. The most visited and trusted sites are Amazon and Flip-kart, according to the survey. After this analysis, we conclude that India has a vast potential for growth of a multibillion-dollar industry of E-Commerce as the top players of the market have more than 100% growth yearly. Their valuations are crossing billions of dollars. There are many recommendation systems in the market, but the sector we have chosen to make is not yet touched for a recommendation. If you think, every individual in one day in his life needs to go to the market with a list, making the same as digital so one can carry it all along will surely help. Moreover, no matter how attentively you make a grocery list, one always forgets one or two items. Giving recommendations will help them choose the right pair of products and not forget any. To provide an ultimate experience to the user, the algorithm we have used is K nearest neighbor, which is extremely fast and versatile. The computation time is less. It also provides a reliable and accurate result, which made it our ideal option. The overall power of prediction certainly depends on k. The more value of k is, the better result it gives so that the user experience will be smoother and better over time. The algorithm is seamlessly integrated, and for the optimum result, we have taken care of any backdoor or limiting scenario affecting K.

References

1. B.M. Sarwar, G. Karypis, J.A. Konstan, J. Riedl et al., Item-based collaborative filtering recommendation algorithms. **1**, 285–295 (2001)
2. H. Yildirim, M.S. Krishnamoorthy, A random walk method for alleviating the sparsity problem in collaborative filtering, in *Proceedings of the 2008 ACM conference on recommender systems* (Vol. 95, pp. 131–138). ACM (2008)
3. K. Kalaivendhan, P. Sumathi, An efficient clustering method to find similarity between the documents. Int. J. Innov. Res. Comput. Commun. Eng. **1** (2014)

A Supervisory Control and Data Acquisition System Filtering Approach for Alarm Management with Deep Learning

Isaac Segovia Ramírez(✉), Pedro José Bernalte Sánchez, and Fausto Pedro García Márquez

Ingenium Research Group, University of Castilla-La Mancha, Ciudad Real, Spain
{Isaac.segovia,pedro.bernalte,faustopedro.garcia}@uclm.es

1 Introduction

The necessity of advances in renewable energies lead to novel technological improvements. The wind energy capacity has risen with high growth in recent years, reaching the 60.4 GW of new installations in 2019 [1]. Wind turbines (WTs) produce electricity by the conversion of wind energy received by the blades and transmitted to the shaft and generator. The variable physical efforts and hard environmental working conditions increment the probability of critical failures. The 15% of the failures may cause the 75% of the downtimes being the 60% of the total failures produced in the gearbox, yaw, blades and electrical systems [2, 3].

The operation and maintenance costs are estimated between 10 and 20% of total costs of energy, being the unplanned maintenance activities between 30 and 60% of the total maintenance costs WTs [4]. Cost reduction strategies are focused on the reduction of maintenance costs and the improvements in specific components reliability [5]. A suitable maintenance management plan is required to ensure the designed performance of the WT components and its proper life cycle using the optimal amount of material and human resources [6, 7]. The size and increasing capacity of current WTs need new monitoring and controlling systems to ensure the viability of the electricity generation.

Condition monitoring systems (CMS) measure the reliability of the system, determining the real state of the wind energy conversion systems [7, 8]. CMS allows the acquisition of reliable information about WT components and the detection or prediction of failures employing several techniques, e.g. thermography, acoustic emission or vibration [9, 10]. The supervisory control and data acquisition (SCADA) system collect the data from several types of CMS providing signals and alarms to the operators [11, 12]. The alarm is a warning message indicating that a type of anomaly, failure or irregularity is modifying the suitable behaviour of the WT [13, 14]. One of the most relevant issues in the WT maintenance is the false alarm detection [15, 16]. False alarms are activated although the WT does not present a real malfunction due to failures in the data transmission or issues in the control model, incrementing the maintenance costs for unnecessary

F. P. García Márquez (ed.), *International Conference on Intelligent Emerging Methods of Artificial Intelligence & Cloud Computing*, Smart Innovation, Systems and Technologies 273, https://doi.org/10.1007/978-3-030-92905-3_10

maintenance activities [17, 18]. Due to the amount of CMS, control and emergency systems, the volume of the generated alarms and the possibility of false alarms, it is necessary new algorithms with better capabilities to obtain reliable information about the WT [19, 20].

The main methodologies for fault detection in WTs employ mathematical models considering the real behaviour of the WT. These conventional computing methods are designed based on predefined procedures or rules that provide a precise definition of the task. The identification of critical components is obtained by quantitative analysis with binary decision diagrams and graphic representation of fault trees [21]. Machine learning algorithms apply computational methods to find patters not identical to those defined by learning [22]. Support vector machine (SVM) is a supervised machine learning algorithm formed by statistical learning and minimization of probabilities employing data regression [23]. Artificial neural networks (ANNs) are models of biological neural system imitating the learning procedures of human brain with processing components interconnected called neurons. Garrett [24] defines an ANN as:"A computational mechanism able to acquire, represent and compute mapping from one multivariate space of information to another, given a set of data representing that mapping". ANNs are able to manage discontinuous problems using input patterns defined by the user and related to a target output, with the aim of detecting these patterns in other untrained datasets. ANNs can learn and identify complex connections without restrictions in the input variables. The training is one of the most critical phases since it must be verified by human experts [25].

This machine learning technique is applied in several application fields, e.g., image processing, pattern recognition, forecasting. In the wind energy field, ANNs are employed to forecast, predict and control the energy production. Several authors are focused on the false alarm detection using ANNs. Chacón, Segovia and García [26] analyse and filter the false alarms produced in the bearing system. Kusiak and Verma [27] used a ANNs to develop models for predicting bearing faults. The authors minimize false alarms with statistic algorithms. Marugán, Chacón and García [28] developed ANN for false alarm detection and presented a real case study with SCADA data and vibration system. The authors employed redundant dataset. Adouni et al. [29] employed a novel ANN architecture to increment the fault diagnosis and detection of false alarms. Neural networks in the state of the art are normally over-dimensioned, and the data used in both training and testing are redundant, increasing the computational load and reducing the validity of the results [30].

This paper proposes a novel approach is based on the reduction of parameters of ANNs for WT maintenance management. The alarm dataset is studied with Pareto chart to choose the critical alarm and for ensuring proper periods for the study it is determined a filtering process. Different techniques are employed to reduce the signal dataset, e.g., correlation and principal component analysis (PCA). This filtering process reduces the redundant information and increases the accuracy, of the ANN. A real case study with data from a real WT is presented, comparing the ANN results from initial dataset and filtered scenario. All the procedure is applied in the original data, obtained the critical alarm and the related dataset.

This paper is divided in different sections: Section 2 defines the approach providing details about the algorithms and ANN definition and Section 3 explain the real case study and the results obtained after the application of the approach developed in this work.

2 Approach

The novelty presented in this work is based on the data filtering process in the SCADA dataset to increase the reliability of ANN and reduce the redundant data. The application of valuable information in the ANN increases the reliability of the analysis. Figure 1 shows flowchart of the proposed approach.

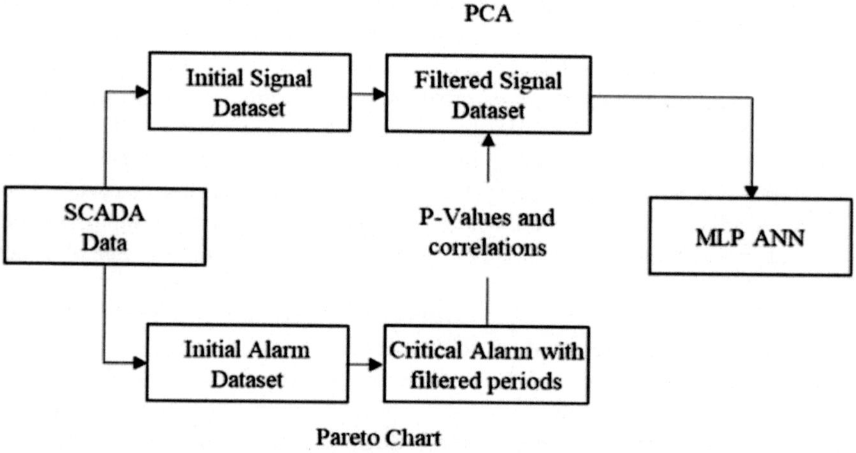

Fig. 1. Flowchart of the approach

Several types of alarms are produced in the SCADA system and the identification of the critical alarm is fundamental. The operators may define this critical alarm directly or it is required different selection criteria. This study considers the number of activations, the maximum and the average alarm period, and the average period without alarms as relevant parameters. Pareto chart is also applied because it is a graphic tool that identify the frequencies of the parameters. The period between alarm activations is decisive and it is considered one day before the failure as security range, being reduced period discarded to ensure enough data for the analysis [31]. It is required a proper range between alarm activations to ensure enough data to introduce in the ANN. The alarm activations that do not comply with this condition are excluded. The probability to find pattern that activate the alarm in periods closer to the alarm activation is higher, and it has more relevance in the study. The range definition showed in Fig. 2 shows the range with more and less probability of causing the alarm activation.

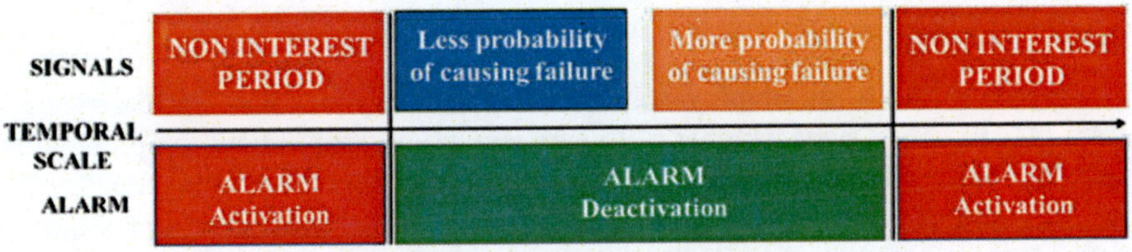

Fig. 2. Diagram of the approach

Different signals are provided to SCADA systems, e.g., temperature, electric behavior, and the reduction of this signal dataset in It is necessary the determination of a dataset of signals related to the critical alarm selected for this study. The reduction of number of the signals is developed with p-values and correlation. PCA develops a multivariate analysis to reduce the dataset dimensions maintaining the patterns and most important characteristics.

The P-values shows the probability of the hypothesis and it provides information about the correlation between signals. The test statistic Z is showed in (1).

$$z = \frac{\hat{n} - n_0}{\sqrt{\frac{n_0(1-n_0)}{n}}} \tag{1}$$

being $\hat{n}$ the proportion of the sample; n_0 the population percentage detected in the hypothesis, and n the sample size. The P-value is defined by (2):

$$Pvalue = \text{Prob}(z \geq ts|Hypothesis.true|) = cdf\,(ts) \tag{2}$$

Being cdf the distribution function of the test static and ts the observed value of the test statistic. It is determined that the p-value threshold is lower than 0.05 [32].

PCA is a mathematical method to reduce the dimensionality of the dataset. The initial large dataset is transformed into a set of smaller variables with the same patterns and information than the original [33]. The initial dataset X is defined with $n \cdot p$ dimensions, being p variables structured in columns and n observations in rows. The principal components are a weighted average of the initial variables and the new dataset is constructed with them. The principal component is determined with (3):

$$Y_{ij} = w_{1j}X_{1j} + w_{2j}X_{2j} + \ldots + w_{pi}X_{pj} \tag{3}$$

being w_{1j}, w_{2j},... w_{pi}, the coefficient weights of the linear correlation defined by PCA. The matrix W of weights is obtained from the covariance matrix S, identified with (4):

$$Sij = \frac{\sum_{k=1}^{n} (x_{ik} - \overline{x_i}) \cdot (x_{jk} - \overline{x_j})}{n-1} \tag{4}$$

The correlation between the original data and the componenti^{th} i^{th} is defined with (5):

$$r_{ij} = \frac{u_{ij} \cdot \sqrt{l_i}}{s_{ij}} \tag{5}$$

Being u_{ij} the eigenvector matrix, l_i is a singular value decomposition of Sij.

2.1 Artificial Neural Network

Several types of ANN are employed in the wind industry analysis. Multi-Layer Perceptron (MLP) neural network does not make any assumptions about the probabilistic information of the classes of patterns compared to other probability-based modality. This type of neural network detects the non-linear connections between input and output and increments the viability and suitability of the output. The MLP architecture is variable but it is possible to identify different types of layer: input, hidden and the output layer, see Fig. 3. The weighs are parameters between 0 and 1 that associate the input with the outputs, and it is recommended a suitable training dataset to ensure reliable weights between layers. This work uses a MLP ANN with 10 hidden layers.

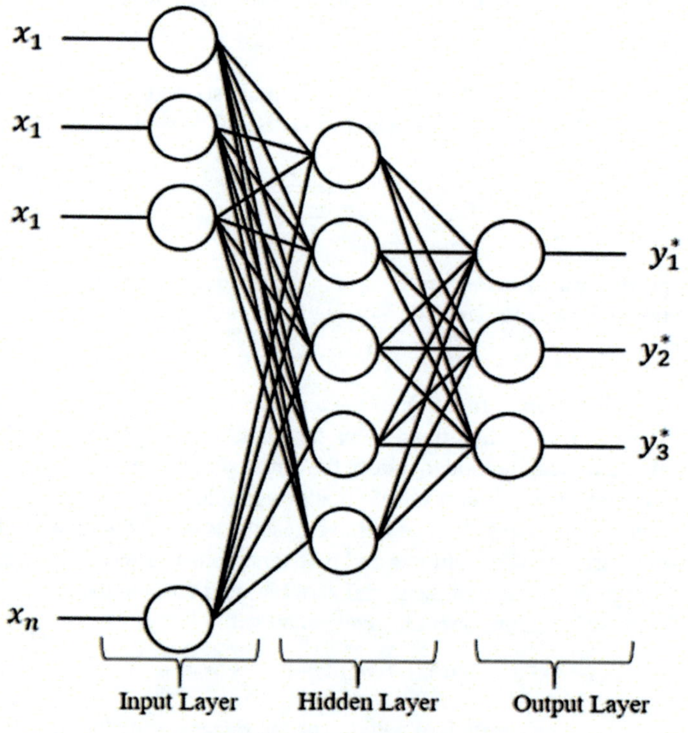

Fig. 3. MLP ANN diagram

The connections between neurons are quantified with a weight parameter. The output O_i is defined in (6):

$$O_i = \sigma(\sum_{S=1}^{N} W_{rs} \cdot x_s + T_i^{hidden\_layer}) \tag{6}$$

being the transfer function characterized by $\sigma()$; N the number of input neurons; W_{rs} is the matrix of weights between neurons; x_s shows the inputs and; $T_i^{hidden\_layer}$ the tolerance of the hidden neurons.

After different tests and combinations, the MLP ANN designed for this work is formed by 20 hidden layers with a 70% of the dataset employed for the training, 15% for the validation and the rest 15% for testing.

3 Real Case Study and Results

The case study is formed by SCADA data from real WT. The study period is one year with a data acquisition rate of 1 min. The WT alarm system has 200 different alarms, and the signal dataset presents more than 90 signals, obtaining more than 180 million of data. The critical alarm is established according to different criteria: the number of activations, see Fig. 4, the maximum period of the alarm and the difference between alarms without filtering process. The accumulated data is represented with the blue line.

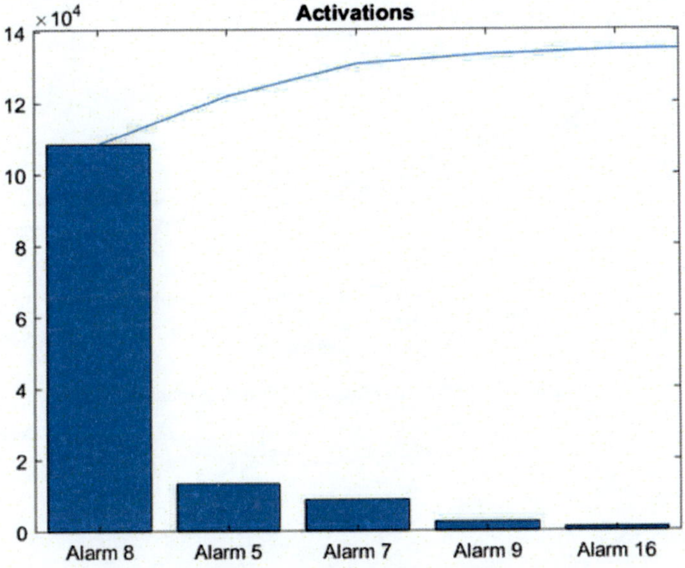

Fig. 4. Maximum alarm activations

The critical alarm for this study is an alarm about discrepancy in generator rotation. As it is mentioned in previous section, the range of interest is stablished in one day before the failure. The critical alarm is activated 25 times but only 12 activations accomplish the filtering requirement.

The original dataset is analysed using correlation and p-values. Different thresholds are applied, and the new signal dataset is formed by only 11 signals of the initial 96. Once it is determined the related signals, PCA is used to decrease the data volume. The 99% of the initial dataset is defined with two principal components.

Initial and filtered dataset are introduced in the MLP neural network to test the validity of the approach. The designed MLP is formed by ten hidden layers. The performance of the ANNs train, validation and test is showed for both situations: Fig. 5 shows initial dataset with no filtering process and Fig. 6 dataset after the application of the approach. The ANN with the filtering process represents a better performance with a smaller number of epochs. This dataset with the application of the approach provides better stabilization of the network, due to reduced dataset for filtering redundant information.

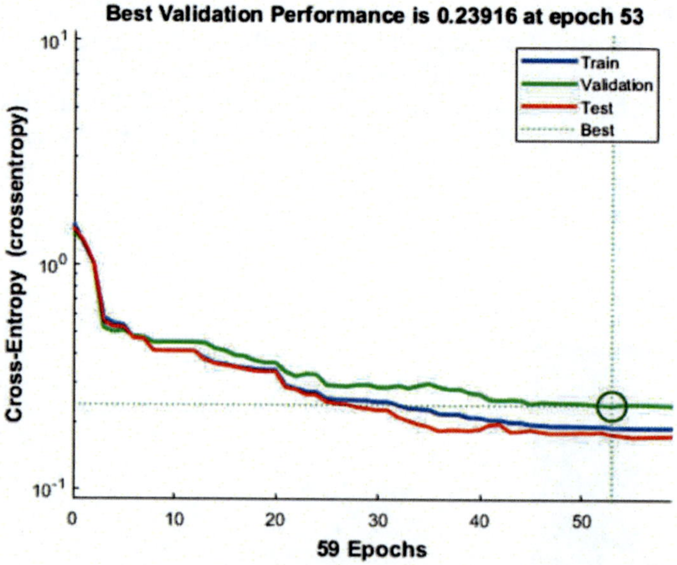

Fig. 5. ANN performance with initial dataset

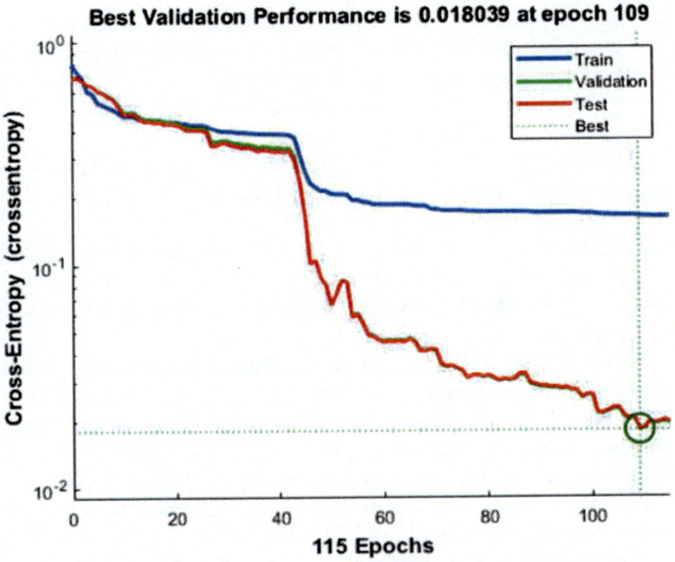

Fig. 6. ANN performance with dataset filtered with the approach

With the confusion matrix of the ANN, it is possible to observe that the number of proper cases identified by the network is elevated and it is proved the capacity to classify alarm activations. Comparing with the initial results, it is concluded that the 6% of the alarms may be considered as false.

Acknowledgements. The work reported herewith has been financially by the Dirección General de Universidades, Investigación e Innovación of Castilla-La Mancha, under Research Grant ProSeaWind project (Ref.: SBPLY/19/180501/000102)

References

1. F.Z. Joyce Lee, *Global wind report*; Global Wind Energy Council (2020)
2. F. García Márquez, A. Pliego Marugán, J. Pinar Pérez, S. Hillmansen, M. Papaelias, Optimal dynamic analysis of electrical/electronic components in wind turbines. Energies **10**, 1111 (2017)
3. P. Tchakoua, R. Wamkeue, M. Ouhrouche, F. Slaoui-Hasnaoui, T.A. Tameghe, G. Ekemb, Wind turbine condition monitoring: State-of-the-art review, new trends, and future challenges. Energies **7**, 2595–2630 (2014)
4. C.A. Walford, *Wind turbine reliability: Understanding and minimizing wind turbine operation and maintenance costs*. Sandia National Laboratories (2006)
5. A. Pliego Marugán, F.P. García Márquez, J. Lorente, Decision making process via binary decision diagram. Int. J. Manag. Sci. Eng. Manag. **10**, 3–8 (2015)
6. F.G. Marquez, An approach to remote condition monitoring systems management. (2006)
7. F.P. Garcia Marquez, C.Q. Gomez Munoz, A new approach for fault detection, location and diagnosis by ultrasonic testing. Energies **13**, 1192 (2020)

8. A.H. Butt, B. Akbar, J. Aslam, N. Akram, M.E.M Soudagar, F.P. García Márquez, M. Younis, E. Uddin, Development of a linear acoustic array for aero-acoustic quantification of camber-bladed vertical axis wind turbine. Sensors 20, 5954 (2020)
9. Z. Liu, C. Xiao, T. Zhang, X. Zhang, Research on fault detection for three types of wind turbine subsystems using machine learning. Energies **13**, 460 (2020)
10. C.Q. Gómez Muñoz, F.P. García Marquez, B. Hernandez Crespo, K. Makaya, Structural health monitoring for delamination detection and location in wind turbine blades employing guided waves. Wind Energy **22**, 698–711 (2019)
11. J.M.P. Pérez, F.P.G. Márquez, A. Tobias, M. Papaelias, Wind turbine reliability analysis. Renew. Sustain. Energy Rev. **23**, 463–472 (2013)
12. F.P.G. Márquez, A.M.P. Chacón, A review of non-destructive testing on wind turbines blades. Renew. Energy (2020)
13. A. Pliego Marugán, F.P. Garcia Marquez, B. Lev, Optimal decision-making via binary decision diagrams for investments under a risky environment. Int. J. Prod. Res. **55**, 5271–5286 (2017)
14. F.P. García Márquez, I. Segovia Ramírez, B. Mohammadi-Ivatloo, A.P. Marugán, Reliability dynamic analysis by fault trees and binary decision diagrams. Information **11**, 324 (2020)
15. I. Segovia Ramirez, B. Mohammadi-Ivatloo, F.P. Garcia Marquez, Alarms management by supervisory control and data acquisition system for wind turbines. Eksploatacja I Niezawodnosc-Maintenance Reliability **23**, 110–116 (2021)
16. A. Pliego Marugán, F.P. García Márquez, Advanced analytics for detection and diagnosis of false alarms and faults: A real case study. Wind Energy **22**, 1622–1635 (2019)
17. A. Pliego Marugán, A.M. Peco Chacón, F.P. García Márquez, Reliability analysis of detecting false alarms that employ neural networks: A real case study on wind turbines. Reliab. Eng. Syst. Saf. 191, 106574 (2019)
18. F.P.G. Márquez, A new method for maintenance management employing principal component analysis. Struct. Durab. Health Monit. **6**, 89 (2010)
19. I.S. Ramirez, F.P.G. Marquez, In Supervisory control and data acquisition analysis for wind turbine maintenance management, in *International Conference on Management Science and Engineering Management*, 2020; Springer, pp. 470–480
20. F.P. Garcia Marquez, A. Pliego Marugan, J.M. Pinar Perez, S. Hillmansen, M. Papaelias, Optimal dynamic analysis of electrical/electronic components in wind turbines. Energies **10**, 1111 (2017)
21. F.P. García Márquez, I. Segovia Ramírez, A. Pliego Marugán, Decision making using logical decision tree and binary decision diagrams: A real case study of wind turbine manufacturing. Energies **12**, 1753 (2019)
22. A.A. Jiménez, L. Zhang, C.Q.G. Muñoz, F.P.G. Márquez, Maintenance management based on machine learning and nonlinear features in wind turbines. Renew. Energy **146**, 316–328 (2020)
23. S. Sridhar, K.U. Rao, R. Umesh, K. Harish, In Condition monitoring of induction motor using statistical processing, in *2016 IEEE Region 10 Conference (TENCON)*, 2016; IEEE: pp. 3006–3009
24. J. Garrett, J. Where and why artificial neural networks are applicable in civil engineering (1994)
25. A.A. Jiménez, C.Q.G. Muñoz, F.P.G. Márquez, Dirt and mud detection and diagnosis on a wind turbine blade employing guided waves and supervised learning classifiers. Reliab. Eng. Syst. Saf. **184**, 2–12 (2019)
26. A.M.P. Chacón, I.S. Ramírez, F.P.G. Márquez, False alarms analysis of wind turbine bearing system. Sustainability **12**, 7867 (2020)
27. A. Kusiak, A. Verma, Analyzing bearing faults in wind turbines: A data-mining approach. Renew. Energy **48**, 110–116 (2012)

28. A.P. Marugán, A.M.P. Chacón, F.P.G. Márquez, Reliability analysis of detecting false alarms that employ neural networks: A real case study on wind turbines. Reliab. Eng. Syst. Saf. **191**, 106574 (2019)
29. A. Adouni, D. Chariag, D. Diallo, M. Ben Hamed, L. Sbita, Fdi based on artificial neural network for low-voltage-ride-through in dfig-based wind turbine. ISA Trans. **64**, 353–364 (2016)
30. S. Han, J. Pool, J. Tran, W. Dally, In Learning both weights and connections for efficient neural network, in *Advances in neural information processing systems*, pp. 1135–1143 (2015)
31. M. Schlechtingen, I. Ferreira Santos, Comparative analysis of neural network and regression based condition monitoring approaches for wind turbine fault detection. Mech. Syst. Signal Process. **25**, 1849–1875 (2011)
32. R.J. Feise, Do multiple outcome measures require p-value adjustment? BMC Med. Res. Methodol. **2**, 8 (2002)
33. F.G. Marquez, An approach to remote condition monitoring systems management, in *IET International Conference on Railway Condition Monitoring*, pp. 156–160 (2006)

Routing Vehicles on Highways by Augmenting Traffic Flow Network: A Review on Speed Up Techniques

Jayanthi Ganapathy[1](✉), Fausto Pedro García Márquez[2], and Medha Ragavendra Prasad[1]

[1] Sri Ramachandra Engineering and Technology, Sri Ramachandra Institute of Higher Education and Research, Chennai 6000 116, India
jayanthig@sret.edu.in
[2] Ingenium Research Group, University of Castilla-La Mancha, 13071 Ciudad Real, Spain

1 Introduction

Until today over the 40 decade of years, transportation research is active and dynamic in all nations across the world. The variation in travel time and delay in travel faced by commuters is the adverse effect of traffic congestion. Therefore, it is essential to manage congestion as it cannot be avoided but can be mitigated. The common problem that was addressed by various researchers in transportation is finding the shortest path between desired source and destination [1]. Although this problem has been solved by many researches from theoretical view point, accurate and realistic shortest path that could overcome travel delay in a time varying road network is yet to be focused [2]. Travel to a location is represented by path traversed between source and destination with cost of travel represented in terms of speed, distance, time delay etc. Thus, path computation on time varying network is a function of space and time. Path computation involves (1) Temporal path and (2) Spatial path.

1.1 Temporal Path

In real road network, temporal variation in traffic volume has strong influence towards travel delay. Variation in travel time is the effect of congestion in time varying network. The computation of shortest path considering travel time information like earliest arrival time and latest departure time [3–6] alone is insufficient. Hence, temporal instances of traffic information are highly required in analysing traffic at preceding instances. Edge weight augmented with temporal information would help in realizing real traffic conditions more effectively.

1.2 Spatial Path

Road network is a spatial graph G = (S, R) where S is set of vertices representing arterial junctions and R is set of edges representing road segment connecting the junctions.

F. P. García Márquez (ed.), *International Conference on Intelligent Emerging Methods of Artificial Intelligence & Cloud Computing*, Smart Innovation, Systems and Technologies 273, https://doi.org/10.1007/978-3-030-92905-3_11

Spatial relation "connected" as defined in RCC-8 [7] is used to identify the topological connectivity of road segments R intersecting at arterial junctions S of the network. Identification of critical node (arterial junction) in the spatial network based on temporal variation of traffic with analysis on congestion patterns is required in real time scenarios [2]. Traffic congestion incurred due to time varying travel time needs much consideration in solving path computation problem. Traffic information at different time instances carry useful information about road network. Moreover, traffic information has to be analyzed at different time instances rather than arrival time and departure time alone. Computation on static network does not yield accurate results as traffic congestion varies with time. Therefore, time dependent travel time computation over such network is demanded [2–6]. When road network is congested, static information like distance, pre computed travel time alone are not enough. Rather, traffic condition in previous time instances are required to assess congestion. In this view, logical analysis of temporal traffic information is required for congestion management. Moreover, logical analysis of temporal traffic information is useful in analysing traffic in previous instance or interval. Allen's Interval relation [8] can be used in achieving logical analysis of temporal information as traffic information in preceding time instances is more significant in estimating the traffic in succeeding instances. Sequencing spatial and temporal traffic information is useful in analyzing traffic in previous instance or interval [7, 8]. Sequential pattern mining can be used in achieving analysis of temporal traffic information as traffic information in preceding time instances is more significant in estimating the traffic in succeeding instances [7]. Travel to a location is represented by path traversed between source and destination with cost of travel represented in terms of speed, distance, time delay etc. In this way, when congestion is detected the path is re-routed. In a spatially connected network, the road segments (a path) that are connecting the source and destination in preceding time instances are not considered in succeeding instance for re-routing by finding an alternate path to a congested route.

The road network is modelled using Randell's RCC-8 connected relation which enables logical analysis of spatial connectivity of road network in finding the alternate path. In the interest of reducing travel delay, SP-STAR algorithm is proposed for traffic congestion management considering the temporal and spatial factors of the time varying urban transport system. The temporal and spatial factors are logically analyzed by representing the time varying network using Allen's Interval relations and Randell's RCC-8 relations respectively. This paper is organized as follows. Section 2 presents the static shortest path algorithms. Section 3 presents a review on Time dependent Shortest Path Problems. Section 4 explores the research challenges and future directions.

2 Static Shortest Path Problem (S-SPP)

Route Planning on Transportation network can be categorized under two broad sub divisions (i) Static Shortest Path Problem(S-SPP) and (ii) Time dependent Shortest Path Problem (TD-SPP). In early years, the Shortest Path Problem (SPP) was well studied on a static network. Graph theory is the fundamental concept behind SPP. Many algorithms guaranteed correctness and were formally proved from theoretical view point. The primal theory behind SPP was well formulated in late 1950's [1, 9, 10]. The theory behind these basic solutions is being active until today as most of the recent works are extension to these fundamental approaches. The existence of shortest path from a given vertex to every other vertex on a static network is proved [1]. Hence, the algorithm solved Single Source Shortest Path (SSSP) problem. The algorithm is valid as long as edge cost is non-negative. As Dijkstra's algorithm iterates on all vertices, when applied on large graph its computation complexity is $O(n^2)$ which is high. In contrast, [9, 10] solves negative edges and detects negative cycle in the path. It fails to find shortest path when negative edge cycle is detected. Later Floyd, in 1962 formulated all—pair shortest path to find shortest path between every pair of vertices on a weighted graph with both positive and negative edge weights [11]. The complexity of various static shortest path algorithms is shown in Table 1. To overcome the limitation of these algorithms heuristic search was formulated that works on the principle of best-first search strategy [12].

Table 1. Static shortest path problem.

Problem type	Algorithm	Complexity#
Label Setting Algorithm	**DIKQ** [1]	$O(V^2)$
	DIKB [13]	$O(E + VC)$
	DIKBM [14]	$O(E + V(C/\alpha + \alpha))$
	DIKBA [14]	$O(E\beta + V(\beta + C/\beta))$
	DIKBD [14]	$O(E + V(\beta + C/\beta))$
	DIKF [15]	$O(E + V \log(V))$
	DIKAF-Heap [15]	$O(\log V/\log\log V)$
	DIKH [16]	$O(E \log(V))$
	DIKR [17]	$O(E + V \log(C))$
Label Correcting Algorithm	Dynamic Programming [9, 10]	$O(EV)$
	APSP [11]	$O(V^3)$

#E is number of Edges; V is number of Vertices; C, α, β are constants

Unlike adaptation of greedy strategy by Dijkstra's algorithm, A* works by informed search by which it has the ability to keep track of distance it has travelled in reaching the goal node. The goal directed search is based on a heuristic function. It fails when search technique does not follow an admissible heuristics [12]. This simple A* search was extended to work with landmarks termed as ALT [18] and has achieved better performance compared to other variation in A*. This technique is combination of A*, landmark and Triangle in-equality. It has pre-processing stage wherein it selects required number of landmarks and computes distances between them using triangle in-equality property. The computation of A* combined with chosen landmark and triangle in-equality was proved to exhibit constant time lower bound [12, 18]. In contrast to ALT, a reach based method was proposed [19]. The computations use Euclidean distance measure and reach value of all vertices. Although it outperforms ALT given single landmark, fails to perform when sixteen landmarks were given. The critical drawback is the algorithm suffers from specific assumption, complex pre-processing routines and inability to extend further in dynamic scenarios. Several pre-processing techniques were proposed to speed—up Dijkstra algorithm. Static shortest path with specialized technique is shown in Table 2. Hierarchical properties of road network were well studied for ordering of vertices in hierarchy of road network comprising of suburban streets, motor ways, urban streets and arterials. Static highway hierarchy algorithm was proposed in which a local search is performed between vertices in close proximity and highway edge is created if the edge lies in the path between source and destination vertex and does not lie close to the source or destination. Data preprocessing using wavelet transforms has its significance in pattern recognition in major engineering domains [20].

Table 2. Static shortest path problem with specialised technique.

Algorithm	Speed up technique
A^* [12]	Goal-directed informed search
ALT (A^* + Landmark + Triangle inequality) [18]	Augmented goal directed search with landmarks satisfying triangle inequality
SHARC [3]	Augmented hierarchies with edge flag replacement
Goal directed ALT [21]	Hierarchical search on ALT
Hierarchical Bi-directional Search [22, 23]	Augmenting least important vertex in highway hierarchy

Hierarchical algorithm with bidirectional search technique was proposed [22] to reduce the search space in highway hierarchies. Contraction hierarchies iteratively contract least important vertex in highway hierarchy thereby replacing shortest path passing through vertex with shortcuts. An extension to Dijkstra's algorithm was proposed with pre-processing based on hierarchical properties of road network [23]. In addition, complexity in pre-processing of edge in contraction hierarchies is further reduced by introducing edge flag termed SHARC that targets only important edges in pre-processing [3]. Generalized pre-processing and speed up technique for Dijkstra's algorithm on dense graph was reported [21] in which speed up performance is achieved by introducing goal directed ALT to hierarchical search. Real road network is time varying which means traffic information keeps changing. In this view, Time varying network has two ways of defining SPP (1) Fastest path problem and (2) Minimum cost path problem, where fastest path is based on travel time while minimum cost path is based on distance, arc length etc., In both the cases, traffic information or edge cost is a function of time. Although static shortest path algorithms paved solution, they stand far apart when applied to dynamic or time—dependent scenarios in real road network. Static SPP has proven solution but they cannot be applied to real road network unless it has efficient pre-processing on dynamic update of edge cost.

3 Time Dependent Shortest Path Problem (TD-SPP)

In reality travel time or edge cost varies with respect to time. In a road network, the travel delay that occurs due to traffic congestion varies with time. Thus, transit of vehicle on a path is time—dependent. Static shortest path algorithms are based on assumption that edge cost is constant between any two points [24] by which it fails to analyze the dynamic characteristics of time varying networks e.g. road networks. Research on time-dependent shortest path was analyzed to derive an iterative approach as an extension to Bellman's optimality condition [10]. This iterative method computes path to single destination in discrete time steps. Similarly Dreyfus, proposed dynamic label setting approach, a generalization of static Dijkstra's algorithm with an implicit assumption that links on network upholds First-in-First-Out (FIFO) property. The algorithm fails otherwise. The same conclusion was met by several other researchers [25–27]. Nevertheless, algorithm for non-FIFO links based on waiting policies was proposed but it fails when waiting is not allowed anywhere along the path [27].

The problem of computing fastest path from all nodes to one destination considering all possible departure times is solved by [28, 29]. All—to—one fastest path is computed using 3-queue data structure which is an extension to label correcting with 2—queue data structure introduced by [30]. Dynamic adaptation of static shortest path named Decreasing Order of Time (DOT) is formulated using 3- queue data structure. It is implemented using simple two-dimensional array data structure and proved to be optimal compared to dynamic label correcting algorithm [28, 29]. Shortest path algorithm based on Bellman's optimality was proposed for a network with time- dependent edge cost [31]. The proposed algorithm computes path between every pair of origin and destination nodes for each time step. Correctness of algorithm on backward search from destination node is arrived analytically. A survey was presented on FIFO network [32] as special case of time dependent SPP in dynamic scenarios with direction to develop polynomial time algorithms. George et al., 2006 modelled vertices and edges into time series to formulate Time—Aggregated Graph (TAG) [33, 34]. The presence and absence of vertices and edges at any instant of time makes the network time dependent. The objective of TAG is to solve SPP considering time dependency of the network with estimation of SP-STAR time of travel. Greedy strategy was adapted to formulate SP-TAG algorithm to compute shortest path at the time of query. BEST algorithm was formulated to estimate best SP-STAR time of travel over time—aggregated network for any given period of time [33, 34]. The static ALT algorithm proposed by [18] was experimented in time-dependent scenario. The results were promising with dynamic adaptation of ALT where pre-processing of edge cost is negligible when variation in traffic flow on the network is moderate.

A survey was presented on speed techniques to route planning. Algorithm to estimate departure time on large scale network considering dynamic traffic conditions was proposed [35, 36]. They investigated various time dependent scenarios in which they stated and argued that it is impossible to adapt shortcut arcs in static network to time—dependent scenarios. On analysing variants of time—dependent algorithms in SPP, they concluded that correctness of algorithm on time dependent graphs is guaranteed with augmented pre-processing and query subroutines phases. Several search techniques were introduced over static algorithm to work in time-dependent scenarios. The complexity of various TD-SPP is shown in Table 3. Bidirectional search technique is applied to ALT where forward search explores the network while backward search is bounded by number of nodes traversed by forward search. Although this technique is several time faster than Dijkstra's algorithm, it can obtain only near optimal solution due to approximation limits [37]. Static SHARC algorithm was extended to work in time-dependent scenarios [3, 38]. Further, the enhanced time—dependent version has resulted in reduced space complexity with efficient usage of memory in query processing while ensuring correctness of the algorithm [6]. Core routing on bidirectional search was proposed to improve the suboptimal solution, where only subsets of original nodes were searched [39]. This resulted in minimal search space with less complexity in pre-processing of edge cost. When contraction hierarchies were adapted to time—dependent edge weights, high memory requirement was experienced in query processing [4].

Table 3. Time-dependent shortest path problem.

Algorithm	Complexity[$]
Modified Bellman Ford Moore [24]	*Polynomial time*
Dynamic Dijkstra [25]	
Decreasing Order of Time (DOT) [28, 29]	$O(n^2 + nM + mM)$
Time dependent A* [40]	$O(nM + mM + SSP(n,m))$
SP-TAG (Shortest Path -Time Aggregated Graph) & BEST [33, 34]	$SP\text{-}TAG = O(m(logM + \log n)$ $BEST = O(n^2\ mM)$
TDALT [35]	*Faster than Dijkstra in several orders of magnitude*
TCH [4, 5]	
Bidirectional Core Routing [39]	
Bidirectional A* search[37]	
TD-SHARC [38]	
TD-RSPP [41]	
HTNGD [2]	

[$]n: nodes; m: arcs; M: time interval

Time dependent Contraction Hierarchies (TCH) faced high pre-processing cost while the improved version of TCH [5] used approximated edge weights to reduce this effect. Approximated TCH (ATCH) extracts sub graphs with shortcuts. Time dependent search is performed after replacing shortcuts. This approach has significantly reduced memory usage with less effect on query processing time [4, 5, 42]. In view of computation overhead and storage efficiency in time dependent scenarios, speed up technique was proposed for time dependent spatial graphs in which pre-processing is performed offline and fastest path computation done online [43]. The offline process involves non-overlapping partition of graph while online process utilizes heuristic function. This technique has significantly reduced both storage and computational complexity. Alternate path algorithms were formulated to overcome the computational overhead in updating the edge cost between all pair of vertices. iSPQF was proposed in which storage scheme reduces generation of quad-trees at each vertex. SPQF algorithm finds alternative path with reuse of previously computed results on unused edge weight or vertex 'v'. The algorithm runs in $O(n)$ complexity when experimented over single source and all pairs i.e., the path from source to destination without considering vertex 'v' and all sources to all destinations without considering vertex 'v' respectively [44]. Reliable SPP for time dependent scenarios (TD-RSPP) was proposed in [41] to solve forward search and backward search independently. They claimed that TD-RSPP is irreversible in which backward search algorithm from destination to origin cannot be solved using forward search.

Study on complexity of arrival time was reported in which time dependent edge cost was analysed by mapping arrival time to parametric shortest path problem [45]. In most of the recent works reviewed here, efforts taken to extend the speed up technique by pre-processing edge cost [18, 46–50, 13, 51–56] based on travel time is yet to devise solution for mitigating traffic congestion especially for metropolitan transport systems. In this aspect, the use of temporal and spatial information of traffic has its significance in mitigating transport congestion which was not considered in early achievements [4, 35, 38, 42, 57, 58, 52]. The interest of this work is re-routing the path when there is traffic congestion considering the temporal and spatial information of time varying road traffic network. Thus, Shortest Path—Spatio-TemporAl Reroute (SP-STAR) is proposed for managing congestion by augmenting the time varying network using temporal and spatial relations.

4 Research Challenges and Future Directions

Temporal variation in traffic flow essentially captures the recurring and non-recurring congestion in dynamics of physical traffic flow. However, temporal traffic information alone is not sufficient in travel decisions when there is a need for reliable path on a spatially connected road network. In route guidance [54, 56], a path in a travel is said to be reliable when flow rate in successive time instance is made known. This has motivated researchers to focus on influence of spatial characteristics in dynamics of traffic flow as flow rate from neighboring links contribute significant amount of traffic at current location. In this perspective, a fully automated traffic management system is not feasible. Therefore, it is essential to manage traffic flow congestion systematically as it cannot be avoided but can be mitigated. In this view, a speedup technique is necessary to bridge the gap between traffic flow estimation and path routing of vehicular traffic between origin and destination (OD) [59, 20].

In real road network, temporal variation in traffic volume has strong influence towards travel delay. Variation in travel time is the effect of congestion in time varying network. Traffic information at different time instances carry useful information about road network. Moreover, traffic information has to be analyzed at different time instances rather than arrival time and departure time alone. Edge weight augmented with temporal information would help in realizing real traffic conditions more effectively. In a spatially connected network, it is essential to analyze spatial dependency of road segment with respect to upstream and downstream traffic flow thereby, road segments (a path) that are connecting the source and destination in preceding time instances are analyzed in successive time instances for re-establishing connectivity. In this way, when traffic congestion is detected the path is reconnected based on spatial–temporal traffic information.

References

1. E.W. Dijkstra, A note on two problems in connexion with graphs. Numer. Math. **1**(1), 269–271 (1959)
2. M.M. Nejad, L. Mashayekhy, R.B. Chinnam, P. Anthony, Hierarchical time-dependent shortest path algorithms for vehicle routing under ITS. IIE Trans. **48**(2), 158–169 (2016)
3. R. Bauer, D. Delling, SHARC: 'Fast and Robust unidirectional routing. ACM J. Exp. Algorithmics **14**, 2.4–2.29 (2009)
4. G. Batz, R. Geisberger, S. Neubauer, P. Sanders, Time-dependent contraction hierarchies and approximation. Exp. Algorithms 166–177 (2010)
5. G.V. Batz, R. Geisberger, P. Sanders, C. Vetter, Minimum time-dependent travel times with contraction hierarchies. J. Exp. Algorithmics **18**, 1.1–1.43 (2013)
6. E. Brunel, D. Delling, A. Gemsa, D. Wagner (2010) Space-efficient SHARC-routing. , in *Experimental Algorithms, 9th International Symposium, SEA 2010, Ischia Island, Naples, Italy. Proceedings, Springer*
7. D.A. Randell, Z. Cui, A.G. Cohn, A spatial logic based on regions and connection. Knowl. Represent. Reason. 165–176
8. J.F. Allen, An interval-based representation of temporal knowledge. Int. Jt. Conf. Artif. Intell. Morgan Kaufmann **1**, 221–226 (1981)
9. R. Bellman, "Dynamic Programming", Princeton University Press, 1957.
10. R. Bellman,On a routing problem. Q. Appl. Math. (1958)
11. R. Floyd, Algorithm 97: shortest path. Commun. ACM 344–348 (1962)
12. P.E. Hart, N.J. Nilsson, B. Raphael,A formal basis for the heuristic determination of minimum cost paths. IEEE Trans. Syst. Sci. Cybern. **4**(2), 100–107 (1968)
13. R. Dial, Algorithm 360: shortest path forest with topological ordering. Commun. ACM **12**, 632–633 (1969)
14. V. Cherkassky, B. Goldberg, A. Radzik Tomasz, Shortest paths algorithms: theory and experimental evaluation. Math. Program. **73**, 129–174 (1994)
15. M.L. Fredman, R.E. Tarjan, Fibonacci heaps and their uses in improved network optimization algorithms. L Assoc. Comput. **34**(3), 596–615 (1987)
16. T.H. Cormen, C.E. Leiserson, R.L. Rivest, in *Introduction to Algorithms* (MIT Press, Cambridge, MA, 1990)
17. R.K. Ahuja, K. Mehlhorn, J.B. Orlin, R.E. Tarjan, Faster algorithms for the shortest path problem. J. Assoc. Comput. Math. **37**(2), 213–223 (1990)
18. A.V. Goldberg, C. Harrelson, Computing the shortest path: a* search meets graph theory, Technical report (2003)
19. R. Gutman, Reach-based routing: a new approach to shortest path algorithms optimized for road networks, ALENEX/ANALC (2004)
20. C.Q. Gómez Muñoz, F.P. García Márquez, B. Hernandez, K. Makaya, Structural health monitoring for delamination detection and location in wind turbine blades employing guided waves. Wind Energy **22** (2019). https://doi.org/10.1002/we.2316
21. R. Bauer, D. Delling, P. Sanders, D. Schieferdecker, D. Schultes, D. Wagner, Combining hierarchical and goal-directed speed-up techniques for dijkstra's algorithm. J. Exp. Algorithmics 303–318 (2010)

22. R. Geisberger, P. Sanders, D. Schultes, D. Delling, Contraction hierarchies: faster and simpler hierarchical routing in road networks. Exp. Algorithms 319–333 (2008)
23. R. Geisberger, P. Sanders, D. Schultes, D. Delling, C. Vetter, Exact routing in large road networks using contraction hierarchies. Transp. Sci. 388–404 (2012)
24. K.L. Cooke, E. Halsey, The shortest routes through a network with time-dependent inter nodal transit times. J. Math. Anal. Appl. **14**(3), 493–498 (1966)
25. S.E.Dreyfus, An appraisal of some shortest-path algorithms. Oper. Res. **17**(3), 395–412 (1969)
26. D.E. Kaufman, R.L. Smith, Fastest paths in time-dependent networks for intelligent vehicle highway systems application. J. Intell. Transp. Syst. **1**(1), 1–11 (1993)
27. R. Orda Rom, Shortest-path and minimum-delay algorithms in networks with time-dependent edge-length. J. ACM **37**, 607–625 (1990)
28. I. Chabini, A new algorithm for shortest paths in discrete dynamic networks, in *Presented at the 8th IFAC Symposium on Transportation Systems, Chania, Greece*, 1997.
29. I. Chabini, Discrete dynamic shortest path problems in transportation applications: Complexity and algorithms with optimal run time. Transp. Res. Rec.: J. Transp. Res. Board **1645**(1), 170–175 (1998)
30. G. Gallo, S. Pallottino,Shortest paths algorithms. Ann. Oper. Res. **13**, 3–79 (1988)
31. H. Ziliaskopoulos, A. Mahmassani, Time-dependent shortest path algorithm for real-time intelligent vehicle highway system applications. Transp. Res. Rec. **1408**, 94–104
32. B.C. Dean, Shortest paths in FIFO time-dependent networks: theory and algorithms, Technical Report, Massachusetts Institute of Technology, Cambridge, MA, 2004
33. B. George, S. Shekhar,Time-aggregated graphs for modeling, in *Advances in Conceptual Modelling* (2006), pp. 85–99
34. B. George, S. Kim, S. Shekhar. Spatio-temporal network databases and routing algorithms: a summary of results. Spat. Temporal Databases 460–477 (2007)
35. D. Delling, D. Wagner, Time-dependent route planning. Robust Online Large-Scale Optim. **2**, 1–18 (2009)
36. B. Ding, J.X. Yu, L. Qin, Finding time-dependent shortest paths over large graphs, in *Proceedings of the 11th International Conference on Extending Database Technology Advances in Database Technology EDBT 2008*, vol. 8 (2008), p. 205.
37. G. Nannicini, D. Delling, D. Schultes, L. Liberti, Bidirec-tional A* search on time-dependent road networks. Networks **59**(2), 240–251 (2012)
38. D. Delling,Time-dependent SHARC-Routing. Algorithmica **60**(1), 60–94 (2011)
39. D. Delling, G. Nannicini,Core routing on dynamic time-dependent road networks. Inf. J. Comput. **24**(2), 187–201 (2012)
40. I. Chabini, S. Lan, Adaptations of the A* algorithm for the computation of fastest paths in deterministic discrete-time dynamic networks. IEEE Trans. Intell. Transp. Syst. **3**(1), 60–74 (2002)
41. B.Y. Chen, W.H. Lam, A. Sumalee, Q. Li, M.L. Tam, Reliable shortest path problems in stochastic time-dependent network. J. Intell. Transp. Syst. **18**(2), 177–189 (2013)
42. G.V. Batz, D. Delling, P. Sanders, Time-dependent contraction hierarchies, in *Proceedings of the 11th Workshop on Algorithm Engineering and Experiments (ALENEX'09), New York*
43. U. Demiryurek, F. Banaei-kashani, C. Shahabi, *Online Computation of Fastest Path in Time Dependent*, SSTD (2011), pp. 92–111.
44. K. Xie, K. Deng, S. Shang, X. Zhou, K. Zheng, Finding alternative shortest paths in spatial networks. ACM Trans. Database Syst. 1–31 (2012)
45. L. Foschini, J. Hershberger, S. Suri, On the complexity of time-dependent shortest path. Algorithmica (2014)
46. R. Goldberg Werneck, Computing point-to-point shortest paths from external memory, ALENEX/ANALCO (2005)

47. B. Goldberg, Point-to-point shortest path algorithms with preprocessing, SOFSEM (2007), pp. 9–12
48. P. Sanders, D. Schultes, Highway hierarchies hasten exact shortest path queries. Algorithms ESA (2005), pp. 568–579
49. P. Sanders, D. Schultes, Engineering highway hierarchies, Algorithms, ESA (2006), pp. 804–816
50. B. Casey, A. Bhaskar, H. Guo, E. Chung, Critical review of time-dependent shortest path algorithms: a multimodal trip planner perspective. Transp. Rev. **34**(4), 522–539 (2014)
51. A. Benantar, R. Ouafi, J. Boukachour, A combined vehicle loading and routing problem: a case study of fuel logistics. Int. J. Logist. Syst. Manag. **32**(3/4), 346–371 (2019)
52. D. Pavlyuk, Feature selection and extraction in spatiotemporal traffic forecasting: a systematic literature review. Eur. Transp. Res. Rev. **11**(6), 1–19 (2019)
53. R.J. Pemberthy, J. Muriel, A.A. Correa-Espinal, A cross-border, long haul freight transportation problem with transshipments. Int. J. Logist. Syst. Manag. **32**(3), 437–464 (2019)
54. Y. Shi, M. Deng, J. Gong, C. Lu, Y. Xuexi, H. Liu, Detection of clusters in traffic networks based on spatio-temporal flow modeling. Trans. GIS **23**(2), 312–333 (2019)
55. W. Wangyang, W. Honghai, M. Huadong, An auto encoder and LSTM-based traffic flow prediction method. Sensors **19**(2946), 1–16 (2019)
56. Y. Zhuang, R. Ke, Y. Wang, Innovative method for traffic data imputation based on convolutional neural network. IET Intel. Transp. Syst. **13**(4), 605–613 (2019)
57. T. Azad, M.A.A. Hasin, Capacitated vehicle routing problem using genetic algorithm:a case of cement distribution. Int. J. Logist. Syst. Manag. **32**(1), 132–146 (2019)
58. G. Kartikay, C. Niladri, Forecasting through motifs discovered by genetic algorithms. IETE Tech. Rev. **36**(3), 253–264 (2019)
59. G. Jayanthi, P. Jothilakshmi, Traffic time series forecasting on highways - a contemporary survey of models, methods and techniques. Int. J. Logist. Syst. Manag. Inderscі. (2019). https://doi.org/10.1504/IJLSM.2020.10024052,ISSN:1742-7975
60. R.E. Turochy, B.L. Smith, Measuring variability in traffic conditions by using archived traffic data. Transp. Res. Rec. **1804**(2), 168–172 (2002)
61. B.L. Smith, M.J. Demetsky, Investigation of extraction transformation and loading techniques for traffic data. Transp. Res. Rec. **1879**, 9–16 (2004)
62. E.I. Vlahogianni, J.C. Golias, M.G. Karlaftis, Short-term traffic forecasting: overview of objectives and methods. Transp. Rev. **24**(5), 533–557 (2004)
63. B. Ghosh, B. Basu, M.O. Mahony, Multivariate short-term traffic flow forecasting using time-series analysis. IEEE Trans. Intell. Transp. Syst. **10**(2), 246–254 (2009)
64. G. Aakarsh, G. Aman, S. Samridh, S. Varun, Factors affecting adoption of food delivery apps. Int. J. Adv. Res. **7**(10), 587–599 (2019)
65. G. Jayanthi, P. Jothilakshmi, Prediction of traffic volume by mining traffic sequences using travel time based PrefixSpan. IET Intel. Transp. Syst. **13**(7), 1199–1210 (2019). https://doi.org/10.1049/iet-its.2018.5165,PrintISSN1751-956X

False Alarm Detection in Wind Turbine Management by K-Nearest Neighbors Model

Ana María Peco Chacón, Isaac Segovia Ramirez(✉), and Fausto Pedro García Márquez

Ingenium Research Group, University of Castilla-La Mancha, Ciudad Real, Spain

1 Introduction

The reliability and availability of the wind farms is required to maximize the production of energy from the wind [1]. Condition monitoring systems (CMS) provide consistent data about critical Wind Turbine (WT) components [2–4]. The supervisory control and data acquisition (SCADA) system incorporates various types of CMS, including signals and alarms [5–7]. False alarms are produced although the WT presents real health conditions, causing unnecessary maintenance tasks and downtimes [8–10]. The alarm evaluation requires advanced and complex algorithms to reduce false alarms [11, 12]. Machine learning algorithms use computational approaches to learn from the dataset and detect patterns [13–15]. The K-Nearest Neighbor (KNN) algorithm is widely applied in data mining [16]. It is a classification method based on the closest training examples in the feature space [17]. Eyecioglu et al. [18] applied KNN for the prediction of power generation in WT. Several research have been found for the diagnosis and detection of faults by the KNN method [19], but this method have not been applied yet for the analysis of the alarms generation.

The main contribution of this work is the application of a data-based study to predict and detect false alarms with KNN methods. The main objective of this work is to determine a methodology for the detection of false alarms in WTs. This approach can be applied with historical SCADA data and in real-time to improve the reliability of WTs.

2 Approach

The proposed methodology correlates the alarm log of the WT with the SCADA variables. The data acquired from the alarm activations is synchronized in temporal scale with the rest of the SCADA variables. The next step is the classification and prediction with KNN model, being holdout validation and K-fold cross validation (CV) applied and compared. Different types of KNN algorithms are studied, and the best classification method is selected according to the accuracy of the model. The third step applies confusion matrix to classify the results. The misclassifications are analyzed with the comparison between the data from the alarm log and maintenance record.

KNN is an algorithm of nonparametric with lazy learning without suppositions about the main dataset [20]. KNN uses all participating cases in the dataset and categorizes new

F. P. García Márquez (ed.), *International Conference on Intelligent Emerging Methods of Artificial Intelligence & Cloud Computing*, Smart Innovation, Systems and Technologies 273, https://doi.org/10.1007/978-3-030-92905-3_12

cases based on their similarity indices [21]. The objective is the obtention of the highest similarity indices. The selection of the required number of neighbors (K) is important since this number depends on the metrics used for classification. The KNN classifier estimates the distances between the point and the points in the training data set. The most common distance metric is the Euclidean distance (Table 1).

Table 1. Comparison between different KNN methods.

Classifier	Model Flexibility	Distance	Neighbors
Fine	Precise distinctions between classes	Euclidean	1
Medium	Medium distinctions between classes	Euclidean	10
Coarse	Gross distinctions between classes	Euclidean	100
Cosine	Medium distinctions between classes	Cosine	10
Cubic	Medium distinctions slow prediction speed	Minskowski	10
Weighted	Medium distinctions between classes	Euclidean	10

Several distance functions are used to calculate the distance between feature vectors A and B in a feature space, where $A = (x_1, x_2, ..., x_m)$, $B = (y_1, y_2, ..., y_m)$ and m is the dimensionality of the feature space. The most used distance function is the Euclidean distance function [22], see Eq. (1):

$$dist(A, B) = \sqrt{\frac{\sum_{i=1}^{m} (x_i - y_i)^2}{m}} \tag{1}$$

The cosine KNN measures the cosine of the angle between the two vectors [23], see Eq. (2).

$$\cos(A, B) = \frac{\vec{A} \cdot \vec{B}}{\left|\vec{A}\right| \cdot \left|\vec{B}\right|} \tag{2}$$

The cubic KNN uses the distance of Minskowski [24], in this case r is 2 for default.

$$dist\_Minskowsky(A, B) = \left(\sum_{i=1}^{m} |x_i - y_i|^r \right)^{1/r} \tag{3}$$

The aproach studied the KNN classifier for CV of K-fold and holdout validation. K-Fold Cross-Validation is used to analyze the evaluation error with a statistical learning method [25]. This technique divides a set of observations into random K equal-sized folds. All the training data is used during the testing phase, and then it is decided the best training subset [26]. The K-fold algorithm employs a vector as input to the K training data set and the most common class is used to classify the K nearest neighbors. The neighbors are categorized during the training phase according to their distances from the test dataset. The classes of the dataset are evaluated during the test phase. The holdout validation divides the data into training and testing [27].

The n × n confusion matrix, see Table 2, is used to evaluate the efficiency of the classifiers, where *n* is the number of classes. The confusion matrix represents the result of a classification that identifies the real classes from the predicted classes.

Table 2. Confusion matrix.

	True class	
Hypothesis class	True positive (TP)	False positive (FP)
	False negative (FN)	True negative (TN)

The ratio of the total number of correct predictions is the accuracy, Eq. (4).

$$Accuracy = \frac{TP + TN}{TP + TN + FP + FN} \tag{4}$$

The sensitivity, shown in Eq. (5), suggests the proportion of positive cases.

$$Sensitivity = \frac{TP}{TP + FN} \tag{5}$$

The specificity is the ratio of negatives that are properly classified, given by Eq. (6).

$$Specificity = \frac{TN}{TN + FP} \tag{6}$$

3 Case Study and Results

The case study presented in this paper considers SCADA data from the European project OPTIMUS. The predictors used in the method are 38 SCADA signals measured every 10 min for 2 months. The alarm log is the response variable, where two class labels are regarded: alarm activation or no alarm activation. The method with the best results is weighted with the highest accuracy (98,7%) by a 5-Fold CV. This model has a sensitivity of 99,13% and a specificity of 88,52%. This means that the method predicts cases better when no alarm occurs (Table 3).

Table 3. Comparison for different KNN methods.

Classifier type	Holdout validation		5 fold cross validation	
	Accuracy (%)	Misclassifications	Accuracy (%)	Misclassifications
Fine	98.2	78	98.6	124
Medium	97.8	93	98.2	158
Coarse	96.2	163	96.9	265
Cosine	97.9	91	98.1	162
Cubic	97.8	95	98.1	164
Weighted	98.3	74	98.7	117

The power curve is a graph that compares the WT power versus wind speed, being widely applied to detect WT failures [28]. The weighted KNN model with a 5-Fold CV correctly estimated 8524 measurements, and only 117 points have been considered as misclassifications. The incorrect predicted points by the model are shown in Fig. 1b. The classification model predicted points without the alarm activation when real alarms are activated. These points are the FN of the model, shown as blue points. For this particular case study, there are 72 points. The opposite case is the FP points, where alarms are predicted without being activated. There are 45 cases, where these points are shown in orange in Fig. 1.

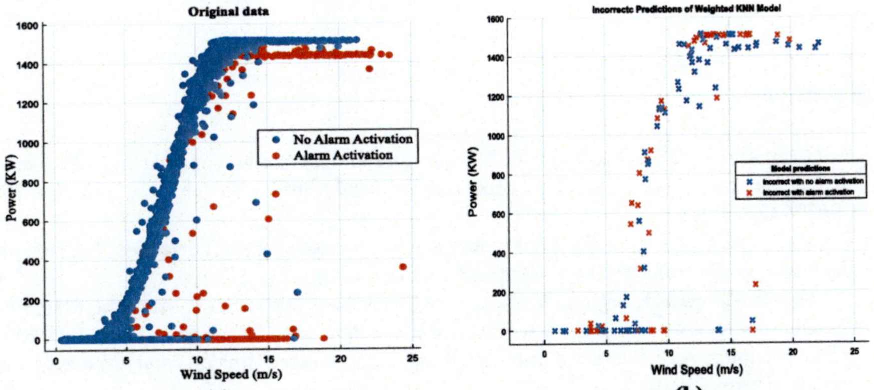

(a) (b)

Fig. 1 **a** Power Curve. **b** Misclassifications of Weighted KNN Model

The main conclusions about the 72 points of FN predicted by the method are:

- The alarm related to the level of turbulence has been activated 14 times, and the method has been unable to detect it.
- There were 6 maintenance activities, although 4 were undetected.

- The alarm was activated for a few seconds in 15 cases and these activations are considered as false alarms produced by the SCADA system.
- There is a desynchronization between the start and end of the alarm respect to the periods predicted by the model. It is concluded that the model is unable to detect the alarm period for this particular case.

The classification model generates false alarms, called FN. There are 45 FN and the alarms occurred in these points just before or after the alarm prediction. Three cases were produced by maintenance activities and they were detected as alarms.

4 Conclusions

False alarm detection is essential to ensure proper wind turbine maintenance management. Advanced analytics using machine learning algorithms are required due to the volume and variety of the generated data. The novelty presented in this paper is the implementation of the KNN method for alarm detection and classification. A real case study with SCADA data from a real wind turbine is analyzed to validate the methodology. The results imply accuracy of 98.7%. The specificity and the sensitivity are 99.13% and 88.52% respectively. For future research, the survey could be included the classification of alarm types and the use of other artificial intelligence models.

Acknowledgements. The work reported herewith has been financially by the Dirección General de Universidades, Investigación e Innovación of Castilla-La Mancha, under Research Grant ProSeaWind project (Ref.: SBPLY/19/180501/000102).

References

1. A.P. Marugán, A.M.P. Chacón, F.P.G. Márquez, Reliability analysis of detecting false alarms that employ neural networks: a real case study on wind turbines. Reliab. Eng. Syst. Saf. **191**, 106574 (2019)
2. F.P. Garcia Marquez, C.Q. Gomez Munoz, A new approach for fault detection, location and diagnosis by ultrasonic testing. Energies **13**, 1192 (2020)
3. F.P. García Márquez, I. Segovia Ramírez, B. Mohammadi-Ivatloo, A.P. Marugán, Reliability dynamic analysis by fault trees and binary decision diagrams. Information **11**, 324 (2020)
4. F.P.G. Márquez, A.M.P. Chacón, A review of non-destructive testing on wind turbines blades. Renew. Energy (2020)
5. F.P. Garcia Marquez, A. Pliego Marugan, J.M. Pinar Pérez, S. Hillmansen, M. Papaelias, Optimal dynamic analysis of electrical/electronic components in wind turbines. Energies **10**, 1111 (2017)
6. A. Pliego Marugán, F. P. García Márquez, and J. Lorente, "Decision making process via binary decision diagram," *International Journal of Management Science and Engineering Management,* vol. 10, pp. 3–8, 2015.
7. A.H. Butt, B. Akbar, J. Aslam, N. Akram, M.E.M. Soudagar, F.P. García Márquez, et al., Development of a linear acoustic array for aero-acoustic quantification of camber-bladed vertical axis wind turbine. Sensors **20**, 5954 (2020)

8. F.P.G. Márquez, A new method for maintenance management employing principal component analysis. Struct. Durab. Health Monit. **6**, 89 (2010)
9. F.P. García Márquez, I. Segovia Ramírez, A. Pliego Marugán, Decision making using logical decision tree and binary decision diagrams: a real case study of wind turbine manufacturing. Energies **12**, 1753 (2019)
10. A. Pliego Marugán, F.P. García Márquez, Advanced analytics for detection and diagnosis of false alarms and faults: a real case study. Wind Energy **22**, 1622–1635 (2019)
11. A. Pliego Marugán, F.P. Garcia Marquez, B. Lev, Optimal decision-making via binary decision diagrams for investments under a risky environment. Int. J. Prod. Res. **55**, 5271–5286 (2017)
12. A.M.P. Chacón, I.S. Ramírez, F.P.G. Márquez, False alarms analysis of wind turbine bearing system. Sustainability **12**, 7867 (2020)
13. A.A. Jiménez, L. Zhang, C.Q.G. Muñoz, F.P.G. Márquez, Maintenance management based on machine learning and nonlinear features in wind turbines. Renew. Energy **146**, 316–328 (2020)
14. I. Segovia Ramirez, B. Mohammadi-Ivatloo, F.P. Garcia Marquez, Alarms management by supervisory control and data acquisition system for wind turbines, in *Eksploatacja i Niezawodnosc-Maintenance and Reliability*, vol. 23 (2021), pp. 110–116
15. C. Q. Gómez Muñoz, F. P. García Márquez, B. Hernández Crespo, K. Makaya, Structural health monitoring for delamination detection and location in wind turbine blades employing guided waves. Wind Energy **22**, 698–711 (2019)
16. Q.P. He, J. Wang, Fault detection using the k-nearest neighbor rule for semiconductor manufacturing processes. IEEE Trans. Semicond. Manuf. **20**, 345–354 (2007)
17. T. Cover, P. Hart, Nearest neighbor pattern classification. IEEE Trans. Inf. Theory **13**, 21–27 (1967)
18. O. Eyecioglu, B. Hangun, K. Kayisli, M. Yesilbudak, Performance comparison of different machine learning algorithms on the prediction of wind turbine power generation, in *2019 8th International Conference on Renewable Energy Research and Applications (ICRERA)* (2019), pp. 922–926
19. A.A. Jimenez, C.Q.G. Muñoz, F.P.G. Márquez, Dirt and mud detection and diagnosis on a wind turbine blade employing guided waves and supervised learning classifiers. Reliab. Eng. Syst. Saf. **184**, 2–12 (2019)
20. D. Wettschereck, D.W. Aha, T. Mohri, A review and empirical evaluation of feature weighting methods for a class of lazy learning algorithms. Artif. Intell. Rev. **11**, 273–314 (1997)
21. X. Wu, V. Kumar, J.R. Quinlan, J. Ghosh, Q. Yang, H. Motoda et al., Top 10 algorithms in data mining. Knowl. Inf. Syst. **14**, 1–37 (2008)
22. L.-Y. Hu, M.-W. Huang, S.-W. Ke, C.-F. Tsai, The distance function effect on k-nearest neighbor classification for medical datasets. SpringerPlus **5**, 1304, (2016)
23. A. Maghari, Prediction of student's performance using modified KNN classifiers, in *Prediction of Student's Performance Using Modified KNN Classifiers. In The First International Conference on Engineering and Future Technology (ICEFT 2018)*, ed. by S.S. Alfere, A.Y. Maghari (2018), pp. 143–150
24. K. Chomboon, P. Chujai, P. Teerarassamee, K. Kerdprasop, N. Kerdprasop, An empirical study of distance metrics for k-nearest neighbor algorithm, in *Proceedings of the 3rd International Conference on Industrial Application Engineering* (2015), pp. 280–285
25. C.-L. Liu, C.-H. Lee, P.-M. Lin, A fall detection system using k-nearest neighbor classifier. Expert Syst. Appl. **37**, 7174–7181 (2010)
26. H. Shahabi, A. Shirzadi, K. Ghaderi, E. Omidvar, N. Al-Ansari, J.J. Clague et al., Flood detection and susceptibility mapping using sentinel-1 remote sensing data and a machine learning approach: hybrid intelligence of bagging ensemble based on k-nearest neighbor classifier. Remote. Sens. **12**, 266 (2020)

27. S. Yadav, S. Shukla, Analysis of k-fold cross-validation over hold-out validation on colossal datasets for quality classification, in *2016 IEEE 6th International Conference on Advanced Computing (IACC)* (2016), pp. 78–83.
28. M. Lydia, S.S. Kumar, A.I. Selvakumar, G.E.P. Kumar, A comprehensive review on wind turbine power curve modeling techniques. Renew. Sustain. Energy Rev. **30**, 452–460 (2014)

Classification Learner Applied to False Alarms for Wind Turbine Maintenance Management

Isaac Segovia Ramirez(✉) and Fausto Pedro García Márquez

Ingenium Research Group, Universidad Castilla-La Mancha, 13071 Ciudad Real, Spain
{isaac.segovia,faustopedro.garcia}@uclm.es

1 Introduction

Wind energy is in continuous expansion, being one of the most cost-competitive renewable energies with a global cumulative wind power capacity up to 651 GW in 2019, see Fig. 1 [1]. Wind energy market was growing 4% each year, but the effects of COVID-19 slowed this trend. This impact is still not quantified, although it is expected to have reached the 76 GW of new installations in 2020 driven by China and USA. There has also been an improvement in the size of Wind Turbines (WTs), reaching the 20 MW and, therefore, an increment in the complexity in maintenance operations due to higher failure rates [2]. The wind energy market must continue growing to play a key role in the global energy generations towards 2030 objectives [3].

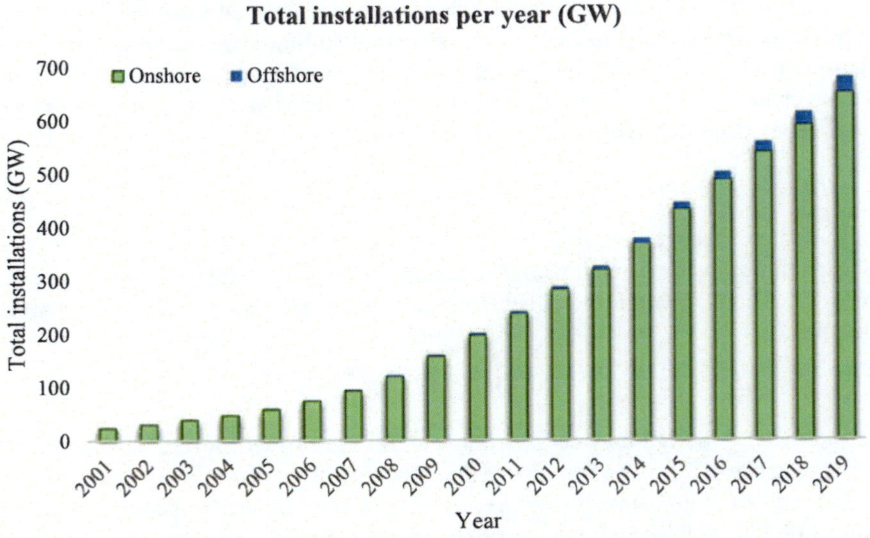

Fig. 1. Total installations onshore and offshore. *Source* Global Wind Energy Council [1].

WTs use the rotor blades to transform wind energy into mechanical energy that is transferred to the generation and convert it into electric energy. The main parts of

F. P. García Márquez (ed.), *International Conference on Intelligent Emerging Methods of Artificial Intelligence & Cloud Computing*, Smart Innovation, Systems and Technologies 273, https://doi.org/10.1007/978-3-030-92905-3_13

the nacelle are the pitch system, generator, rotor and gearbox, among others. The rotor includes the blades, the bearings and a rotor hub. The generator converts the mechanical energy to electrical regarding on the requirements of the grid. The gearbox transforms the torque and adjust the speed of the generator. The pitch system adjusts the angle of attack of the blade in function of the wind direction. Electric, control and yaw systems together with generator and gearbox present elevated failure rate with hight associated downtimes [4]. WT systems operate under hard working conditions, e.g., extreme temperatures or varying wind conditions and loads [5]. The lifetime of WTs is estimated for 20 years and operation and maintenance (O&M) account for 15–25% of lifecycle costs, although these costs are dynamics and influenced by several variables [6]. Condition-based maintenance with the monitoring of critical components is needed to reduce the O&M costs and avoid unnecessary operations and downtimes [7]. Fault pattern recognition and prediction techniques allow the detection of fault generation with the analysis of incoming alarms. Fault diagnosis and condition monitoring have a key role for predictive maintenance and condition-based maintenance of WTs, since the working conditions perform a critical challenge for reliable prognostics and diagnostics.

Condition monitoring systems (CMS) are formed by the combination of sensors and signal processing devices that provide continuous information about the WT components using different techniques [8, 9], e.g., ultrasonic and acoustic techniques, vibration, thermography [10–12]. Supervisory control and data acquisition (SCADA) system collects data about the operating conditions of the WT, and this information is usually divided into alarms and signals to determine the health of the system [13]. Alarms are warnings messages that reveal malfunctions or issues that require maintenance activities [14]. Alarms are produced when predefined conditions or threshold limits are reached, mainly with changes of the working state, malfunction or failure mode activation [15, 16]. Alarms are also characterized as time series events T of n ordered values t_n divided in several subsequence Q produced by a failure system sequence, as it is observed in Eq. (1).

$$T = t_1, \ldots, t_n \tag{1}$$

The pattern recognition aims to detect subsequences Q with similar performance and capabilities in comparison with the interest region. The matching subsequence D is defined in Eq. (2), being R the positive range of the subsequence and M the analysed subsequence.

$$D(Q, M) \leq R \tag{2}$$

The SCADA design may cause a noisy alarm system with several activations and overlapping alarm functions that reduce the reliability of the analysis [17]. The monitoring and analysis of the alarm system is essential due to the increasing activations of false alarms [18, 19]. A false alarm is considered when the alarm is triggered although it is not produced a real failure, reducing the effectiveness of the maintenance management plant and increasing associated costs. Several alarm activations in short periods of time may indicate that the alarm turns out to be false. New techniques and algorithms are needed to acquire reliable information for false alarm recognition [20].

Several research studies are being developed to improve the reliability of pattern recognition methods in huge volume of data [21, 22]. Artificial Neural Networks (ANNs),

Machine Learning (ML) or Deep Learning are the most applied techniques due to pattern recognition and forecasting capabilities [23]. ANNs adjust the weights of the connections between layers and neurons by means of training process to develop forecasting, pattern recognition and function approximation. ANNs have been applied in the alarm processing in several works [24, 25]. Pliego et al. [26] employed an ANN structure in vibration dataset for false alarm detection, reaching a precision more than 80%. Bangalore et al. [27] developed an approach to apply filtered data to the ANN model and avoid false alarms. However, ANNs require advanced training and high computational costs that may reduce the reliability of the results [28]. The K-Nearest Neighbour (KNN) algorithm is a pattern recognition technique for regression, classification and fault detection although it is not currently applied for false alarm detection [29]. The Euclidean distance, shown in Eq. (3) [30], is applied to determine the distance of the K observation to the feature space to develop an estimation.

$$D(X, Y) = \left(\sum_{k=1}^{K} |x_i - y_i|^2 \right)^{1/2} \tag{3}$$

Support Vector Machine (SVM) is widely applied to classification problems and fault detection. SVM applies nonlinear Kernels reducing over-fitting issues and presenting high reliability with reduced dataset. Laouti et al. [31] applied SVM for fault detection and pattern recognition in simulated WTs. Several studies have focused on the application of ML techniques in wind energy. Durbhaka and Selvaraj [32] applied KNN and SVM to classify the type of fault. The decision tree belongs to supervised learning and it is able to identify several categories and patterns from an initial dataset with high reliability [33]. It is usually defined an ensemble of decision trees to increase the accuracy of the classifier. Bagged trees develop new random groups from the initial training datasets to generate the model and define the probabilities of the classes [34, 35]. The Naïve-Bayes classifier considers all the classes independent and develops different estimation about the probability. This technique is widely applied for WT prediction and fault diagnosis [36].

This paper develops a new approach to analyse SCADA data to increase the reliability of alarm detection by means of the application of different algorithms. The main contributions of this paper are summarised as follows:

- The definition of critical alarm regarding on three key factors and the acquisition of the signal dataset related to this alarm through correlations. The dataset introduced in the algorithms presents a significant reduction of the computational load ensuring a proper analysis.
- The application of several types of algorithms to study the filtered dataset and model the behaviour of the alarm with different predictors. A case study with a real WT is presented, where the possible false alarms are identified as misclassification points by at least of two algorithms. The validation is developed comparing the results provided by the algorithms and the real SCADA activations.

This paper is organized as follows: Sect. 2 presents the methodology and the approach; Sect. 3 describes a real case study with real SCADA data studied with the proposed algorithms and it is presented the analysis of the results provided by the methodology. The conclusions are summarized in Sect. 4.

2 Approach

The approach developed in this paper aims to prioritize the alarm distribution obtained with SCADA system to detect possible false alarms. The first phase is the filtering of the SCADA data to reduce the volume of the dataset and the computational load of the algorithms. The alarms are quantified based on three key performance indicators (average alarm rate, maximum alarm rate and the percentage time) to define the critical alarm. Once this alarm is determined, it is required a related dataset to reduce the initial signal dataset and ensure a suitable analysis. The correlation coefficient provides the relation between different variables. The coefficient r, employed in Eq. 4, shows the correlation between the signals x and y.

$$r = \frac{N \cdot (\sum_{n=1}^{N} x \cdot y) - (\sum_{n=1}^{N} x) \cdot (\sum_{n=1}^{N} y)}{\sqrt{(N \cdot \sum_{n=1}^{N} x^2 - (\sum_{n=1}^{N} x)^2) \cdot (N \cdot \sum_{n=1}^{N} y^2 - (\sum_{n=1}^{N} y)^2)}} \tag{4}$$

Five data analysis methods with different versions and variations are applied to reduce the dataset for forecasting and false alarm identification: decision tree, Naïve Bayes, SVM, KNN and ensemble tree. It is applied a cross-validation approach to divide the dataset into sub datasets to increase the reliability and reduce the overfitting. For this study, it is considered the accuracy as fundamental indicator of the performance of the models. The confusion matrix is employed to quantify the performance of the algorithms computing the differences between the predictions and the real cases: the accuracy is the ratio of the total number of proper predictions, being the True Positive (TP) the cases where the alarms are properly categorized; False Positive (FP) or false alarms, where different activations are incorrectly determined as alarms but the WT presents a healthy condition; the False Negative (FN) or missed alarms are cases not detected by the prediction methods and the True Negative (TN) are the alarms rejected properly, see Eq. (5).

$$Accuracy = \frac{TP + TN}{TP + TN + FP + FN} \tag{5}$$

The response of the algorithms produces reliable information to prioritize alarms. The approach developed in this work considers as possible false alarms those points in the period of alarm deactivation not correctly classified by the largest number of algorithms. A misclassification must be detected at least by two different methods to be considered as possible false alarm. The validation of the results is established with the comparison of these misclassification and the alarm triggering by the SCADA system.

3 Case Study and Results

The case study presented in this paper is formed by SCADA data from a real WT of 2 MW. The initial dataset is defined with 273 signals and 96 alarms from one year and a sampling rate of 10 min, obtaining an initial dataset with 200 million of data. The distribution of the 96 alarms regarding on the three key indicators (maximum alarm rate, average alarm activation and number of activations) is shown in Fig. 2.

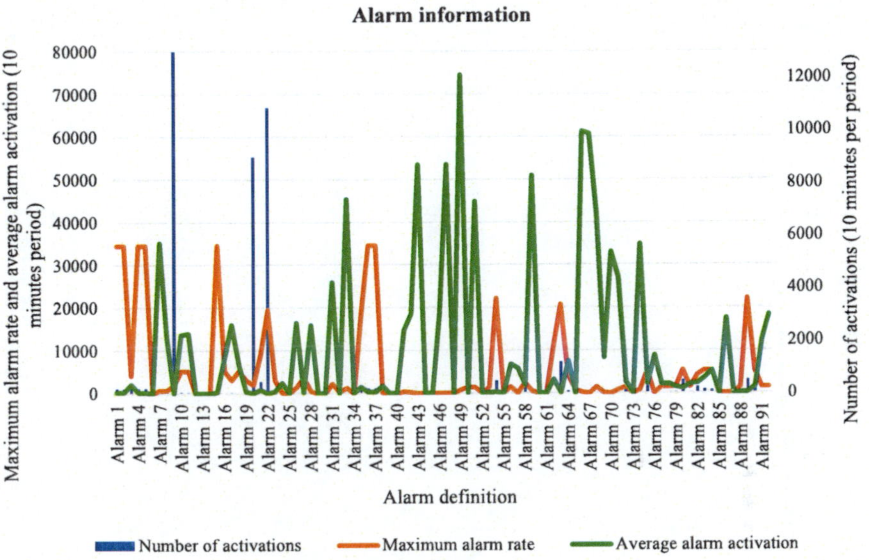

Fig. 2. SCADA alarm description.

To determine the critical alarm, it is also considered the difference between alarm activations since several activations in short periods of times may be classified as false alarms. With this information and the values of the three key indicators, it is possible to determine the critical alarm. For this case study, the critical alarm is related to the over-speed of the generator. This alarm is triggered for predefined wind velocity conditions higher than 25 m/s with slow filter or 30 m/s with quick filter. The main characteristics of this alarm are stablished in Table 1.

Table 1. Alarm information.

Number of data	597,618 (414 days)
Number of activations	37
Average period time of activation	97.89

A new dataset related to the critical alarm it is required to ensure a proper application of the algorithms proposed in this work. The ROC curve by logistic regression is defined to obtain an area under the curve (AUC) around 0.9721 and identify the correlated signals, see Fig. 3. For this particular case study, it is only considered the 3 signals with more correlation with the alarm due to the high computational load of the algorithms.

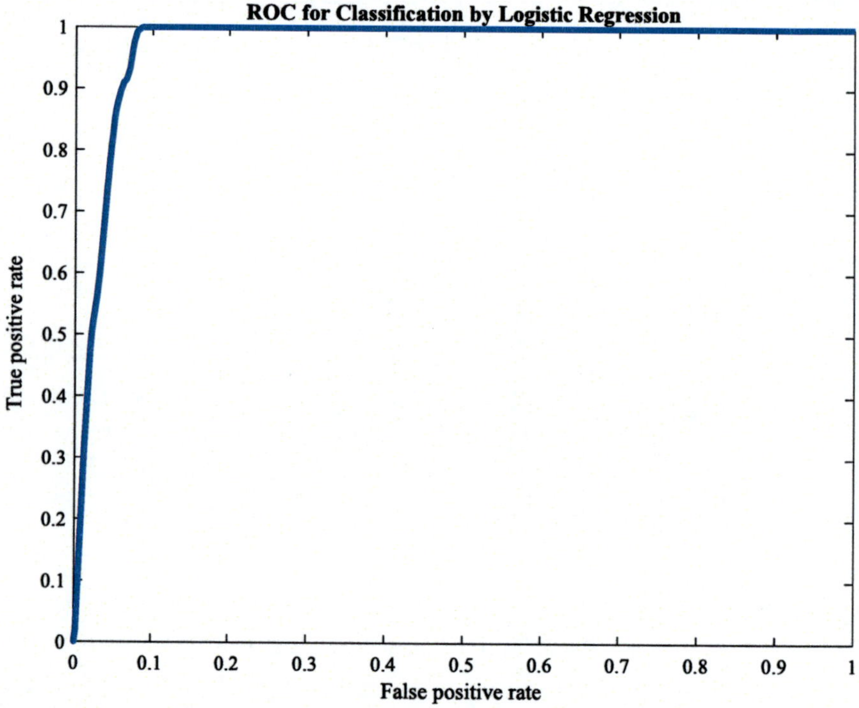

Fig. 3. ROC curve.

Table 2 shows the results of the algorithms and the misclassifications. The accuracy of each algorithm and its version is provided, obtaining constant and suitable values in all cases.

The results provided by Naïve Bayes, linear SVM and coarse KNN are discarded since it is not produced any misclassification for the study. Only 6 points fulfilled that the misclassifications must be detected by two algorithms to be considered as false alarm. These points are compared with the SCADA activation and it is determined that the misclassifications do not coincide with any alarm activation. There is no information about maintenance activities o reparation tasks that may lead to the alarm triggering and they are finally determined as false alarms.

Table 2. Comparison for different KNN methods.

Classifier type	Holdout validation	
	Accuracy (%)	Misclassifications
Fine tree	99.1	104
Medium	99.1	7
Coarse	99.2	0
Logistic regression	99.4	3
Naive Bayes	93.7	NaN
Linear SVM	99.4	0
Fine KNN	98.9	28
Medium KNN	97.4	25
Coarse KNN	97.4	0
Ensemble boosted trees	99.4	6
Ensemble bagged trees	99.3	9

4 Conclusions

The wind energy needs a reduction in the operation and maintenance costs to increase the competitiveness of this renewable energy. Supervisory control and data acquisition system acquired valuable information about several types of condition monitoring systems about the real state of the wind turbine. The false alarm identification is a critical issue in the wind energy industry, being necessary new algorithms and methodologies, such as machine and deep learning, to manage large volumes of data providing a consistent accuracy.

This paper proposes an initial filtering process determining the critical alarm and the related dataset to increase the reliability of the false alarm detection. The robustness of the methodology lies in the application of several types of algorithms to determine the misclassification points considered as possible false alarms. The results obtained from a real case study proves that all the algorithms present an accuracy around the 90% in all the cases. The misclassification points are compared with the real activations of the supervisory control and data acquisition system. It is proposed as future work the application of larger filtered datasets to increase the reliability of the modelling.

Acknowledgements. The work reported herewith has been financially by the Dirección General de Universidades, Investigación e Innovación of Castilla-La Mancha, under Research Grant ProSeaWind project (Ref.: SBPLY/19/180501/000102).

References

1. Joyce Lee, F.Z. *Global wind report*; Global Wind Energy Council: 2020.
2. C. Dao, B. Kazemtabrizi, C. Crabtree, Wind turbine reliability data review and impacts on levelised cost of energy. Wind Energy **22**, 1848–1871 (2019)
3. F.P.G. Márquez, A. Karyotakis, M. Papaelias, *Renewable energies: Business outlook 2050*. (Springer, Berlin, 2018)
4. E. Artigao, S. Martín-Martínez, A. Honrubia-Escribano, E. Gómez-Lázaro, Wind turbine reliability: a comprehensive review towards effective condition monitoring development. Appl. Energy **228**, 1569–1583 (2018)
5. A. Pliego Marugán, F.P. Garcia Marquez, B. Lev, Optimal decision-making via binary decision diagrams for investments under a risky environment. Int. J. Prod. Res. **55**, 5271–5286 (2017)
6. D. Chan, J. Mo, Life cycle reliability and maintenance analyses of wind turbines. Energy Procedia **110**, 328–333 (2017)
7. F.P.G. Márquez, A.M.P. Chacón, A review of non-destructive testing on wind turbines blades. Renewable Energy (2020)
8. C.Q. Gómez Muñoz, F.P. García Márquez, B. Hernández Crespo, K. Makaya, Structural health monitoring for delamination detection and location in wind turbine blades employing guided waves. Wind Energy **22**, 698–711 (2019)
9. A.H. Butt, B. Akbar, J. Aslam, N. Akram, M.E.M. Soudagar, F.P. García Márquez, M. Younis, E. Uddin, Development of a linear acoustic array for aero-acoustic quantification of camber-bladed vertical axis wind turbine. Sensors **20**, 5954 (2020)
10. I.S. Ramirez, C.Q.G. Muñoz, F.P.G. Marquez, in *A Condition Monitoring System for Blades of Wind Turbine Maintenance Management* (Springer, Singapore, 2017), pp 3–11
11. C.Q. Gómez Muñoz, F.P. García Márquez, A new fault location approach for acoustic emission techniques in wind turbines. Energies **9**, 40 (2016)
12. P.J.B. Sánchez, F.P.G. Marquez, in *New Approaches on Maintenance Management for Wind Turbines Based on Acoustic Inspection, International Conference on Management Science and Engineering Management* (Springer, 2020), pp 791–800
13. P. Bangalore, M. Patriksson, Analysis of scada data for early fault detection, with application to the maintenance management of wind turbines. Renew. Energy **115**, 521–532 (2018)
14. F.P.G. Márquez, A new method for maintenance management employing principal component analysis. Struct. Durab. Health Monit. **6**, 89 (2010)
15. A. Pliego Marugán, F.P. García Márquez, Advanced analytics for detection and diagnosis of false alarms and faults: a real case study. Wind Energy **22**, 1622–1635 (2019)
16. F.P. Garcia Marquez, C.Q. Gomez Munoz, A new approach for fault detection, location and diagnosis by ultrasonic testing. *Energies*, **13**, 1192 (2020)
17. A.M.P. Chacón, I.S. Ramírez, F.P.G. Márquez, False alarms analysis of wind turbine bearing system. Sustainability **12**, 7867 (2020)
18. I.S. Ramirez, F.P.G. Marquez, in *Supervisory Control and Data Acquisition Analysis for Wind Turbine Maintenance Management, International Conference on Management Science and Engineering Management* (Springer, 2020), pp 470–480
19. I. Segovia Ramirez, B. Mohammadi-Ivatloo, F.P. Garcia Marquez, Alarms management by supervisory control and data acquisition system for wind turbines. Eksploatacja I Niezawodnosc-Maintenance and reliability **23**, 110–116 (2021)
20. Y. Qiu, Y. Feng, P. Tavner, P. Richardson, G. Erdos, B. Chen, Wind turbine scada alarm analysis for improving reliability. Wind Energy **15**, 951–966 (2012)
21. F.P. Garcia Marquez, A. Pliego Marugan, J.M. Pinar Perez, S. Hillmansen, M. Papaelias, Optimal dynamic analysis of electrical/electronic components in wind turbines. Energies **10**, 1111 (2017)

22. F.P. García Márquez, I. Segovia Ramírez, B. Mohammadi-Ivatloo, A.P. Marugán, Reliability dynamic analysis by fault trees and binary decision diagrams. Information **11**, 324 (2020)
23. A.A. Jiménez, L. Zhang, C.Q.G. Muñoz, F.P.G. Márquez, Maintenance management based on machine learning and nonlinear features in wind turbines. Renew. Energy **146**, 316–328 (2020)
24. B. Chen, Y. Qiu, Y. Feng, P. Tavner, W. Song, Wind turbine scada alarm pattern recognition (2011)
25. A.P. Marugán, F.P.G. Márquez, J.M.P. Perez, D. Ruiz-Hernández, A survey of artificial neural network in wind energy systems. Appl. Energy **228**, 1822–1836 (2018)
26. A.P. Marugán, A.M.P. Chacón, F.P.G. Márquez, Reliability analysis of detecting false alarms that employ neural networks: A real case study on wind turbines. Reliab. Eng. Syst. Saf. **191**, 106574 (2019)
27. P. Bangalore, S. Letzgus, D. Karlsson, M. Patriksson, An artificial neural network-based condition monitoring method for wind turbines, with application to the monitoring of the gearbox. Wind Energy **20**, 1421–1438 (2017)
28. G. Li, J. Shi, On comparing three artificial neural networks for wind speed forecasting. Appl. Energy **87**, 2313–2320 (2010)
29. A.A. Jimenez, C.Q.G. Muñoz, F.P.G. Márquez, Dirt and mud detection and diagnosis on a wind turbine blade employing guided waves and supervised learning classifiers. Reliab. Eng. Syst. Saf. **184**, 2–12 (2019)
30. K.-P. Chan, A.W.-C Fu, in Efficient time series matching by wavelets, in *Proceedings 15th International Conference on Data Engineering* (*Cat. No. 99CB36337*); IEEE (1999), pp 126–133
31. N. Laouti, N. Sheibat-Othman, S. Othman, Support vector machines for fault detection in wind turbines. IFAC Proc. Vol. **44**, 7067–7072 (2011)
32. G.K. Durbhaka, B. Selvaraj, in *Predictive Maintenance for Wind Turbine Diagnostics Using Vibration Signal Analysis Based on Collaborative Recommendation Approach*, 2016 *International Conference on Advances in Computing, Communications and Informatics* (*ICACCI*), IEEE (2016), pp 1839–1842
33. F.P. García Márquez, I. Segovia Ramírez, A. Pliego Marugán, Decision making using logical decision tree and binary decision diagrams: a real case study of wind turbine manufacturing. Energies **12**, 1753 (2019)
34. J.C.-W. Chan, D. Paelinckx, Evaluation of random forest and adaboost tree-based ensemble classification and spectral band selection for ecotope mapping using airborne hyperspectral imagery. Remote Sens. Environ. **112**, 2999–3011 (2008)
35. J. Lee, W. Wang, F. Harrou, Y. Sun, Wind power prediction using ensemble learning-based models. IEEE Access **8**, 61517–61527 (2020)
36. Y. Zhao, D. Li, A. Dong, D. Kang, Q. Lv, L. Shang, Fault prediction and diagnosis of wind turbine generators using scada data. Energies **10**, 1210 (2017)

Agricultural Image Analysis on Wavelet Transform

Rishi Sikka(✉)

Department of Electronics and Communication Engineering, Sanskriti University, Uttar Pradesh, Mathura, India
rishisikka.ec@sanskriti.edu.in

1 Introduction

Due to increasing problems in agricultural field, to manage and identify the difficulties present like crop diseases and other factors leading to reduced growth, it has become a necessity to figure out the accurate and quantitative figures. To get the accurate information about the problems existing in the crops, it is necessary to get the digital images and analyze them to know to the reasons. However, the process to manage and identify the problems is an extensive task. The agricultural digital images [1] do not show the actual problems leading to the use of different products in case of crop diseases, thus the exact and correct crop disease detection is necessary to avoid such problems.

The image processing with computer vision or machine vision is a rapid growing industry with vast number of applications in the fields of medical, research, agriculture, engineering etc. The processing of digital images is far better than the processing done by human vision and is thus advantageous in terms of adaptability, productivity, accuracy etc. and thus has major applications in the field of agriculture. The main applications of image processing using computer technology lies in the identification of plant species, detection of quality, classification of products in agricultural fields [1]. The processes and techniques used in the field of image processing including the machine vision technology includesacquisition of image, processing of the images, pattern recognition and classification [2] etc. these steps of processing the images must give an efficient output as the resultant image so that the important and relevant information can be extracted. Especially the agricultural images must be processed effectively without any kind of blurs or noisy data in the image.

Therefore the techniques used for processing the information in the agricultural images without impacting the actual information and retrieving the best data include noise removal techniques [3], image compression and image enhancement for getting better results. The wavelet transform [4] based approach is used for the filtering of the agricultural images due to its efficiency and better results which help in better transmission and processing of the agricultural images.

F. P. García Márquez (ed.), *International Conference on Intelligent Emerging Methods of Artificial Intelligence & Cloud Computing*, Smart Innovation, Systems and Technologies 273, https://doi.org/10.1007/978-3-030-92905-3_14

2 Methodology

The quality of the agricultural images is not in a form which can be easily accessed and processed by human vision and computer system. To make the images better and extract the relevant information from the images required for transmission and further processing required according to the technologies used in the agricultural field, certain steps of image analysis like removing noise, image enhancement and image compression are the most important techniques required today. These techniques of image processing use the method of wavelet transform for performing denoising, image compression and image enhancement as discussed.

2.1 Image Denoising

The technique to remove the noise from images or data to extract the relevant information is a process known as denoising, which plays an important role in image processing. The noise removal of images is done to prevent the distortion of edges or boundaries area of the images so that information is not lost while processing of the images. Many filtering techniques are available for removing noise from the data based on spatial or frequency domains, where the frequency domain techniques are more advantageous over that of spatial domain methods. Denoising of images in the field of agricultural images is the most initial step to be followed by using wavelet transform where the wavelet coefficients [5] of the image are generated and analyzed in the denoising technology.

The wavelet denoising is performed based on the threshold values and the wavelet coefficients of the image in the wavelet domain. Using the threshold in wavelet transform preserves the coefficients of the image with relevant data and the coefficients with noise are made nil for removing noise from the images. The noise signal is decomposed by setting a threshold value where the wavelet coefficients having value less than the predefined threshold are kept 0 which are the noise signals in the data. The wavelet coefficients remaining are further processed to recover the image by using inverse wavelet transform. The threshold function generated for performing the wavelet transform [6] on the images is as shown below:

$$\widehat{w}_{i,j} = \begin{cases} sign\left[\left|w_{i,j}\right| - \frac{2(1-a)\lambda}{1+e^{b(\left|w_{i,j}\right|-\lambda)}}\right], & \left|w_{i,j}\right| \geq \lambda \\ cw_{i,j} & \left|w_{i,j}\right| \geq \lambda \end{cases}$$

While performing the wavelet transform, the threshold value chosen is a local thresh-oldwhich decreases along with the increase in decomposition. The threshold value chosen is not global threshold in wavelet denoising because even after the wavelet decomposition [7], the signals are visible in the image even after the wavelet transforms. The threshold to be applied on the image is denoted by (Figs. 1 and 2):

$$\lambda_j = \sigma\sqrt{2\log(N)}/\sqrt{\log_2^{j+1}}$$

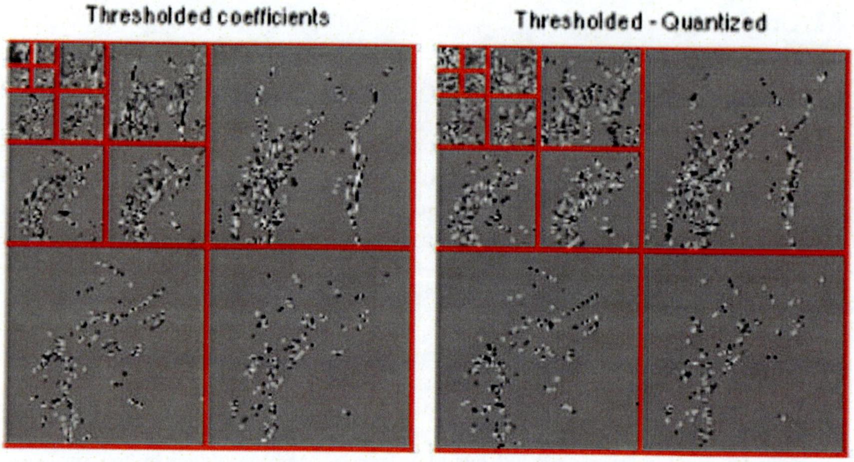

Fig. 1. Denoising by threshold.

Fig. 2. Images with and without noise.

2.2 Image Compression

Image compression of the images is a necessary step to be followed when analyzing the agricultural images due to large data sets which leads to complexity if the size is not reduced. The image compression using the wavelet transform [8] gives an effective resultant image which can be stored and transferred easily. Image compression is achieved by the wavelet transform by setting a threshold value.

If the original image includes multiple parts or components, it becomes a necessity to decompose the image into sub parts. The wavelet transform on the images for compression work in the frequency domain to achieve better results [9]. The sub parts of the images are known as blocks which are further processed using entropy coding which is carried out on the coefficient values of the blocks which lead to compressed data. The encoded data is then extracted and reconstructed to form the image back by decoding the blocks. The processing done for image compression is as shown in the block diagram below in Fig. 3:

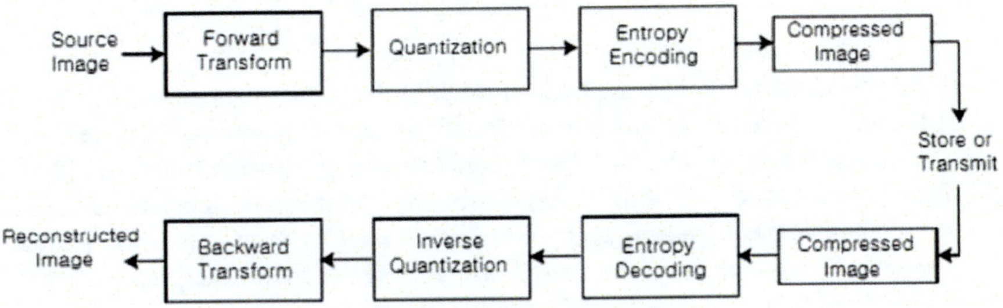

Fig. 3. Image compression.

Entropy encoding being the main step in the image compression by wavelet transforms is performed at the step of quantization where the quantized values are compressed in a way where the data is not lost. The coded values are generated equivalent to each of the quantized value to reduce the size of the data or the image and these coded values are further decoded on reconstruction of the images by inverse quantization. The entropy coding can be performed using various techniques like Huffman coding.

2.3 Image Enhancement

Image enhancement is the step used in image analysis for improving the image quality so that the relevant information can be extracted from the images. The resultant images after image enhancement are improved such that the machine system as well a human vision can gain information from the images easily. The methods of enhancement are based on both spatial and the frequency domain of the images. The frequency domain techniques are much used due to better results which include the processing of the images by convolution using Fourier transforms. The image enhancement on the agricultural images is performed using the technique of 2D stationary dyadic wavelet transform (Fig. 4) [8, 10].

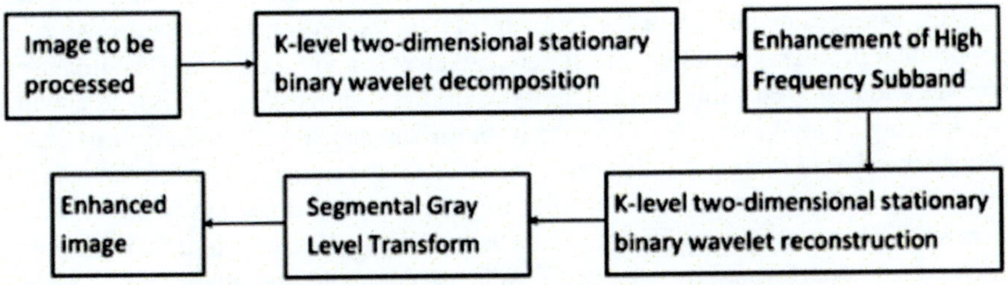

Fig. 4. Image enhancement using 2D wavelet transform.

The nonlinear enhancement is applied as a method on the agricultural images, where the direct function represented by f(x) in the time domain can be further represented as:

$$\hat{f}(x) = (x) - \left[E\left(2^p \frac{d^2 f}{dx^2}\right) - 2^p \frac{d^2 f}{dx^2}\right]$$

The enhancement method is applied on the high frequency components of the sub parts of the image and further the wavelet coefficients reconstruct the image back to original for image analysis. The two dimensional stationary wavelet transform [11] has high frequency components in a direction diagonally and it achieves better information from the images at the high frequency level. The wavelet transform used generates the decomposed components in the image which is better than other techniques used for the processing of same. The modulus of the 2D wavelet transform also has further applications in determining the edges of the images. The image enhancement performed by using 2D wavelet transform is as shown in Fig. 5:

Fig. 5. Enhanced image.

3 Conclusion

In the field of agriculture, recent technical advancements have led to increased research using the techniques of image processing. The agricultural images are processed for identifying the diseases occurring in the crops or plants, the reason for slow growth of the crops or other problems that may lead to reduced growth and production. The images are processed and analyzed using various methods of image processing like image enhancement, filtering of images to remove noise from the images and the compressing images to reduce the size of image. As discussed in the paper the image enhancement using 2D dyadic wavelet transform is done in order to get correct information from the images which are easily accessible by human vision as well as computer systems. The denoising of the images is performed by setting the value of threshold and quantizing in order to remove noise from the images by filtering thus removing the unwanted data from agricultural images which may otherwise lead to suppressing of the actual information. The other technique used in image analyzing of the agricultural images is image compression using 2D wavelet transforms applied on the high frequency components of the images. The paper address that the agricultural images have a large amount of data that is not easy to process, store or transfer. So the image compression using entropy encoding is performed to reduce the size of data. Thus applying these techniques the improved and better results of agricultural images is achieved which are useful for further analysis and processing in using technology in agriculture.

References

1. A. Vibhute, S.K. Bodhe, Applications of image processing in agriculture: a survey. Int. J. Comput. Appl. (2012)
2. J.D. Pujari, R. Yakkundimath, A.S. Byadgi, Image processing based detection of fungal diseases in plants. Procedia Comput. Sci. (2015)
3. P.K. Sethy, B. Negi, S.K. Behera, N.K. Barpanda, A.K. Rath, An image processing approach for detection, quantification, and identification of plant leaf diseases -a review. Int. J. Eng. Technol. (2017)
4. J. Gao, H. Sultan, J. Hu, W.W. Tung, Denoising nonlinear time series by adaptive filtering and wavelet shrinkage: A comparison. IEEE Signal Process. Lett. (2010)
5. F. Luisier, C. Vonesch, T. Blu, M. Unser, Fast interscale wavelet denoising of Poisson-corrupted images. Signal Processing (2010)
6. J. Wang, X. Meng, Study on robust noise reduction algorithm based on wavelet transfrom, in *2011 International Conference on Remote Sensing, Environment and Transportation Engineering, RSETE 2011 - Proceedings* (2011)
7. C.-L. Liu, A tutorial of the wavelet transform. *History* (2010)
8. D. Gupta, S. Choubey, Discrete wavelet transform for image processing. Int. J. Emerg. Technol. Adv. Eng. (2015)
9. S. Majumder, N.L. Meitei, A.D. Singh, M. Mishra, Image compression using lifting wavelet transform. Transform (2010)
10. J. Gilles, G. Tran, S. Osher, 2D empirical transforms. Wavelets, ridgelets, and curvelets revisited. SIAM J. Imaging Sci. (2014)
11. J. Ma, Z. Wang, B. Pan, T. Hoang, M. Vo, L. Luu, Two-dimensional continuous wavelet transform for phase determination of complex interferograms. Appl. Opt. (2011)

Deep Learning Algorithms

Swapnil Raj(✉)

SOEIT, Sanskriti University, Mathura, India
swapnil.cse@sanskriti.edu.in

1 Introduction

Various machine learning approaches are available for a variety of applications growing with the advanced technology. One of the approach of machine leaning known as deep learning is widely used. The increase in hardware technologies are being researched and deep learning is utilized to outperform the conventional approaches. The deep learning approach or algorithm is also known as representation learning [1] and uses the technology working with graphs to develop learning models. The deep learning finds various applications like natural language processing, visual processing, audio processing, etc. Conventionally, the machine learning algorithms differ in terms of data representation. Data representation directly affects the performance; thus better data representation is required. Thus, the machine learning approach uses the feature engineering to build relevant features from the large amount of raw data. But, in comparison to such methods, the deep learning automatically extracts the relevant features for further processing. This leads to effective research by less human efforts and also does not requires extensive research on the domain knowledge.

Deep learning, a subclass of machine learning methods [2] and techniques and advantageous over machine learning in terms of performance as shown in Fig. 1, work with multiple layers of processing the information for computing the hierarchical features of the data deriving the higher level of information from the lower level data. The architecture of the deep learning is similar to that of the artificial intelligence where the process is simulated in the human brain. The human brains are capable of extracting data from different areas. The information is received through eyes and the objects are classified from the data processed through the brain. Thus the deep learning algorithms is equivalent to the processing of human brain. Deep learning works in the manner similar to that of the structure and function of human brain, thus is a subpart of artificial intelligence along with neural networks. It allows the computation of information with multiple layer processing in order to represent the data in a form with multiple layers of abstractions i.e. deep learning methods are the methods with representation of multiple layers which are transformed at each level starting from the level with raw input data by abstraction [3]. These layers in deep learning are generated by the input data using a process of learning known as deep learning models. These deep learning models depend on artificial neural networks [4] and the algorithms like deep belief networks, convolutional neural networks, deep neural networks etc.

Deep learning with neural network is a neural network having complexity due to multiple layers as use of various complex mathematical models are used for processing of

F. P. García Márquez (ed.), *International Conference on Intelligent Emerging Methods of Artificial Intelligence & Cloud Computing*, Smart Innovation, Systems and Technologies 273, https://doi.org/10.1007/978-3-030-92905-3_15

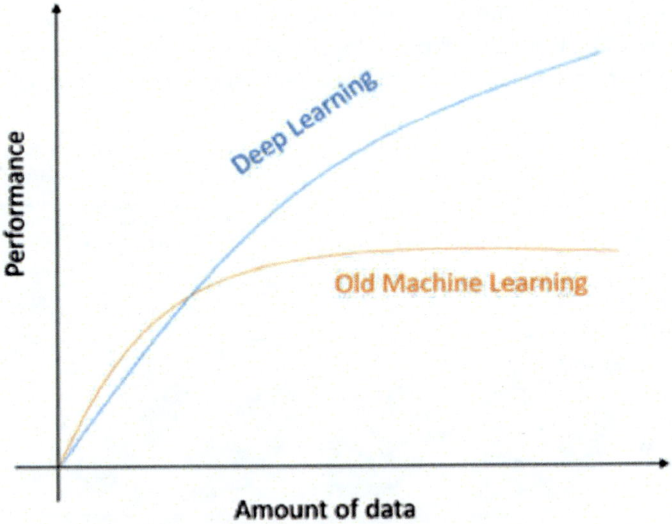

Fig. 1. Deep learning versus machine learning

the input data. The capability of processing large amount of data and their features makes the deep neural network an effective and efficient tool for dealing with unstructured data. The deep neural network [5] is scalable as well as ability to extract features from the data automatically and this is known as feature learning.

2 Deep Learning Algorithms

The algorithms or methods of deep neural networks are updated class under artificial neural networks which support significantly reduced cost of computation, increased processing speed as well as advancement in machine learning. They are capable of creating larger and complex neural networks and for this purpose various algorithms are used as discussed in detail.

1. *Convolutional neural network*
   The deep learning algorithm that can assign weights to the input data and differentiate them from each other is known as convolutional neural networks. This algorithm is also known as shift invariant artificial neural network [6] depending on the characteristics and architecture.. CNN is applied to various applications such as computer vision, audio and visual processing, NLP, etc. The effective processing of two dimensional data extracted from the raw data depends on the shared weights of the network. This results in faster performance and operations. Altogether these properties or layers of CNN enable the achievement of better generalization with a requirement of reduced memory storage and larger training thus making the networks powerful. The CNN has applications [8] in various fields specifically in image classification, recognition, NLP and image analysis in medical field. A CNN [7] consists of two

layers being an input and an output layers along with many hidden layers. These hidden layers include convolutional layers, pooling layers, fully connected layers, normalization layers, and the final convolutional layer. The layers of CNN as shown in Fig. 2.

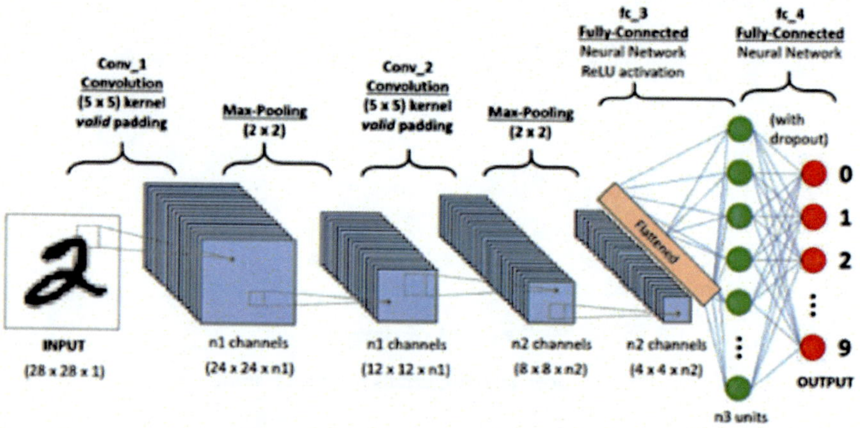

Fig. 2. CNN sequence for classification

In the convolutional layers, a CNN utilizes various kernels to convolve the whole image as well as the intermediate feature maps for generating various feature maps. The pooling layer following the convolutional layer is used to reduce the dimensions of feature maps and network parameters. Similar to convolutional layers, pooling layers are also translation invariant, because their computations take neighboring pixels into account. The fully connected layers convert the 2D feature maps into a 1D feature vector for further feature representation. Fully-connected layers perform like a traditional neural network and contain about 90% of the parameters in a CNN. It enables to feed forward the neural network into a vector with a predefined length. CNN is trained in two stages known as forward stage and backward stage. The forward stage represents the input along with the current parameters in the layers. The prediction output computes the loss cost which is used in the backward stage to further compute the gradient of each of the parameters. These gradients are used in forward computation and similarly after going through several iterations, the training of network is fulfilled.

2. *Recurrent neural network*
   Recurrent neural network or RNN [9] is the subfield of artificial neural network in which the input to current level is received from the output of previous level i.e. the levels are dependent on each other. The most important part of RNN is the hidden state to solve the complexity of hidden layers. It also has a memory known as internal state to store all the data that can be repetitively used to produce the output hence reducing complexity of the parameters. This is an advantage over other algorithms and it can be used in various applications like speech or character recognition. The

RNN algorithms are used with CNN to extend the pixel neighborhood. RNN uses the sequential form of data available in the network. This manner of data utilization is effective in various applications that conveys the relevant information from the sequential structure of data.

RNN can be considered as short term memory unit that has input layer, hidden layer and an output layer as illustrated in Fig. 3. One of the important part of recurrent neural networks is recursive neural networks which can be created by using the same set of data again in topological order. It is a linear architectural variant of the recurrent neural networks and facilitates branching of features related to hierarchy thus resulting in imitation of network architecture. An example of recursive neural networks is a structure corresponding to linear chain and is also used in NLP by using the libraries like Tensor Flow [10], MxNET etc. However, an issue of sensitivity to the gradients is seen in TNN. The sensitivity diminishes with time which leads to loss of prior data after new data is introduced. Due to this issue, the long short term memory (LSTM) is introduced in deep learning by providing memory blocks. The memory blocks having memory cells are used to store the temporary states of the network and also controls the flow of data.

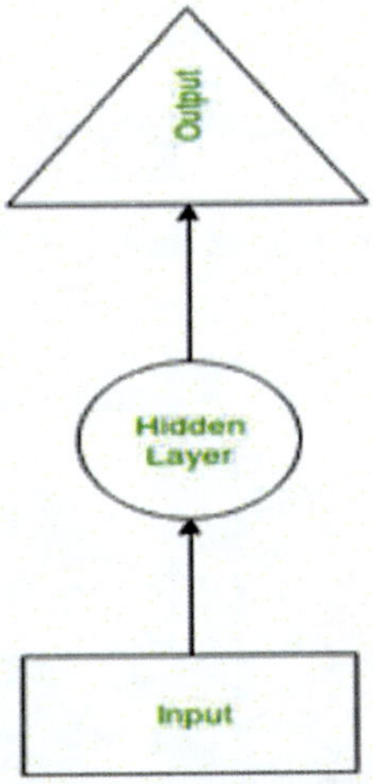

Fig. 3. Recurrent neural network

3. *Deep belief networks*
An algorithm of deep neural networks known as deep belief networks [11] are models made up of multiple layers with hidden variables. The upper two layers in this algorithm are symmetrically connected and the lower layer receives the connection from upper layers by the top-down approach as shown in Fig. 4. A DBN algorithm is when trained on the unsupervised dataset, it can reconstruct its input and the layer of this reconstructed input is known as feature detectors. The algorithm is then further trained with supervision for performing the classification step. Deep

belief networks or DBN uses unsupervised machine learning model like restricted Boltzmann machines [12] to get desired output. In this algorithm, the hidden layers act as visible layers for the next layer for connecting the layers as in Fig. 5. The most common and useful feature of DBN is that the connection between the layers are present but not included as a part of the layer. This algorithm has applications in the areas which require less data labeling, have less structured systems and the results are based on iterative processes and random variables.

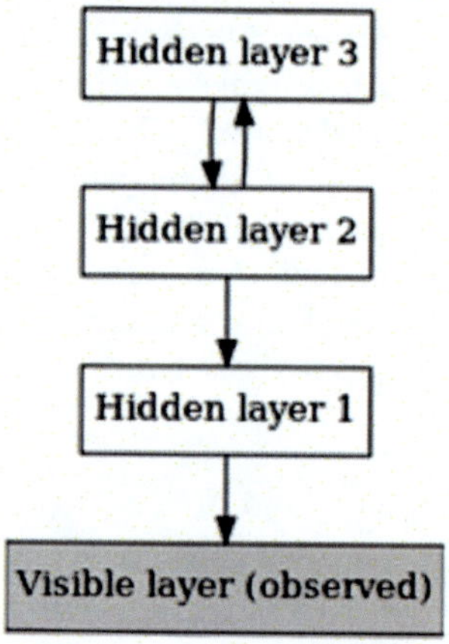

Fig. 4. Deep belief network architecture

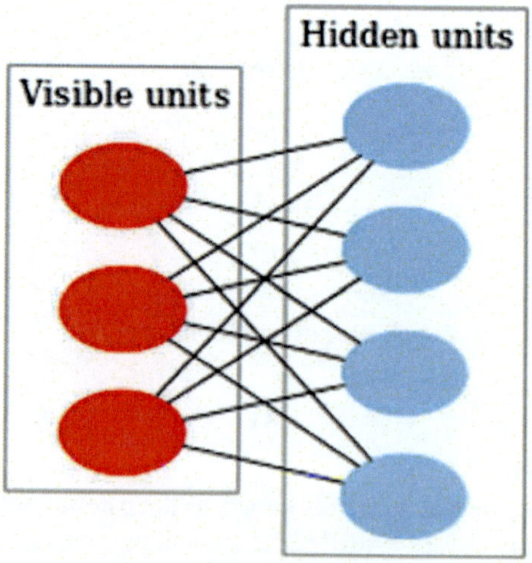

Fig. 5. Visible units of DBN

Deep belief network is a generative model where the single directional connections are formed on the topmost two layers of the network. The lower layers are meant to receive the inputs from the above layers for generating information in form of vectors. The last layer also known as visible unit or layer is used to represent the states of the data vector in the input units or the hidden units. This approach used the unsupervised method to reconstruct the inputs where the layers are in form of feature detectors. Also the training process in the deep belief networks performs the task of classification.

This network has several hidden layers or units that are seen through the visible layer for the next sub network in the sequence. Deep belief network (DBN) provides probability distribution of the data by efficiently using layer by layer strategy. The use of layers is advantageous in providing initialization of the network along with showcasing the difficulty occurring during parameter selection. The other being in case of unsupervised learning where no labels are required while training. However, upon being used in the computer vision tasks, the deep belief networks do not consider the two dimensional structure of the input data or images.

4. *Deep neural networks*
   Deep neural network (DNN) has multiple layers between the input and output layers of the network. The DNN finds the correct mathematical manipulation to turn the input into the output, whether it be a linear or non-linear relationship. The network moves through the layers by calculating the probability of each output. The extra layers enable composition of features from lower layers, potentially modeling complex data with fewer units than a similarly performing shallow network. Deep neural networks also known as hierarchical networks or deep structured learning is an algorithm using machine learning and artificial neural networks having multiple layers between the input and output layers. A deep neural network [13] is a type of neural network having a level of complexity in terms of datasets and having more than two layers for computation as shown in Fig. 6. The processing or the computation of data using deep neural networks involves models using sophisticated mathematical approaches for computing data in a complex manner. In the DNN algorithm, the networks move between the layers for converting the input into output irrespective of the relationship being linear or non- linear.

To model the nonlinear relationships complex data, the DNN generates models where the object can be represented in form of a layered composition of primitives. These layers are extra layers and collect features from the lower layers which model complex data and thus perform networking. The applications of deep neural networks are found in the areas of fraud detection [14], medical analysis, image recognition, military, image classification etc.

3 Conclusion

Deep learning is a growing technique having a better output efficiency by applying deep learning techniques and algorithms discussed in the paper. The deep learning and neural networks algorithm discussed in the paper have various useful applications in the field of

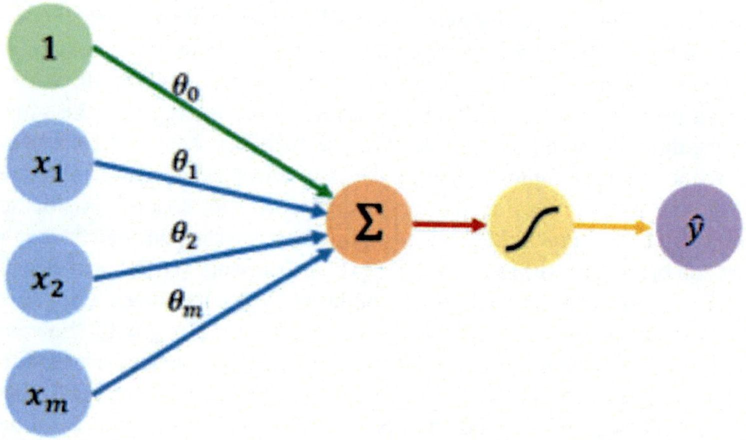

Fig. 6 Deep neural networks

image recognition, image analysis, speech recognition etc. by providing effective results which are advantageous over machine learning. This paper addresses the most useful algorithms in deep learning which are Convolutional neural networks, recurrent neural networks, Deep belief networks, deep neural networks. The deep learning algorithms provide advancement for the large and fast complex computations along with larger training data. The training and compression of neural networks is done by reduced cost and in a simplified manner using the discussed models or algorithms. Different deep learning algorithms helps to improve the learning performance, broaden the scopes of applications, and simplify the calculation process. However, the extremely long training time of the deep learning models still remains a major problem for the researchers. Furthermore, the classification accuracy can be drastically enhanced by increasing the size of training data and model parameters.

References

1. L. Deng, A tutorial survey of architectures, algorithms, and applications for deep learning, in *APSIPA Transactions on Signal and Information Processing* (2014)
2. J. Schmidhuber, Deep learning in neural networks: an overview. Neural Netw. (2015)
3. K. Simonyan, A. Zisserman, Very deep convolutional networks for large-scale image recognition, in *3rd International Conference on Learning Representations, ICLR 2015 - Conference Track Proceedings* (2015)
4. R.E. Neapolitan, R.E. Neapolitan, Neural networks and deep learning, in *Artificial Intelligence* (2018)
5. D. Erhan, C. Szegedy, A. Toshev, D. Anguelov, Scalable object detection using deep neural networks, in *Proceedings of the IEEE Computer Society Conference on Computer Vision and Pattern Recognition* (2014)
6. H. Mostafa, Supervised learning based on temporal coding in spiking neural networks. IEEE Trans. Neural Netw. Learn. Syst. (2018)

7. Y. Kim, Convolutional neural networks for sentence classification, in *EMNLP 2014 - 2014 Conference on Empirical Methods in Natural Language Processing, Proceedings of the Conference* (2014)
8. A. Bhandare, M. Bhide, P. Gokhale, R. Chandavarkar, Applications of convolutional neural networks. Int. J. Comput. Sci. Inf. Technol. (2016)
9. A.L. Caterini, D.E. Chang, Recurrent neural networks, in *SpringerBriefs in Computer Science* (2018)
10. M. Zhang, S. Rajbhandari, W. Wang, Y. He, DeepCPU: serving RNN-based deep learning models 10x faster, in *Atc* (2018)
11. R. Sarikaya, G.E. Hinton, A. Deoras, Application of deep belief networks for natural language understanding. IEEE Trans. Audio, Speech Lang. Process. (2014)
12. A. Fischer, C. Igel, An introduction to restricted Boltzmann machines, in *Lecture Notes in Computer Science (including subseries Lecture Notes in Artificial Intelligence and Lecture Notes in Bioinformatics)* (2012)
13. K. Yun, A. Huyen, T. Lu, Deep neural networks for pattern recognition, in *Advances in Pattern Recognition Research* (2018)
14. E.W.T. Ngai, Y. Hu, Y.H. Wong, Y. Chen, X. Sun, The application of data mining techniques in financial fraud detection: a classification framework and an academic review of literature. Decis. Support Syst. (2011)

Deep Reinforcement Learning

Mrinal Paliwal(✉)

SOEIT, Sanskriti University, Mathura, India
mrinalpaliwal.cse@sanskriti.edu.in

1 Introduction

Reinforcement learning also known as semi- supervised learning model is a method that allows the software agent to take actions and further interact with the environment to maximize the result. Being a part of deep learning in machine learning, reinforcement learning follows the Markov Decision Process (MDP) [1] to achieve a complex objective and to maximize the value which represents the long term objective. The reinforcement learning is modeled as Markov Decision Process as shown in Fig. 1. The interaction in reinforcement learning where the controller gets an input of the reward and the state of the system linked with the previous transaction. The action calculated by the controller is transferred back to the system, which makes a new transition to new state and so on.Reinforcement learning is a part of many fields like, control system, simulation based optimization, statistics, genetic algorithms [2] etc.

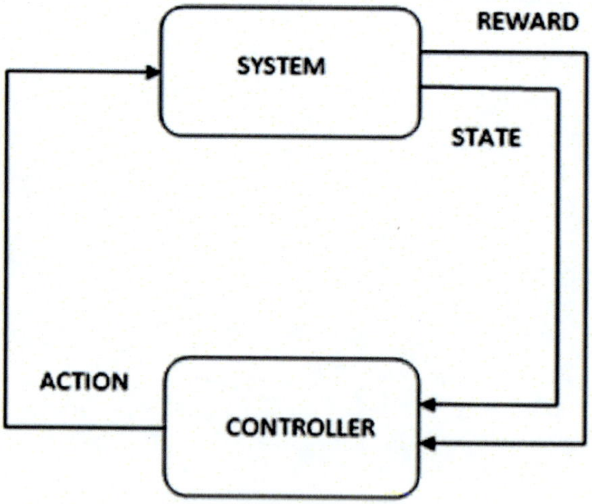

Fig. 1. Reinforcement learning

F. P. García Márquez (ed.), *International Conference on Intelligent Emerging Methods of Artificial Intelligence & Cloud Computing*, Smart Innovation, Systems and Technologies 273, https://doi.org/10.1007/978-3-030-92905-3_16

2 Learning Models of Reinforcement Learning

1. *Q- learning:*
   Q- learning is an algorithm in the model- free approach [3] of deep reinforcement learning with the aim of learning the policy to determine the action to be taken in any certain situation. Since the model of the environment is not required the problems can be handled by using the stochastic transitions and rewards thus not requiring adaptation to the environment.The actions performed in Q- learning generated maximum of the rewards known as Q- value and the strategy can be formolutated as:

$$Q(s,\ a) = r(s,\ a) + \gamma \max Q(s',\ a)$$

The Q-value is the output by being at the state s and action a with an immediate reward r(s,a) in addition to the maximum Q-value generated from the nest state s′. Q(s′,a) depends on the the next state s″ and requires a coefficient gamma to incerase or reduce the contribution of the rewards and is shown as follows:

$$Q(s,a)\,\gamma Q(s',a) + \gamma^2 Q(s'',a) \ldots\ldots\ldots \gamma^n Q(s''^{\ldots n},a)$$

2. *Markov decision process:*

The deep reinforcement learning can be described as markov Decision process(MDP) [4] [5] which consistes of parameters(S, A, P, R, y) where S is the state space, A is the action set, P is the transition probability function denoted as:

$$\mathcal{P}^a_{ss'} = \mathbb{P}[S_{t+1} = s' \mid S_t{=}s,\ A_t = a]$$

R is the rewrad function denoted as:

$$\mathcal{R}^a_s = \mathbb{E}[R_{t+1} \mid S_t{=}s,\ A_t = a]$$

and y is the discount factor $\gamma \in [0, 1]$.The policy defines the behavior of the agent as defined:

$$\pi(a|s) = \mathbb{P}[A_t = a|S_t = s]$$

The markov decision process is based on the present state and has no link to the past states thus the action taken by the agent id pre determined for all the different actions to be performed. MDP is shown in Fig. 2

3 Algorithms

There are two important methods in deep reinforcement learning based on value functions [4] and policy search.

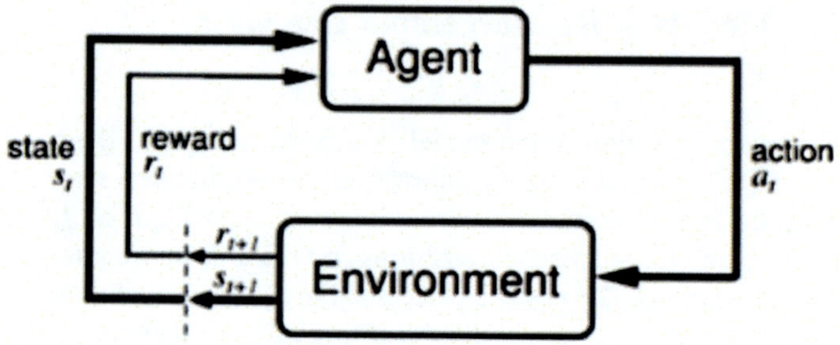

Fig. 2. Markov decision process

1. *Value Functions:*
The methods of value functions are based on the estimation of values [3] in the state known as state value function which is the return expected. The state value function can be defined as:

$$V^{\pi}(\mathrm{s}) = \mathbb{E}[R|\mathrm{s}, \pi]$$

The optimal state value function is defines as:

$$V^{*}(s) = \max_{\pi} V^{\pi}(\mathrm{s}) \ \ \forall \mathrm{s} \in S.$$

Generalized policy iteration which is a combination of policy improvement and policy evaluation is used to estimate the best policy. The policy evaluation [6] enhances the estimate of value function which is achieved by reducing the errors generated by policy. On improvement of estimation, the policy automatically improves by updating the value function.

2. *Policy search:*
In the policy search methods, the value function model need not be maintained and the optimal policy can be searched directly. A parameterized policy [7] is considered, where the parameterscan be updated hence the expected return can be maximized by the use of either gradient free optimization or gradient based optimization being sample efficient where large number of parameters are present.

Policy gradients [8] can be used to significantly improve the poly of parameterization. The computation of return needs averaging of the policy parameterization which requires deterministic approximations on a model based setting. For the gradient based learning method, reinforce rule is used as an estimator for the gradient and it can be used to compute the gradient of expectation over that of the function f of a variable X with respect to parameters and is defined as:

$$\nabla_{\theta}\mathbb{E}_{X}[f(X;\theta)] = \mathbb{E}_{X}[f(X;\theta)\nabla_{\theta}\log p(X)].$$

3. *Actor-critic Methods:*
The value functions can be combined with representation of policy and the resultant method is known as actor-critic method as shown in Fig. 3. The actor is the policy which uses feedback from the critic which is a value function [9], where the value function is used as a gradient making this method a subpart of policy gradient methods.

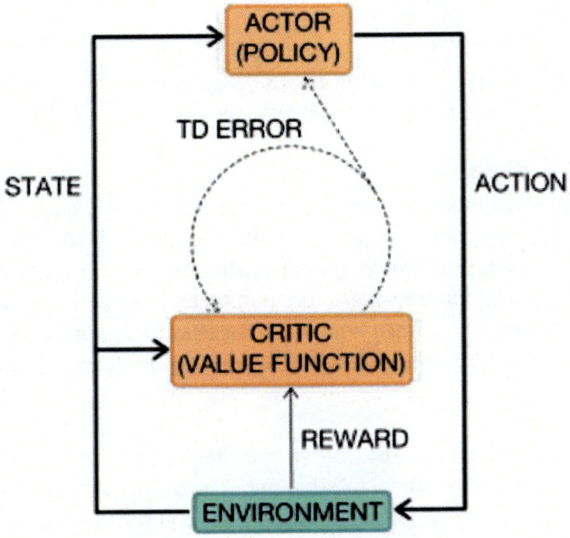

Fig. 3. Actor critic methods

4. *Model based:*
The model based Reinforcement learning (refer to Fig. 4) allows the simulation of environment without any interaction with the environment inany direct manner. A virtual model of the real environment can be created and the operations further required are performed in that specific virtual environment. The model based methods [10] reduce the interactions with the real environment thus reducing the time consumption. For example, in case of robotics, the use of this method reduces the time and wear and tear of the hardware.

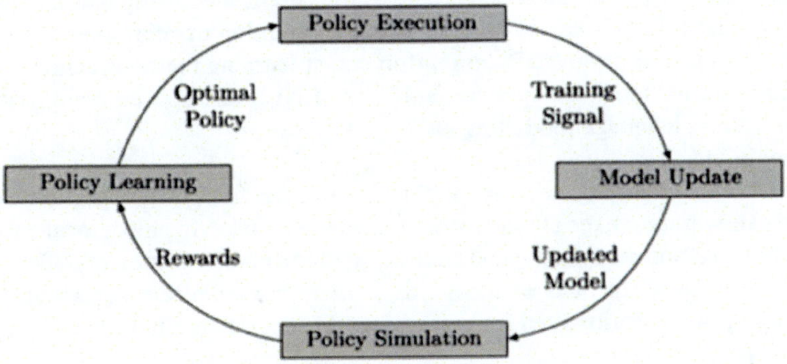

Fig. 4. Model based reinforcement learning

4 Applications

Deep reinforcement learning has wide range of applications in the fields of robotics, web system, gaming, NLP, computer vision etc.

1. *Web System Configuration:*
   Due to enormous number of parameters in a web system reconfiguration [11] of parametres done by tuning of the parameters by using deep reinforcement learning where the reconfiguration is done by Markov Decision Process. The system configuration is considered as state, the parameteres as actions and the difference between the two response times which are target and measured. The computational complexity is reduced by using a combination of deep learning and reinforcement learning.
2. *Gaming:*
   Games like chess, Othello, Go etc. prove to be an excellent testbed for reinforcement learning algorithms and artificial intelligence. The best application of deep reinforcement algorithm learning in gaming is by using an algorithm named AlphaGo Zero which is neither supervised nor unsupervised learning but performs policy improvement and evaluation. It is build with a combination of deep convolutional neural networks, reinforcement learning and supervised learning.
3. *Robotics:*
   Reinforcement learning plays an important role in the field of robotics to develop techniques for controlling robots by automatic processes like sensing, planning, contolling.Deep reinforcement techniques provide ability in robotics for achieving optimal behaviour by the use of trial- error methods with its environment. The performance of robots are easily known by getting feedbacks in form of scalar objective function.
4. *Natural Language Processing:*
   The area of NLP has now shifted to neural networks from statistical approaches due to challenges which cannot be overcome by the latter. Hence deep learning has been playing vital role in areas of NLP [12] and an addition of reinforcement learning made a significant progress.Deep reinforcement learning methods achieve state of the outcome for language specific problems of NLP like sppech recognition, text classification, language modelling etc.
5. *Recognition:*
   Reinforcement learning has vast applications in the field of recognition as it can significantly improve the efficiency of image classification in object localization and detection. Motion analysis, visual control, situation and environment understanding are the best areas in field of recognition where the deep reinforcement learning algorithms are be formulated for performing this operation efficiently.
6. *Healthcare:*
   Due to the rise in popularity of healthcare field in providing personalized medicines many challenges have occurred. Deep reinforcement learning algorithms systematically optimizes the potential of growth in this area by using the algorithms effectively. Decision support applications are widely used in the medical field however it has many limitations due to use of existing models or techniques. Application of deep

reinforcement learning in the field of healthcare is capable of achieving optimized treatment strategies that require large states and action spaces by using Markov Decision Process (MDP). This strategy overcomes many limitations and thus reduces the computational complexity.

5 Conclusion

Deep learning models are capable of extracting complex representations of data from input with high dimensional data and thus overcome the traditional machine learning methods. The reinforcement learning methods cannot optimize the policies efficiently due to data inefficiency. Thus a combination deep learning and reinforcement learning is an useful approach and has become a valuable tool for overcoming challenges faced. This paper discusses the reinforcement learning framework using the deep learning methodology with a combination of both supervised and unsupervised learning networks. The deep reinforcement learning methods use architectures like deep convolutional neural networks which are appropriate for the challenges and problems with partial observable environment of Markov Decision process. The deep reinforcement learning due to its efficient methods and algorithms have wide range of applications in areas like robotics, machine vision, NLP, medical, gaming, videos etc. which have been addressed in the paper. The deep reinforcement learning has the significant use in real world applications as direct approach of policies and values is done. Thus the approach of deep reinforcement learning addressed in the paper overcomes many challenges and is a progressing field.

References

1. Z. Wei, J. Xu, Y. Lan, J. Guo, X. Cheng, Reinforcement learning to rank with markov decision process, in *SIGIR 2017 - Proceedings of the 40th International ACM SIGIR Conference on Research and Development in Information Retrieval* (2017)
2. H. Iba, C.C. Aranha, Introduction to genetic algorithms, in *Adaptation, Learning, and Optimization* (2012)
3. S. Gu, T. Lillicrap, U. Sutskever, S. Levine, Continuous deep q-learning with model-based acceleration, in *33rd International Conference on Machine Learning, ICML 2016* (2016)
4. C. Szepesvári, Algorithms for reinforcement learning, in *Synthesis Lectures on Artificial Intelligence and Machine Learning* (2010)
5. M.T.J. Spaan, Partially observable markov decision processes, in *Adaptation, Learning, and Optimization* (2012)
6. S. Racanière et al., Imagination-augmented agents for deep reinforcement learning, in *Advances in Neural Information Processing Systems* (2017)
7. S. Gu, E. Holly, T. Lillicrap, S. Levine, Deep reinforcement learning for robotic manipulation with asynchronous off-policy updates, in *Proceedings - IEEE International Conference on Robotics and Automation* (2017)
8. R. Houthooft et al., Evolved policy gradients, in *Advances in Neural Information Processing Systems* (2018)
9. R. Lowe, Y. Wu, A. Tamar, J. Harb, P. Abbeel, I. Mordatch, Multi-agent actor-critic for mixed cooperative-competitive environments, in *Advances in Neural Information Processing Systems* (2017)

10. A. Nagabandi, G. Kahn, R.S. Fearing, S. Levine, Neural network dynamics for model-based deep reinforcement learning with model-free fine-tuning, in *Proceedings - IEEE International Conference on Robotics and Automation* (2018)
11. J.P. O'Doherty, S.W. Lee, D. McNamee, The structure of reinforcement-learning mechanisms in the human brain, in *Current Opinion in Behavioral Sciences* (2015)
12. W.Y. Wang, J. Li, X. He, Deep reinforcement learning for NLP, in *ACL 2018 - 56th Annual Meeting of the Association for Computational Linguistics, Proceedings of the Conference Tutorial Abstracts* (2018)

Different Texture Segmentation Techniques: Review

Rishi Sikka(✉)

Department of Electronics and Communication Engineering, Sanskriti University, Mathura, Uttar Pradesh, India
rishisikka.ec@sanskriti.edu.in

1 Introduction

The term segmentation basically relates to the partition of images or the objects based on certain characteristics like area of interest, background etc. the partition or division of an image into a number of small parts is done for highlighting the required or important features from the image for further processing. For this image processing has a feature known as image segmentation [1] whose main aim is to provide information to be processed in a simple manner. The image segmentation of the images basically works in a way to separate the line or curves or boundaries from the objects in the image and thus the labeling of each pixel depending on the intensity such that same level is given to the pixels having same features or characteristics.

To extract the meaningful information from the image and process it accordingly, features or texture of the images are required. Texture of the image is the is the means to derive the information from the image apart from the gray values and thus segmentation of the image is done on the basis of texture thus known as texture based segmentation (Fig. 1) [2]. Texture of an image relates to the pattern in the image which is same in intensity level determining the surface properties of an image and can separate the non textured part from the image. Texture basedsegmentation is difficult for a single dimensional image thus 2D images are used for thepurpose since texture can be easily determined in such kind of images.

The methods of image segmentation are divided on the basis of its processing into two parts. One of the approaches is edge based segmentation [3] and the other is region based segmentation [4]. The steps as shown in Fig. 2 to determine the segments from the image is done by following few basic steps like reading image from the database, grayscale conversion, filtering image, extracting region of edge [5] from the image and displaying segmented image as shown in the figure.

2 Image Segmentation Techniques

1. *Edge detection:*
   Edges in an image are important features to obtain the useful information from the image or separate the main objects from the background or classifying the objects.

F. P. García Márquez (ed.), *International Conference on Intelligent Emerging Methods of Artificial Intelligence & Cloud Computing*, Smart Innovation, Systems and Technologies 273, https://doi.org/10.1007/978-3-030-92905-3_17

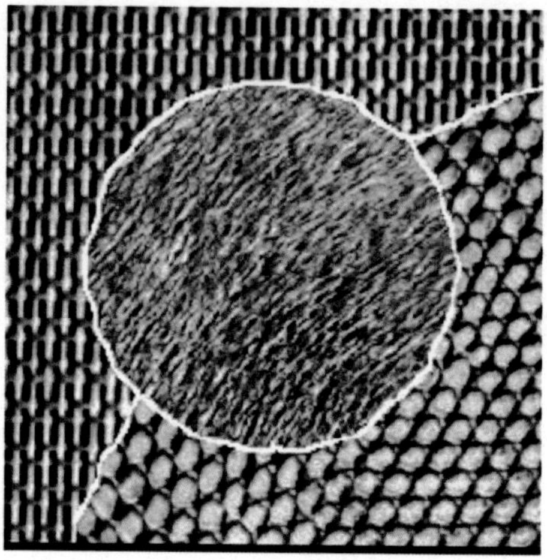

Fig. 1. Textures in segmentation

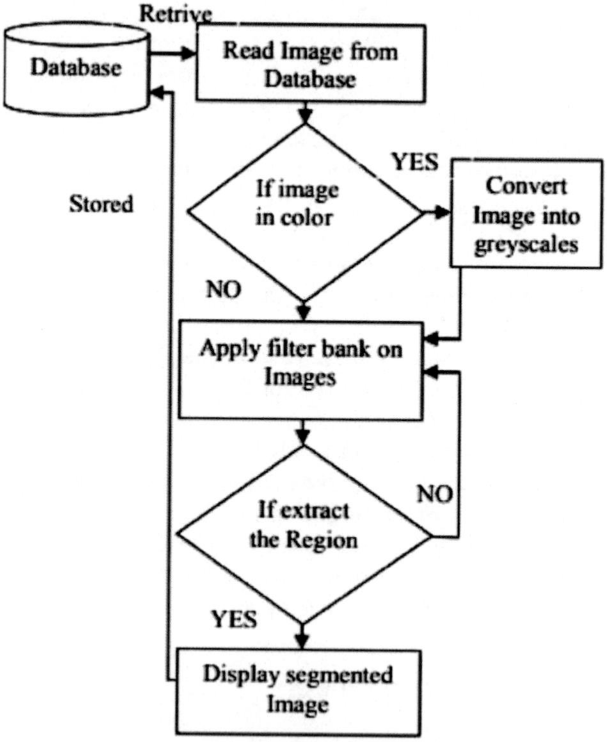

Fig. 2. Steps of segmentation

For this purpose the method known as edge detection is used in image processing and is the base technique to be followed for texture based segmentation. Various approaches are available in edge detection like canny edge detection, Prewitts edge detection, OSTUs algorithm, SOBEL algorithm [6] etc. the edges in the image being a significant part is the part of the image where there is an immediate change in the intensity level. For separating the edges from the image based on the texture three steps are followed using the process of edge detection.

i *Filtering of an image:* The images taken as raw input in the system are full of unwanted information known as noise like Gaussian noise [7] thus degrading the quality of edges to be detected. So filters like wiener or low pass filters are used for removing the noise signals from the images.

ii *Enhancement*: After removing the noise from the images it becomes necessary to pre process the image or enhances an image in a form where retrieving useful information becomes easy which is done by enhancing the intensity value of the required regions and this is performed by the computation of gradient of the image. The gradient of the image provides the strength of the image so calculating the gradient plays an important role and the gradient can be expressed as:

$$\text{Gradient Vector} : \nabla I = \left[\frac{\partial I}{\partial x}, \frac{\partial I}{\partial y}\right]^{\mathrm{T}}$$

iii *Detection*: Thresholding [8] is the method used for the detection of edges or the boundaries in the image. The edge detection based on different values of threshold is as shown in Fig. 3.

Fig. 3. Edge detection based on threshold

2. *Gabor filter*:

Gabor filters are the linear filters used in face recognition by detecting the edges to be used for feature extraction. The Gabor filters [9] represent the frequency and the orientations in a similar way to that of human vision system. The impulse response of the Gabor filters is defined by the product of sine wave as shown in Fig. 4 with that of the Gaussian function. Due to the product of sine wave and Gaussian function, the Gabor filter's Fourier transforms is the convolution of Gaussian and harmonic function's convolution. Since the images used for texture segmentation are 2D images, a 2D filter is used and the Gabor function g(x,y) is defined as:

$$g(x, y) = \frac{1}{2\pi\sigma_x\sigma_y} \exp\left[-\frac{1}{2}\left\{\frac{x^2}{\sigma^2 x} + \frac{y^2}{\sigma^2 y}\right\} + 2\pi j W_x\right]$$

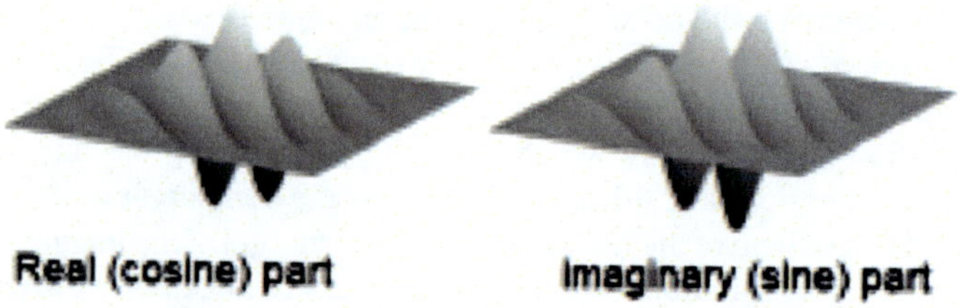

Fig. 4. Sine and cosine waves by Gabor Filter

3. Histogram based segmentation:
   Texture segmentation done by utilizing the histogram of the images is done by applying the algorithm to the clusters in the image to get more significant output. This is done in order to divide the cluster of the images into small parts and continued until no division of clusters can be done. The approaches based on histogram are efficient when multiple frames are to be segmented by merging the parts segmented so that valuable information can be taken as the resultant. The segmentation based on histogram [10] is as shown in Fig. 5.

Fig. 5. Histogram based segmentation

The histogram is applied based on the pixels, objects or a stable environment thus resulting in the applications of video tracking etc. the texture as well non texture regions from the images can be separated easily by using the local spectral histograms. The approach of histogram leads to more accurate regions or the boundaries which can be further localized by localization.

4. *Supervised and unsupervised segmentation*:
The segmentation techniques based on supervised segmentation [11] algorithms utilize the data gained from the previous steps of pre processing or training and use the data as input for further processing. The properties like accuracy, robustness or the speed of the segmentation processes greatly impact the applications using the process of segmentation which can be improved significantly by using supervised segmentation. The unsupervised segmentation [12] algorithms do not require any prior knowledge about the texture or the information about the data extracted earlier. Here the segmentation of the images is done by using clustering process like Fuzzy Clustering. The clustering of the images is done starting from the lowest levels and the segmentation is performed at the final level where the texture from the images is separated.
5. *Region based segmentation*:
The texture segmentation based on the regions of the images uses the approach of region growing. This approach is also known as pixel based texture segmentation because it uses the initial pixel value for processing. In this the neighboring pixels of the image are determined and extra pixels are added to make the region much better for segmentation. The algorithm is similar to that of data clustering algorithm. The main aim of the approach of region based segmentation is to partition an image into regions directly. The region based texture segmentation has two steps to be followed namely region splitting and region merging as discussed below:

Region splitting: the process of growing of the region starts from the initial pixel value rather than growing total of the image thus in this step the image divided into parts thus known as region splitting.

Region merging: after region splitting the step of merging the regions is followed by considering small matrix of the region having similar values. Thus the merging of regions with similar features is done.

The image after region based segmentation [13] is as shown in Fig. 6.

Fig. 6. Region based segmentation

3 Conclusion

Texture segmentation is the process of partitioning an image into small segments in a way that entire image is segmented and a set of small parts of image based on certain characteristics is done. The pixel values in eat set is similar thus defining the different texture in the image. The paper discusses about various methods or approaches that are used for texture segmentation like edge based segmentation, histogram based segmentation, region based approach, Gabor filter based segmentation and supervised-unsupervised segmentation. The techniques addressed in the paper are important aspects required in image analysis for segmenting an image on the basis of texture. The techniques have their own advantages over each other and the best technique used is by the Gabor filters that use decomposition of the image by wavelet transforms and the intensity filters to achieve better accuracy in terms of texture segmentation. The paper hence concludes various techniques used for texture segmentation.

References

1. D. Oliva, M. Abd Elaziz, S. Hinojosa, Image processing, in *Studies in Computational Intelligence* (2019)
2. Z. Guo, L. Zhang, D. Zhang, A completed modeling of local binary pattern operator for texture classification. IEEE Trans. Image Process. (2010)
3. C.J. Taylor, Towards fast and accurate segmentation, in *Proceedings of the IEEE Computer Society Conference on Computer Vision and Pattern Recognition* (2013)
4. K. Butchiraju, B. Saikiran, Region-based segmentation and object detection. Int. J. Innov. Technol. Explor. Eng. (2019)
5. S.F. Eskildsen et al., BEaST: Brain extraction based on nonlocal segmentation technique. Neuroimage (2012)
6. C. Choi, A.J.B. Trevor, H.I. Christensen, RGB-D edge detection and edge-based registration, in *IEEE International Conference on Intelligent Robots and Systems* (2013)
7. J.V. Manjón, P. Coupé, L. Martí-Bonmatí, D.L. Collins, M. Robles, Adaptive non-local means denoising of MR images with spatially varying noise levels. J. Magn. Reson. Imaging (2010)
8. R. Yogamangalam, B. Karthikeyan, Segmentation techniques comparison in image processing. Int. J. Eng. Technol. (2013)
9. J.L. Raheja, S. Kumar, A. Chaudhary, Fabric defect detection based on GLCM and Gabor filter: a comparison. Optik (Stuttg). (2013)
10. V. Rajinikanth, M.S. Couceiro, RGB histogram based color image segmentation using firefly algorithm. Procedia Comput. Sci. (2015)
11. D. Pathak, P. Krahenbuhl, T. Darrell, Constrained convolutional neural networks for weakly supervised segmentation, in *Proceedings of the IEEE International Conference on Computer Vision* (2015)
12. K. Greff, A. Rasmus, M. Berglund, T. H. Hao, J. Schmidhuber, H. Valpola, Tagger: deep unsupervised perceptual grouping, in *Advances in Neural Information Processing Systems* (2016)
13. A.V. Vo, L. Truong-Hong, D.F. Laefer, M. Bertolotto, Octree-based region growing for point cloud segmentation. ISPRS J. Photogramm. Remote Sens. (2015)

Fully Protected Image Algorithm for Transmitting HDR Images Over a WSN

Mukesh Pathela(✉), Tejraj, Arjun Singh, and Sunny Verma

Department of Electronics & Communication Engineering,
Dev Bhoomi Institute of Technology, Dehradun, Naugaon, India

1 Introduction

This is only Multimedia communication which has grown extensively over the last few years over a Wireless Sensor Network. With the increased demand of communication, the vulnerability of the network poses a major challenge as the vital information can be compromised. If unsecured medical image transmission is done over WSN then hackers can easily.

get access to them and can use it for unethical purpose. The same can happen in case of other applications also. Smart cities have also seen an increase in demand for fast and secured transmission of data.

The major challenge is to provide a secured platform without compromising the quality and security of the data. High Definition data transmission is another big challenge with the introduction of HD in almost all the Multimedia gadgets. Reading an HD data then adding security and transmitting it over a network without compromising QOS is what the paper concentrating on. For the purpose of research, the High Definition image dataset has been created from the images available over the internet. The data set is first read and then converted into linear form and encrypted using a newly designed algorithm before transmitting over the network. Hackers are aware of the known algorithms of Encryption like Genetic, Arnold and more. There is a need to design a more robust Algorithm for Encryption and also design the routing in such a way that without compromising the QOS the data is transmitted over the network.

The paper focuses on designing a new Algorithm for secured transmission of HD images over a network.

HDRI

A single image taken from the camera usually do not contains wide range of light intensity. This problem can be resolved by merging images taken at various exposure values. The overexposed images works good for the darker regions while the underexposed images tone down the intensity in the extra bright areas. To recover an image with a high dynamic range these image sets are combined. HDR or HDRI(High-Dynamic-Range-Imaging) is an enhancement of the image's dynamic range in order to show the image details in both shadows and highlights. The dynamic-range can be defined as the ratio between the physical measurement's minimum and maximum values. It is generally

F. P. García Márquez (ed.), *International Conference on Intelligent Emerging Methods of Artificial Intelligence & Cloud Computing*, Smart Innovation, Systems and Technologies 273, https://doi.org/10.1007/978-3-030-92905-3_18

noticed that between the highlights and the shadows there is a broad range of luminance. Radiance picture format (.hdr,.pic) was used for the HDR format for the first time in 1989 [1] since it was used in the graphics community, primarily for HDR photography and image-based lighting [2, 3]. To get low dynamic from high dynamic images tone map is used.

Tonemap

Light spanning from snow to sunlight is very fast, it spans to approximately order ten of absolute range. Our output devices like printer and screens can span shades at the maximum order two. The difference between the absolute range and the supported range leads to the issue of reproducing the image maintaining its quality. So, lot of efforts have been put in on the issue of tone reproduction [4] and major portion of this work made use of the perceptual model to control the operator [5]. Such methods of [5] were also applied for the dynamic as well as the interactive settings [6]. The other work concentrated on dynamic range compression by spatially adjusting the mapping from scene to luminance display while preserving local contrast around the same time [7]. Finally, computational models of the visual system of human beings may also direct spatially varying maps [8].

2 Historical Background

A method that is primarily based on encryption and faster channel coding for redundancy was proposed in paper [9] for a comfortable information transmission over WI-FI channels. In this approach, safety of the transmitted data is obtained by means of the BB equation which is based totally on encryption and decryption algorithms. Redundancy is achieved as a channel coding for the transmission of information over wireless channels via rapid codes. The results for image communication show that turbo coding under SNR environments is an effective form of channel coding. The paper proposes a mixed SPIHT/OFDM coding scheme for the communication of image/video over the wireless fading path [10]. In the proposed approach, a set of rules for cost manipulation is used for photo coding and an adaptive OFDM modulation system for the communication coding scheme is used. The device is analyzed through a Nakagami-m fading channel that is uncorrelated in accordance with the channel and compressed bit circulation characteristics, and the coding scheme is changed with the assistance of a source controller as well as a channel encoder [11]. When receiving terminals, an adjustable retransmission controller blended with CRC is often used to deliver the latest reported packet consistent with its importance. The effects display that the proposed method is efficient in reaching right photograph satisfactorily, particularly in low sign noise ratio for photograph wireless transmission system. One greater method in which key data of the image is taken and extracted first and is transmitted over correct channel condition to make certain that the majority the errors are shifted to the standard part of the picture [12]. Simulations are performed with one of a kind sign to Noise Ratios (SNR) beneath changing thresholds. it's miles proven that the transmission pleasant is very tons improved. The identical kind of idea is used where key statistics from the photo is extracted and transmitted handiest whilst the channel is right to make sure that most errors arise inside the less crucial photograph information [13] The performance results suggests that the proposed method can significantly improve transmitted photograph satisfactory. The intention proposed in [14] is to reduce the overall machine power intake according to changing channel situations and characteristics of the input photograph. in this scheme, the sources are allotted among source encoder, channel encoder, and amplifier's transmission. The simulation consequences show that the proposed scheme can extensively shop the entire device power consumption if as compared with the equal transmission protection schemes having same device constraints. For image transmission over Rayleigh fading channels, a modulation and channel coding method that is adaptive is implemented [15]. Depending on the importance of the compressed bit movement and channel position, the modulation degree and coding rate are agreed upon. The proposed system aims to achieve a BER target and prevent over-protection, thereby maximizing the efficiency of the bandwidth, which will increase the efficient transmission rate. Simulation findings show that the spectral output maintaining excessive peak sign-to-noise ratio (PSNR) overall performance and fantastic quality are substantially advanced by the proposed scheme. The proposed method [16] gives modern lossless photograph transmission beneath influence of slight fading with none sort of channel coding. It makes use of a fast analysis and synthesis algorithm, in which 3 instances much less real additions are there as compared to JPEG2000 with wireless OFDM. A at ease technique for transmission of scientific pix over wi-fi channels [17] And in this fully algorithms based on Chaos and Brahmagupta Bhãskara (BB) equation are proposed for encryption as well as decryption and lossless encoding is used to codify the encrypted clinical pics. In addition, to correct transmission errors over noisy wi-fi networks, turbo coding is proposed..

3 Methodology

Reading HDR images

There are various method for reading HDR images. Matlab has an in-built method to read images. For this research in-built function of Matlab was used to read HDR images and for Flat HDR images [18] following method was used.

fl← open file in read mode

[filetype,noofbits] ← get format of fl

/* 8 or 16 bit, rgbeprxyze type */

fsensor← get information of fl

/* get wavelength and spectral */

flluminant← get Illumination of fl

himage← make image readable

/* read 8 bit or 16 bit data, reshape and create new Low intensity image */

[himage, Gluminance] ← convert high dynamic to low using Reinhard global tonemap algo [19]

After reading an HDR image and tone mapping Following algorithm is used to encrypt and transmit it over a WSN –

Assumptions

Transport Protocol ← TCP
Routing Protocol ←DSDV
MAC Layer ← 802.11 g/n
Bandwidth← 20 GB
Area← 25 Km (areal)
No of nodes← 100
Image size← 256X256

Arnold's Cat Map

Encryption using Arnlod's Map62

Step1. Ii← convert rgb to grey
Step2. imgX ← initialize zero matrix of image size
Step3. imgX2 ← initialize zero matrix of image size
Step4. [r c] ← size(Ii)
Step5. **for i ← 1 to r** % loop through all the pixels to generate cat map
Step6. **for j ← 1 to c**
Step7. Imgi← get new i coord (m+n) mod NoPixel
Step8. Imgj ← get new j coord (m+2n) mod NoPixel
Step9. imgX(Imgi,Imgj,:) ← I(i,j,:)
Step10. imgX2(i,j,:) ← imgX(i,j,:);
Step11. **endfor**
Step12. **endfor**
Step13. **for i ← 1 to r** % loop through all the pixels to generate cat map
Step14. **for j ← 1 to c**
Step15. imgX2(i,j,:) ← I(i,j,:)+245
Step16. **endfor**
Step17. **endfor**
Step18. encryptedImage ← bitxor(imgX2,Ii)

For decryption follow reverse of loop 5 to 18.

Genetic Algorithm

Step1. I← rad image
Step2. Igray ← rgb to gray(I)
Step3. Split I to N vectors of length L,
where L=128
Step 4. Generate two matrices M1 and M2 using -

$$M1=\sum_{i=0}^{r}\sum_{j=0}^{c}(-1)^{(i+j)}\times I(i,j)/256*L$$

$$M2=\sum_{i=0}^{r}\sum_{j=0}^{c}(-1)^{(i+j+1)}\times I(i,j)/256*L$$

M1 and M2 are initially zero matrices and size is same as the original image.

```
Step5. x ← M1, y ←M2
Step6. Perform Crossover and mutation
        for i← 1 to r
  for j← 1 to c
        k1← j*random
        j1 ← i*rand()
     x(i,j) ← y(k1,j1)
     v(i,j) ← 255 – x(i,j)/256
   end
  end
```

Decrypted image is re-constructed reversing the encryption steps.

Proposed FPI Algorithm

```
Step 1: Deploy nodes
Step 2: I ← Read image
Step 3: if RGB then
     Igray ← convert I to gray scale
     end if
Step 4: for i←1 to rows
     for j← 1 to columns
        pixelValue← image(i,j)
           newPixelValue ←
        (pixelvalue + i^columns)/256
        tImage(I,j) ←newPixelValue
     end j
  end i
```

Step 5: Ilinear ← Convert tImage to linear matrix

Step 6: M ← OQPSK(Ilinear)

Step 7: [p1 p2 p3 p4] ← divide M into four packets

Step8: Transmit packets one by one over the network

At Receiver End

Step 9: Mc ← [p1 p2 p3 p4] //convert into one

Step 10: DM ← demodulate(Mc)

Step 11: Im ← reshape(DM) //reconstruct image

Step 12: Iret ← retrieve image reversing step 4

Step 13: Calculate MSE

Step 14: Calculate PSNR

Step 15: If MSE, PSNR not acceptable

Send error message to sender

Explanation

Encryption is not a very new and vulnerable also, however it helps in protecting vital information transmitted over the network to some extent. HDR images are first converted into readable format and then tone mapped to make them low intensity to support the devices. Encryption of the image is done with the use of the newly designed algorithm and then converted into linear form for transmission, the data is then modulated using OQPSK(Offset Quadrature Phase Shifting) the modulated data received is then divided into four packets and transmitted over the network making it almost impossible for a hacker to decipher the data. Since the data is divided into packets the speed is also not compromised and the data is transmitted smoothly without compromising its quality as well. At the receiver's end the packets are again combined into one and the demodulated, reshaped into original size of the data and then finally decrypted using the reverse steps of the Encryption algorithm.

4 Simulation Results

Performance of two algorithms along with that of proposed algorithm is evaluated. As compared to the genetic algorithm and Arnold Cat map algorithm, the proposed algorithm shows improvement in PSNR and MSE is also reduced. Following tables shows the performance of previous and proposed algorithms (Tables 1, 2 and 3).

Table 1. Shows the results obtained after Genetic Algo

Image	Type	PSNR	MSE
	HDR	82.1145	0.0210192

Table 2. Shows the results obtained after Arnold Cat Map

Image	Type	PSNR	MSE
	**HDR**	84.2680	0.0193221

Table 3. Proposed method

Image	Type	PSNR	MSE
	**HDR**	91.5954	0.0007247

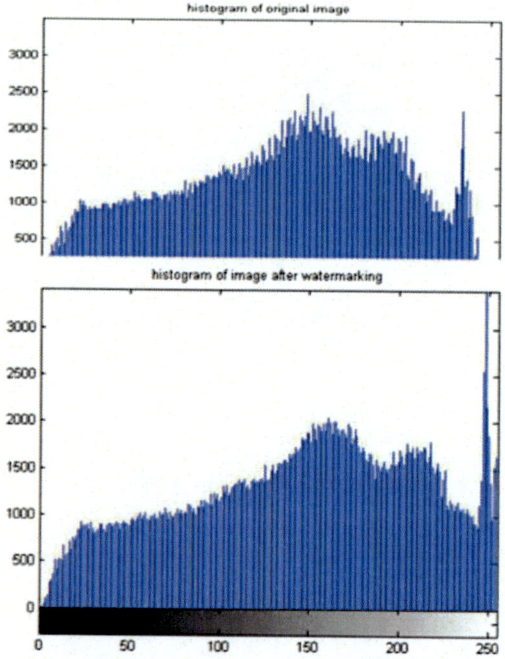

5 Conclusions

The proposed work concludes with the secured transmission of High-Dynamic-Range images (HDRI) using a newly designed algorithm for encryption. The results obtained using the proposed algorithm indicates that the quality of the image transmitted is not compromised. Mean Square Error (MSE) and Peak Signal to Noise Ratio (PSNR) acquired by suggested techniques show a significant improvement over traditional studies and also show that the technique is successful in achieving reasonable image quality. In future the work can be carried over Wi-max and can also be applied in medical images transmission.

References

1. V. Santhi, N. Rekha, S. Tharini, A hybrid block based watermarking algorithm using DWT-DCT-SVD techniques for color images, in *International Conference on Computing, Communication and Networking, 2008. ICCCn 2008* (2008), pp. 1,7, 18–20
2. G. Ward-Larson, R.A. Shakespeare, *Rendering with Radiance* (Morgan Kaufmann, San Francisco, 1998)
3. P.E. Debevec, Rendering synthetic objects into real scenes: bridging traditional and image-based graphics with illumination and high dynamic range photography, in *SIGGRAPH 98 Conference Proceedings, Annual Conference Series, ACM SIGGRAPH* (1998), pp. 45–50

4. E. Reber, R.L. Michell, C.J. Carter, Traditional and Image-based Graphics with Illumination and High DyE. Oxygen Absorption in the Earth's Atmosphere," Technical Report TR-0200 (420–46)-3, Aerospace Corp., Los Angeles, Calif., Nov. 1988. (Technical report with report number)
5. A. McNamara, A. Chalmers, T. Trocianko, Visual perception in realistic image synthesis. Comput. Graph. Forum **20**(4) (2001)
6. Tumblin and Rushmeier, Tone reproduction of realistic images, 1993. IEEE Comput. Graph. Appl.
7. Cohen et al., The gaussian watermarking game. IEEE Trans. Inform. Theory **48**, 1639–1667 (2001)
8. Tumblin and Turk, LCIS: a boundary hierarchy for detail-preserving contrast reduction, 1997, College of Computing, Georgia Institute of Technology, Atlanta, GA 30332–0280
9. M. Padmaja, S. Shameem, Secure image transmission over wireless channels Publication Year: 2007, in *IEEE International Conference on Computational Intelligence and Multimedia Applications*, vol. 4 (2007), pp. 44–48
10. M. Khedr, M. Sharkas, A. Almaghrabi, O. Abdelaleem, A SPIHT/OFDM with diversity technique for efficient image transmission over fading channels, in *IEEE International Conference on Wireless Communications, Networking and Mobile Computing* (2007), pp. 480–483
11. J. Zhang, C. Chen , C. Lv, A robust image transmission strategy over wireless channels, in *IEEE International Conference on Service Operations and Logistics, and Informatics*, vol. 1 (2008), pp. 606–609
12. H. Zhang, A cross layer selective approach for wireless image transmission system over fading channels, in *8th International Symposium on Antennas, Propagation and EM Theory* (2008), pp. 1508–1511
13. H. Zhang , A. Gulliver, A channel selective approach to wireless image transmission over fading channels, in *IEEE Pacific Rim Conference on Communications, Computers and Signal Processing* (2009), pp. 720–724
14. Z. Tang, T. Qin, W. Liu, Energy-minimized adaptive resource allocation for image transmission over wireless channel, in *2010 International Conference on Intelligent Control and Information Processing* (*ICICIP*), pp. 398–403
15. R.A. Zrae, M. Hassan, M. El-Tarhuni, Wireless image transmission using joint adaptation of modulation and channel coding, in *IEEE Symposium on Computers and Communications (ISCC)* (2011), pp. 218–223
16. M. Sabelkin; F. Gagnon, Combined source-channel transform for image transmission over wireless channel, in *Wireless Telecommunications Symposium(WTS)* (2011), pp. 1–4
17. K. Praveen Kumar, M.N.S. Swamy, K.D. Rao, A high secure approach for medical images transmission over wireless channels, in *Annual IEEE India Conference (INDICON)* (2012), pp. 841–846
18. Rahman et al., A multiscale retinex for bridging the gap between color images and the human observation of scenes. IEEE Trans. Image Process. **6**(7) (1997)
19. F. Xiao, J.M. DiCarlo, P.B. Catrysse, B.A. Wandell, Stanford University Stanford, California; "High Dynamic Range Imaging of Natural Scenes"

Gabor Wavelets in Face Recognition and Its Applications

Mr. Manoj Ojha(✉)

Department of Electronics and Communication Engineering, Sanskriti University, Mathura, Uttar Pradesh, India
manoj@sanskriti.edu.in

1 Introduction

Face recognition is a computer application [1] to intelligently identify an individual by using the image in its digital form by comparing patterns and analyzing the data. The algorithms used in this recognition extract the features from an individual's face and compare them to the information stored in the database in order to find the best match. Various algorithms are available and perform different operations. One of the methods includes identifying the facial features by extraction [2] of exclusive characteristics from digitized image; other includes analyzing the shape, size or positionscertain features of the face. The identification and extraction of these features are used for matching the features of other images. Popular algorithms used for facial recognition includes principal Component Analysis [3], wavelet transforms [4], Linear Discriminate analysis [5] etc. the principal component analysis uses Eigen value for face recognition and is an effective algorithm.

Wavelet transforms [6] are an effective substitute of Fourier transform in various applications of signal processing used in computer vision, speech analysis, etc. where the wavelet expansions are applicable. The most basic application of wavelet transform in face recognition lies in use of wavelet coefficients in form of features where it can detect the edges of facial images or features.The wavelet transforms are applicable to both continuous and discrete signals. The continuous wavelet transforms are defined as following where the elements of Xwt(T,S) are known as wavelet coefficients which are associated with frequency in the time domain.

$$X_{WT}(\tau, s) = \frac{1}{\sqrt{|s|}} \int_{-\infty}^{\infty} x(t)\psi * \left(\frac{t-\tau}{s}\right) dt.$$

The discrete wavelet transforms is a transform where wavelets are transformed discretely. The filter analysis in DWT is as shown in Fig. 1.

Face recognition being one of the most important applications in the image processing using Gabor wavelets [7] have been amount to wide researches based on holistic and analytic approaches.

F. P. García Márquez (ed.), *International Conference on Intelligent Emerging Methods of Artificial Intelligence & Cloud Computing*, Smart Innovation, Systems and Technologies 273, https://doi.org/10.1007/978-3-030-92905-3_19

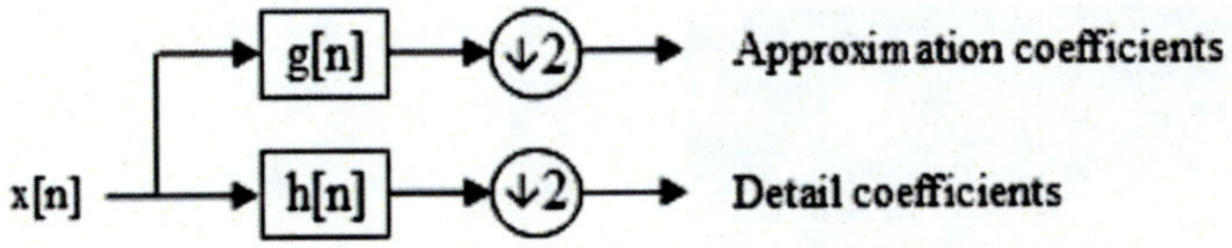

Fig. 1. Filter analysis of DWT

Analytic approach:

Many features points are extracted from the face like landmark points of a face being eyes, mouths and nose which are also known as fiducial points. The distance between the fiducial points and many other local points along with the angle between these points and quantitative measurements extracted from the facial features are used for face recognition. Analytic approaches are advantageous since different angles for the view of faces and a flexible deformation of the main feature points extracted can be used for face recognition.

Holistic approach:

Unlike analytic approach where the feature points only from specific parts of face are extracted, the holistic approach uses the feature points from all the areas of the face. The pre-processing techniques used in holistic approach include the normalization of face size as well as rotation so as to make the face recognition process more robust and useful. The two important techniques used in holistic approach are the principal component analysis (PCA) using the Eigen values and vectors and the other approach known as linear discriminant analysis (LDA) which is based on fisher values.

2 Methodology

Among all the wavelet transforms the one providing optimal resolution in spatial and frequency domains is the Gabor wavelet transform as shown in Fig. 2 as it has multi resolution and multi orientation properties. It is a representation of behavior in the field of construction of facial features by selecting energized points in the wavelet response as feature points which are extracted by using local characteristics of the face. In this paper the image processing methods use 2D wavelet transforms as a partial differential of the given order. The Gabor wavelets are used for detecting corners and edges of the face for recognition process as it is more efficient than that of the Haar wavelet [8] or Gaussian function. The Gabor wavelet can be defined as following, where x and y are the orientation and scale of Gabor wavelet,

$$\phi_{(x,y)}(z) = \frac{\left\|k_{x,y}\right\|^2}{\sigma^2} e^{\frac{-\left\|k_{x,y}\right\|^2 \|z\|^2}{2\sigma^2}} \left[e^{ik_{x,y}z} - e^{-\frac{\sigma^2}{2}}\right]$$

Kx,y is the wave vector defined as,

$$k_{x,y} = k_y e^{i\phi_x}$$

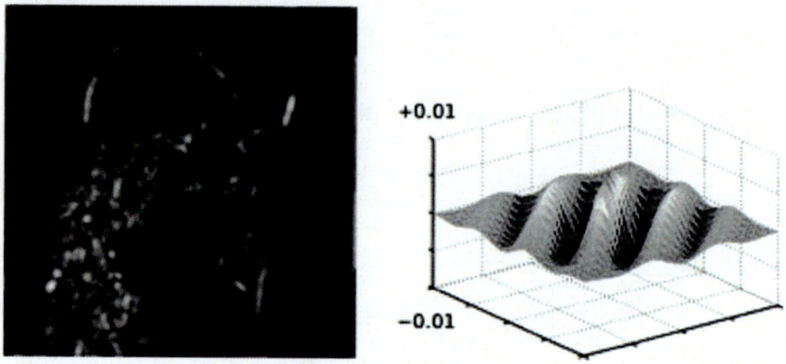

Fig. 2. Gabor transforms

where, Kx,y = Kmax/f, and Kmax is the maximum frequency with f being the space factor in the frequency domain.

Gabor wavelets as shown in Fig. 3 used in many applications of face recognition require analyzing the multi-orientation and multi-resolution images [9], which is defined as follows:

$$\{\varphi_{discrete}(f_u, \theta_v, \gamma, \eta)(x, y)\}$$

$$f_u = \frac{f_{\max}}{\sqrt{2}^u}, \theta_v = \frac{v}{V}\pi, u = 0, \ldots \ldots U - 1, v = 0, \ldots \ldots V - 1$$

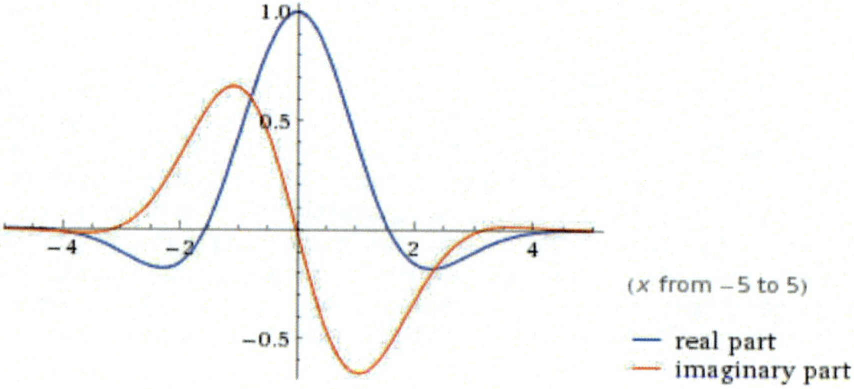

Fig. 3. Gabor wavelet

where fmax is the maximum central frequency, 2 is the space factor between the central frequencies, fu is the orientation of Gabor wavelets and Qv is the scale of Gabor wavelets.

Gabor filters are the linear filters used in face recognition by detecting the edges to be used for feature extraction. The Gabor filters [10] represent the frequency and the orientations in a similar way to that of human vision system. The impulse response of the Gabor filters is defined by the product of sine wave with that of the Gaussian function. Due to the product of sine wave and Gaussian function, the Gabor filter's Fourier transforms is the convolution of Gaussian and harmonic function's convolution.

The real and imaginary functions of the Gabor filter are as defined below (Fig. 4):

$$g(x, y; \lambda, \theta, \psi, \sigma, \gamma) = \exp\left(-\frac{x'^2 + \gamma^2 y'^2}{2\sigma^2}\right)\cos\left(2\pi\frac{x'}{\lambda} + \psi\right)$$

$$g(d, y; \lambda, \theta, \psi, \sigma, \gamma) = \exp\left(-\frac{x'^2 + \gamma^2 y'^2}{2\sigma^2}\right)\sin\left(2\pi\frac{x'}{\lambda} + \psi\right)$$

Fig. 4. Real and imaginary part of Gabor filter

3 Application of Gabor Transform

1. *Facial expression classification:*
   Using the Gabor transform for classifying the facial expressions [11] is one of the most important applications. In the case of facial expression, the classification is done by generating the features from the filtered image and training a framework using neural networks called two layer perceptron. The Gabor wavelet is more preferred for expression classification because it overcomes the challenges faced by the geometry based method.
2. *Texture classification:*
   The features of the texture depend on the local spectrum of the image that is obtained by the Gabor filters. Texture classification [12] can be achieved using Gabor wavelets in which the features are searched by calculation of the variance and the mean of the images which are filtered by Gabor. Further a vector of the features is generated for an image.
3. *Face reconstruction:*
   Due to non-orthogonality, the capability of Gabor wavelets to explore the characteristics and features of the data is limited. For the purposes of object representation a processing the face, an image of the face can be reconstructed using a set of coefficients of Gabor wavelets. The quality of the face reconstruction highly depends on the number of the Gabor wavelets used for the processing.

4. *Iris recognition:*
One of the most important applications of Gabor wavelets is in the field of iris recognition [13]. majorly used for biometrics. The use of Gabor filters and zero crossing representation of the wavelet transforms is done on the iris by using two approaches where one being the single circle of iris and other being the approach based on annular region. The wavelet transform is used for translation, invariation and rotations.
5. *Edge detection:*
Gabor transforms for edge detection is important to be used as a partial differential operator of the first order. For edge detection, the convolution is performed for the two directions which are perpendicular to each other. The edged are obtained by getting the local maxima which is greater than the threshold value due to the presence of the small edges and noise in the images. The local maxima are defined as following:

$$M(x, \sigma_D) = \sqrt{I_x^2(x, \sigma_D) + I_y^2(x, \sigma_D)}$$

6. *Corner detection:*
In corner detection, the aim is to find the points on the images known as corners which represent the stable locations of the image for image matching or template matching. The use of Gabor wavelets lies in the use of Log-Gabor filter where the corners can be represented in terms of information of the localized frequency and also Gabor filters are used to smoothen out the edgy corners to get accurate results.

4 Conclusion

Use of wavelet transforms for face recognition specifically, Gabor wavelet transforms is advantageous in terms of dilation, rotation or translation. The Gabor transforms can also abstract features from the training data and generalizes the features in order to code or analyze maximum information from the image sets. Face recognition being one of the most important applications in the image processing using Gabor wavelets have been amount to wide researches based on holistic and analytic approaches. The applications of face recognition using Gabor wavelet transforms is present widely in fields like surveillance, military and security, biometrics etc. This paper discusses about the Gabor wavelet transform used for face recognition and also the applications of Gabor wavelet transform in various face recognition systems not limited only to the classification or surveillance but also reconstruction, texture and expression classification, iris recognition and many more. This growing application of face recognition along with wavelet transforms is the basic step taken towards more advanced technologies used for interpreting actions of human, human machine interaction, human behavior and smart environments.

References

1. Z. Cao, Q. Yin, X. Tang, J. Sun, Face recognition with learning-based descriptor, in *Proceedings of the IEEE Computer Society Conference on Computer Vision and Pattern Recognition* (2010)

2. J. Kumari, R. Rajesh, K.M. Pooja, Facial expression recognition: a survey. Procedia Comput. Sci. (2015)
3. G.M. Zafaruddin, H.S. Fadewar, Face recognition using eigenfaces, in *Advances in Intelligent Systems and Computing* (2018)
4. K. Choudhary, N. Goel, A review on face recognition techniques, in *International Conference on Communication and Electronics System Design* (2013)
5. S. Liao, A.K. Jain, S.Z. Li, Partial face recognition: alignment-free approach. IEEE Trans. Pattern Anal. Mach. Intell. (2013)
6. J. Olkkonen, *DISCRETE WAVELET TRANSFORMS - THEORY AND APPLICATIONS Edited by Juuso Olkkonen.* 2011.
7. Á. Serrano, I.M. de Diego, C. Conde, E. Cabello, Recent advances in face biometrics with Gabor wavelets: a review. Pattern Recognit. Lett. (2010)
8. M. Alwakeel, Z. Shaaban, Face recognition based on haar wavelet transform and principal component analysis via levenberg-marquardt backpropagation neural network. Eur. J. Sci. Res. (2010)
9. A. Kar, D. Bhattacharjee, M. Nasipuri, D.K. Basu, M. Kundu, High performance human face recognition using Gabor based pseudo Hidden Markov model. Int. J. Appl. Evol. Comput. (2013)
10. T. Wu, M.S. Bartlett, J.R. Movellan, Facial expression recognition using Gabor motion energy filters, in *2010 IEEE Computer Society Conference on Computer Vision and Pattern Recognition - Workshops, CVPRW 2010* (2010)
11. V. Triyanti, Y. Yassierli, H. Iridiastadi, Basic emotion recogniton using automatic facial expression analysis software. J. Optimasi Sist. Ind. (2019)
12. Y. Xu, X. Yang, H. Ling, H. Ji, A new texture descriptor using multifractal analysis in multi-orientation wavelet pyramid, in *Proceedings of the IEEE Computer Society Conference on Computer Vision and Pattern Recognition* (2010)
13. K. Nguyen, C. Fookes, A. Ross, S. Sridharan, Iris recognition with off-the-shelf CNN features: a deep learning perspective. IEEE Access (2017)

Harris Corner Detection for Eye Extraction

Rishi Sikka(✉)

Department of Electronics and Communication Engineering, Sanskriti University, Mathura, Uttar Pradesh, India
rishisikka.ec@sanskriti.edu.in

1 Introduction

One of the important aspects in the field of face recognition is extraction of eyes, which is considered as a stable feature among all the facial features. Thus detection of eyes or eye extraction plays a vital role in the applications of face recognition or other fields using eyes like iris recognition, eye movement tracking, gaze tracking etc. during the detection of facial features,extracting the position of eyes done in the beginning is advantageous since the location of other features can be done with ease depending on the position of eyes on the face. The detection of eyes [1] is further sub divided into two parts namely eye contour detection and eye position detection where the need of eye contour detection lies in the applications like video conferencing or any interfaces requiring vision assistance. For eye contour detection [2] also the need to detect the position of eyes on the face is the important and initial step.

The approaches used for face recognition earlier defined the position of eyes manually so as to reduce the timing constraints and extracting other features easily. But since the modern technologies are real time based applications, giving the manual data about eyes is not possible so the eye detection has become an important aspect for the face recognition and thus approaches to efficiently detecting and location the eyes has become a necessity. Detecting eyes and extracting the information has become a need in all the automatic facial features detection and further in all the applications using these features for their working and performance.

Today the extraction of eyes can be done by using various algorithms which are based on the features, templates or the appearance and among all these algorithms the feature based approaches are widely used due to less time consumption and efficient results. Oneof the major factors in the process of eye extraction [3] or detection is the detection of corners which are defined as the intersection point of two edges or a point of intersection having two dominant edges. Corner detection is the approach used for detecting or locating the corners and similar features and is used in various applications like object recognition, face recognition, imageregistration, detection of motion etc. many corner detection approaches are available including Moravec corner detection, Harris corner detection, Forstner detection, SUSAN corner detection etc.

Among all the approaches used in corner detection the most significant is Harris corner detection [4]. It is an operator for corner detection which is commonly used in field of computer vision for extracting the corners and inferring the image features.

F. P. García Márquez (ed.), *International Conference on Intelligent Emerging Methods of Artificial Intelligence & Cloud Computing*, Smart Innovation, Systems and Technologies 273, https://doi.org/10.1007/978-3-030-92905-3_20

Harris algorithm uses the methodology of calculating the curvature and the gradient in order to detect the corners. It is widely used due to easy approach, highly stable extraction of corners and efficient results. The algorithm of Harris corner detection is an improvement over the Moravec corner detection method as Harris algorithms uses the differential rather than using shifting patches. The Harris operator calculates the auto correlation matrix and the gray values extracted are converted into corners. The algorithm processes in few steps as below:

1. *Conversion of grayscale to corners:*
   Let f(x,y) be the gray values of the pixels and after conversion to the corners by changing the gray level intensity, the change is expressed as following:

$$E_{u,v}(x, y) = \sum_{u,v} w_{u,v}[f(x + u, y + v)]^2$$

2. *Calculation of spatial derivative*:
   By using Taylor expansion on the above formula, the converted gray scale intensity defined earlier can be expressed as:

$$f(\Delta x, \Delta y) \approx \sum_{(x,y)\in W} (I_x(x, y)\Delta x + I_y(x, y)\Delta y)^2.$$

   Which is further expressed in its matrix form as following:

$$f(\Delta x, \Delta y) \approx (\Delta x \ \Delta y)M\begin{pmatrix}\Delta x\\ \Delta y\end{pmatrix}.$$

3. *Structure tensor:*
   Where M is known as structure tensor and expressed as following:

$$M = \sum_{(x,y)\in W} \begin{bmatrix} I_x^2 & I_x I_y \\ I_x I_y & I_y^2 \end{bmatrix} = \begin{bmatrix} \sum_{(x,y)\in W} I_x^2 & \sum_{(x,y)\in W} I_x I_y \\ \sum_{(x,y)\in W} I_x I_y & \sum_{(x,y)\in W} I_y^2 \end{bmatrix}$$

4. *Response calculation:*
   The smallest Eigen value of the tensor structure is calculated in this step using approximation and defined as following:

$$R = \mathrm{Det}(M) - k(trM)^2$$

5. *Non—maximum suppression:*
   A local maximum is calculated as corners for finding the optimal values to indicate corners in the image by using a 3*3 template.

2 Methodology

The steps for implementing Harris corner detection in this paper are described according to the Fig. 1. The steps include pre-processing of the image, registration of image, morphological operations [5], applying corner detection algorithm [6] as explained above, and thus extracting the eyes from the image as final result.

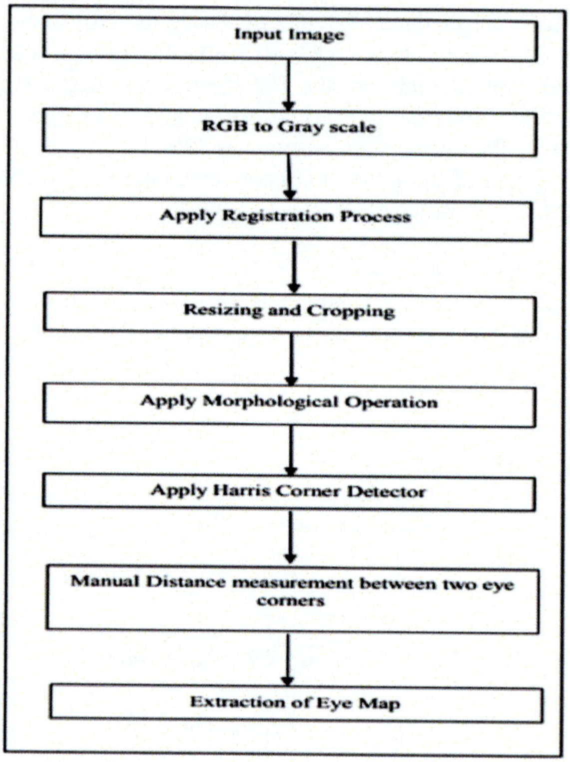

Fig. 1. Steps of eye extraction

1. *Pre-processing of the image:*
   For processing of the images, conversion of images from color to grayscale to binary form is a necessity and the resultant image is known as binary image. The grayscale image is converted to binary by the replacement of pixels with a greater luminance to a level 1 and rest to 0. It is done in order to perform image registration on the input images which is possible only on the binary images.
2. *Image registration:*
   After the pre processing of images, the next step is for registration of the images [7]. Image registration being a process to align multiple images of the original image where, one image is taken as the reference image and transformations and other processes are performed on the other images in alignment with the reference image. The image analysis tasks require image registration as it is a crucial step where the information like change detection, image restoration, fusion of images etc. are combined altogether. The alignment done in the image registration collects all the images in same format for further processing using morphological operations.
3. *Morphological operations:*
   Morphological operations like dilation and erosion are applied to the image after the image registration is done. Dilation [8] is the step to grow the objects present in the image by adding extra pixels depending on the structure element used for

processing the image. Erosion [8] is the process of removing the pixels from the edges or boundaries of the image by comparing the pixels to the structuring elements.

4. *Harris corner detection:*
Corners being the point of intersection of the edges or the junction of the edges where the brightness level is different from rest of the image parts [9]. The important features that are needed to be extracted from an image are the corners which are also known as interest points and are invariant to rotation, brightness or conversion. They are the most prominent features of an image for applications like image restoration or image reconstruction. Harris detection from the corners is applied to the image preprocessed and achieved after morphological operations are applied to it from better extraction of the needed features from the image. Harris detection for the corners is processed by using autocorrelation of the intensity values or the gradient values in the image. The use of gradient values is done for the processing of corners in the images. The gradient of the image and the standard deviation is denoted as:

$$I_x, I_y, \sigma_i$$

The coefficients of the autocorrelation matrices are denoted as;

$$\tilde{A},\ \tilde{B},\ \tilde{C}$$

For each pixel at (i,j), coefficients of the autocorrelation matrix are calculated as follows:

$$\begin{aligned} A(i, j) &\leftarrow I_x^2(i, j) \\ B(i, j) &\leftarrow I_x(i, j)I_y(i, j) \\ C(i, j) &\leftarrow I_y^2(i, j) \end{aligned}$$

Which are further convolved with the Gaussian function:

$$\begin{aligned} \widetilde{A} &\leftarrow \text{gaussian}(A, \sigma_i) \\ \widetilde{B} &\leftarrow \text{gaussian }(B, \sigma_i) \\ \widetilde{C} &\leftarrow \text{gaussian}(C, \sigma_i) \end{aligned}$$

The autocorrelation matrix yields two real positive Eigen values which are further analyzedand the corner response function is defined by Harris function as below:

$$R_H = \lambda_1\lambda_2 - \kappa \cdot (\lambda_1 + \lambda_2)^2 = det(M) - \kappa \cdot trace(M)^2$$

where k has a value of approximately 0.04. This function calculates the interest points or the corners in the image.

The final corners detected by the Harris corner detection is threshold [10] to focus on the most relevant and significant corners. The resultant image after applying all the operations and Harris corner detection is shown in Fig. 2.

Fig. 2. Original image and resultant image

3 Conclusion

Eye detection or eye extraction is the important factor required in today's technology used in surveillance, security or biometrics which requires facial features for access or functioning. The paper addresses about the importance of detecting corners in an image for image analysis required in the fields of face recognition, iris recognition, and vision tracking etc. having the eye extraction as main factor. The corner detection techniques having many approaches to be applied on images but the paper discusses about Harris corner detection which has many advantages over other approaches like Moravec corner detection as Harris approach works with differential rather than using the shifting patches. It calculates the curvature and gradient in order to detect the corners. The approach of Harris corner detection gives an accurate and reliable result and the use of gradient in this approach makes it a stable approach.

References

1. M.F. Peterson, M.P. Eckstein, Looking just below the eyes is optimal across face recognition tasks. Proc. Natl. Acad. Sci. U. S. A. (2012)
2. A.B. Roig, M. Morales, J. Espinosa, J. Perez, D. Mas, C. Illueca, Pupil detection and tracking for analysis of fixational eye micromovements. Optik (Stuttg). (2012)
3. F. Song, X. Tan, X. Liu, S. Chen, Eyes closeness detection from still images with multi-scale histograms of principal oriented gradients. Pattern Recognit. (2014)
4. N. Dey, P. Nandi, N. Barman, D. Das, S. Chakraborty, A comparative study between moravec and harris corner detection of noisy images using adaptive wavelet thresholding technique. Int. J. Eng. Res. Appl. (2012)

5. T. Rajpathak, R. Kumar, E. Schwartz, Eye detection using morphological and color image processing. Camera (2010)
6. H. Cho, P.E. Rybski, W. Zhang,"Vision-based bicyclist detection and tracking for intelligent vehicles, in *IEEE Intelligent Vehicles Symposium, Proceedings* (2010)
7. M. Al Najjar, M. Ghantous, M. Bayoumi, Image registration. Lect. Notes Electr. Eng. (2014)
8. B.G. Batchelor, F.M. Waltz,"Morphological image processing, in *Machine Vision Handbook* (2012)
9. E. Rosten, R. Porter, T. Drummond, Faster and better: a machine learning approach to corner detection. IEEE Trans. Pattern Anal. Mach. Intell. (2010)
10. J. Malik, R. Dahiya, G. Sainarayanan, Harris operator corner detection using sliding window method. Int. J. Comput. Appl. (2011)

Human Computer Interface Using Electrooculogram as a Substitute

Laxmi Goswami(✉)

Department of Electronics and Communication Engineering, Sanskriti University, Uttar Pradesh, Mathura, India
laxmigoswami.ee@sanskriti.edu.in

1 Introduction

Recently, human computer interface (HCI) [1] is a field with increasing demand due to the benefits of HCI in the advanced technologies. HCIs have found its way in the medical field being used in the technologies specially meant to be used as assistive methods like technologies for disabled, assisting robotic arms, robot controlled wheelchairs, moving cursor with blinking eyes etc. These technologies using HCI systems need a bridge to cover the gap between the real world and the virtual environment created to mimic the reality [2]. Eyes of a person act as a bridge between these two worlds since eyes are the features of a human body which are not usually affected by the nervous system of a human being and thus can act as substitute for nerves used to do physical movements required for controlling many tasks.

The movement of eyes can be captured and processed by many devices or systems like electroencephalogram, electrooculogram, magneto encephalograph, eletromyogram etc. Generally these methods or devices are used by the HCI systems to serve as a tool or bridge between the human and the computers or machines for translating the electrical signals produced by the human brain due to certain physical activity to commands that are readable by a machine in order to perform the certain task decoded by the electrical signals. In eletromyogram, the signals are generated by the movement of skeletal muscles and are in many areas like prosthetics, electric wheel chairs etc., however the analysis of EMG [3] signals is difficult. The electroencephalogram or the EEG [4] signals are the representation of electrical signals produced in the brainfor developing control systems for performing any activity but the analysis of EEG signals iscomplex and hence disadvantageous. Among all the available tools the most prominently used tool is electrooculogram due to its robust performance and simplicity in its architecture thus overcoming the complexity and other challenges using other systems.

The electrooculogram signal used in the process of electrooculography [5] is a signal that captures the eye movements and the value of signals changes according to the change in the movement. In this process the changes in the signal are captured by the difference in the potential between the cornea and the retina of the eyes and this potential is generated by the use of electrodes. There is an increase in the potential value when in the eye movement the cornea is closer to the electrode and the decrease in the potential is seen

F. P. García Márquez (ed.), *International Conference on Intelligent Emerging Methods of Artificial Intelligence & Cloud Computing*, Smart Innovation, Systems and Technologies 273, https://doi.org/10.1007/978-3-030-92905-3_21

when the cornea is far from the electrode which means it moves in the opposite direction. When the gaze of the person is straight i.e. there is no eye movement in any direction, the potential in the electrodes remain constant and the value does not change. EOG is not only used for eye movement but also for opening and closing of eyes can be stored in form of signals.

2 Method for Eog in Eye Movement

The measurement of eye movement by the use EOGs require certain steps as shown in Fig. 1 like measurement of the signals, processing of the signals, feature extraction [6], and recognition.

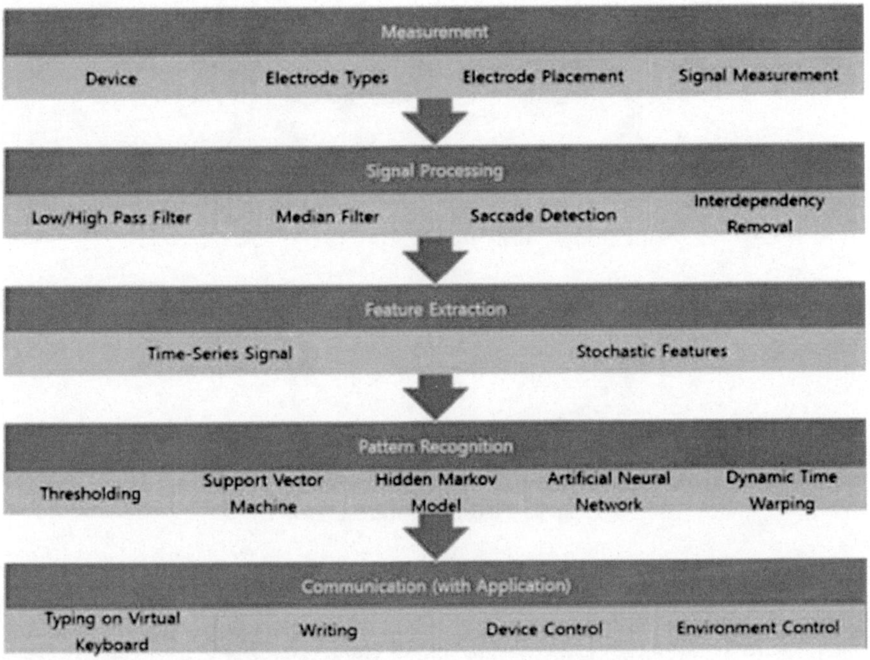

Fig. 1. Steps for eye movement measurement

1. *Measurement of signals*
   For the measurement of signals, initially the placement of electrodes or sensors near the eye area is done. The data in form of signals are acquired by the system depending on the placement of electrodes. The electrodes are placed on the left and right of the yes to capture movement in both directions and an electrode used as reference electrode is placed at the forehead or the middle of eyes or placing two electrodes one above and the other below the eye. The placement of electrodes [7] as shown in Fig. 2. The signals captured might contain certain noise or unwanted data due to the

quality and placement of the electrodes used for capturing signals. The amplitude calculated by the eye movement is directly proportional to the magnitude of the movement of the eyes and the EOG signal is stored by using an amplifier.

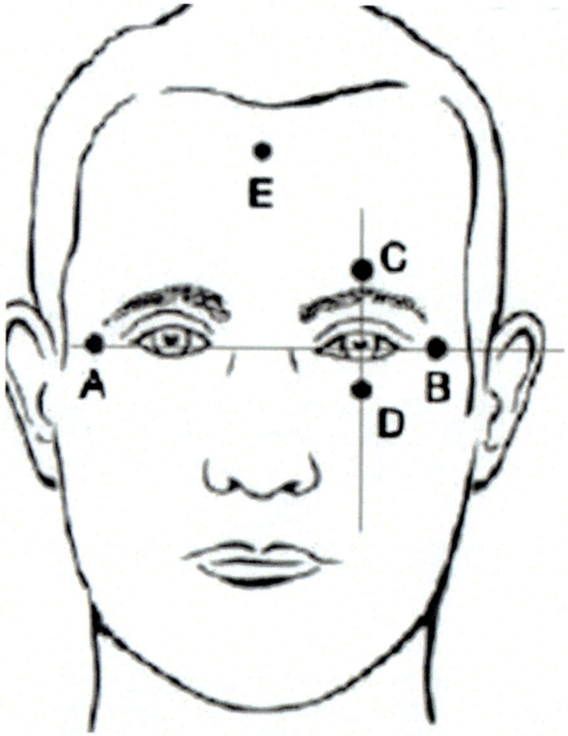

Fig. 2. Placement of electrodes

2. *Processing of signals*
The signals produced by the eyes are extracted from the sensors or the electrodes in form of potential and are further stored for processing. The noise signals or other unwanted signals like tremor [8] are removed from the stored data so that only significant signal may remain for further processing. Certain more signals like crosstalk may also be present in the captured signals and thus needed to be removed. These unwanted data or artifacts are removed using low pass or high pass and median filters thus generating a noise free signal containing only the relevant information. Polynomial regression is also a technique used for removal of noise signals from the captured signals and is a better option over filters. Wavelet transformations are also used in place of the high pass filters to avoid the diminishing of edges in the signals stored.

3. *Feature extraction*
   One of the most important steps in the process of using electrooculography signals for processing the eye movements is feature extraction. The features in the EOG signals are extracted on the basis of time domain and are compared with most common eye movement directions known. There are many techniques used for feature extraction from the signals captured by the electrooculography. The techniques like areaunder curve [6] (AUC) where the AUC value is the addition of amplitude under both + ve and –ve curves, peal amplitude value (PAV) where the PAV is the measurement of EOG signals at a maximum point at both the channels, peak amplitude position (PAP) which is the measurement of EOG signal at the maximum position of the highest point in both the channels. Similarly many techniques are available for the extraction of feature from the electrooculogram signals.
4. *Recognition*
   The features extracted by the various techniques mentioned in the above step are used to determine the direction of the movement of human eye. These extracted features are further used as an input for classification and recognition. The best approach that can be used for recognition in the systems using EOG can be Hidden Markov Model [9] (HMM). The HMM is applied on EOG signals to determine the data extracted by the feature extraction process and generates command signals for the task to be performed according to the eye movement. The eye movement as shown in Fig. 3.

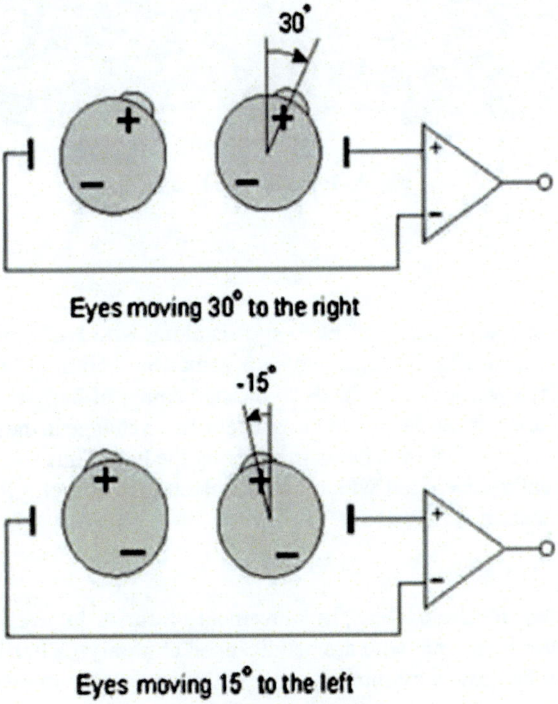

Fig. 3. Eye movement

3 Applications of EOG in HCI

Human computer interface has a vast and increasing role in the advanced technologies recently. The use of electrooculogram especially in the field of medical has found to be of great application altogether with the HCI. The applications are as discussed:

1. *Robotic wheelchair*
   Paralyzed people today require the robotic wheelchairs or automatic wheelchairs that can be operated by the movement of eyes and this control of the wheelchair is done by capturing the eye movements by electrooculogram thus controlling movement as well as sped of wheelchair. The speed is controlled by the short or long eye movements, blinks of eye, or opening and closing of eyelids. Many applications have been found where the direction of the wheel chairs is directly based on the eye movement's direction. The robotic wheelchair [10] is as shown in Fig. 4.

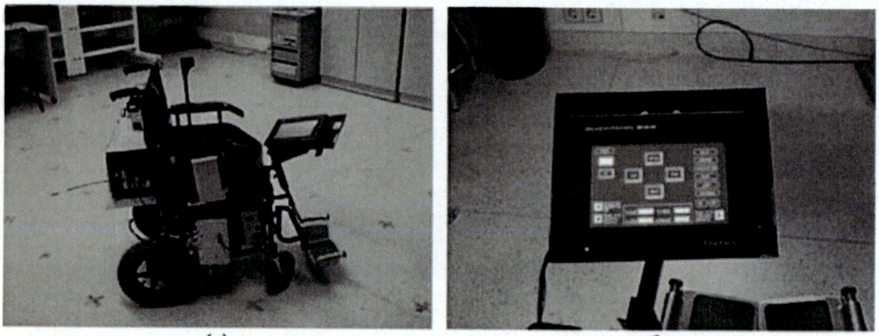

(a) (b)

Fig. 4. Robotic wheelchair

2. *Gaming*
   Another important application of EOG signals along with HCI is found in gaming systems as shown in Fig. 5. For example a game for shooting [11]was developed using balls. The eye movement leads to the movement of balls to the left or right direction and the downwards movement of eyes causes change in the color of the ball. The upward movement of eyes helps in shooting the ball. Thus similar applications have been recently developed which use the electrooculography for detecting eye movement by human computer interface.

3. *Cursor control*
   A control system for controlling the movement of cursor in any direction instead of using a mouse by people who are handicapped or paralyzed has been developed. The cursor can be moved by moving the eyes in the specific direction as required. The blinking eyes control the task to be performed which is otherwise done by the mouse used by a hand. The movement of cursor depends on the threshold of the eye

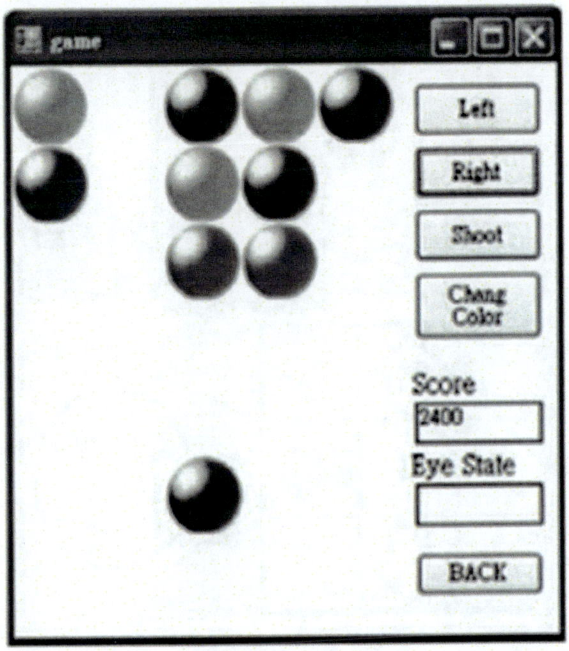

Fig. 5. Gaming system

movement and the minimum threshold do not causes any movement thus preventing the unwanted movement of curve due to eyes.

4. *Dance game*
   In this application the user is required to operate according to the instructions on the screen displaying the game. The user is asked to follow only the instructions below the arrow and if the actions are followed otherwise it leads to termination of the game. The noise signals were removed to improve the precision of this game. The Fig. 6 shows the dance game [12].

4 Conclusion

The electrooculography is the means to detect eye movement for controlling certain tasks when used altogether with human computer interface. Recently the applications of HCI along with EOG have been a wide topic of research due to its vast applications and better functionalities. Various techniques are available to detect the eye movements like EEG, EMG and EOG but EOG is the most preferred one due to its robust performance and simplicity in performing tedious tasks in complex applications. The applications of EOG with human computer interface have been discussed in the paper which successfully proves EOG to be of great help in medical as well as other areas requiring EOG as a substitute for many purposes. The HCI systems using EOG have been tested as discussed in the paper. Hence the paper concludes that the electrooculography when used with the

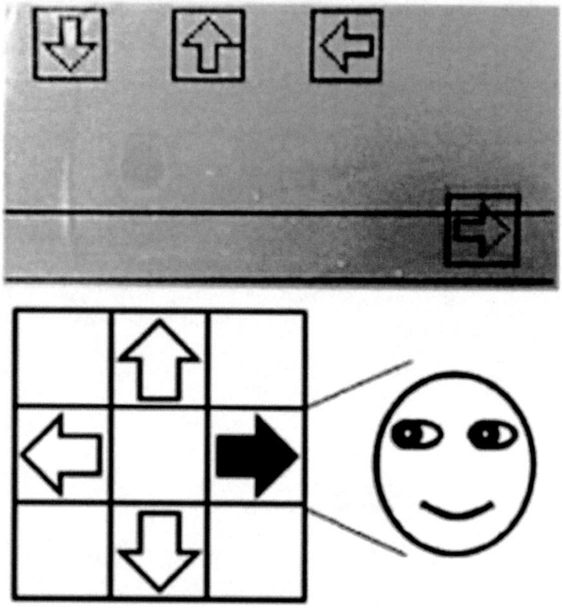

Fig. 6. Dance game

HCI systems or applications prove to be the best approach used for the communication between real and the virtual world where the virtual worlds emulates the real human situations and the tasks are performed by using the tools like EOG as a bridge between the two worlds.

References

1. S. Amiri, R. Fazel-Rezai, and V. Asadpour, "A review of hybrid brain-computer interface systems," *Advances in Human-Computer Interaction.* 2013.
2. D. Friedman, R. Leeb, G. Pfurtscheller, and M. Slater, "Human-computer interface issues in controlling virtual reality with brain-computer interface," *Human-Computer Interact.*, 2010.
3. R. M. Enoka, "Electromyography (EMG)," in *Encyclopedia of Movement Disorders*, 2010.
4. C. M. Michel and M. M. Murray, "Towards the utilization of EEG as a brain imaging tool," *NeuroImage.* 2012.
5. A. Bulling, J. A. Ward, H. Gellersen, and G. Tröster, "Eye movement analysis for activity recognition using electrooculography," *IEEE Trans. Pattern Anal. Mach. Intell.*, 2011.
6. S. Aungsakul, A. Phinyomark, P. Phukpattaranont, and C. Limsakul, "Evaluating feature extraction methods of electrooculography (EOG) signal for human-computer interface," in *Procedia Engineering*, 2012.
7. A. López, F. J. Ferrero, M. Valledor, J. C. Campo, and O. Postolache, "A study on electrode placement in EOG systems for medical applications," in *2016 IEEE International Symposium on Medical Measurements and Applications, MeMeA 2016 - Proceedings*, 2016.
8. M. Sanjeeva Reddy, B. Narasimha, E. Suresh, and K. Subba Rao, "Analysis of EOG signals using wavelet transform for detecting eye blinks," in *2010 International Conference on Wireless Communications and Signal Processing, WCSP 2010*, 2010.

9. Sahoo, Singh, and Mukhopadhyay, "A Hidden Markov Model for Collaborative Filtering," *MIS Q.*, 2012.
10. F. Aziz, H. Arof, N. Mokhtar, and M. Mubin, "HMM based automated wheelchair navigation using EOG traces in EEG," *J. Neural Eng.*, 2014.
11. L. Y. Deng, C. L. Hsu, T. C. Lin, J. Sen Tuan, and S. M. Chang, "EOG-based Human-Computer Interface system development," *Expert Syst. Appl.*, 2010.
12. M. R. Kim and G. Yoon, "Control Signal from EOG Analysis and Its Application," *Int. J. Electr. Comput. Energ. Electron. Commun. Eng.*, 2013.

Image Fusion Using Wavelet Transforms

Manoj Ojha(✉)

Department of Electronics and Communication Engineering, Sanskriti University, Uttar Pradesh, Mathura, India
manoj@sanskriti.edu.in

1 Introduction

In recent technologies using machine or computer vision [1], the field of image processing for analyzing the images is the approach which leads to effective application of the computer vision or machine vision in various fields like medical imaging, satellite imaging, remote sensing [2] etc. for these applications in image analysis the most effective method used is image fusionwhich is a method of fusing multiple images together to produce single image with all the relevant information. Image fusion is the approach applied to the image processing techniques when it requires both the information including spatial as well as spectral from single image. It is needed in the satellite imaging or remote sensing applications of image processing.

Image fusion which is basically a method where two or more i.e. multiple images are bound together to form a single informative image i.e. each image has certain information in them which are combined together. The fusion of the images [3] is needed when the images captured by different techniques related to the same object are available having different level of information available in each of them. The example for such a method can be imaging with multiple focus of a single image, in remote sensing few images give good spectral information and in medical images MRI or CT are used for information gathering.

The steps in image fusion start from the pre processing of the images beginning with registration of images which is a process to align the multiple images of same scene or objects. The registration of images is done by considering one image as the reference image or fixed image and aligning other images to the reference image by applying transformations to them. The next step in image fusion is combining the images together to form a single image by applying fusion techniques to all the images generating the features required from the multiple images. The most basic technique for fusion of images is done by calculating average of the pixels in the images but since it lacks in terms of quality of the information gathered in the image, complex methods like PCA [4] or wavelet transforms can be used. The approaches applied to the images for fusion depends on the spatial value of the images and is the transformations of the images is done from spatial then the fusion of those transformed images is done by using wavelet transforms.

The paper discusses about the image fusion done by the wavelet transforms particularly discrete complex wavelet transforms (DCWT) [5]. This technique or algorithm

F. P. García Márquez (ed.), *International Conference on Intelligent Emerging Methods of Artificial Intelligence & Cloud Computing*, Smart Innovation, Systems and Technologies 273, https://doi.org/10.1007/978-3-030-92905-3_22

depends on the gradient values of the images. The images are registered first by using a point method image registration when the images are multimodal images. The tool used for image fusion known as wavelet transform is used to partition the given image on the basis scale value. The combination of wavelet transformation of the images is shown in Fig. 1.

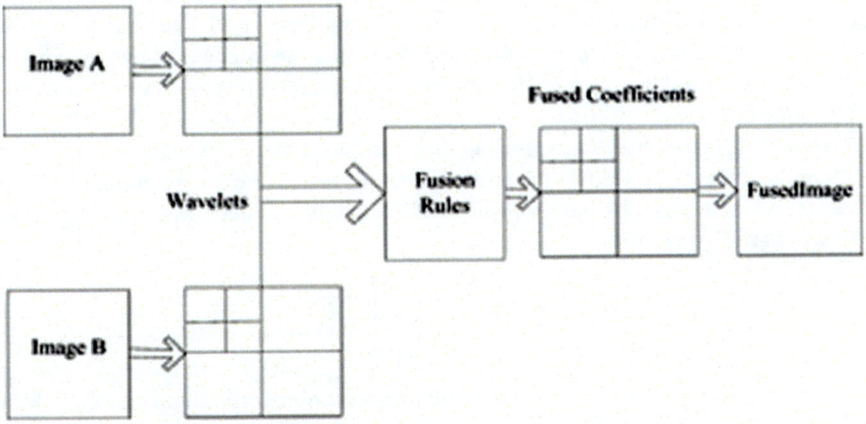

Fig. 1. Fusion of different wavelet transforms of the images

Wavelet transformation are an effective substitute of Fourier transform [6] in various applications of signal processing used in computer vision, speech analysis, etc. where the wavelet expansions are applicable [7]. The most basic application of wavelet transform in face recognition lies in use of wavelet coefficients in form of features where it can detect the edges of facial images or features. The wavelet transforms are applicable to both continuous and discrete signals. The continuous wavelet transforms are defined as following where the elements of Xwt (T, S) are known as wavelet coefficients which are associated with frequency in the time domain.

$$X_{WT}(\tau, s) = \frac{1}{\sqrt{|s|}} \int_{-\infty}^{\infty} x(t)\psi^{*}\left(\frac{t-\tau}{s}\right)dt.$$

The discrete wavelet transforms is a transform where wavelets are transformed discretely. The filter analysis of the discrete wavelet transforms is as shown in the Fig. 2

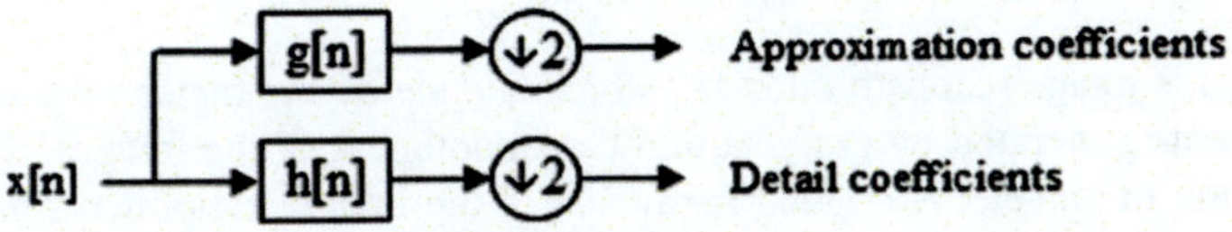

Fig. 2. Filter analysis

2 Methodology

The images used in the image fusion are registered images which are obtained from the same scene or belong to the same objects. The images registered are first broken down or decomposed into multiple smaller parts with the help of wavelet transform. The sub division of the images is done by selecting the images with same resolution at same point of level. The fusion of the images is done on the sub divided images having high frequency level and the resultant image is again obtained by the use of wavelet transforms. The wavelet transform used for subdividing the image is known as forward wavelet transform and the one used for obtaining the resultant image is known as inverse wavelet transform.

The method of wavelet transforms applied for image fusion [5] depends on the use of gradients of image as well as the smoothness of image. The image gradient at pixel location (x,y) is obtained by the application of 2D directional derivative on the pixels in the image as shown below:

$$\nabla I(x, y) = \begin{bmatrix} Gx \\ Gy \end{bmatrix} = \begin{bmatrix} \frac{\partial I(x,y)}{\partial x} \\ \frac{\partial I(x,y)}{\partial y} \end{bmatrix}$$

where I(x,y) is the gray scale image to which wavelet transformation is applied. Magnitude of gradient and the decreasing rate of approximations can be generated by the following:

$$G = |\nabla I(x, y)| = \sqrt{Gx^2 + Gy^2} \approx |Gx| + |Gy|$$

The smoothness of image is calculated by using the histogram values of the area in the neighboring to (x,y). "z" is considered as a variable to denote gray levels and p(zi) as the histogram corresponding to the variable in gray level, where i- 1,2, L−1, where L represents total no of levels of gray in image.

The variance of "z" is represented in terms of mean "m" by:

$$\sigma^2(z) = \sum_{i=0}^{L-1} (z_i - m)^2 p(Z_i)$$

$$m = \sum_{i=0}^{L-1} z_i p(Z_i)$$

The smoothness of the pixels in the adjoining of (x,y) can further be calculated as:

$$R(x, y) = 1 - \frac{1}{1 + \sigma^2(z)}$$

Next step is image reconstruction [8] which is done by the fusion of the images or the components generated by gradient and the smoothness of the images. The inverse transformation of wavelet is applied for fusion of the image components having both high and low frequency components.

The process of image fusion from raw data and applying wavelet decomposition is as shown in Fig. 3.

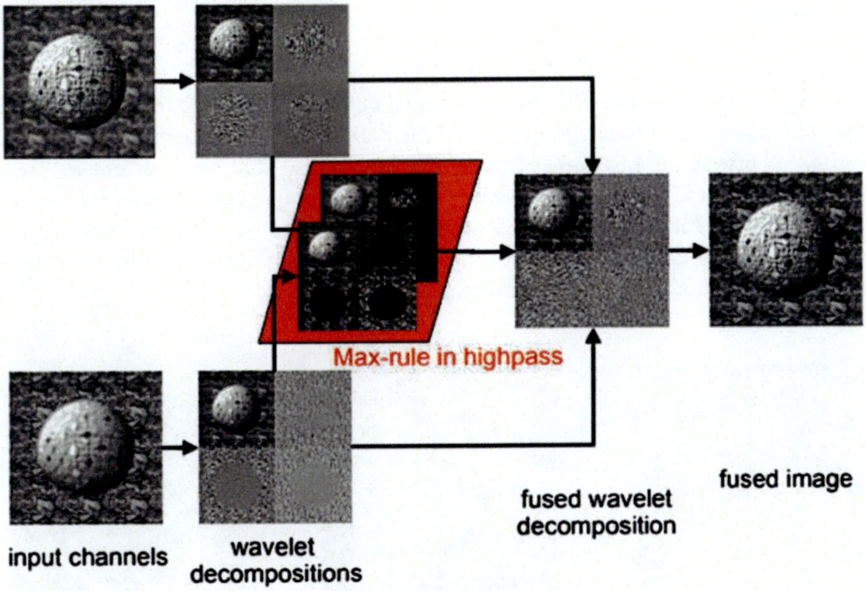

Fig. 3. Resultant image

3 Applications

1. Remote sensing
   One of the applications where image fusion is used is in field of remote sensing [9]. The image fusion is found in several domains of remote sensing where images with multi-resolution are needed. The images like panchromatic images and multispectral images are available in the field of remote sensing. The panchromatic images are the images which are collected using the broad wavelength range in the form of black and white image. The multispectral images in remote sensing are the images which are acquired in more than one spectral interval having same background area but spectral band is different.

   The fused image of the remote sensing image is as shown below in Fig. 4

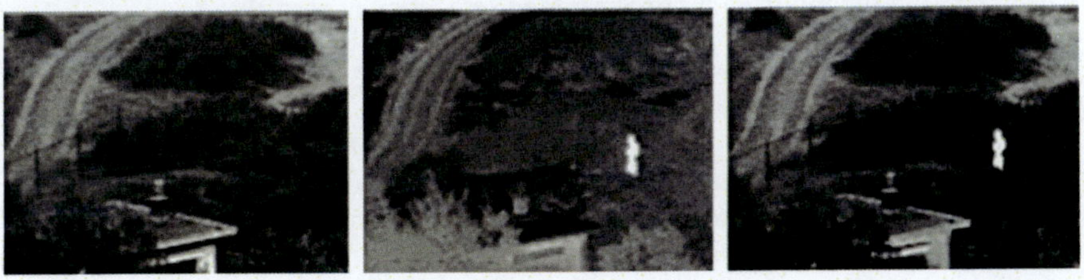

Fig. 4 Image fusion for remote sensing

2. Medical

Image fusion being used in the field og medical is used on multiple images [10] of the registered patient and the merged image of the patient gives the important information which is otherwise hard to gather from multiple images. The fusion of images colled from the MRI and CT scans of the pateints is done for diagnosing the cause accurately. It is mostly used for the patient diagnosed from cancer as the relevant information is required by oncologists to provide relevant therapy to the patient. The fused image of the MRI of the patient is as shown in Fig. 5.

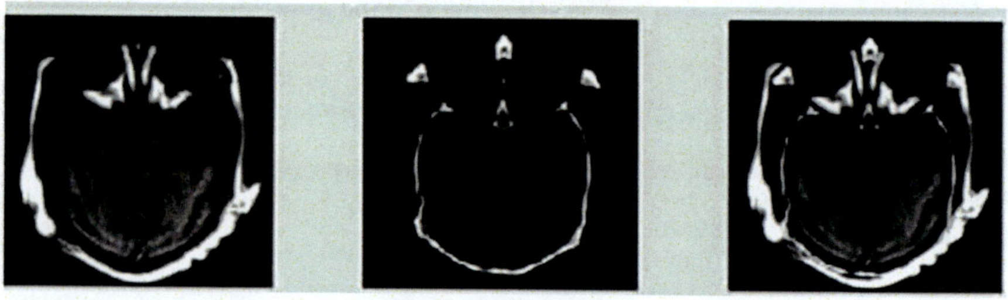

Fig. 5 Image fusion of medical imaging

4 Conclusion

In recent technologies, the field of image processing for analyzing the images is the approach which leads to effective application of the computer vision or machine vision in various fields like medical imaging, satellite imaging, remote sensing etc. the paper addresses the approach of image fusion applied on the images for the applications of medical and remote sensing. Image fusion is the approach applied to the image processing techniques when it requires both information from the spatial as well as spectral information generated in a single image. The use of gradient and smoothness are calculated and combined together by the decomposed wavelets transform and hence giving an image with uniform intensity and reduced noise. The approach discussed in the paper can be applied to the multimodal images due to its efficient result.

References

1. C. J. Du, Q. Cheng, "Computer vision," in *Food Engineering Series*, (2014)
2. H. Ghassemian, A review of remote sensing image fusion methods. Inf Fusion. (2016). https://doi.org/10.1016/j.inffus.2016.03.003
3. A. de Juan, A. Gowen, L. Duponchel, C. Ruckebusch, Image fusion, in *Data Handling in Science and Technology*, (2019)
4. T. V. Šibalija, V. D. Majstorović, Novel approach to multi-response optimisation for correlated responses. FME Trans. (2010)

5. H. Demirel, G. Anbarjafari, IMAGE resolution enhancement by using discrete and stationary wavelet decomposition. IEEE Trans. Image Process. (2011). https://doi.org/10.1109/TIP.2010.2087767
6. A. Muthukrishnan, J. Charles Rajesh kumar, D. Vinod Kumar, M. Kanagaraj, Internet of image things-discrete wavelet transform and Gabor wavelet transform based image enhancement resolution technique for IoT satellite applications. Cogn. Syst. Res. (2019). doi: https://doi.org/10.1016/j.cogsys.2018.10.010.
7. G. Anbarjafari, S. Izadpanahi, H. Demirel, Video resolution enhancement by using discrete and stationary wavelet transforms with illumination compensation, *Signal*. Image Video Process. (2015). https://doi.org/10.1007/s11760-012-0422-1
8. F. Wübbeling, PET image reconstruction, in *Correction Techniques in Emission Tomography*, (2012)
9. J. Zhang, Multi-source remote sensing data fusion: Status and trends. Int. J. Image Data Fusion (2010). https://doi.org/10.1080/19479830903561035
10. A. Sotiras, C. Davatzikos, N. Paragios, Deformable medical image registration: a survey. IEEE Trans. Med. Imaging (2013). https://doi.org/10.1109/TMI.2013.2265603

Review on Traction Control System

Arvind Kumar(✉)

Department of Mechanical Engineering, Sanskriti University, Uttar Pradesh, Mathura, India
arvindkumar.me@sanskriti.edu.in

1 Introduction

Traction control was first used on the Turbo Supra in 1994 and extended in 1997 to include the six Camry and Avalon cylinder versions. The Traction Control System's purpose is to prevent wheel spin due to acceleration. The maximum torque that can be transmitted to the wheels is determined by the friction coefficient between the tires and the road. When torque reaches that amount, it is possible that the wheels would rotate. TRAC operating conditions may include: slippery road surfaces, loose gravel surfaces, cornering acceleration and fast acceleration. The TRAC system, once activated, reduces engine torque and drives wheel speed as needed to bring the vehicle under control, improving vehicle stability when slippery roads start, accelerate or turn. While both the Supra and Camry/Avalon TRAC system control engine torque and drive wheel braking, how this is accomplished varies and the two systems are therefore covered separately in this section [1, 2].

In order to provide traction control, TRAC ECU/ECM and ABS work together. To determine the speed of each wheel and vehicle speed, the ABS/TRAC/ECU monitor signals from the four speed sensors. When determining the slippage:

- The ABS/TRAC ECU enables the actuator solenoids and the pump motor to apply hydraulic pressure on the drive wheels to the brakes.
- The ECM controls the throttle position sensor and rejects up to five cylinders of fuel injection to reduce the torque of the engine.
- The ECM prohibits the automatic transaxle movement.
- The slip indicator light is switched on to notify the TRAC driver and a signal is sent to the ECM.

F. P. García Márquez (ed.), *International Conference on Intelligent Emerging Methods of Artificial Intelligence & Cloud Computing*, Smart Innovation, Systems and Technologies 273, https://doi.org/10.1007/978-3-030-92905-3_23

Traction control is a control system to increase vehicle stability and safety in automotive applications. In internal combustion engine vehicles (ICVs), well-known vehicle control systems such as anti-slip regulation (ASR), electronic stability program (ESP) and anti-lock brake system (ABS) are used. Traction control prevents the vehicle from sweeping when accelerating on a loose surface, reduces engine output until the vehicle is able to move without skidding the wheels, and provides maximum stability when turning especially on wet or icy roads. When a wheel is skidding, a conventional difference does not usefully distribute the torque to the wheels. All the power is applied to the skidding wheel, not to the more tractable wheel. By applying a brake to that wheel, an electronic traction control system prevents wheel from skidding, allowing the difference to apply power to the other wheel. The control scheme consists of a device estimating the condition of the road surface and a traction controller regulating the slip of the wheel at the desired values. In the literature, several control strategies were proposed based primarily on sliding mode controllers, fuzzy logic, and adaptive schemes for controlling four wheel vehicles moving in sliding surfaces. Such control schemes are motivated by the non-linear and uncertain nature of the system. A lot of work has been developed recently in defining traction control algorithms for electric vehicles (EV). The generation of torque in EV is very fast and accurate, both to accelerate and decelerate. The inverter ensures the torque control of each wheel, and a mechanical differential gear is not required. The automatic torque and speed control of each of the four separate wheels allows the EV to operate more efficiently and prevent slippage. In addition, efficient torque control makes it possible to increase the vehicle's energy efficiency. Applying traction control to motorcycles is not as widely used as four wheel vehicles, probably due to the control system's high cost. MotoGP (e.g., Yamaha and Ducati) uses traction control systems to improve the ability to drive the motorcycle during competitions. The wheel speed is measured of Ducati motoGP bike using hall-effect gear tooth sensors, in a manner similar to that used in this work. 8 pick-up points (the bolts) are used by the Ducati motoGP (Fig. 1). The Yamaha M1 motoGP has sensors for redundancy on either side of the wheel. The solid disk used is a magnetic ring element in which for more data points and accuracy a strip of small magnets is embedded than a toothed ring [2–5].

1. When is it useful?
   Traction control is helpful if vehicle is tried to speed up in conditions of low friction. Such may involve rocky, frozen, rough, rainy or poorly maintained paths. Several concrete examples of the utility of traction control include:

   - When the driver is trying to speed up in a hill where the surface is gravely and loose. Vehicles wheels spin without traction control and it start sliding backwards.
   - When the driver hits a slushy ground patch that causes you to lose traction on your wheels. Your vehicle slows down as a result and starts fishtailing.
   - Two of the vehicles wheels are approaching an icy road segment, causing them to turn and lose traction.As a consequence, the vehicle is spinning out of balance abruptly.
   - Driving through a puddle vehicle lose traction. As a consequence, vehicle can't keep its speed, leaving the driver and vehicle in danger of being hit by other vehicles.

- At a green light, if the driver try to accelerate on a slick road with traffic coming from behind.

  Traction control's usefulness is not reserved for off-road adventure. Variable temperatures and seasonal changes often lead to rapid weather changes and can take a tough toll on road conditions. Traction control can provide with additional support for safe driving in a variety of situations.

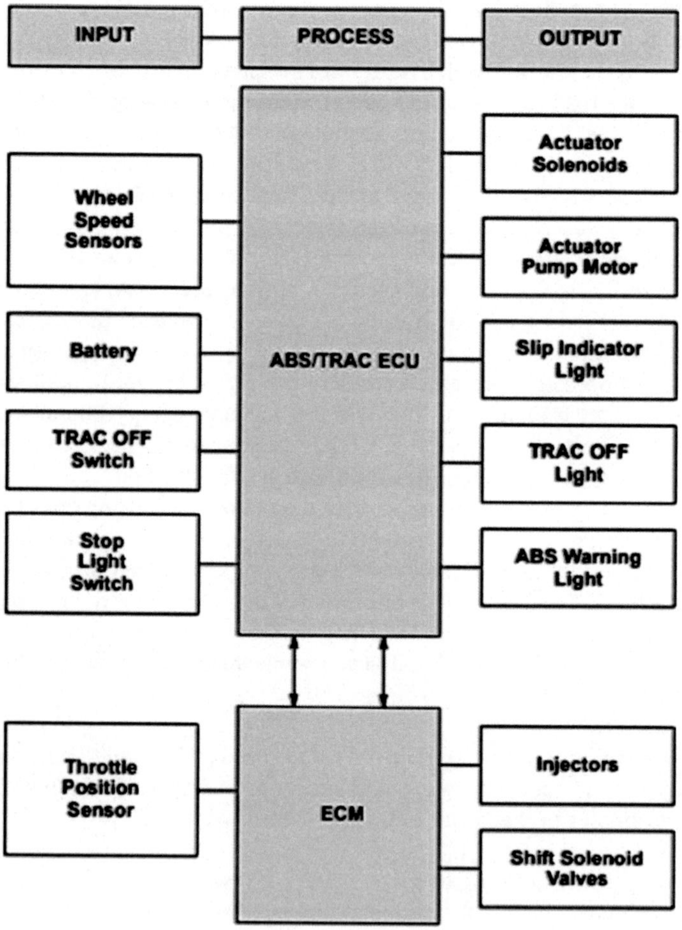

Fig. 1. Traction control system

2. Working

   Traction control operates similar to ABS and is often seen as an addition to existing ABS systems. Both systems work to solve opposite wheel slippage or wheel locking related problems. In fact, the traction control feature uses the same components as ABS in most modern vehicles, including wheel speed sensors (sensors that measure the wheel's rotational speed), hydraulic modulator (that applies the brakes) and ECU. Adding traction control to ABS requires adding a different valve to the hydraulic brake modulator. It is therefore relatively simple to install traction control on a vehicle that already has ABS.

The traction control uses individual wheel-speed sensors in order to measure the difference in rotational speed of each wheel. On each wheel, these sensors are located. When the ECU senses that one wheel spins faster than the other (an indicator that the wheel loses traction), it sends a message to the hydraulic brake-modulator (attached to the ECU) and automatically lowers the wheel speed that lowers the slip. Traction control mechanisms have different ways to reduce individual wheel rotational speed. Others "pump" the brake to the tire with less traction, while others pair wheel braking with lower engine speed. In the vehicle that uses less engine power to control slipping wheel rotation, when traction control is active, the driver may experience a gas pedal pulsation. This sound is identical to the sensation when ABS is working with the brake pedal pulsate (Fig. 2). The traction control systems operate to track ground speed and measure the rotational speed of the wheels of the vehicle once the wheels have regained traction [6, 7].

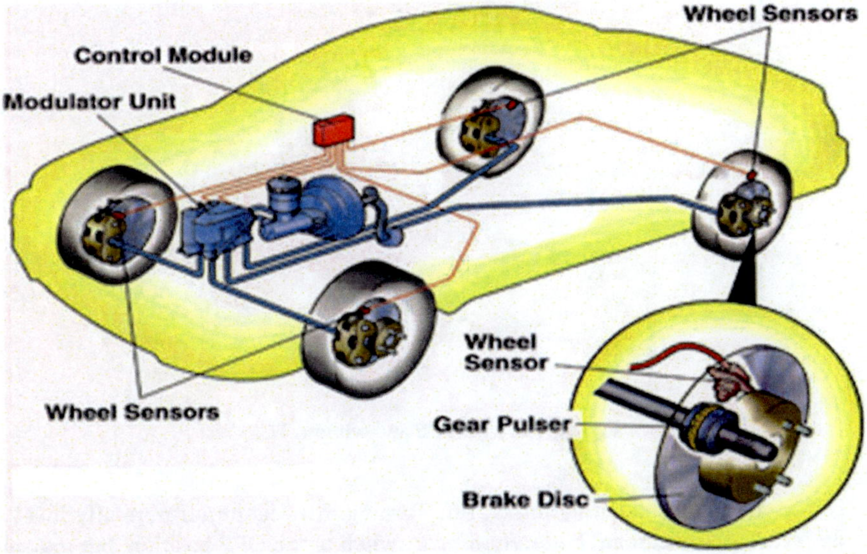

Fig. 2. Working of TCS

3. Effectiveness of traction control
Tests have shown that traction control is effective in reducing wheel slippage when accelerating under low-friction conditions, although this effect is more noticeable in four-wheel drive vehicles than in front-wheel drive vehicles. Furthermore, the same study found that traction control systems incorporating reductions in engine power to wheels with less traction are associated with slightly better stability, but that traction control systems are suitable for improving a vehicle's acceleration performance. The

efficacy of traction control has not been well established in minimizing or avoiding road accidents. Furthermore, owing to the repeated labeling of ABS and ESC, it is reasonable to suggest that driving a vehicle with this trio greatly reduces the risk of crashing. To illustrate, ESC has been found to decrease the risk of fatal single-vehicle crashes by nearly 50%, and reduce risk of rollovers by about 75%.However, due to the way drivers react to the safety system, there may be risks associated with traction control. For example, it has been argued that some of the conflicting data about the effectiveness of ABS is the result of how drivers act on both the functioning sound of their ABS setups and the recovered control of the steering. First, drivers uncertain about the sounds and sensations associated with the proper functioning of ABS, and this potential for confusion also has traction control consequences. Traction control systems make scraping sounds when it operates and the gas pedal will pulsate [8].

When drivers are unfamiliar with how their traction control operates, they might be mistaken for signals that traction control is somehow faulty, so drivers might disengage it. Secondly, because traction control allows drivers to retain steering power by stopping wheels from spinning, drivers must be careful not to exaggerate their steering commands in conditions where traction control operates, as these excessive commands can make the vehicle more difficult to control (Fig. 3).

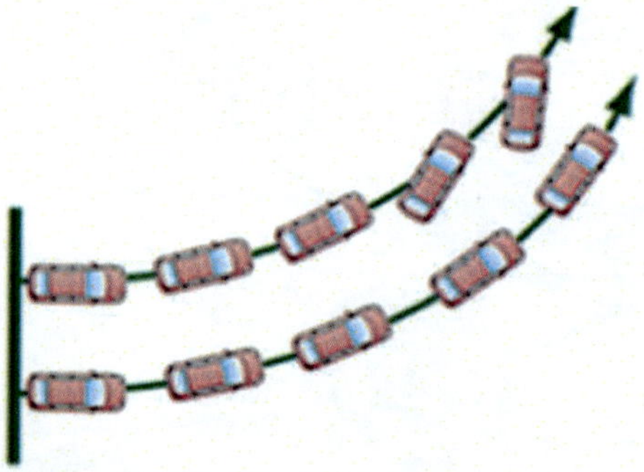

Fig. 3. Car with TCS and without TCS

The traction control can be turned off. Turning this feature off normally involves simply pressing a dashboard button/turning a switch to the OFF position, but may also involve pulling a fuse manually from inside or outside the vehicle. The manual of your owner will explain how to disengage the power of traction.

There are circumstances in which driver may want to switch off the traction control. Of starters, if driver already stuck in the snow, sometimes the only way forward is by "blasting out," i.e. pushing down heavily on the accelerator and making the wheels spin in hopes of gaining some traction. Unless traction control is off, the wheels will be stopped from turning and driver probably stay trapped. If driver turn off traction control, remember to turn it back on as soon as the situation is resolved which led to it being disengaged. Traction control systems was initially introduced on high-end vehicles in 1987, though early versions of traction control systems were equipped with some powerful rear-wheel drive vehicles in the early 17s [9, 10].

2 Conclusion

Traction control is generally equipped on any vehicle that has ABS because traction control was designed and built from existing ABS technology in terms of availability. Both ABS and traction control have been around for a long time compared with other safety features, and are available in the high-end and economical markets on a range of vehicles. Nevertheless, it doesn't automatically mean it has traction control just because the car has ABS. Only some older vehicles may have ABS. If driver is uncertain about ABS-equipped vehicle has traction control as well, he should check the manual of your driver. Traction control is usually offered as part of a larger protection kit, instead of as a stand-alone system. ABS, ESC and traction control are generally packaged together to equip drivers with the latest, most complementary braking technology. It can cost between $250.00 and $450.00.traction control system has come a long way forward for providing safety to every segment of vehicle, still there is a lot of improvement in this system.

References

1. P. Seiniger, K. Schröter, J. Gail, Perspectives for motorcycle stability control systems. Accid. Anal. Prev. (2012). https://doi.org/10.1016/j.aap.2010.11.018
2. K. L. Pfeiffer, Traction control, in *The Shakespearean International Yearbook*, (2018)
3. P. G. Dearman, Integrating traction protection and control, in *12th IET International Conference on Developments in Power System Protection, DPSP 2014*, (2014), pp. 1–3
4. P. Urda, J. A. Cabrera, J. J. Castillo, A. J. Guerra, An intelligent traction control for motorcycles, in *The Dynamics of Vehicles on Roads and Tracks - Proceedings of the 24th Symposium of the International Association for Vehicle System Dynamics, IAVSD 2015*, (2016), pp. 809–822, doi: https://doi.org/10.1201/b21185-86
5. J.-S. Hu, D. Yin, Y. Hori, F.-R. Hu, Electric vehicle traction control, *IEEEMIA*, no. December 2011, pp. 1–11, (2012)
6. M. M. Abdelhameed, M. Abdelaziz, N. E. Elhady, A. M. Hussein, Development of integrated brakes and engine traction control system, (2014). doi: https://doi.org/10.1109/REM.2014.6920246
7. C. Marathe, R. Annamalai, U. S. Karle, K. P. Venkatesan, "Development of adaptive traction control system," in *SAE Technical Papers*, vol. 5 (2013). doi: https://doi.org/10.4271/2013-26-0085
8. M. Richardson, Hybrid vehicles—The system and control system challenges, (2010). doi: https://doi.org/10.1049/ic.2010.0251
9. M. Spiryagin, P. Wolfs, F. Szanto, C. Cole, Simplified and advanced modelling of traction control systems of heavy-haul locomotives. Veh. Syst. Dyn. (2015). https://doi.org/10.1080/00423114.2015.1008016
10. D. Dujic et al., Power electronic traction transformer-low voltage prototype. IEEE Trans. Power Electron. (2013). https://doi.org/10.1109/TPEL.2013.2248756

Various Algorithms Used for Image Compression

Rishi Sikka(✉)

Department of Electronics and Communication Engineering, Sanskriti University, Mathura, Uttar Pradesh, India
rishisikka.ec@sanskriti.edu.in

1 Introduction

Images are expressed in form of pixels in an array for several of image analysis used in machine or computer vision and various other fields. The applications of image analysis in growing technology require a large amount of data to be stored and transferred for further use [1]. It is not possible to keep such large data stored and it becomes complex and difficult to transfer such data due to huge size so to overcome this issue a technique called compression of images is used. In image analysis this compression is known as image compression. Image compression is a kind of algorithm where the image gets reduced in size by encoding the images in a less number of bits. This reduction by encoding may lead to loss of relevant information, so it becomes important for the algorithms to be such as to save the data and transfer them in a much better way.

Image compression is one of the important techniques of image processing having many applications in image analysis due to its storage reduction and easy transfer of image information. The main objective of image compression is to keep all the information in an image safe while reducing the bit size of the image. Due to the rapid increase in technologies the large amount of data are used in a better manner by using various techniques available for compressing images. The research done in the field of image compression includes various steps like pre processing of the images, decision of the images, quantization and transformations, compression and decompression time.

Many algorithms are available to perform image compression including lossy [2] and lossless compression [3]. The images that are required to be compressed are the grayscale images with a pixel value ranging from 0 to 255. The initial stage of image compression includes converting the image from its spatial domain to an easy form for encoding the values in form of coefficients thus allowing a large amount of data to be compressed thus reducing huge computational costs and preventing the loss of data while transferring. The steps followed in image processing for performing image compression is as shown in Fig. 1.

Image compression is divided into two types, lossy compression [4] and lossless compression [5]. The lossy compression is applied on the images to reduce the size of image by eliminating the bits of the data which is not required or irrelevant data. The images might not always contain the data which is useful for image analysis, so that

F. P. García Márquez (ed.), *International Conference on Intelligent Emerging Methods of Artificial Intelligence & Cloud Computing*, Smart Innovation, Systems and Technologies 273, https://doi.org/10.1007/978-3-030-92905-3_24

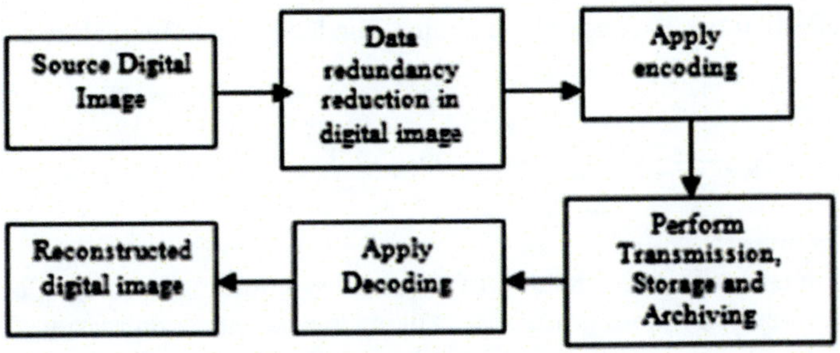

Fig. 1. Steps of image compression

irrelevant information is eliminated to reduce the bit size of the image in order to reduce the storage and complexity while transfer of data. For compressing the images, the steps involved in lossy compression include transformation and quantization of the images. Discrete cosine transform (DCT) [6] is used for the transformation, where the image is converted from its spatial domain to frequency domain for representing the image in a compact and reduced form. The step of quantization [7] of the image as shown in Fig. 2 is performed in order to represent the data of the image in a reduced form by the encoding the image followed by compression of the image. The lossy compression can be performed by methods like discrete cosine transform (DCT), wavelet compression etc.

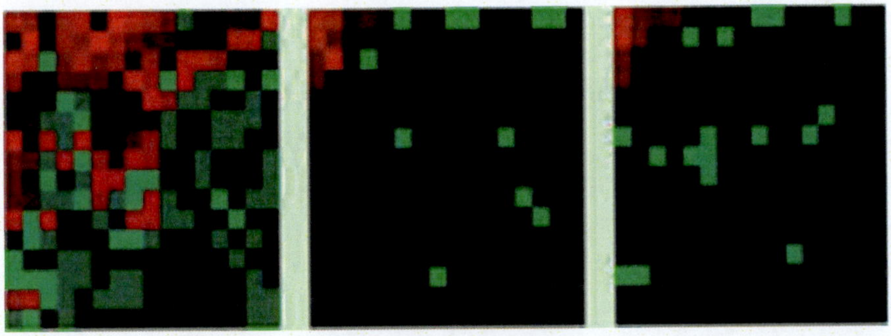

Fig. 2. Quantization of image

The lossless image compression is the technique where the data is not lost or eliminated but the same image as the initial is generated by reducing the bit size. The data in form of documents or executable files are reduced in bit size by compression techniques and then the data is decompressed to produce the same data but in a reduced form. The decompressed data may or may not be the exact replica of the raw data but there is no loss of information in the image but in terms of resemblance the images remain same. The lossless compression is also known as noiseless compression techniques as no noise signals are found in the images. The various techniques used in lossless compression of images include methods like arithmetic coding, run length encoding, Huffman encoding etc.

2 Various Algorithms of Image Compression

1. *Wavelet compression*
To fulfill the aim of storing the data of an image in a minimum storage, the technique applied for image compression is that of using the wavelet compression. The method of wavelet compression [8] can be found in either lossy or lossless compression. The use of wavelet compression is found suitable for images having components with high frequency, where the wavelet compression represents the transients of the components and these transients are reduced in size to reduce the bit size of the image to make it suitable for storage as well as transfer. For the processing of images using this algorithm a wavelet transform is applied to the image to generate the coefficients equivalent to all the pixel values of the image and then the coefficients are compressed. This step is known as transform coding following which the quantization of the coefficients is done and further the quantized values encoded by following either entropy or the run length encoding. The wavelet compression steps are as shown in Fig. 3.

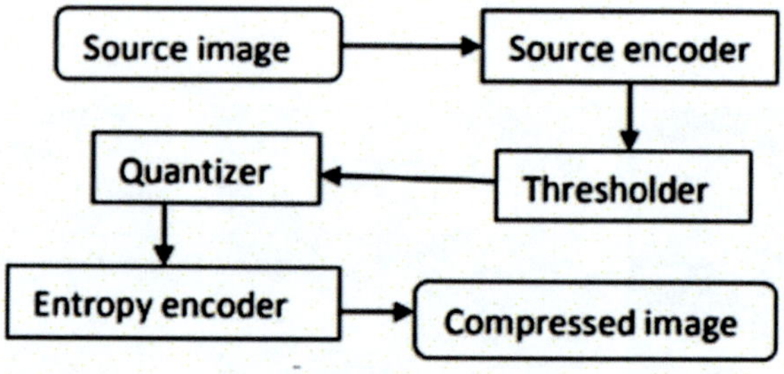

Fig. 3. Wavelet compression

Embedded zero wavelet (EZW) [9] is one of the best technique used in wavelet compression and is an image compression algorithm being a lossy compression technique.

2. *DCT based image compression*
Discrete cosine transform applied on the images for image compression is a lossy image compression technique. It uses two steps for compression of images including quantization and entropy coding of the coefficients. Quantization being the method to reduce the possible values of a image data by reducing the bits required for it. The entropy coding in the image compression is perform on the coefficients of the image generated after the quantization of the image. It represents the quantized data in a compact form where the data has been quantized by different levels of quantization. To measure the amount or the level of quantization obtained from the quantization

matrix, the step of entropy is used. The step used in image compression by following the technique of discrete cosine transform [10] is shown in Fig. 4.

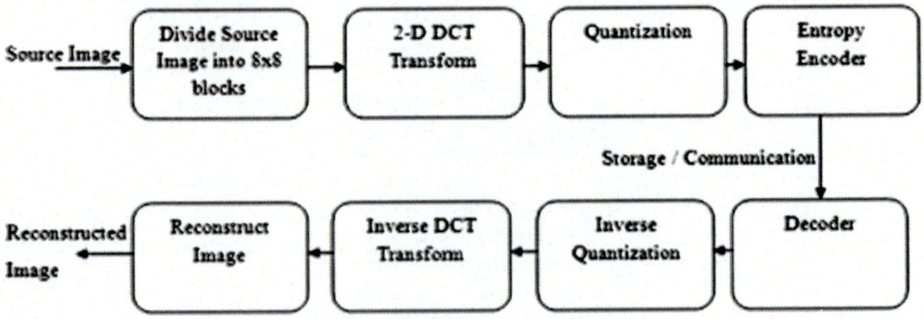

Fig. 4. DCT Compression

3. *Fractal compression*
   The type of image compression used in the images in the medical is the fractal compression which is a lossy compression. This approach or method as shown in Fig. 5 is used largely for the images with textures and natural images. The fractal algorithm [11] first partitions the image into parts in form of blocks known as range blocks. The image parts or the blocks generated are converted to fractal codes to recreate the image that has been encoded. The fractal code for the blocks known as range blocks consists of the coefficients produced by the quantization of the data. The decoding of the image is followed by the step of applying a compression transform on the blocks until all the decoded blocks are obtained known as domain blocks. The range and the domain blocks are as shown in Fig. 6.

4. *Huffman Compression*
   One of the most basic algorithms for image compression is known as Huffman encoding technique which is useful for the images and video compression and wok on the pixel intensity values of the data. The algorithm of Huffman encoding [12] depends on short code words and long code words. The input image in this algorithm is reduced to an ordered histogram value which determines the probability of the occurrence of the specific intensity pixel value and the total number of the pixels in the image. The applications of this algorithm are majorly in encoding of the data in the field of communication like music etc. The Huffman encoding as shown in Fig. 7 is used as the algorithm in lossless JPEG compression also which is majorly used in medical purposes. The medical equipments like MRI machines, ultrasound or CT machines etc. require this lossless algorithm to save the data without losing any information.

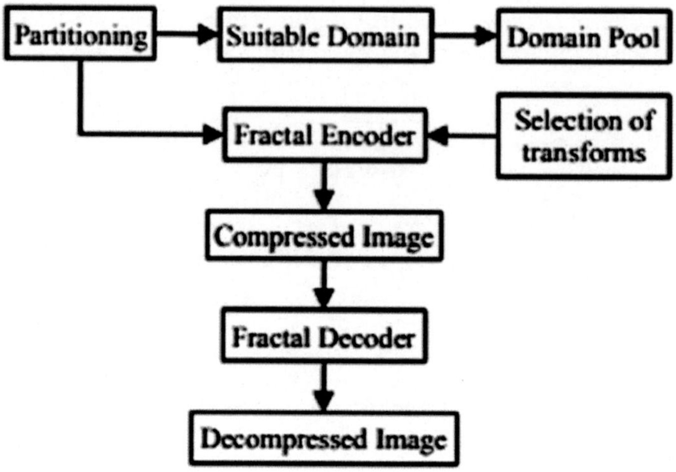

Fig. 5. Fractal compression

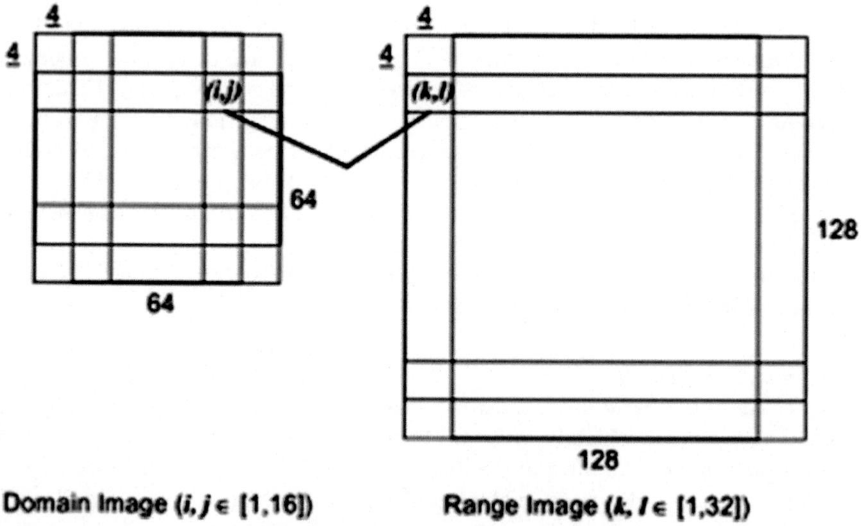

Fig. 6. Range block and domain block

3 Discussion

The technique of image compression in the image analysislies in the compressing of images or data so as to store the data in less space and reduce the complexity while transferring large amount of data. The main advantages are as discussed below:

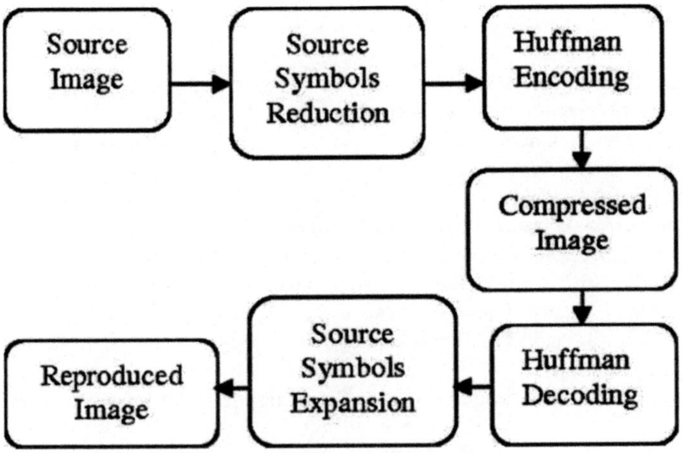

Fig. 7. Huffman compression

1. *Loss of data*
   The reduction of data or images leads to elimination of certain parts of the images or the data which can be recovered in the final image or data after the decompression. Therefore the image compression algorithm has the capability to restore all the data without losing any information and only reduces the size of the data.
2. *Reduction of size*
   The most significant advantage of image compression is the reduction of size which can be problematic in many applications due to complexity while transferring and storing data with large size.
3. *Speed of devices*
   Many electronic devices have to load the data in a fast speed which is possible only by image compression. Since the large amount of data is present in the devices the algorithms must be followed so that the faster loading of data and transfer from one device to another can be done efficiently.

4 Conclusion

Image compression being the technique to reduce the size of the images or the data is order to reduce the storage and the easy manner to transfer the data from one place to another. Various image compression techniques have been developed in research field which have been discussed in the paper. The image compression depending on the types of being lossy or lossless compression further utilizes various algorithms depending on the type of applications. The paper discusses about few important algorithms like wavelet transform compression and discrete cosine transform compression which are both lossy algorithms of image compression. The fractal compression is used in medical fields and belongs to the lossy algorithm of image compression. The Huffman coding is a type of lossless image compression algorithm used in medical fields and instruments where the loss of any kind of data or information is fatal like MRI, CT, or ultrasound images.

References

1. O. Rippel, L. Bourdev, Real-time adaptive image compression, in *34th International Conference on Machine Learning, ICML 2017*, 2017
2. L. Theis, W. Shi, A. Cunningham, F. Huszár, Lossy image compression with compressive autoencoders, in *5th International Conference on Learning Representations, ICLR 2017 - Conference Track Proceedings*, 2019
3. H. Malepati, Lossless Data Compression, in *Digital Media Processing*, 2010
4. S.E. Marzen, S. DeDeo, The evolution of lossy compression. J. R. Soc. Interface (2017)
5. A.J. Hussain, A. Al-Fayadh, N. Radi, Image compression techniques: a survey in lossless and lossy algorithms. Neurocomputing (2018)
6. R.J. Cintra, F.M. Bayer, A DCT approximation for image compression. IEEE Signal Process. Lett. (2011)
7. M.H. Horng, Vector quantization using the firefly algorithm for image compression. Expert Syst. Appl. (2012)
8. D. Gupta, S. Choubey, Discrete wavelet transform for image processing. Int. J. Emerg. Technol. Adv. Eng. (2015)
9. G. Chopra, A.K. Pal, An improved image compression algorithm using binary space partition scheme and geometric wavelets. IEEE Trans. Image Process. (2011)
10. R. A.M, K. W.M, E. M. A, W. Ahmed, jpeg image compression using discrete cosine transform - a survey. Int. J. Comput. Sci. Eng. Surv. (2014)
11. J. Wang, N. Zheng, Y. Liu, G. Zhou, Parameter analysis of fractal image compression and its applications in image sharpening and smoothing. Signal Process. Image Commun. (2013)
12. M. Sharma, Compression Using Huffman coding. IJCSNS Int. J. Comput. Sci. Netw. Secur. (2010)

Wavelet Transformation for Digital Watermarking

Laxmi Goswami(✉)

Department of Electronics and Communication Engineering, Sanskriti University, Uttar Pradesh, Mathura, India
laxmigoswami.ee@sanskriti.edu.in

1 Introduction

In the modern technology, increased used of digital data and information along with the transmission of digital data through internet has become a growing field. Many processing fields like image processing and signal processing are utilized for protecting the digital data while storing as well as transmitting the data in real time. The data and information available in digital form require to be secured against illegal transfer and redistribution thus protecting all the available information from many fraud activities. A wide research has been carried out in the field for safe guarding digital data like encrypting solutions and products, strong authentication, advanced digital signatures [1], SSL certificates [2] etc. One of the methods to secure digital data and information includes the use of digital watermark [3] or digital signature which is added to the information in either visible or invisible mode and further is received at the end of receiver for validating the authenticity of the data.

Digital watermarking is a technique to secure the data or information by embedding the watermark within the digital form of the data. The watermarks in the digital data determine the ownership of the authorized data. The digital watermarking is used to safeguard data and information which has become hard to protect due to increase in internet traffic and digitization of the documents. It is used for copyright protection [4] of the data like images, audio, video, texts etc. available in form of digital data. The use of digital watermark is done to avoid the situations of copyright infringements or violations of the rule of copyright ownership. Also watermarking is a method to reduce the piracy of data as ownership is been declared by the use of watermarks. Various research works are going in the field of protecting the data by seeking encryption methods which efficiently secure information along with many data hiding techniques which gives a robust result.

Many techniques are available for digital watermarking working on the basis of spatial domain as well as frequency domain. The frequency based techniques[5] are mostly used in the field of watermarking due to better results for human visual and machine vision systems with the help of spectral coefficients. The techniques which are frequency based gives robust results on using wavelet transform [6] as the method of watermarking. Many wavelet transform techniques are available like Haar wavelet transform [7], Symlet wavelet transform [8], Coiflet wavelet[9] transform etc. These

F. P. García Márquez (ed.), *International Conference on Intelligent Emerging Methods of Artificial Intelligence & Cloud Computing*, Smart Innovation, Systems and Technologies 273, https://doi.org/10.1007/978-3-030-92905-3_25

watermarking process consists of processing like encoding and decoding of the data or the information to which the techniques are applied. The encoding process works by decomposition of the data into frequency domain and adding the watermark to the decomposed image. The decoding process of the data or the information works using the inverse wavelet transformation by using 2D inverse wavelet transform [10]. The process of digital watermarking using the wavelet transforms is as shown in Fig. 1.

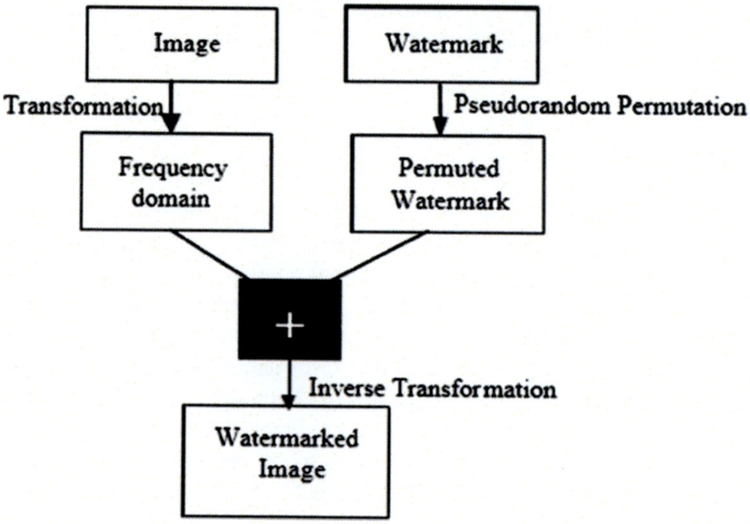

Fig. 1. Watermarking by wavelet transforms

2 Wavelet Transform Techniques

The techniques used fordigital watermarking based on the frequency domain are efficient as they give better and efficient results for human visual and machine vision systems. The wavelet transforms based on frequency domain are classified into categories like discrete wavelet transform (DWT) and discrete cosine transform (DCT). Discrete cosine transform (DCT) [11] aremethods that converts a signal into frequency terms and generate the frequency coefficients. The frequency components are robust when compared to the spatial components due to its efficiency in comparison to the spectral coefficients. The discrete wavelet transform used for processing the images decompose the images and make a multi resolution perspective of the image. The method of creating a multi resolution of the image is advantageous in the process of watermarking enables the watermark to be embedded in any of the frequency bands and inverse wavelet transform is applied to process the watermarking on all the frequency bands being high or low frequencies as shown in Fig. 2.

Discrete wavelet transform depends on the wavelets of varying frequencies and represents the data in other form by securing all the information in the data. DWT

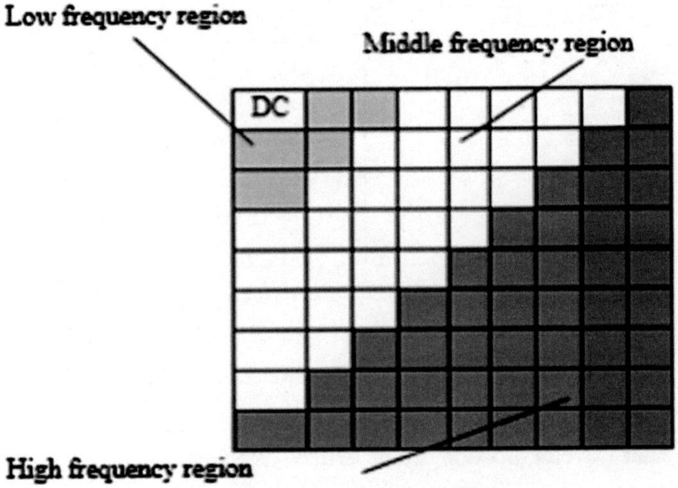

Fig. 2. Frequency coefficients of DCT

being an important and useful process in the field of image analysis provides with a better visual quality of the images. The discrete wavelet transform can be used in both frequency as well as time domain adding to the advantages. In the method of discrete wavelet transforms the images are decomposed in sub parts based on different frequency bands denoted as LL, LH, HL, HH as shown in the Fig. 3.

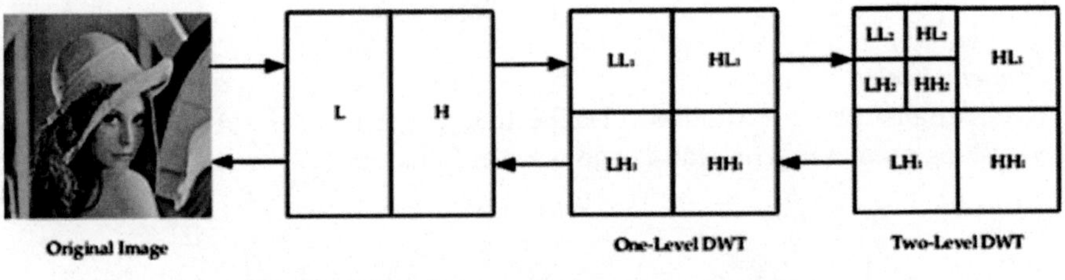

Fig. 3. DWT decomposition and reconstruction

Various techniques are used in the discrete wavelet transform like Haar wavelets, Coiflet wavelets, Symlet wavelet transforms etc. as discussed further.

1. Haar wavelet
   Haar transform is widely used in processing of images and recognition, mostly for 2D signals due to their square shaped or wavelets like structure [12] as shown in Fig. 4.

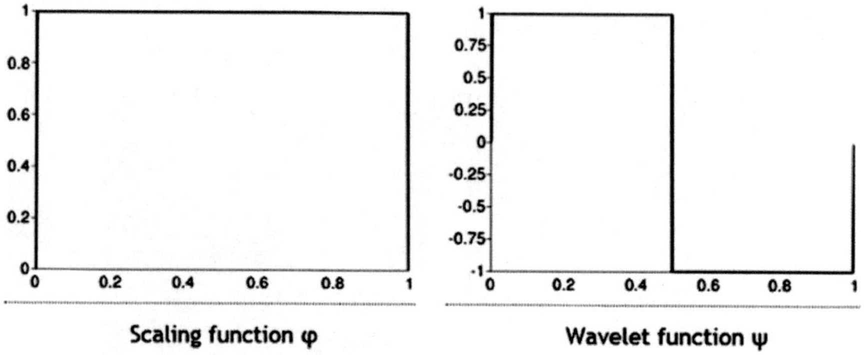

Fig. 4. Haar wavelet

The Haar wavelet function can be defined as below:

$$\psi(t) = \begin{cases} 1 & 0 \leq t < \frac{1}{2} \\ -1 & \frac{1}{2} \leq t < 1 \\ 0 & \text{otherwise} \end{cases}$$

The scaling function is defined as:

$$\varphi(t) = \begin{cases} 1 & 0 \leq t < 1 \\ 0 & \text{otherwise} \end{cases}$$

2. *Symlet wavelet*
   Symlet transforms are similar to Daubechies wavelet transform having maximum vanishing moments. The scaling and wavelet functions [12] are as shown in Fig. 5.

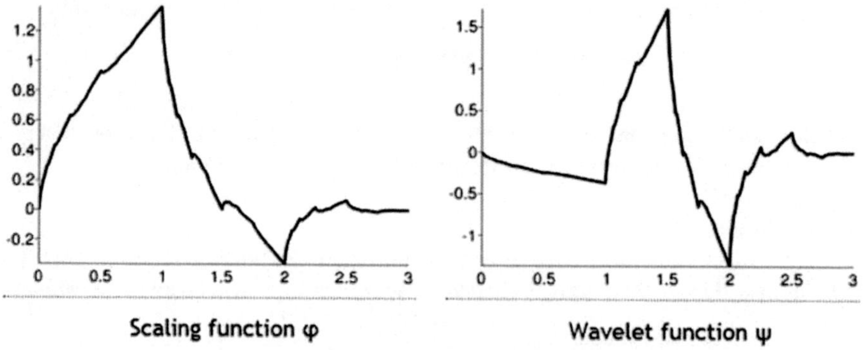

Fig. 5. Symlet wavelet

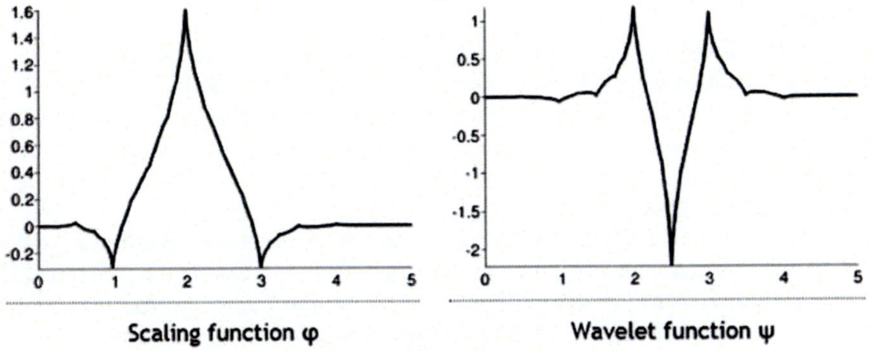

Fig. 6. Coiflet wavelet

3. *Coiflet wavelet*
   Coiflet waveforms are in discrete form having the scaling function and wavelet function with vanishing moments [12] as shown in Fig. 6.

The image after digital watermarking applied to the image is as shown in Fig. 7 along with the digital watermark which is embedded in the image and the original image.

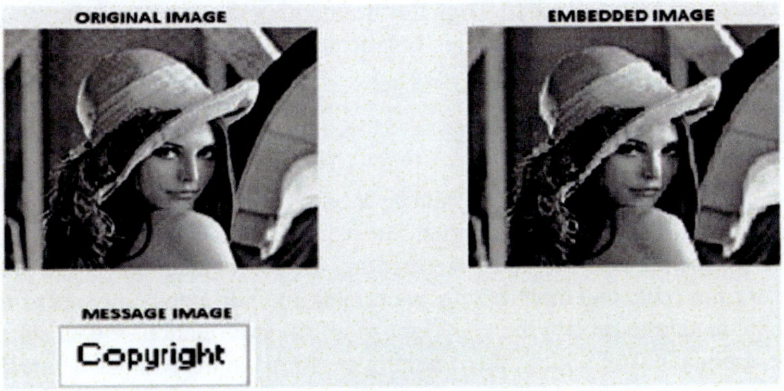

Fig. 7. Digital watermarking

3 Applications of Digital Watermarking

1. Data authentication
   The data or the information available on the internet can be embedded with a digital watermark to avoid any change of data made illegally thus maintaining the actual work. If the information is tampered, due to the watermark the pixel values of the data do not change and the data if copied without permission will not contain the digital watermark thus making the data unauthorized. The example to this is cryptography [13] which includes digital signatures for authentication the data.

2. Medical
   The medical reports of the patient like CT scans, MRI or X-ray needs to be imprinted with patient's details. It is done along with a digital watermarking technique to prevent the reports from getting tampered[14]. Also the digital watermarking makes it easy for the reports to be identified for providing the required treatment to right patient.
3. Copyright protection
   To protect the data from misuse and unauthorized access the owner embeds the data with digital watermark as it provides the users identity depicting the copyright [15] of the information and thus if it is illegally used elsewhere the owner can seek remedy by proving the copyright ownership through the digital watermark used.
4. Source tracking
   The embedded data or information when transferred by the owner to some other source, it contains a different watermark for each of the source where data is being transferred. The different watermarks embedded for different sources provide facility to track[16] the data being shared and thus protecting the data from being lost or misused further.
5. Broadcast monitoring
   The digital watermarking used in the content to be transmitted makes sure about the authenticity of the data transmitted. The timing of the broadcast and the content along with duration can be easily monitored by the digital watermarking done to the contents to be broadcasted. Digital watermarking can automatically monitor the information to be transmitted during broadcast.

4 Conclusion

Various methods and techniques are available for protecting the illegal and unauthorized access to the data shared through internet. The paper discusses the technique of embedding the information with a digital watermarking to prevent such issues and protecting all the data from fraud and theft. Digital watermarking is an important area of research which is applicable to protect the copyright of all the information or data stored digitally. The applications of digital watermarking are found in many fields where the need to safeguard data is required like broadcasting, medical, copyright protection, ownership etc. Many techniques are available for digital watermarking based on spatial and frequency domain, among which the techniques discussed in the paper are based on frequency domain known as discrete wavelet transforms (DWT). Discrete cosine transform can also be used but due to certain advantages of DWT over DCT, discrete wavelet transform is more preferred. The methods like Haar wavelet, Symlet wavelet, and Coiflet wavelet transforms are generally used in digital watermarking to protect the ownership of data in electronic form.

References

1. A. Biryukov et al., Digital signature standard, in *Encyclopedia of Cryptography and Security* (2011)
2. M. Almishari, E. De Cristofaro, K. El Defrawy, G. Tsudik, Harvesting SSL certificate data to identify web-fraud. Int. J. Netw. Secur. (2012)
3. M. Barni, S. Katzenbeisser, Digital watermarking, in *Handbook of Financial Cryptography and Security* (2010)
4. M. Hemis, B. Boudraa, Digital watermarking in audio for copyright protection, in *Proceedings—ICACSIS 2014: 2014 International Conference on Advanced Computer Science and Information Systems*, (2014)
5. P. Parashar, R. K. Singh, A survey: digital image watermarking techniques. Int. J. Signal Process. Image Process. Pattern Recognit. (2014)
6. V. S. Jabade, "Literature review of wavelet based digital image watermarking techniques. Int. J. Comput. Appl. (2011)
7. J. Musić, I. Knezević, E. Franca, Wavelet based watermarking approach in the compressive sensing scenario, in *Proceedings—2015 4th Mediterranean Conference on Embedded Computing, MECO 2015 - Including ECyPS 2015, BioEMIS 2015, BioICT 2015, MECO-Student Challenge 2015*, (2015)
8. G. Zhao, H. Li, H. Y. Huang, M. H. Gao, A new digital audio watermarking algorithm based on DWT, in *2011 2nd International Conference on Artificial Intelligence, Management Science and Electronic Commerce, AIMSEC 2011—Proceedings*, (2011)
9. K.R. Kakkirala, S.R. Chalamala, Block based robust blind image watermarking using discrete wavelet transform, in *Proceedings—IEEE 10th International Colloquium on Signal Processing and Its Applications. CSPA* **2014**, 2014 (2014)
10. D. Gupta, S. Choubey, Discrete wavelet transform for image processing. Int. J. Emerg. Technol. Adv. Eng. (2015)
11. D. Rosiyadi, S. J. Horng, P. Fan, X. Wang, M. K. Khan, Y. Pan, Copyright protection for e-government document images. IEEE Multimed. (2012)
12. P. Fryzlewicz, Wavelet methods, *Wiley Interdisciplinary Reviews: Computational Statistics.* (2010)
13. P. Gupta, Cryptography based digital image watermarking algorithm to increase security of watermark data, Int. J. Sci. Eng. Res. (2012)
14. S. Mehta, R. Nallusamy, R. V. Marawar, B. Prabhakaran, A study of DWT and SVD based watermarking algorithms for patient privacy in medical images, in *Proceedings—2013 IEEE International Conference on Healthcare Informatics, ICHI 2013*, (2013)
15. S. G. Hombal, K. N. Prasad, Digital copyright protection: issues in the digital library environment. DESIDOC J. Libr. Inf. Technol. (2012)
16. J. Nin, S. Ricciardi, Digital watermarking techniques and security issues in the information and communication society, in *Proceedings—27th International Conference on Advanced Information Networking and Applications Workshops, WAINA 2013*, (2013)

Singular Value Decomposition Based Image Compression

Laxmi Goswami(✉)

Department of Electronics and Communication Engineering, Sanskriti University, Mathura, Uttar Pradesh, India
laxmigoswami.ee@sanskriti.edu.in

1 Introduction

With increased technology, advancement in digital images and video has increased tremendously. The digital products like cameras, cell phones etc. enable the storage of large amount of data in form of images and videos which is beneficial in many areas of applications like medical, security, surveillance, etc. The applications require a better quality of data to be stored to getbetter results which leads to use of high definition images and videos further utilizing high amount of storage. To store and transmit large amount of data complex systems are needed and requires more cost which is disadvantageous in the increased technology. For overcoming such difficulties compression of images is done to get a reduced sized image, which can be done by using either lossy compression techniques or lossless compression techniques.

The compression techniques [1] are known as lossy techniques [2] because certain data is removed from the images which are not of much relevance and thus reduce the size by losing some information from the images. This loss is however not visible and does not leads to any transformation in the visibility of the images. Thus making the images tolerable and easy to store and transmit. Whereas the lossless techniques [3] are the ones where pixel size is reduced to reduce the size of the images. No information is removed from the images and thus it is lossless as compared to the lossy techniques.

The technique addressed in the paper is lossy compression technique which uses singular value decomposition method along with it to get better image compression. The Eigen vectors of the image to be reduced in size is calculated and then further processed by portioning the image in form of small blocks [4]. The method of singular value decomposition used alone gives data with redundancy and hence is not beneficial alone for image compression. To overcome these disadvantages of the method certain other techniques can be merged together to get better results. Hence to compress the images, Eigen values and partitioning of the images in form of blocks is done along with the use of singular value decomposition [5].

Compression of images is important for storage as well as transmission of the images and the aim of using compression is to produce an image in a compact form thus making it reliable for further processing. Compression is done to remove the high frequency components in order to get a smooth image by the low frequency components [6]. The

F. P. García Márquez (ed.), *International Conference on Intelligent Emerging Methods of Artificial Intelligence & Cloud Computing*, Smart Innovation, Systems and Technologies 273, https://doi.org/10.1007/978-3-030-92905-3_26

LL sub band which is the approximation band is useful for getting the smooth image and is useful component in the compression of images. Other sub bands contains high frequency components and represents the vertical, horizontal and the diagonal details which are not required in the compression of images to get more accurate results.

Image compression methods are based on reducing redundancies in the data representation. The redundancies present in images can be divided into three categories. The coding redundancy refers to the use of a number of bits greater than absolutely necessary to represent the intensities of the pixels of the image. The interpixel redundancy is associated with the relation or similarity of neighboring pixels. Finally, the psycho visual redundancy concerns the imperceptible details of the human visual system.

2 Image Compression

The Fig. 1 represents the method of image compression applied on the images. The figure shows the two steps followed being image compression methodology and decompression methodology. The compression algorithm has as input an image and for output the compressed image is represented by the singular value decomposition matrices in a binary file [7]. The decompression algorithm has as input the binary file containing the decomposition matrices and, as output, an image.

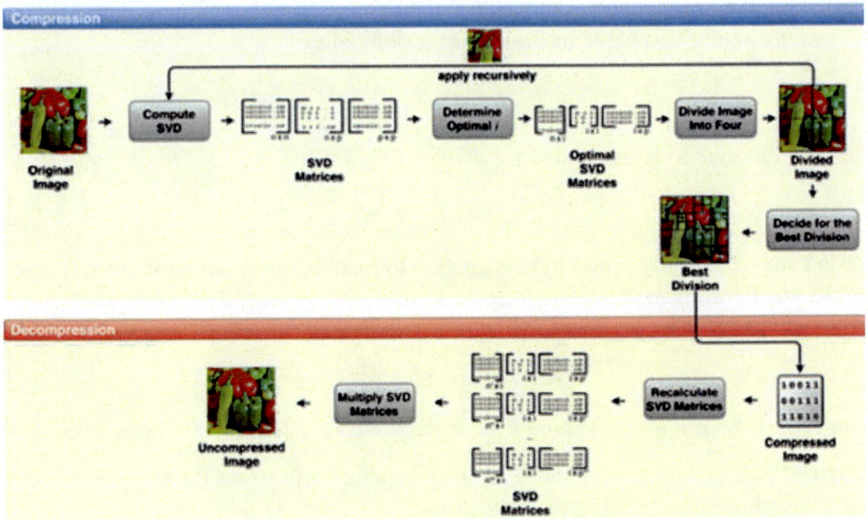

Fig. 1. Compression and decompression

3 Methodology

The steps followed in the methodology are as following:

The singular value decomposition (SVD) for the entire image is calculated as the first step. The singular value decomposition of the real matrix can be given by:

$$A = U\sum V^T$$

where U is a real unit matrix, Σ is a diagonal rectangular matrix with diagonal nonnegative real numbers and V is a real unit matrix. Furtherthe real matrix can be represented as:

$$A = U\sum V^{\mathbf{T}} = \begin{bmatrix}\mathbf{u}_1\ \mathbf{u}_2\ \ldots\ \mathbf{u}_n\end{bmatrix}\begin{bmatrix}\sigma_1 & 0 & \cdots & 0\\ 0 & \sigma_2 & \cdots & 0\\ \vdots & \vdots & \ddots & \vdots\\ 0 & 0 & \cdots & \sigma_n\end{bmatrix}\begin{bmatrix}\mathbf{v}_1^{\mathbf{T}}\\ \mathbf{v}_2^{\mathbf{T}}\\ \vdots\\ \mathbf{v}_n^{\mathbf{T}}\end{bmatrix}$$

$$= \mathbf{u}_1\sigma_1\mathbf{v}_1^{\mathbf{T}} + \mathbf{u}_2\sigma_2\mathbf{v}_2^{\mathbf{T}} + \cdots + \mathbf{u_n}\sigma_n\mathbf{v}_n^{\mathbf{T}}$$

The use of arrays in the singular value decomposition makes the matrix larger in size than the size of the original matrix. The use of lossy compression [8] in the paper leads to calculation of Eigen values in the matrix and then removing some Eigen values to make the size of the matrix smaller.

Thus, the optimized value of the matrix is defined as:

$$\text{opt\_ value} = \arg\max_i(\alpha - 1)(\lambda_{\text{i}}) + (\alpha)(\delta_i)$$

where the variance λi is calculated as:

$$\lambda_i = \lambda_{i-1}^{-} + \sigma_i^*$$

Next step is to apply linear transformation to map the original interval to a range of 0 to 255 and defining as:

$$g = \frac{g_{\max} - g_{\min}}{f_{\max} - f_{\min}}(f - f_{\min}) + g_{\min}$$

where f_{min} is the minimum value of the matrix and f_{max} is the maximum value of the matrix and g_{min} and g_{max} are 0 and 255. The variable f represents the original value and g the value obtained after the transformation.

The next step in the methodology being the partitioning of the image in form of blocks of a certain size but the blocks are all of the same size. Theses blocks are applied with the image compression technique individually on all the blocks to compress the entire image. This process is performed until the desired size of the image is achieved. The step followed here produces the images in form of a quad tree decomposition [9] with the nodes at inner side representing the singular value decomposition.

The Fig. 2 represents the steps followed in the compression of the images, where the (a) represents the original image in binary form. (b) represents the partitioning of

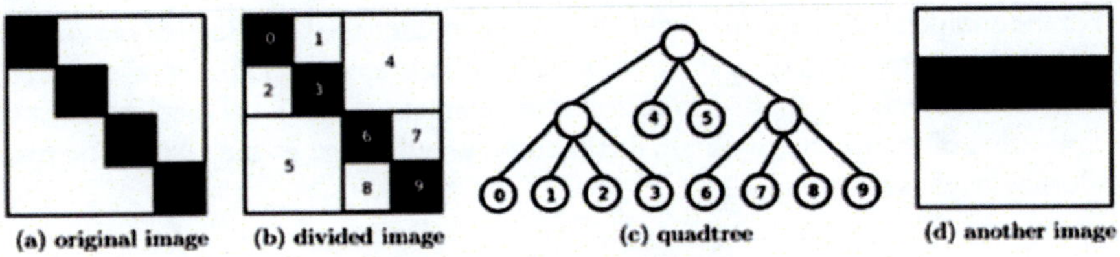

Fig. 2. Steps of compression

the image in blocks, (c) represents the quad tree of the images in the blocks and (d) represents the image after application of singular value decomposition.

The partitioning of the image is done in form of blocks which is chosen if the optimized value satisfies the below condition.

$$opt\_value > \frac{\sum_{j=0}^{4} opt\_value_j}{4}$$

where opt_value is the optimal value for the entire image and opt_value_j is the optimal value for the j block.

For colored images, decomposition [10] is calculated separately for each color channel and the compressed image is stored in a binary file. The protocol used to store the colored image in form of global header and local data is shown in Fig. 3. The binary file starts at a global header. The header stores data containing the width and height of the images and the number of channels of the image. They determine the size of each node in the tree, which is necessary to correctly retrieve the data in the decompression step.

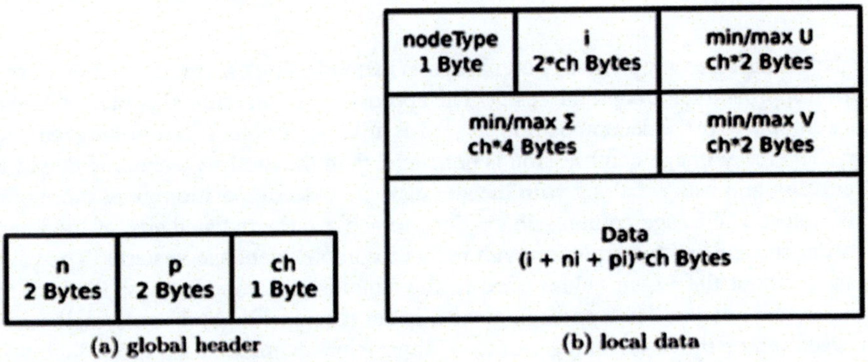

Fig. 3. Decomposition of colored images

The first information in the header is internal or leaf node. For internal nodes 4 sub bands are availableand for the leaf nodes, besides the node type, the rank value r, the minimum and maximum values of each of the matrices are present.

For the compressed image file is read for reconstructing the tree. Each of the singular value decomposition matrixes a linear transformation is applied on the image and nest transformation inverse is applied to restore the decimal values. Final step includes the multiplication of the matrix to obtain the uncompressed image as the result. The result of colored image is shown in Fig. 4.

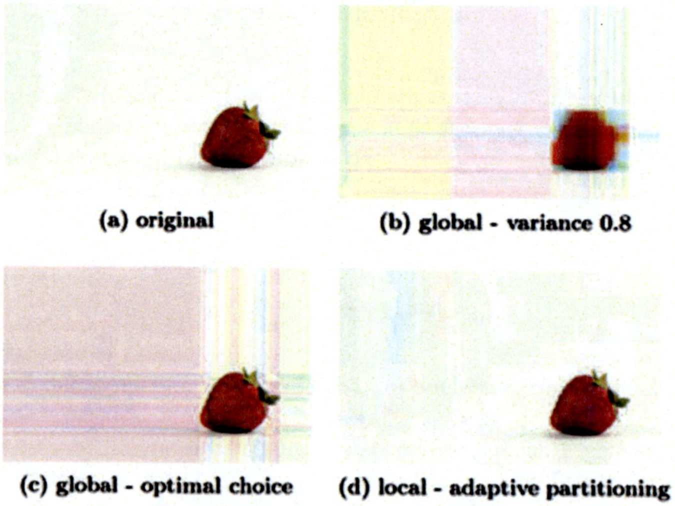

Fig. 4. Compression of colored image

4 Conclusion

The paper addresses a lossy image compression method using the singular value decomposition approach for an optimal result. The approach uses an optimal choice of Eigen-values computed in the decomposition, as well as utilizing the block partitioning methodology. The lossy image compression is beneficial than the lossless technique due to its better result and using it along with the singular value decomposition gives more optimized value. The image compression is done to achieve the reduced size of the image for better storage and transmission with reduced complex data and systems. The paper discusses about the Eigen values used in the singular decomposition applied on the blocks of the images. The approach gives a better result with good compression ratio thus overcoming the disadvantages of the compression technique used individually.

References

1. O. Rippel, L. Bourdev, Real-time adaptive image compression, in *34th International Conference on Machine Learning, ICML 2017*, (2017)
2. L. Theis, W. Shi, A. Cunningham, F. Huszár, Lossy image compression with compressive autoencoders, in *5th International Conference on Learning Representations, ICLR 2017—Conference Track Proceedings*, (2019)
3. H. Malepati, "Lossless Data Compression," in *Digital Media Processing*, 2010.
4. I. K. Kim, J. Min, T. Lee, W. J. Han, J. H. Park, Block partitioning structure in the HEVC standard. IEEE Trans. Circuits Syst. Video Technol. (2012)
5. L. Cooper, Singular-value decomposition. J. Empirical Generalisations Mark. Sci. (2013)
6. K. Gregor, F. Besse, D. J. Rezende, I. Danihelka, D. Wierstra, Towards conceptual compression, in *Advances in Neural Information Processing Systems*, (2016)
7. M. Lebrun, A. Leclaire, An implementation and detailed analysis of the K-SVD image denoising algorithm. Image Process. Line. (2012)
8. A. Furqan, M. Kumar, Study and analysis of robust DWT-SVD domain based digital image watermarking technique using MATLAB, in *Proceedings—2015 IEEE International Conference on Computational Intelligence and Communication Technology, CICT 2015*, (2015)
9. S. Aouat, S. Larabi, Object retrieval using the quad-tree decomposition. J. Intell. Syst. (2014)
10. R. Gutjahr et al., Material decomposition and virtual non-contrast imaging in photon counting computed tomography: an animal study, in *Medical Imaging 2016: Physics of Medical Imaging*, (2016)

Wavelet Transform for Signature Recognition

Manoj Ojha(✉)

Department of Electronics and Communication Engineering, Sanskriti University, Uttar Pradesh, Mathura, India
manoj@sanskriti.edu.in

1 Introduction

The biometric systems work with the principle of recognition of physical human traits used to authenticate the identity of person. The human traits used for authentication depends on the biometrics of facial features, iris, finger prints etc. biometric recognition [1] also uses certain other characteristics like voice, signature or other behavioral possessions. Among the available behavioral traits of human signature recognition is an important field in the area of research as it provides the better results in authentication of a person by using biometric recognition systems. Biometric authentication has many disadvantages as many of the biometrics of a human can be forged and leads to risks thus signature recognition becomes advantageous over the physical characteristics used in authentication of the identity of the person. The signature authentication [2] is a secure form of biometrics as the electronic approach of signature recognition not only depends on the signature captured but also measures the way of creating a signature with characteristics like pressure or the speed and way of forming the signature which cannot be copied by other humans. The approach of signature recognition [3] is widely used in many areas having vast application like that of legal works, banks, user verification in surveillance, security, applications used in mobile systems etc.

Recognition of certain features from the images or real time input given electronically uses the field of image analysis and image processing along with machine vision. Many methods and techniques are available for the processing of images depending on the spatial or the frequency based approaches and provide with an efficient result. The wavelet transforms being a tool used to represent the signals in a way to showcase the characteristics of the signal as a function of both frequency as well as time. The wavelet transforms are used for compressing the data, reducing noise etc. and frequency based wavelet transforms are used by working on a specific frequency band based on the features or the characteristics depicted by the signal. The frequency based discrete wavelet transform approach gives a robust result in the area of signature recognition as they work well with the systems using machine vision and image processing. The discrete wavelet transform extracts the relevant features from the input signatures along with the certain specific characteristics like pressure and speed used to create the signature. Various image processing techniques like pre processing of the data, image registration, and extraction of relevant features, matching and recognition are used in the processing of the signature for signature recognition. The discrete wavelet transforms [4] is used

F. P. García Márquez (ed.), *International Conference on Intelligent Emerging Methods of Artificial Intelligence & Cloud Computing*, Smart Innovation, Systems and Technologies 273, https://doi.org/10.1007/978-3-030-92905-3_27

in the step of feature extraction for the processing of signature recognition. Discrete wavelet transform depends on the wavelets of varying frequencies and represents the data in other form by securing all the information in the data. DWT being an important and useful process in the field of image analysis provides with a better visual quality of the images. The discrete wavelet transform can be used in both frequency as well as time domain adding to the advantages.

2 Discrete Wavelet Transform

Discrete wavelet transform depends on the wavelets of varying frequencies and represents the data in other form by securing all the information in the data. DWT being an important and useful process in the field of image analysis provides with a better visual quality of the images. The discrete wavelet transform can be used in both frequency as well as time domain adding to the advantages. The image used for signature recognition being 2D signals, two dimensional discrete wavelet transform is applied on the signature by applying transformation in terms of rows and columns. In the method of discrete wavelet transforms the images are decomposed in sub parts [5] based on different frequency bands denoted as LL, LH, HL, HH, where the a low pass band demoted as low-low(LL) contains the approximations of the raw input image and other frequency bands for extracting the details of the image in horizontal, vertical and diagonal manner as shown in Fig. 1.

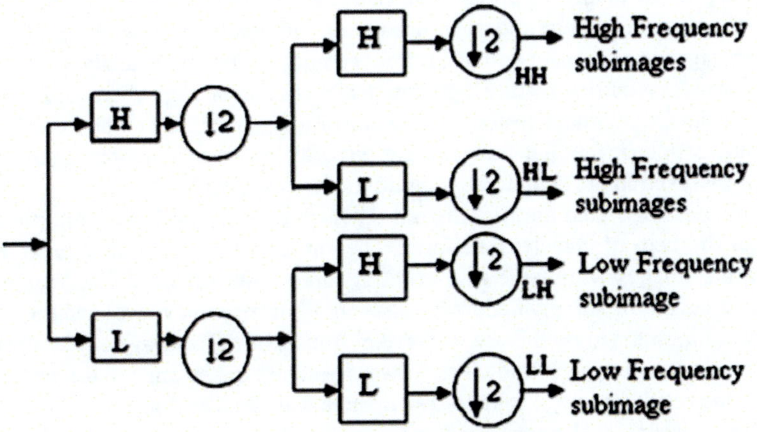

Fig. 1. Frequency bands in DWT

The wavelet decomposition using shifting and scaling parameters [6] is performed for better analysis of the signature images used for signature recognition. The wavelet transform used can be expressed as:

$$\psi_{a,b}(t) = \psi\left(\frac{t-b}{a}\right)$$

$$W_{\psi}x(a,b) = \frac{1}{\sqrt{a}} \int_{-\infty}^{+\infty} x(t)\psi*\left(\frac{t-b}{a}\right)dt$$

where the denotations of 'a' and 'b' are the shifting and scaling parameters respectively.

The discrete wavelet transform of the image involves using the low pass filters having a impulse response with the convolution of the signals as shown below:

$$y[n] = (x^{*}g)[n] = \sum_{k=-\infty}^{\infty} x[k]g[n-k]$$

3 Methodology

The processing of signature recognition includes many steps like acquiring data, pre-processing of the images, signature image registration, feature extraction and verification. The data is acquired by giving a signature as raw input to the system electronically and converting it in the digital form. The pre processing step of the signature recognition is done by removing noise from the signature captured as an image by the electronic means. The noise id removed by using a filter. After the pre processing of signatures, the next step is for registration of the signature. Signature registration or image registration [7] being a process to align multiple images of the original image where, one image is taken as the reference image and transformations and other processes are performed on the other images in alignment with the reference image. The image analysis in signature recognition tasks require signature registration as it is a crucial step where the information like change detection, image restoration, fusion of images etc., are combined altogether. The alignment done in the signature registration collects all the signatures in same format. The steps involved in the signature recognition are as shown in Fig. 2.

The images registered are first broken down or decomposed into multiple smaller parts with the help of wavelet transform. The sub division of the images is done by selecting the images with same resolution at same point of level. The fusion of the images is done on the sub divided images having high frequency level and the resultant image is again obtained by the use of wavelet transforms. The wavelet transform used for subdividing the image is known as forward wavelet transform and the one used for obtaining the resultant image is known as inverse wavelet transform.

The two dimensional image used for signature recognition uses two dimensional discrete wavelet transform on the signature by applying transformation in terms of rows and columns. In the method of discrete wavelet transforms the images are decomposed in sub parts based on different frequency bands denoted as LL, LH, HL, HH, where the a low pass band demoted as low-low (LL) contains the approximations of the raw input image and other frequency bands for extracting the details of the image in horizontal, vertical and diagonal manner as shown in Fig. 3.

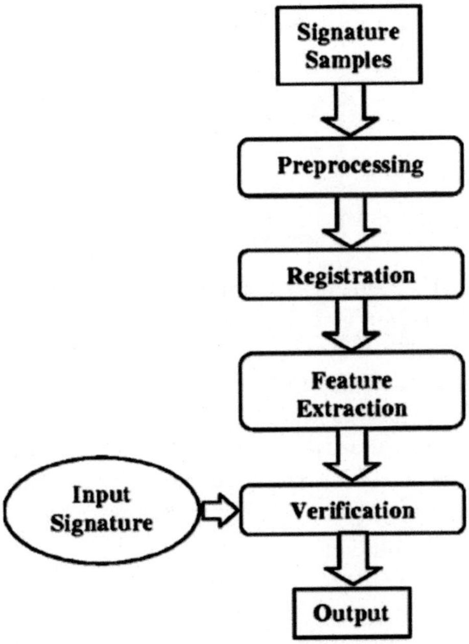

Fig. 2. Signature recognition process

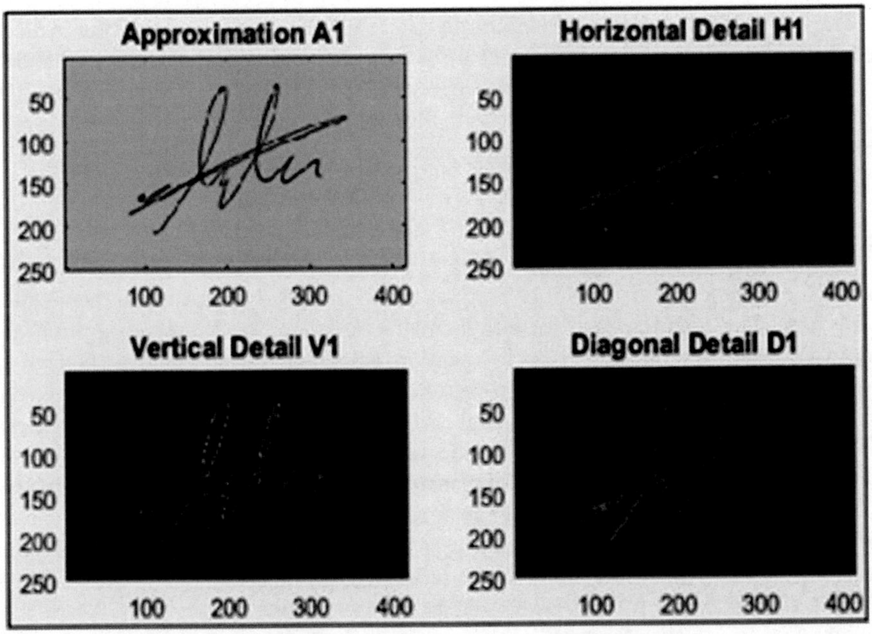

Fig. 3. Signature image in different frequency bands

The signature in the binary form is a matrix represented as below, where the pixels of the signature are shown:

$$\Gamma = \begin{bmatrix} \delta_{1,1} & \delta_{1,2} & \dots & \delta_{1,250} \\ \delta_{2,1} & \delta_{2,2} & \dots & \delta_{2,250} \\ \vdots & \vdots & \vdots & \vdots \\ \delta_{80,1} & \delta_{80,2} & \dots & \delta_{80,250} \end{bmatrix}$$

Feature extraction in the signature recognition [8] is one of the important steps that extract the relevant information and characteristics from the input given in form of a signature. The discrete wavelet transform is applied to extract the perfect characteristics which are further used for processing of the features like matching the features extracted with the data stored in the database and comparing it to verify or authenticate the signature of the user. Discrete wavelet transform uses vectors for the processing of images for extracting the features. The matrix is converted into a vector form for extracting features from the signature image. The vector is as shown below:

$$A = \begin{bmatrix} \delta_{1,1} \\ \delta_{2,1} \\ \vdots \\ \delta_{79,250} \\ \delta_{80,250} \end{bmatrix}$$

The last step being verification of the signature [9] in signature recognition matches the features extracted by the feature extraction step and compares it to the pre store data in the database. The output image is the reconstructed image [10] of the raw input data entered in the system and given as an output after being processed by the discrete wavelet transform in two dimensional. The output reconstructed image is as shown in Fig. 4.

4 Conclusion

The field of biometrics uses a lot of features and characteristics for the authentication of the user. The features used include facial features like iris, face recognition and certain other features like finger print recognition, palm recognition and many behavioral features like voice or the signature recognition. Signature recognition is mostly used due to its accuracy in the field of security and authentication of the user without any risks. The paper addresses the feature extraction in the processing of the signature in the image analysis by 2D discrete wavelet transform. The frequency based DWT is used in signature recognition to give better results as it works on many frequency bands. The paper discusses about the vector form of the input image which is used in feature extraction step and the results in form of a reconstructed image.

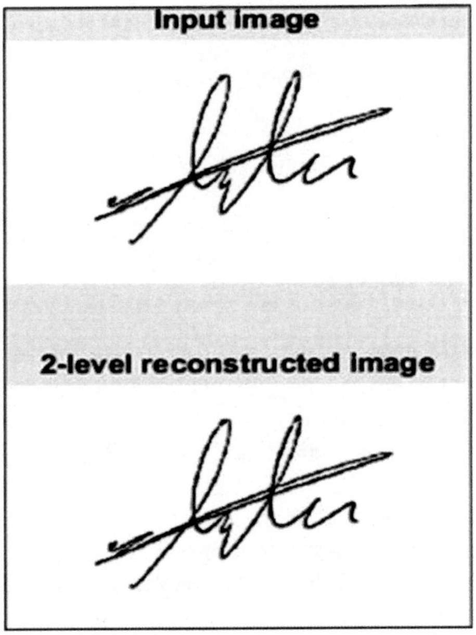

Fig. 4. Results

References

1. J. A. Unar, W. C. Seng, A. Abbasi, A review of biometric technology along with trends and prospects. Pattern Recognit. (2014)
2. X. Su, H. Tong, P. Ji, Activity recognition with smartphone sensors. Tsinghua Sci. Technol. (2014)
3. D. A. Reid, S. Samangooei, C. Chen, M. S. Nixon, A. Ross, Soft biometrics for surveillance: an overview, in *Handbook of Statistics*, (2013)
4. R. W. Goodman, Discrete wavelet transforms, in *Discrete Fourier and Wavelet Transforms*, (2016)
5. H. Demirel, G. Anbarjafari, IMAGE resolution enhancement by using discrete and stationary wavelet decomposition. IEEE Trans. Image Process. (2011)
6. K. Schneider, O. V. Vasilyev, Wavelet methods in computational fluid dynamics. Annu. Rev. Fluid Mech. (2010)
7. M. Al Najjar, M. Ghantous, M. Bayoumi, Image registration. Lect. Notes Electr. Eng. (2014)
8. S. Sen, A. Dutta, N. Dey, Feature extraction, in *SpringerBriefs in Applied Sciences and Technology*, (2019)
9. D. Gálvez-López, J. D. Tardós, Bags of binary words for fast place recognition in image sequences. IEEE Trans. Robot. (2012)
10. R. E. J. Yohanes, W. Ser, G. Bin Huang, Discrete wavelet transform coefficients for emotion recognition from EEG signals, in *Proceedings of the Annual International Conference of the IEEE Engineering in Medicine and Biology Society, EMBS*, (2012)

Object Detection with Compression Using Wavelets

Ruchi Sharma(✉)

Department of Electronics and Communication Engineering, Vivekananda Global University, Jaipur, India
sharma.ruchi@vgu.ac.in

1 Introduction

The field of image processing is a wide field of research since decades which includes the processing of images to gather the required and relevant information. One of the most important areas in image analysis is in the detection of certain objects or parts in the images which are needed to be used in various applications. The detection of objects [1] is an area in computer or machine vision used along with image processing for detecting the objects in still images as well in videos so that it fulfills the requirements of real time applications. Detecting the objects and locating them in images and videos is a very useful approach in many fields to reduce the time consumption along with reduced manpower. The object detection is found in application [2] in the areas of medicals, security and surveillance, automation systems, agriculture etc. These vast areas of applications require an efficient result which is achieved by using reliable techniques.

Many approaches and techniques are available for detection of objects in the images using image processing. One of the most important and reliable techniques is with the use of wavelet transforms in the images. Wavelet transforms [3] is one of the extensively used algorithm or technique that easily handles the complex images or videos for detecting the required objects in the images. The approach of wavelet transform is used for image analysis with multiple scale and easily applied to the two dimensional data. The applications of wavelet transforms is majorly found where complex data is present like medical area, detection of objects for security, enhancement of images, compression of data [4], denoising the audio signals etc. the use of wavelet transforms give reliable results for all kind of complex audio, visual or pattered data using different scales for decomposing the data and the results are achieved by reconstructing the data.

The wavelet transforms works in both time and frequency domains depending on the components in the data. The use of transforms like Fourier transforms [5] is an important domain working with the frequency components but due to certain disadvantages as it does not involve the time based components, it is not reliable for use in complex real time applications. So, wavelet transforms providing both the time as well as frequency components [6] of the data are a powerful tool. It works with low and high frequency components, removes noise signals to improve the quality of the images or the data, compresses the images to reduce the size for storage related problems by decomposing the images in sub parts. In the process of detecting objects with the approach of wavelet

F. P. García Márquez (ed.), *International Conference on Intelligent Emerging Methods of Artificial Intelligence & Cloud Computing*, Smart Innovation, Systems and Technologies 273, https://doi.org/10.1007/978-3-030-92905-3_28

transforms certain other algorithms like segmentation of the images is used to get more efficient and accurate results. The objects present in the images lack in clarity sometimes and problems like overlapping occurs which reduce the accuracy of the end results, so techniques which clarify the details of the objects to be detected is followed.

Image segmentation is the useful approach for overcoming the stated problems. It breaks the images in parts which are a set of similar pixel values and these individual set is known as object of the image. This approach simplifies the data in the image for easy analyzing by detecting the lines, points, curves, edges or boundaries in the object thus making the object detection easy and accurate. The approach of segmentation uses various methods like edge detection, k- means clustering, histogram based methods, region based approaches etc. Among various approaches of segmentation edge based approach [7] gives efficient results by detecting the edges of the objects on the basis of intensity values at the boundaries or the edges of the object to be detected. The segmentation of images which are based on the texture of the image uses the compression based approaches in image segmentation as it reduces the size of the data by compressing the data using lossy compression which is beneficial for storing as well as transmission of the data.

Compression of images is done to remove the high frequency components in order to get a smooth image by the low frequency components. The LL sub band which is the approximation band is useful for getting the smooth image and is useful component in the compression of images. Other sub bands contains high frequency components and represents the vertical, horizontal and the diagonal details which are not required in the compression of images to get more accurate results [8]. The sub bands on the basis of low and high frequency components are as shown in Fig. 1.

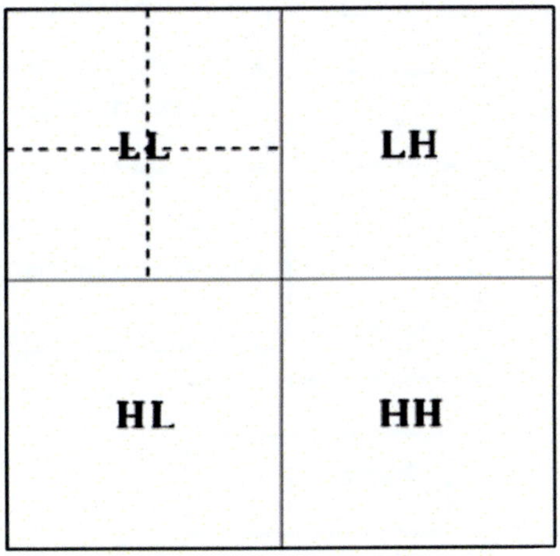

Fig. 1. Frequency sub bands

2 Methodology

The images used for detection of the objects are colored images as the approach used is capable of processing the complex images. The steps used for detecting objects include the compression of images by using certain algorithms of wavelet transform like Haar transform and wavelet decomposition for multi dimensional images. The compression of the images is measured by using metrics like compression ratio or signal to noise ratio. The decomposed images are then processed further by segmentation of the images to retrieve the images required by detecting the edges or the boundaries to detect the objects efficiently. The segmentation is done by using the algorithm of edge detection. The method used in the detection of objects is as shown in Fig. 2.

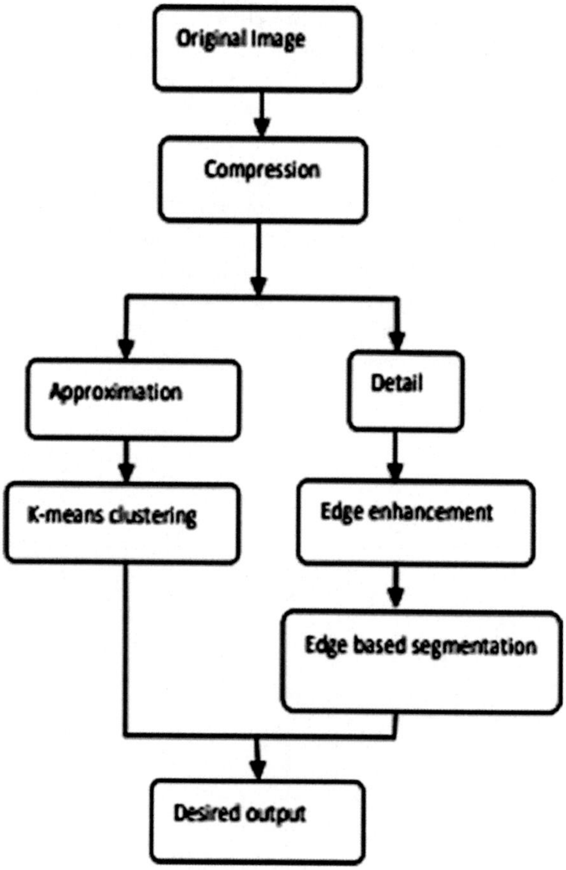

Fig. 2. Steps for object detection

1. Wavelet transform
The wavelet transform represents the real as well a complex function of the orthonormal functions which are produced by the wavelets majorly square wavelets. Many approaches like Haar transform and dual tree complex transforms are used for compressing the images. The approach of Haar transform [9] is widely used in processing of images and recognition, mostly for 2D signals due to their square shaped or wavelets like structure. The compression performed by Haar transforms provides the information of the low frequency components as well as the details of components representing high frequency components. The Haar decomposition levels are as shown in Fig. 3.

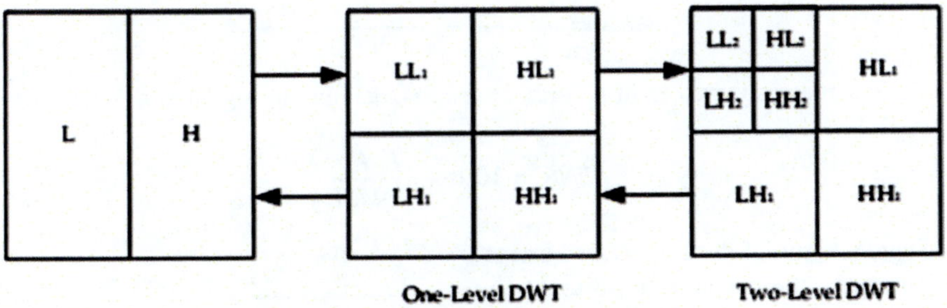

Fig. 3. Decomposition levels of Haar transforms

The Daubechies wavelet transform [10] is also used for decomposition and extracts the required features from the images for object detection. The extracted features by using the Daubechies wavelets are as shown in Fig. 4. The image is decomposed by two dimensional wavelets to form the set of required features for detecting the objects.

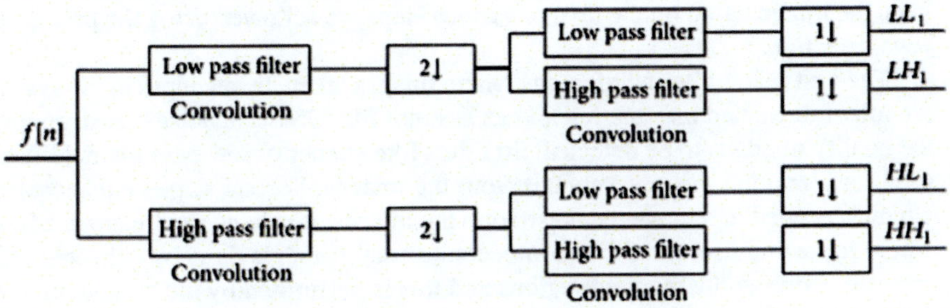

Fig. 4. Daubechies wavelet decomposition

2. Measurement of compression
The level of the compression caused by use of wavelet transform like Haar and wavelet decomposition on the two dimensional images is further measured by compression ratio or signal to noise ratio. The compression ratio is also known as

compression power and measures the size reduced by the use of algorithms for compression. The compression ratio is expressed as:

$$\text{Compression Ratio} = \frac{\text{Uncompressed Size}}{\text{Compressed Size}}$$

Signal to noise ratio is used to measure the quality between the original data and the compressed data to check the whether the compressed image is of good quality or degraded. To calculate the signal to noise ratio of the images the mean squared error is calculated as:

$$MSE = \frac{\sum_{M,N} [I_1(m,n) - I_2(m,n)]^2}{M * N}$$

The mean squared error is further used to calculate the signal to noise ratio by:

$$PSNR = 10 \log_{10}\left(\frac{R^2}{MSE}\right)$$

3. Segmentation of the compressed image
To extract the meaningful information from the image and process it accordingly, features or texture of the images are required. The methods of image segmentation are divided on the basis of its processing into two parts. One of the approaches is edge based segmentation [11] and the other is region based segmentation. The steps to determine the segments from the image is done by following few basic steps like reading image from the database, grayscale conversion, filtering image, extracting region of edge from the image. The edge detection is used for segmentation of the compressed images to detect and locate the objects in the images. Edges of image are important features to obtain the useful information from the image or separate the main objects from the background or classifying the objects. For separating the edges from the image based on the texture various steps are followed using the process of edge detection.
The first being filtering of image where images taken as raw input in the system are full of unwanted information known as noise like Gaussian noise thus degrading the quality of edges to be detected. So filters like wiener or low pass filters [12] are used for removing the noise signals from the images. Second step is enhancement which is needed to remove the noise from the images or enhance an image in a form where retrieving useful information becomes easy which is done by enhancing the intensity value of the required regions and this is performed by the computation of gradient of the image. The last step uses thresholding for the detection of edges or the boundaries in the image. The image after segmented is as shown in Fig. 5.

Fig. 5. Segmented image

3 Conclusion

Object detection is a wide area of research in image processing and computer vision having many applications in real time. The paper discusses about the approach used to detect the object in the images efficiently using the wavelet transform, the compression of images is done by using the Haarwavelet transform and the wavelet decomposition performed in the two dimensional images to compress the complex images so that they can be easily stored and transferred. The compressed images are checked for quality by measuring the compression using compression ratio and the signal to noise ratio. The compression quality must be such that the qualities of the images are not degraded. The paper uses image segmentation techniques to detect and locate the images by using edge based detection so that the accurate results of the objects to be detected can be achieved. The segmentation of images removes the noise that may be present along with image compressed by wavelet transform. Efficient and reliable results can be achieved by using the techniques of image compression and decomposition by wavelet transforms and further using edge detection.

References

1. M. Al Najjar, M. Ghantous, M. Bayoumi, Object detection. Lect. Notes Electr. Eng. (2014)
2. C. Mertz et al., Moving object detection with laser scanners, *J. F. Robot.*, (2013)
3. C.-L. Liu, A tutorial of the wavelet transform. History (2010)
4. O. Rippel, L. Bourdev, Real-time adaptive image compression, in *34th International Conference on Machine Learning, ICML 2017*, (2017)
5. M. J. Baker et al., Using fourier transform IR spectroscopy to analyze biological materials, Nat. Protoc. (2014)

6. J. Gilles, Empirical wavelet transform. *IEEE Trans. Signal Process.*, (2013)
7. C. J. Taylor, Towards fast and accurate segmentation, in *Proceedings of the IEEE Computer Society Conference on Computer Vision and Pattern Recognition*, (2013)
8. J. Li, An improved wavelet image lossless compression algorithm, Optik (Stuttg). (2013)
9. I. Ram, M. Elad, I. Cohen, Generalized tree-based wavelet transform. IEEE Trans. Signal Process. (2011)
10. A. Khare, M. Khare, Y. Jeong, H. Kim, M. Jeon, Despeckling of medical ultrasound images using Daubechies complex wavelet transform. Sig. Process. (2010)
11. A. Khoreva, F. Galasso, M. Hein, B. Schiele, Classifier based graph construction for video segmentation, in *Proceedings of the IEEE Computer Society Conference on Computer Vision and Pattern Recognition*, (2015)
12. C. Harris, M. Stephens, A combined corner and edge detector, (2013)

A Review on Deep Learning Models

Dushyant Singh(✉)

Department of Computer Science and Engineering, Vivekananda Global University, Jaipur, India
dushyant.singh@vgu.ac.in

1 Introduction

Reinforcement learning also known as the semi- supervised learning models is a method that allows the software agent to takes action and further interact with the environment to extend the result. Being a part of deep learning in machine learning, reinforcement learning follows the Markov Decision Process (MDP) [1] to achieve a complex objective and to maximize the value which represents the long term objective. The reinforcement learning is modeled as Markov Decision Process as shown in Fig. 1. The interaction in reinforcement learning where the controller gets an input of the reward and the state of the system linked with the previous transaction. The action calculated by the controller is transferred back to the system, which makes a new transition to new state and so on. Reinforcement learning is a part of many fields like, control system, simulation based optimization, statistics, genetic algorithms [2] etc.

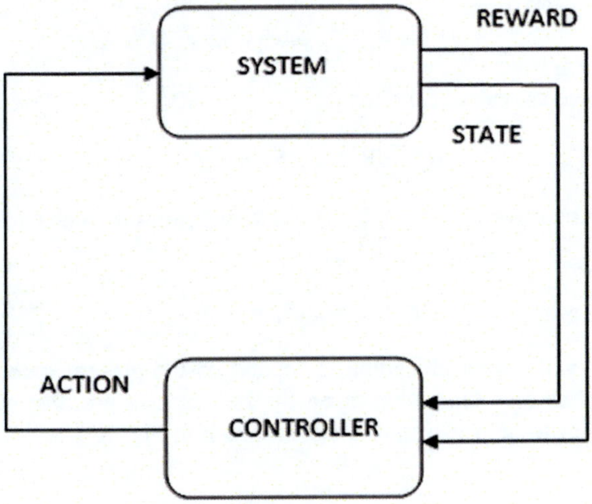

Fig. 1. Reinforcement learning

F. P. García Márquez (ed.), *International Conference on Intelligent Emerging Methods of Artificial Intelligence & Cloud Computing*, Smart Innovation, Systems and Technologies 273, https://doi.org/10.1007/978-3-030-92905-3_29

2 Learning Models of Reinforcement Learning

1. Q-learning
   Q-learning is an algorithm in the model- free approach [3] of deep reinforcement learning with the object of learning the policy to determine the action to be taken in any certain situation. Since the model of the environment is not required the problems can be handled by using the stochastic transitions and rewards thus not requiring adaptation to the environment.

   The actions performed in Q- learning generated maximum of the rewards known as Q- value and the strategy can be formolutated as:

$$Q(s, a) = r(s, a) + \gamma \max_a Q(s', a)$$

   The Q-value is the output by being at the state s and action a with an immediate reward r(s,a) in addition to the maximum Q-value generated from the nest state s'. Q(s',a) depends on the the next state s" and requires a coefficient gamma to incerase or reduce the contribution of the rewards and is shown as follows:

$$Q(s, a) \rightarrow \gamma Q(s', a) + \gamma^2 Q(s'', a) \ldots\ldots\ldots \gamma^n Q(s''^{\ldots n}, a)$$

2. Markov decision process
   The deep reinforcement learning can be explanined as the markov Decision process (MDP) [4, 5] which consistes of parameters (S, A, P, R, y) where S is the state space, A is the action set, P is the transition probability function denoted as:

$$\mathcal{P}^a_{ss'} = \mathbb{P}[S_{t+1} = s' | S_t = s, A_t = a]$$

   R is denoted as rewrad function

$$\mathcal{R}^a_s = \mathbb{E}\left[R_{t+1} | S_t = s, A_t = a\right]$$

   and y is a discount factor $\boldsymbol{\gamma} \in [0, 1]$. The policy defines the behavior of the agent as defined:

$$\pi(a|s) = \mathbb{P}[A_t = a | S_t = s]$$

   The markov decision procedure is depend on the present state and has no link to the past states thus the action taken by the agent id pre determined for all the different actions to be performed. MDP is shown in Fig. 2.

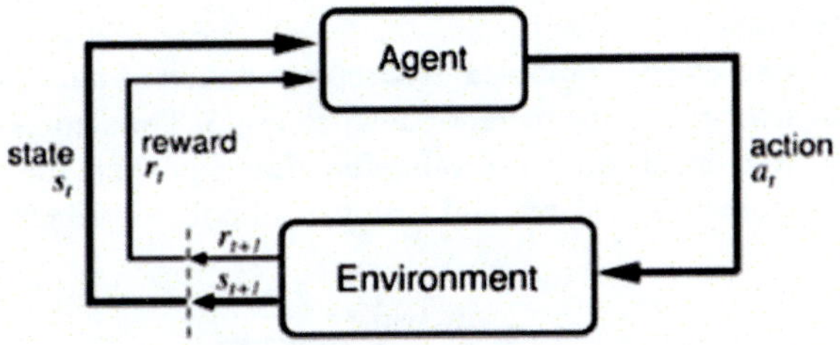

Fig. 2. Markov decision process

3 Algorithms

There are two important methods in deep reinforcement learning based on value functions [4] and policy search.

1. Value functions
The methods of value functions are based on the estimation of values [3] in the state known as state values function which is the return expected. The state values function that is defined as:

$$V^{\pi}(\mathrm{s}) = \mathbb{E}[R|\mathrm{s}, \pi]$$

The state optimal value function is defines as:

$$V^{*}(\mathrm{s}) = \max_{\pi} V^{\pi}(\mathrm{s})\, \forall \mathrm{s} \in \mathrm{S}.$$

Generalized policy iteration which is a combination of policy improvement and policy evaluation is used to estimate the best policy. The policy evaluation [6] enhances the estimate of value function which is achieved by reducing the errors generated by policy. On improvement of estimation, the policy automatically improves by updating the value function.
2. Policy search
In the policy search methods, the value function model need not be maintained and the optimal policy can be searched directly. A parameterized policy [7] is considered, where the parameters can be updated hence the expected return can be maximized by the use of either gradient free optimization or gradient based optimization being sample efficient where large number of parameters are present.

Policy gradients [8] can be used to significantly improve the poly of parameterization. The computation of return needs averaging of the policy parameterization which requires deterministic approximations on a model based setting. For the gradient based learning method, reinforce rule is used as an estimator for the gradient and it can be utilized to compute the gradient of belief over that of the function f of the variable X with acclaim to parameters and is defined as:

$$\nabla_{\theta}\mathbb{E}_{X}[f(X;\theta)] = \mathbb{E}_{X}[f(X;\theta)\nabla_{\theta}\log p(X)].$$

3. Actor-critic methods
The value functions can be combined with representation of policy and the resultant method is known as actor-critic method shown in Fig. 3. The actor is the policy which uses feedback from critic which is a value function [9], where the value function is used as a gradient making this method a subpart of policy gradient methods.

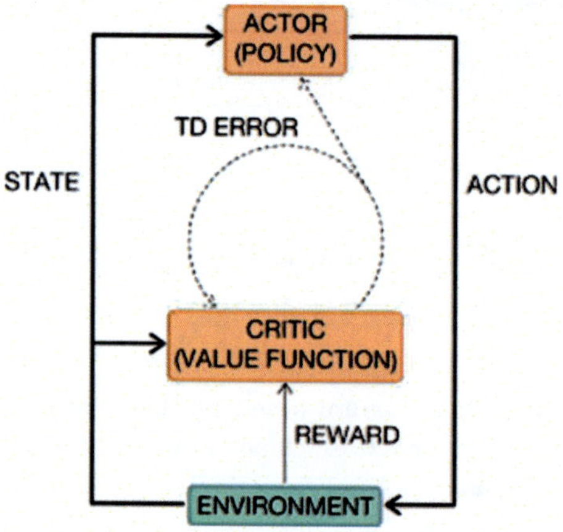

Fig. 3. Actor critic method

4. Model based
Figure 4 shows the model depend on reinforcement learning, the model depend Reinforcement learning allows the simulation of environment without any interaction with the environment in any direct manner. A virtual model of the real environment can be created and the operations further required are performed in that specific virtual environment. The model based methods [10] reduce the interactions with the real environment thus reducing the time consumption. For example, in case of robotics, the use of this method reduces the time and wear and tear of the hardware.

4 Applications

Deep reinforcement learning has wide range of applications in the fields of robotics, web system, gaming, NLP, computer vision etc.

1. Web system configuration
Due to enormous number of parameters in a web system reconfiguration [11] of parametres done by tuning of the parameters by using deep reinforcement learning where the reconfiguration is done by Markov Decision Process. The system configuration is considered as state, the parameteres as actions and the difference between the two response times which are target and measured. The computational complexity is reduced by using a combination of deep learning and reinforcement learning.

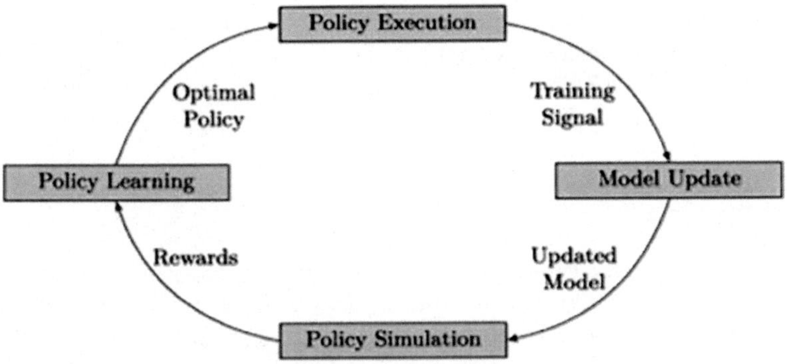

Fig. 4. Model based reinforcement learning

2. Gaming
Games like chess, Othello, Go etc. prove to be an excellent testbed for reinforcement learning algorithms and artificial intelligence. The best application of deep reinforcement algorithm learning in gaming is by using an algorithm named AlphaGo Zero which is neither supervised nor unsupervised learning but performs policy improvement and evaluation. It is build with a combination of deep convolutional neural networks, reinforcement learning & supervised learning.
3. Robotics
Reinforcement learning playthe main role in the field of robotics to develop techniques for controlling robots by automatic processes like sensing, planning, controlling. Deep reinforcement techniques provide ability in robotics for achieving optimal behaviour by the use of trial- error methods with its environment. The performance of robots are easily known by getting feedbacks in form of scalar objective function.
4. Natural language processing
The area of NLP has now shifted to neural networks from statistical approaches due to challenges which cannot be overcome by the latter. Hence deep learning has been playing vital role in areas of NLP [12] and an addition of reinforcement learning made a significant progress. Deep reinforcement learning methods achieve state of the outcome for language specific problems of NLP like sppech recognition, text classification, language modelling etc.
5. Recognition
Reinforcement learning has vast applications in the field of recognition as it can significantly improve the efficiency of image classification in object localization and detection. Motion analysis, visual control, situation and environment understanding are the best areas in field of recognition where the deep reinforcement learning algorithms are be formulated for performing this operation efficiently.
6. Healthcare
Due to the rise in popularity of healthcare field in providing personalized medicines many challenges have occurred. Deep reinforcement learning algorithms systematically optimizes the potential of growth in this area by using the algorithms effectively. Decision support applications are widely used in the medical field however it has many limitations due to use of existing models or techniques. Applications of

reinforcement learning in the field of healthcare is capable of achieving optimized treatment strategies that require large states and action spaces by using Markov Decision Process (MDP). This strategy overcomes many limitations and thus reduces the computational complexity.

5 Conclusion

Deep learning models are capable of extracting complex representations of data from input with high dimensional data and thus overcome the traditional machine learning methods. The reinforcement learning methods cannot optimize the policies efficiently due to data inefficiency. Thus a combination deep learning and reinforcement learning is an useful approach and has become a valuable tool for overcoming challenges faced. This paper discusses the reinforcement learning framework using the deep learning methodology with a combination of both supervised and unsupervised learning networks. The deep reinforcement learning methods use architectures like deep convolutional neural networks which are appropriate for the challenges and problems with partial observable environment of Markov Decision process. The deep reinforcement learning due to its efficient methods and algorithms have wide range of applications in areas like robotics, machine vision, NLP, medical, gaming, videos etc. which have been addressed in the paper. The deep reinforcement learning has the significant use in real world applications as direct approach of policies and values is done. Thus the approach of deep reinforcement learning addressed in the paper overcomes many challenges and is a progressing field.

References

1. Z. Wei, J. Xu, Y. Lan, J. Guo, X. Cheng, Reinforcement learning to rank with markov decision process, in *SIGIR 2017—Proceedings of the 40th International ACM SIGIR Conference on Research and Development in Information Retrieval*, (2017)
2. H. Iba, C. C. Aranha, Introduction to genetic algorithms. Adapt. Learn. Optimization (2012)
3. S. Gu, T. Lillicrap, U. Sutskever, S. Levine, Continuous deep q-learning with model-based acceleration, in *33rd International Conference on Machine Learning, ICML 2016*, (2016)
4. C. Szepesvári, Algorithms for reinforcement learning, in *Synthesis Lectures on Artificial Intelligence and Machine Learning*, (2010)
5. M. T. J. Spaan, Partially observable markov decision processes, in *Adaptation, Learning, and Optimization*, (2012)
6. S. Racanière et al., Imagination-augmented agents for deep reinforcement learning, in *Advances in Neural Information Processing Systems*, (2017)
7. S. Gu, E. Holly, T. Lillicrap, S. Levine, Deep reinforcement learning for robotic manipulation with asynchronous off-policy updates, in *Proceedings—IEEE International Conference on Robotics and Automation*, (2017)
8. R. Houthooft et al., Evolved policy gradients, in *Advances in Neural Information Processing Systems*, (2018)
9. R. Lowe, Y. Wu, A. Tamar, J. Harb, P. Abbeel, I. Mordatch, Multi-agent actor-critic for mixed cooperative-competitive environments, in *Advances in Neural Information Processing Systems*, (2017)

10. A. Nagabandi, G. Kahn, R. S. Fearing, S. Levine, Neural network dynamics for model-based deep reinforcement learning with model-free fine-tuning, in *Proceedings—IEEE International Conference on Robotics and Automation*, (2018)
11. J. P. O'Doherty, S. W. Lee, D. McNamee, The structure of reinforcement-learning mechanisms in the human brain. Curr. Opin. Behav. Sci. (2015)
12. W. Y. Wang, J. Li, X. He, Deep reinforcement learning for NLP, in *ACL 2018—56th Annual Meeting of the Association for Computational Linguistics, Proceedings of the Conference Tutorial Abstracts*, (2018)

Real Time Based Target Detection Method

Baldev Singh(✉)

Department of Computer Science and Engineering, Vivekananda Global University, Jaipur, India
baldev_singh@vgu.ac.in

1 Introduction

Tracking of the videos is an important part of computer vision and leads to major applications fields like surveillance, human computer interaction (HCI) [1], object tracking, human tracking, object identification, robotics, video games, traffic management, biometrics, automation systems, security, defence services, industrial applications, navigation systems, etc. Visual tracking is considered as an alternative method to the conventional techniques of detection or tracking that includes radars [2].

The videos during tracking of objects or humans are analysed in various steps for proper identification. The first step relates to stationary object detection where no movement of the objects is found. The second step relates to the classification of the objects [3] to make sure the object identified to be the correct object while determining the properties of the objects and the third step is the tracking of object after classification. The objects are tracked on various frames and various parameters like velocity, location and speed are considered while tracking the objects on the frames. An efficient object tracking[4] is achieved while continuously updating the detection of the objects along with the object parameters to carefully and correctly identify and track the objects. The flow of object tracking is as shown in the Fig. 1.

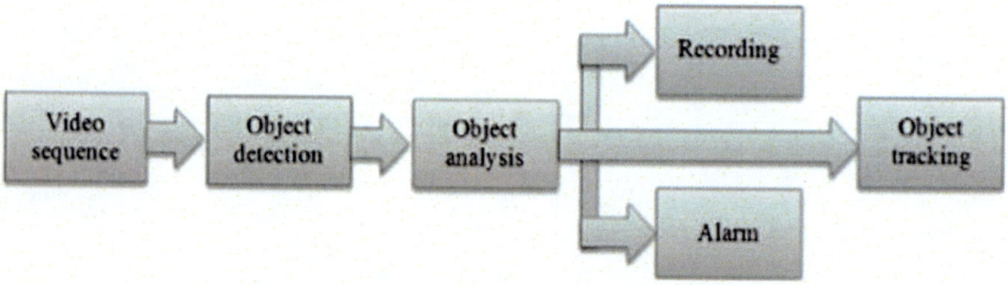

Fig. 1. This flow represent the object tracking

The first step in tracking of the objects in videos is the detection of target which means the identification of objects from several frames of the videos captured by a camera. This step also determines the pixels of the images or the videos related to the objects and form clusters of the pixels. Various algorithms are available for fulfilling this task such as optical flow [5], temporal frame difference [6], machine learning techniques

F. P. García Márquez (ed.), *International Conference on Intelligent Emerging Methods of Artificial Intelligence & Cloud Computing*, Smart Innovation, Systems and Technologies 273, https://doi.org/10.1007/978-3-030-92905-3_30

[7], image processing techniques like subtraction of background [8] in the images, etc. The techniques which are beneficial are mentioned in the Fig. 2 in order to detect the objects in the videos. When the object detection task is to be performed, the images or videos are captured by the camera and these images or videos are further analysed in order to recognise the contents of the images/videos. The objects to be identified are determined in the image sequences or video frames captured by the camera. As soon as the target is pointed in the first frame, the features of the target are extracted to determine the target in the other n number of frames captured.

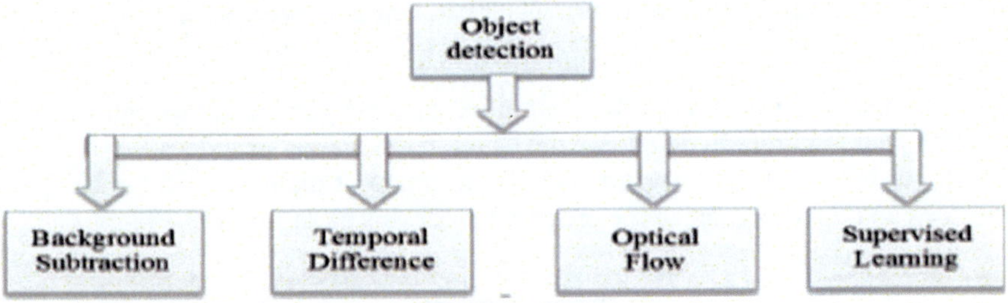

Fig. 2. Object detection techniques

2 Experimental Setup

The proposed system for detection of target and warning about the presence of the target is a simple, low cost and reliable system due to the use of low cost and low power consuming components like camera. The images and videos captured by the camera are transferred to a computing unit basically a computer for further processing of the images and the videos. The software of Python is used in the proposed system, thus making the system economic and efficient to be used. Python is used for processing of the images in both the stationary and moving situation of the target. The camera is utilized to capture the images and the videos for both the situations, when the target is stationary and when the target is in moving.

The video signals captured by the camera that contains the prey data and information which is further move to a computer having software programming (PYTHON) [9] via a serialanchorage. The chosen focal and light camera lens is modified manually and the resolution is modified to provide a wider range of action in distant, poor weather and thus clearer images or videos. After the observation algorithm is performed completely, the output videos is recorded & an alarm is started to warn about the target detected in the visuals. Figure 3 shown in thestructure of the functioning of the proposed systems.

The weighted gain filters[10] is utilized in the methodto reject and remove the noise present in the images and the videos captured by the camera. The noise is added to the images and videos while acquiring them from the source being the camera and the noise is removed in sequence to magnify the image standard. The noise in the image pixels is eliminated by using the morphological transformation filter in the python language and

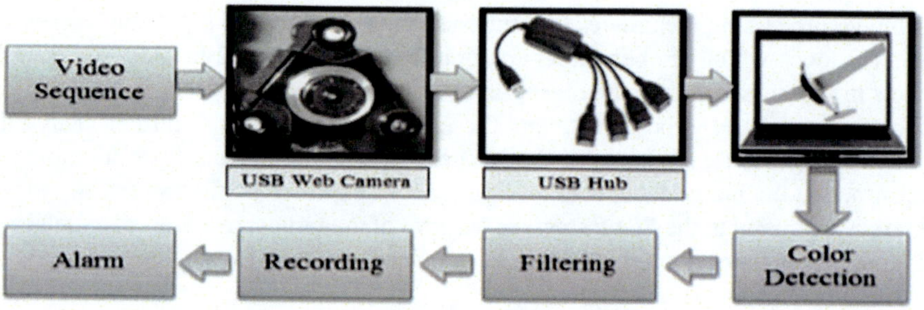

Fig. 3. This block diagram represent the proposed system

the foreground object pixel is further multiplied by a designed factor to compensate for the target signal in order to determine the target in the images or videos.

Color & background subtraction depend detection of the objects in the images are performed and the flowchart of the proposed process is as shown in the Figs. 4 and 5.

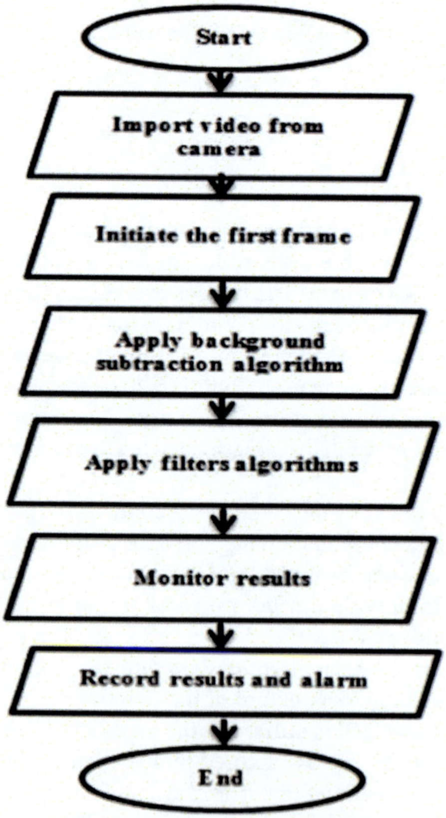

Fig. 4. Background subtraction

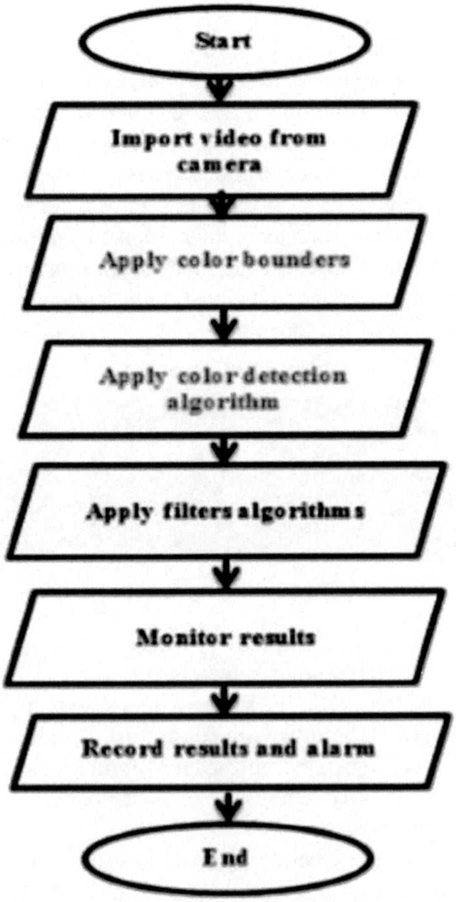

Fig. 5. Color subtraction

The suggested detection algorithm begins by importing video from a USB camera or a video file that has been captured. In determining the identification protocol methods, the two algorithms differ from each other. In the programme algorithm, the upper and lower limits of the target colour are adjusted in the color-based detection process. According to the principle of hue, saturation and value colour representation, target borders are determined. The algorithm identifies target data by colour boundaries and eliminates all unnecessary data from any frames, excluding target data. The first frame is captured in the background subtraction detection process, then the nth frame is captured and subtracted from the original frame. Frame information is separated into foreground and context as a consequence of the subtraction process. Moving targets are identified by foreground data and this process is replicated in each nth frame. A collection of filters are worked to maximise the target signal after the two algorithms have begun and the target is sensed. Filtering performance is delivered to the stages of recording and alert since the goal disappears, the software is ended.

3 Results

1. *Background subtraction depend detection technique*
   The test is conducted on a moving target in order to determine the efficiency of object detection using the python-based programming approach. The background subtraction detection technique determines the reference frame and then subtracts the corresponding frame from it. If there is a moving object inside the frames it will appear as a white photo such as in the case of the yellow object in Fig. 6.This technique is sensitive to changes in the local environment as it is caused by any transformation in the background. The most important constraint is that there must be a previous knowledge of the meaning.

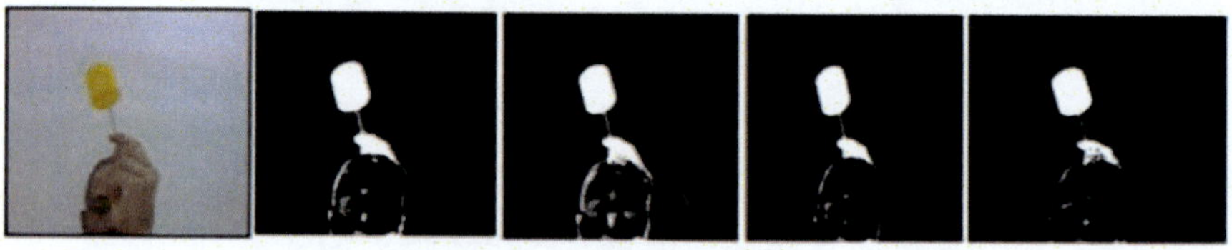

Fig. 6. Moving target

The output of background color subtraction depend on detection technique for moving target is shown in Fig. 6. The image with a yellow object is the actual image captured by the cameras. The other images are the outputs of a sequence of filter to magnify the detected objects output images. Figure 6 shows aadequate result in cases of a rounded object only &the static of background.

2. *Color based detection technique*
   The strategy of color-based detection is favoured where only the target is supposed to appear. The desired colour is initially set prior to its application in the detecting algorithm. Fig. 7 shows the output of the common objects in the past cases but with colors based detection techniques.

Fig. 7. Color based detection

It is clear from Fig. 7 that only the object is appeared in the results and the surrounding environment is removed. When the colour of the target is well known and the background has no resemblance to this colour, this method is applicable. A white plane flying in the blue skies is an example of this. The results of color depend detection techniques shown in Fig. 7. The first picture is the actual image that the camera captures and the rest of the pictures are the result of a collection of filters to increase the image output of the object.

4 Conclusion

The paper addresses the experimental target observation & warning system depend on Python programmed algorithms for detecting the target in both moving and stationary conditions. The proposed system has the ability to detect stationary and moving targets by using two detection algorithms for target detection known as background subtraction and color based techniques. Test groups are performed to determine the efficiency of the proposed algorithm and to reflect the reliance of the proposed method on the visual monitoring of the targets. The built method with the algorithm of context subtraction shows appropriate results as it is only valid for moving goals. The tracking algorithm based on colour produces distinctive effects since it can be used concurrently with stationary and moving targets. For greater performance, the use of hybrid algorithms for target detection using two or more detection algorithms is recommended. Also, the designed algorithms can be easily uploaded on embedded board for real time applications.

References

1. J. S. Olson and W. A. Kellogg, *Ways of knowing in HCI*. 2014.
2. S.M. Patole, M. Torlak, D. Wang, M. Ali, Automotive Radars: A review of signal processing techniques. IEEE Signal Process. Mag. (2017). https://doi.org/10.1109/MSP.2016.2628914
3. J. O'Rourke and G. T. Toussaint, "Pattern recognition," in *Handbook of Discrete and Computational Geometry, Third Edition*, 2017.
4. M. Al Najjar, M. Ghantous, and M. Bayoumi, "Object tracking," *Lect. Notes Electr. Eng.*, 2014, doi: https://doi.org/10.1007/978-1-4614-1857-3_6.
5. A. Dosovitskiy *et al.*, "FlowNet: Learning optical flow with convolutional networks," 2015, doi: https://doi.org/10.1109/ICCV.2015.316.
6. C. Dicle, B. Yilmaz, O. Camps, and M. Sznaier, "Solving temporal puzzles," 2016, doi: https://doi.org/10.1109/CVPR.2016.635.
7. M. Guarascio, G. Manco, and E. Ritacco, "Deep learning," in *Encyclopedia of Bioinformatics and Computational Biology: ABC of Bioinformatics*, 2018.
8. G.E. Setyawan, B. Adiwijaya, H. Fitriyah, Sistem Deteksi Jumlah, Jenis dan Kecepatan Kendaraan Menggunakan Analisa Blob Berbasis Raspberry Pi. J. Teknol. Inf. dan Ilmu Komput. (2019). https://doi.org/10.25126/jtiik.2019621405
9. "Python?," *Seibutsu-kogaku Kaishi*, 2019, doi: https://doi.org/10.1201/b19104-3.
10. N. He et al., Weighted average current control in a three-phase grid inverter with an LCL filter. IEEE Trans. Power Electron. (2013). https://doi.org/10.1109/TPEL.2012.2219322

Business Analytics: Trends and Challenges

Lisha Yugal(✉)

Department of Computer Science and Engineering, Sanskriti University, Mathura, Uttar Pradesh, India
lisha.soet@sanskriti.edu.in

1 Introduction

The current review writing revolves around the area of data analytics and big data. Big data is exponentially increasing its foot in the field of business analytics. It has huge functionality not limited to improved safety, IPR i.e. Intellectual property rights, privacy, cyber law, security, preservation, optimizing and enhancing business processes, improves health sector, advancing Villages, 2-tier Cities and temporary habitats which is built according to their current strengths and the benefits which can be directly availed by the enterprises and also by introducing a new horizons where conventional methodologies, processes and services are meant to get a better version in electrical, electronics, smart devices and obviously a better technology with technological advancements. Now-a-days smart devices are used to even store information then it is mandatory to use optimize information for various machines and devices. The trends for using the processed information comes in use by financial forecasting in banking sectors and enterprises, the use of optimized machine and electronic device for better performance, it forecasts the stock market at least to its nearest, it detects the customer behavior so fraud detection becomes a easy task, online businesses and market screening, market smartness and intelligence, it also improve the research in the various fields such as science and technology.

BA & I is referred to the advancements of the new skills, methodologies, technologies, practices, investigation and study for continuous repetitive discovery of exploring new things and investing the previous models so to we would be able to extract the information of our need and get a clear insight for business plans which gives success with flying colors and also according to the analysis and statistical performance as shown in Fig. 1. BA & I basically focuses on creating new business model, methodologies, practices, terminologies with the help of getting tip and tricks from the previous methodologies. Everything in this works on the factual data i.e. old datasets and statistical methods. Its performance is all based on the quality of the data. The business intelligence has a slight difference in order to work it took consistent set of various metrics alike the business analytics. BI measures the past performance and methodologies and its impacts on the new business guide which is all dependent on previous data and statistical procedures and practices [1].

Business analytics works extensively with analytical& predictive modeling where numerical analysis and fact-driven managerial system is required to take decisions. It

F. P. García Márquez (ed.), *International Conference on Intelligent Emerging Methods of Artificial Intelligence & Cloud Computing*, Smart Innovation, Systems and Technologies 273, https://doi.org/10.1007/978-3-030-92905-3_31

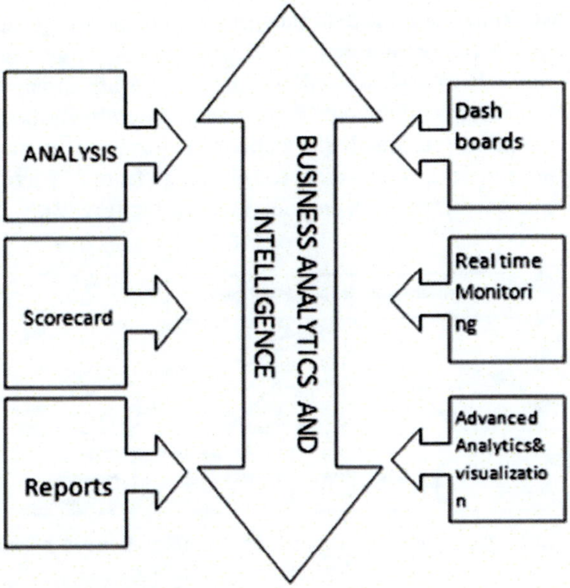

Fig. 1. Business analytics and intelligence works as a master mind in various fields and allows user to get entertain by the plurality of effective results

is basically dependent on management science [2]. Analytics is given in the form of input for social decisions taken by humans or it can also work fully automated even in order to take decisions. Whereas Business intelligence is all about query making, report building, online analytical processing i.e. OLAP and notifications comes in the form of alerts [3].

In this digital era business analytics is the most trending thing after internet as all the data that is generated have a expensive value, and people these days are keen to know that how the data is generated and in these days if anyone knows how to manage data is earning gold. It is getting famous even in business or in academic careers. In basic terms business analytics is the science and art of exploring knowledge and discovering important insights. For this we would use the related methods like machine learning algorithms, deep learning algorithms, statistical models, mathematical models, network sciences methods with the vast variety of knowledge expert data in fast timings. It enables us to take fast and efficient decisions. This era is continuously reserving space for data being called as gold so much soon as we are living in a world where data has large volumes, data is heterogeneous and data has high number of frequency. We can basically say that data analytics is identification of the new or novel information or sets or patterns or words or phrases.

For example, we can say that with evolution of time the evolution of terminology also takes place business intelligence can be named by business analytics, customer intelligence is named by customer analytics, Web mining is named by Web analytics, and knowledge discovery is named by data analytics [4].

Even a naive user knows that the data analytics used in these days required excessive computations as the size of unstructured data is way more than the structured data. The unstructured data consists of high voluminous data with wide variety of data and with the high frequency or velocity of the data that is coining iteratively. So, to manage these data hard & fast tools, practices, methods, techniques, formulas, algorithms are applied so to get the analytics project cover in order to gain a perk from it. There is no limitations to the usage of life sciences, medical, engineering, law etc (Fig. 2).

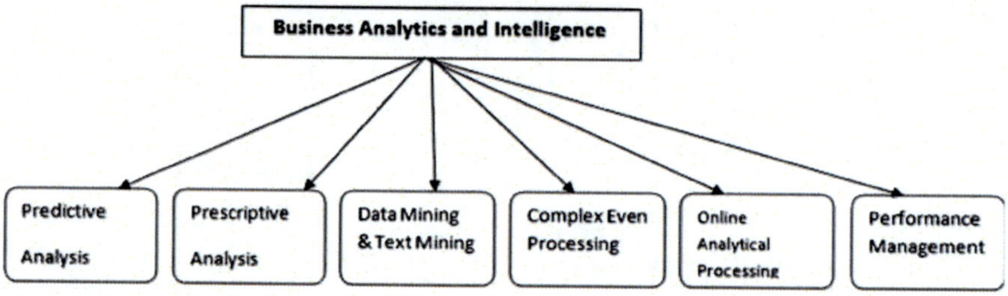

Fig. 2. Numerous techniques come up in making the formation of business analytics and intelligence

BI Versus Big Data

Big Data is way different in various ways as suggested:

1. The information which is stored in big data is more safe, secure, portable and easily adaptable. As it is not stored in a central server rather it is stored in RAID i.e. redundant array inexpensive disks which is basically a distributed file system as to eliminate the dependency on one server it is done.
2. The big data provides the voluminous amount of data to all its users i.e. it is the most responsive approach towards data science as to get done analysis of data we need large amount of data. The big data offers the solutions by charring the process functions with the data not the data with the functions.
3. The problem arises when we analyze that the unstructured data is growing day by day and in all the sectors invariably. The data can be of various types irrespective of structured, semi-structured and unstructured. The unstructured data is generated more than that of the structured data.
4. Data which is pre-processed by Big Data uses various methodologies to come up with the structured one. Data can heterogeneous, overlapped, conical or come from various line sources. Hence, we can say that the enterprises uses various technical advancements in order to give the efficient and reliable results in one time.
5. Big data uses parallel crowd driven processing and various distributed methods, it is used for improving the processing speed. Many instructions are executed parallel with Mass Parallel processing, as we have various tasks which are distributed into set of parallel execution subsets, and finally it integrated and reviewed results. Due to this the managerial department takes the voluminous data in order to ensure the fast delivery of decisions [5].

1.1 Difference Between Analytics and Analysis

Although both the words are contrast to each other and can be used in either place but when it come to its meaning in science and technology there is no replacement of them or it can be again interchangeable as analytics cannot be similar to analysis. In the regular terminology of the analysis it is referred as the method of separating the problem modules in small subsets so that the problem can be viewed from the ground level and basic level. It is mainly used for the investigation of complex and difficult models whose assassination cannot be possible or feasible as a whole so to break it down in a simplified manner in a more understandable or descriptive sub-components and components. After the realizations and examinations are done in the ground level, all the decomposed structure is put back again to its initial structures, that is known as the synthesis. Whereas Analytics shows variety of experimentations, practices, methods, technologies, and similar tools in order to create new knowledge for solving difficult problems. Analytics is a multidimensional, multifaceted, multi-horizon and multidiscipline methodology in order to understand or address the undesirable and complex situations. Analytics uses the data and various mathematical models like no other thing can use to just make sense of things we do in our day to day life even without knowing why we are doing so or anything which is illogical, feels like logical after this. Analytics includes analysis at various stages of its discovery process, it also comes with synthesis and subsequent implementation.

2 Discussions

Business Analytics have filled the gap between the global business comprising IP patterns with their clients. Magnanimous examples are present where business analytics have proved to be at its best are not limited to the world tour, world customer relationship and world inventory. Business analytics comprises plurality of the fields not limited to BI and e-businesses, better governances, research and development in sciences, environment and surroundings provides security and safety. Some of the border areas which business analytics have benefitted are:

1. Improves E-Commerce and Market Intelligence
2. Improves E-governance
3. Improves Research in science and technology
4. Improves health care and public health.

Technologies using BI & BA

1. Big Data
2. Artificial Intelligence
3. Deep Learning
4. Neural Networks
5. The internet of things
6. Micro-segmentation.

While reviewing the research work of The impact of Business Intelligence on the quality of decision making—a mediation model by Bernhard Wielder, Maria-Louise, it was noted that the main objective of the research was how directly or indirectly the aspects of business intelligence and analytics can provide new data driven insights in order to maintain the managerial quality of decision making. Here the analysis of partial least square regression i.e. PLS and mediation identifies that how business intelligence is going to affect the business enterprises. Here the results discloses the utmost important part with the help of business intelligence management quality in order to become by data quality and information quality which supports the order for managing proper business intelligence. They have investigated that if high quality and complex bi methods are implemented then the desired outcomes are also in accordance with the quality check. While implementing the large scale business intelligence solutions may sometimes give the outstanding results, but they have also discussed that the quantity doesn't matter but the quality matters, especially in the case of data insights [3].

In the paper they have discussed the relevancy for academic purpose as well as for industrial application for the benefiting their enterprises by business intelligence and business analytics. They systematically analyzes their effects on the managerial decisions and orderly working of the organization. The advancements in mediation analysis and specific path modeling have been discussed. They showed that a hypothesis will be rejected if a non-signification path of mediation is given and vice-versa. PLS can be used at verge for small sample of the enterprises as there were no valid and reliable tools and instruments to develop a business model so they have created their own.

Whenever we compare the linking between business analytics and decision making a path of model has been analyzed, there study have various limitations. They have followed some four decision rules to create constructive structure so the user can avoid the misspecification. As the research model does not include any reflective items or 'shadow' which covers all the five constructive models, so here selector indicators cannot cover each constructive structure nor can it assess the convergent validity. Then for the second step, they only work on large databases not less than 50 employees, thereby cannot be used in small firms. Then for the third step they utilizes perceived measures to get involves the key variables in their research area whereas the mathematical measures can be proved complemented for same specific decisions. At last, they haven't even checked the erogeneity as they mentioned that they could not get a suitable instrument variable for this construct to measure the erogeneity for the information processes [6].

There study might have several limitations, but they have foreseen the future work in their research. Primarily they have allowed learning a vast knowledge of business analytics and its impact so that the reader may get the investigations according to the context, so that the companies could predict the variables in their businesses and investments. Secondly, they have focused to improve their prior model by enabling them to small industry too so they could use the business analytics understanding in a broader context. Thirdly, they have discussed the work culture, environment and strategic decisions to be handled with the top management teams and the main managerial and organizational team with the help of business analytics in their future research. Atlast, they have discussed that how the researches should have faith in extracting, predicting and improving models of business analytics, and should allow to do some research on the endogeneity as it could take the it empirical to inconsistent and inappropriate estimators so that validity of the research models can be limited [7].

In this entrepreneurial era the electronics and digital devices are the next basic amenities after food, shelter and clothes. And to learn new technologies like big data, ML, AL are more necessary to get a survival in a blooming techno world. Data is the new oil as you cannot oil from spilling and spreading is the same you cannot stop data from generating, millions of peta bytes of data are generated daily and on each and every click of a single person various data is stored at various places. Data mining is the application of data analytics but it is highly useful in the field of business analytics too. Because the business tycoon are not created in a night they have researched on the particular data driven insights that exactly the clients and the users wanted from an enterprise and invest their time, labor and mind in that particular field only. The business tycoons hire the data analyst and then the data-driven insights are been checked by them and the unstructured data is given to them in the form of wide datasets and those data scientists have to mine data for them by managing the whole data and categorizing in order to get the results in succinct way. They also have to manage, clean, extract, load, transform, prepare and draw meaningful drawings and inference for the organizations in order to take decisions that can predict the risk and the profit involved in them widely. In both the papers spawning of data is done which involves new technologies like big data and many more. Big data tools are indirectly required for this purpose. As the need of data analytics and business analytics are interdependent. Data is the raw entity for enterprises like bricks for building home, so the data should be trained well as it the requirement. There is no scope that this field will leave its greener grass as soon as the data is growing at its pace [8] (Fig. 3).

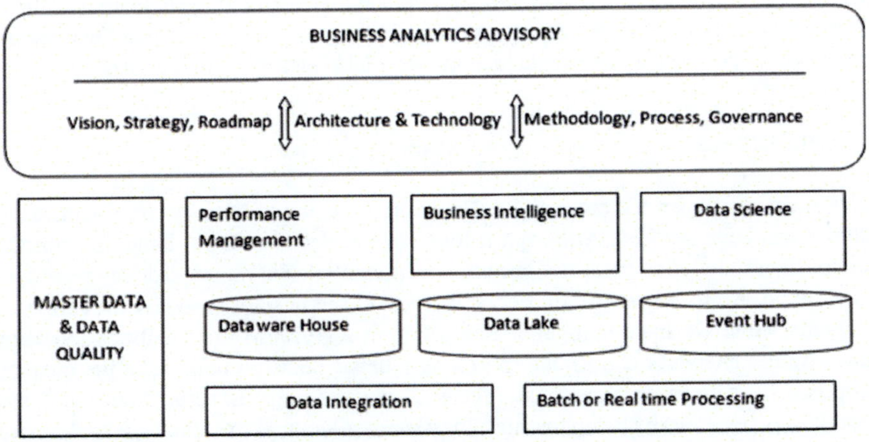

Fig. 3 BA and CPM are important subparts in order to achieve the objective of the corporate and manage the strategy of its execution with the way of its decision making

BA and CPM helps to gain vast knowledge from them and it also creates a business strategy and managerial decisions for making changes. It helps to track our performance management and provides a whole range of strategic consulting services [9].

- An audit is designed so that the assessment can be done of how far the enterprises are in BI & CPM.
- Every little goal is also treated as a milestone is referred to as a vision and strategy for business analytics.
- Architecture in terms of underlying data warehouse architecture, data-modelling, data integration, data quality, flows of data between BA & CPM components, integrated meta & master data and so on.
- a new version is defined for every new procedures for governing business analytics and their regarding performance managerial systems.
- With every organization the performance of business models varies and sometimes they are set to integrated according to their performance in data governance and corporate governance.
- There is a requirement for help in creating physical and abstract team of technical employees who can manage all the activities.
- Procedures and related metrics are vividly used to manage all the schedules and the portfolios of the following projects and each project is in the form of the structured and aligned way with the help of the reusable phrases, templates, words, abbreviations, milestones, documentations and the help given at the time of delivery. To allow execution of business analytics we define the portfolio management to be in metrics and can help in taking managerial decisions. The success of any business model is dependent on the methodology.
- At the time of training and guiding of cooperative management, team leads and members are the major part of the initiatives taken by the management so that the concepts of business analytics and performance managerial systems are made [10].

3 Conclusions

With this review paper, a clear insight of the current status of business intelligence and analytics has been marked. Although it took several decades to get hands-on practice in the business analytics field, the growth is exponential. So, the enterprises regardless of their exposure to the world are becoming sire to the business models for satisfying the clients. Here we have compared two different algorithms and methods by which show business analytics is utilizing in getting a better decision making in the business enterprises have been demonstrated. The PLS methodology is better than the 5 step methodology because the 5step methodology is not tolerant to the small size enterprises and today is the entrepreneurial era. Skipping the small enterprises need doesn't make any sense. The business intelligence and the analytics have inculcated both the technological advancement and the managerial outlook in an enterprise. Using electronic and digital services by big data and AI. Shearing data is generated and shared each single day. The unstructured information can be extracted and can be utilized in planning the data insights according to our needs. Working on machines make humans realizes that it is far easy to program and rule the world. So, we can clearly say that managing data is managing gold for this era. In this paper various factors are identified which leads to successful sets of outcomes in business analytics. This paper shows a theoretical structure and related impact on small and medium enterprises for successful business analytics. Indeed business analytics is playing a vital role in decision making for big enterprises as well as for small enterprises.

References

1. D. Delen, S. Ram, Research challenges and opportunities in business analytics. J. Bus. Anal. (2018). https://doi.org/10.1080/2573234X.2018.1507324
2. M. Suresh Babu, K. Bhavana Raj, D. Asha Devi, Future trends of business intelligence and big data analytics in ubiquitous environment. Int. J. Eng. Adv. Technol. (2019)
3. B. Wieder, M. L. Ossimitz, The impact of business intelligence on the quality of decision making—A mediation model (2015). https://doi.org/10.1016/j.procs.2015.08.599
4. D. Appelbaum, A. Kogan, M. Vasarhelyi, Z. Yan, Impact of business analytics and enterprise systems on managerial accounting. Int. J. Account. Inf. Syst. (2017). https://doi.org/10.1016/j.accinf.2017.03.003
5. M. Kowalczyk, J. P. Gerlach, Business intelligence & analytics and decision quality - Insights on analytics specialization and information processing modes (2015)
6. G.S. Richards, Business intelligence and analytics research. Int. J. Bus. Intell. Res. (2016). https://doi.org/10.4018/ijbir.2016010101
7. A.L. Maxwell, S.A. Jeffrey, M. Lévesque, Business angel early stage decision making. J. Bus. Ventur. (2011). https://doi.org/10.1016/j.jbusvent.2009.09.002
8. J. Bridge, J. C. Dodds, Managerial decision making (2018)
9. D. Trietsch, K.R. Baker, PERT 21: Fitting PERT/CPM for use in the 21st century. Int. J. Proj. Manag. (2012). https://doi.org/10.1016/j.ijproman.2011.09.004
10. A. K. Bhunia, L. Sahoo, A. A. Shaikh, Project management. Springer Optimization and Its Applications (2019)

Home Automation: IoT

Alka Singh[1(✉)] and Poonam Ponde[2]

[1] Department of Information Technology, Noida Institute of Engineering and Technology, Greater Noida, India
alka@niet.co.in
[2] Department of Computer Science and Engineering, Nowrosjee Wadia College, Pune, India
poonamponde@nowrosjeewadiacollege.edu.in

1 Introduction

Automation being new normal in this digital era is like giving a try to each and every new idea to become a new reality and it create a environment which makes our lives more efficient and robust by providing us the automated appliances by which standard of living automatically increased as human efforts are worth using in various other tasks thereby it acts as energy efficient and save a lots of time of the user. Home automation system's basic aim is to remotely access the electronic and electrical appliances at your home at a single click or by sensors and they work on a complete software as a user interface that can be any application or any webpage. Smart homes are also direly important like digitalization in this era. The home automation system is equally helpful to the disabled people as the user interface is designed according to the user which can do many of their tasks easier and fulfill their requirements [1].

The referred smart home is the basically known as ubiquitous computing i.e. universal computing comprises the innovation, smartness into the user's home for healthcare, comfort, safety, energy conservation and security. Saving human effort and providing efficient and secure work is the real success for this era. The smart home automation systems leads to enhancing the conventional safety and security measures by numerous new technologies not limited to artificial intelligence, Internet of things and machine learning [2].

Automation is blended in our life like lemon in lemonade. We start our day with the innovative alarm system and end our day with an advanced lighting system, automation has now-a-days established its roots deeply in everybody's day to day lives. Even the refrigerators, washing machines to appliances in home security, all can be controlled with a server connected with the WIFI or GSM in the smart home automation system.

Figure 1 illustrate the basic layout of home automation. Automatic home appliances are essential contents for the IOT as they require the internet for accessing or observing them remotely. The current scenario for home automation is more for security and for living a better life. Home automation is not just limited to the versatile appliances control, automatic computerized lights, robotized house cleaning, and portable warnings

F. P. García Márquez (ed.), *International Conference on Intelligent Emerging Methods of Artificial Intelligence & Cloud Computing*, Smart Innovation, Systems and Technologies 273, https://doi.org/10.1007/978-3-030-92905-3_32

through email/SMS, remote control for monitoring CCTV and booking machines but many more these are merely sense organs for the home automation. There are sensors inbuilt in the system for numerous kinds of detections like temperature, humidity, light, moisture, dampness, oxygen level, recognizing any intruder etc. Basic technologies in home automation are: Bluetooth, WIFI, zigbee, X10 and Z-wave [3].

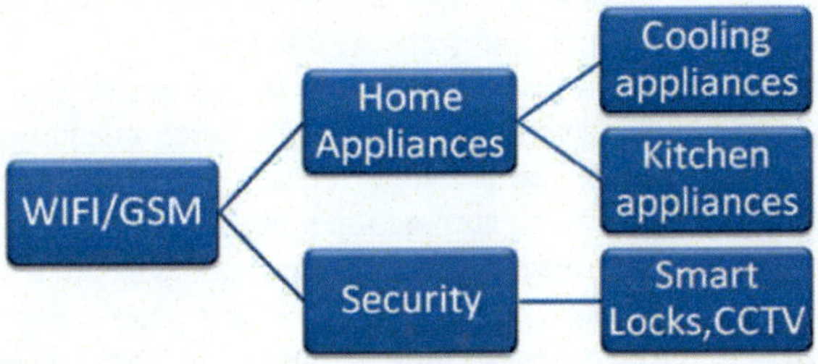

Fig. 1. All the Home appliances and security appliances i.e. cooling appliances, kitchen appliances, entertainment appliances and smart locks, CCTV all are connected with the WIFI/GSM in order to give remote access to the user.

Wireless Home automated system and home security system are the dual aspects of this research work. The current scenario of home automation starts from the outdoors of any home wherever a movement is sensed by the sensor it is send to the owner of the house by SMS or email and if anything contradicting is detected in that case alarms can also be raised in the intruder's direction. The home automated system is also adopted with the functionality of sending messages to the concerned security personnel feeded by the owner in cases of emergency or any sort of disconnection from the owner. On the another edge of the coin if the owner or the concerned person identifies that the following person is nit the intruder instead it is some unexpected guest then it can also add that person in a guest list and allow him to enter in his house by opening the gate and then automatic arrangements can be made for the welcoming of the guest by lighting up some light and warming lights for the guest and even some fresh aroma air fresheners can also be used, as everything is connected through microcontrollers and wifi. At a new instance we can say that the user can do all these arrangements for himself too by just setting up some arrangements in the system for all the world class comfort for him in order to lead an easy and happy life. He can set the feature that if the doorstep detect that the owner is coming it should automatically opens the door and when he enters in its house the lighting should welcomes him every day with a warming light so that the user feels good and enlightened and can also switch on the geyser for boiling water or can open television for the user with his favorite show. The home automation and home security are the compliment terms which are solving problems related to them traditionally. The board or stage of home is only required to have internet, home automation can be done even when the phone is not directly connected to the internet [4, 5].

2 Literature Review

In Home Automation and Security System by Surinder Kaur et al. concluded that to provide the ease to the users the home automation system need the various different methodologies including Arduino, GSM and android based applications and user interface. They have shared various different methodologies used with the components like GSM, arduino and android applications with their designs, flowcharts, implementations in association with their advantages and weaknesses [6].

In Home Automation: Past, Present and Future by Sharda R. Katre and Dinesh V. Rojatkar gives this paper shows a general overview of smart home projects that are masterminded by their expected services. It also covers the survey of various technologies which emphatically support the home automation systems in a reliable way. It also recognizes some future aspects for service providers like remote insurance administrations that work socially and majorly this works for the elderly people and disabled beings. They concluded that the future aspect of home automation would be broadly extended [2].

IoT based Home Automation System with Pattern Recognition by Ritvik Iyer and Antara Sharma demonstrates a generalized summary of home automation current and upcoming projects. They also covered a review on various technologies for consistency and reliability. It contains various new aspects of home automation for smart home research. The future healthcare service provider will consider the smart home powerful method for giving remote social insurance administrations, particularly to the elderly and disabled people. In future home automation will smarter and it would be extended to the large-scale environment [7].

Research Paper on Android based Home Automation using Raspberry pi by Ms. Akanksha Rajendra Surve and Mr. Sampat Vaidya demonstrates an android based home automation which provides an easy and attractive user interface and it makes the system far better in terms of flexibility and security. They have used mobile devices for remotely accessing the home automation system. They have used WIFI in order to establish the communication between the raspberry and the smart phones in which android applications are installed. There three major modules are relays, WIFI and raspberry pi. They have make a simpler project in order to cover the hardships of the conventional application [8].

Home Automation Using Internet of Things by Vinay Lohan and R. P. Singh presents a Home Automation system (HAS) using Intel Galileo that delivers the integration of cloud networking, wireless communication, so that it can provide the user with remote control of various lights, fans, and appliances within their home and storing the data in the cloud. There system automatically changes according to the sensors' data. There system is designed to be low cost and expandable so that it can allow a variety of devices to be controlled [9].

3 Methodology

The system that is proposed in this literature is designed to overcome the shortcomings of the existing systems. The proposed system is better secured, consistent and flexible. This system is carried out using raspberry pi and smart phones in which android applications are present so that the user can control the appliances and relay circuit. The WIFI or GSM signals should be framed strongly so that multiple devices can be controlled. Relay circuitry can also control the appliances. The foremost importance for home automation is providing cheap, secured and reliable automation which can be used to control the home appliances remotely. With the facility of android application a good range for scalability is present. By implementing this the user would be able to save the electric-power as well as man-power.

Each passing week comes with the enhancements in the existing appliances. The architecture shown in Fig. 2 consists of WIFI, relay circuits, sensors, android applications and raspberry pi which is a small size computer which manages the network so the access could be reliable. According to the configuration to the user's home system the WIFI and raspberry pi could set communication [10]. The signal of WIFI and the system should be strong, scalable and flexible.

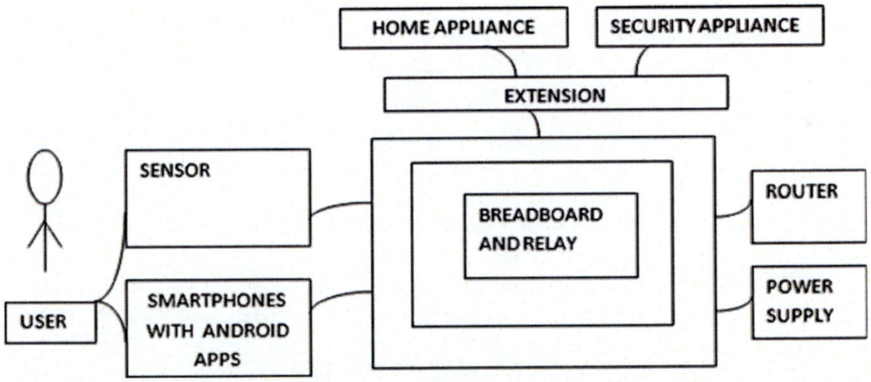

Fig. 2. Architecture of the proposed system: Here breadboard and relay plays an important role in controlling appliance remotely as well as closely in accordance with the user.

The medium to communicate with the devices is WIFI. The home wifi is configured according to the user's needs and it makes the services secure and consistent. Interfaces are used for configuring wifi with raspberry pi. The serial data (it sends the data bits one after another over a single channel) is connected to a raspberry pi circuitry extracting it from the wifi. Here the whole and soul of this research is the raspberry pi circuitry which is a micro computer as small as credit card size with plurality of functions. Raspberry pi will be configured for each and every appliances and the relay circuitry will take inputs for switching and functioning of devices. While thinking of users naive to the technology the installation process of home automation should be easy. The research shows 4 major modules namely: Raspberry pi, Android Application, Wifi and Relay circuit.

1. *Design*

The IR sensors will detect the entry of any intruder, trespasser or even any guest according to your known persons list. It is shown in Fig. 3. Starting from the beginning of any physical activity happens then the IR sensor is tricked into two senses whether it detects the entry of any intruder or did not. If it did not detect anything the system would remain close and save the energy but if it detects any unusual activity or person the alarm will be on with the siren or buzzer and the owner will get the message on his phone too irrespective of the distance the owner is on.

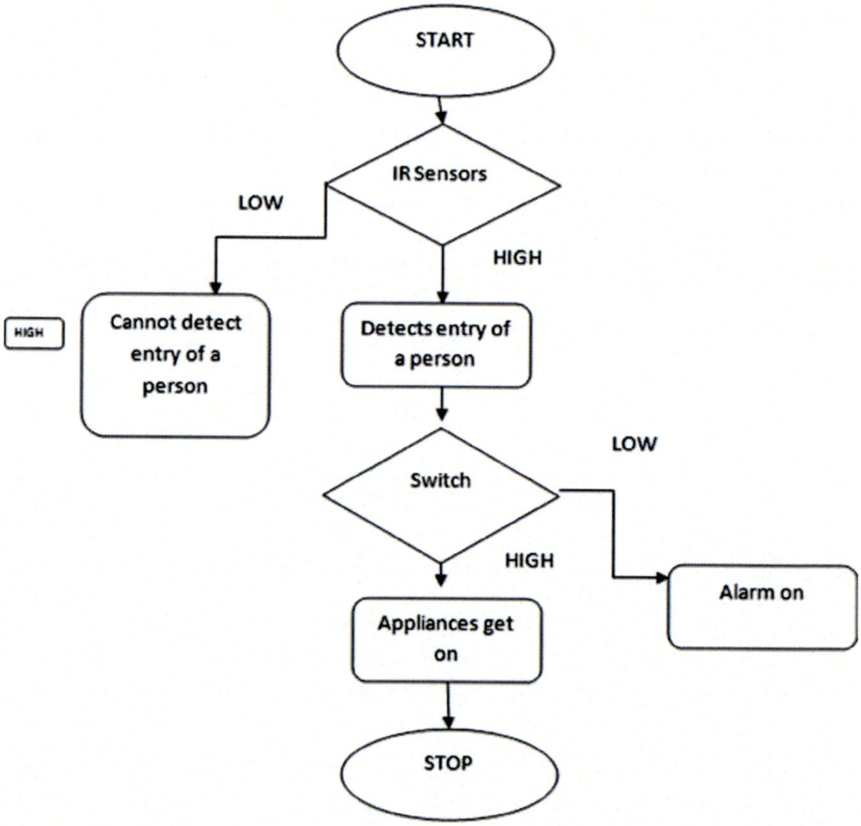

Fig. 3. Flowchart of Arduino based home automation system: A trespasser/intruder/criminal can be caught by this real time system as the alarm will buzz off and the information of entry will be informed to the owner via SMS or e-mail.

The only way to get a robust system we need to have a powerful communication system with the owner and the automated system. There are two possible communication types:

- Input communication
- Output communication

The flow chart of input communication module is shown in the Fig. 4. In Input communication, the user sends new settings or configuration to the home automation system with the help of SMS or android applications. It is used to create a secure environment as whenever any person tries to enter the home of the owner without his intervention, then to establish communication with the user the automated system sends a SMS or e-mail to the user each time whether he is present in the home at that time or not, it doesn't matter.

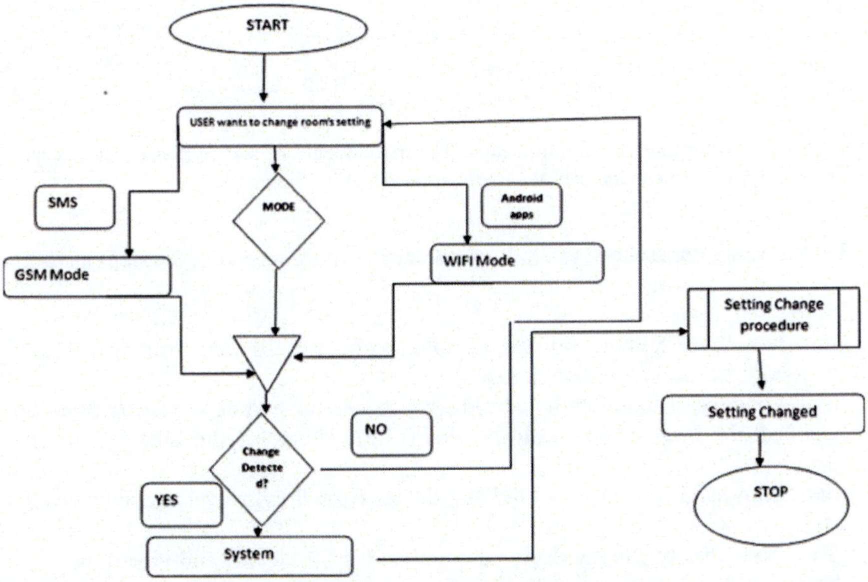

Fig. 4. Flowchart for input communication: In this diagram the user is trying to change the configuration of the system using SMS or an Android application.

Whereas in output communication is established in two modes namely SMS and Android applications. In this an alert message from the system has been sent to the user that the intruder is using the GSM mode. Then the work of GSM starts, it sends the information about the unknown entry in the home to the user or owner on his Smartphone.

As the system in output communication continuously monitors the input the output communication can even change its setting here as shown in Fig. 5. The setting can be changed according to the configuration. To control appliances via mobile or smart phones is the easy task [3]. They can be turned ON/OFF according to the wish of the user.

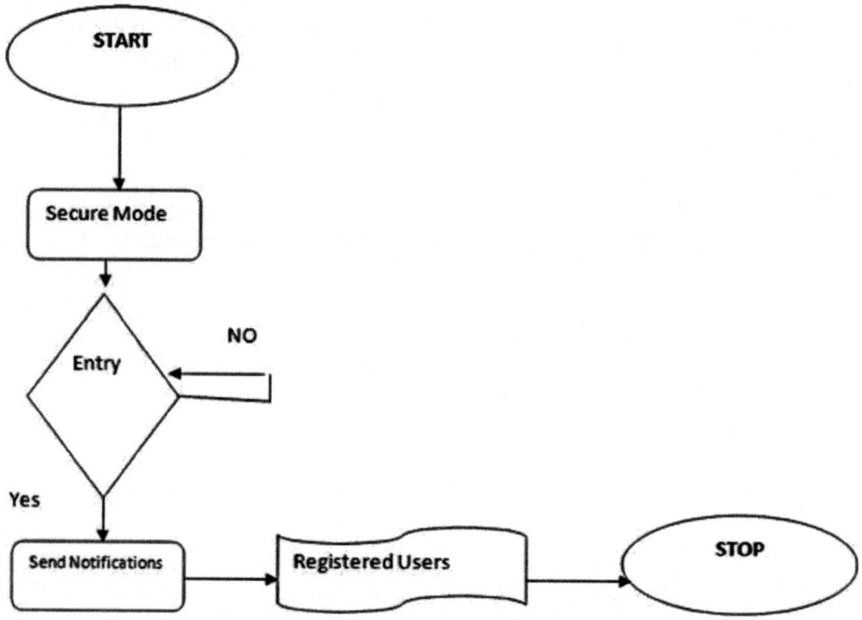

Fig. 5. Flowchart for output communication: The mode would be secure as each time whenever any unauthorized person is detected the notification is sent to the owner.

The above demonstrated system implements in accordance with the steps as follows:

(a) Start
(b) Initiate with the System, no devices hold any control till the system initiation.
(c) Secondly, the Wi-Fi module initiates.
(d) Smart phones installed with android applications have to be opened by the user.
(e) Connect the Smartphone with the WIFI, by waiting for the Smartphone accept device.
(f) Send command of control in ON/OFF device from the Smartphone, then check the status.
(g) Then accordingly ON the device if command is ON device and vice versa.
(h) Stop.

2. *Instrument*

There are majorly 5 modules which are used in this research namely: Android installed smart phones, IR sensors, Raspberry pie, Relay circuitry and WIFI.

2.1. *Android installed smart phones*:
Using android user interface is created for the use of naive users of technology, so that they can easily understand the interface and use it accordingly. Android

provides a plurality of facilities in order to get a desired work done and it provides a user friendly interface with the help of dialogs, notification, and menus. The user interface should complete the both the requirements of handling and monitoring the status of the home automation system.

2.2. *IR sensors*:
An infrared sensor i.e. IR sensors are an electronic component which is used to detect and measure the infrared radiations in its surrounding environment. It is used here to sense the detection of the intruder.

2.3. *Relay*:
A relay, closely appears as a switch that is operated with electric energy. Relays are used to control a circuit with low power signal or where several circuits must be controlled by one signal. The relay circuit is given the output which is generated by the raspberry pi. Due to this the relay will turn on and the device will work. NPN transistors are used in relays. According to the purpose of our application the relays can be selected.

2.4. *Raspberry pi*:
The Raspberry pi is a small credit card size computer whose cost is less but it performs many functions. It is a small single board computer. Raspberry pi automates home appliances and allows users to control them easily through the Internet from anywhere over the world [11].

4 Discussions

The proposed system uses a wireless communication method which can be a radio frequency or infrared waves powered by a real time system. Various other systems exist which use a combination of various other methods to overcome the shortcoming shown by the conventional systems. The software or application which is also the user interface plays a major role in deploying the technological advancements in home automation easily. The conventional systems were not famous just because it does not have any user interface and every new technology is not easily understandable by the users if not explained properly. But in current scenario smart phones are the first choice of any new user as android based user interfaces are easily understandable and have a wide variety of applications and scope for enhancements. Smartphone are easily portable for accessing home appliances from anywhere around the world. The proposed systems that have been demonstrated have certain common features as the system uses a communication technology. The proposed system has a relay circuitry for controlling and monitoring the electrical home appliances. All the home appliances should work on the common command system that will be used to issue commands to the control circuits. The proposed system comprises the security features to ensure that the person entering in the home is the authorized one. The proposed system uses WIFI and GSM in order to maintain a local and common network between the appliances and can monitor and control the appliances. Hence, the main aim of the proposed system is to provide a secure and easy home automation system to the user in order to live a peaceful and worthy life.

5 Conclusion and Implication

This paper presents a home automation system in which Arduino is used as microcontroller to design and implement. This system uses WIFI in order to control and monitor the appliances available in home remotely. This system is a smart system as it can mould itself according to the environment the user wants to create in his home or user sets iteratively in the home. By this system users would be able to save energy as well as human efforts, as the dependency of switching ON or OFF of the appliances reduces for both the normal person or even for the elderly or disabled ones. This paper uses various kinds of automation methods which use Arduino, GSM and Android in order to give designs, implementation and flowcharts. The major aim regarding home automation focuses on providing ease to the people for controlling various devices in less time, money and energy. Other than that it also helps the user for protecting their lovely homes from burglars by using an alarm whenever any unusual activity is sensed by the sensors. In regard to future work we expect to make this system even more advanced by adding LCD monitor which counts the number of people entering the home with their temperature, color, iris scan so that we can know better about the intruders. And for later we would extend this system to automate a large scale environment, such as enterprises and factories.

References

1. S. Palaniappan, N. Hariharan, N.T. Kesh, V. S, A. Deborah S, Home automation systems—A study. Int. J. Comput. Appl. (2015). https://doi.org/10.5120/20379-2601
2. S.R. Katre, D.V Rojatkar, Home automation: Past, present and future. Int. Res. J. Eng. Technol. 343–346 (2017). https://doi.org/10.14257/ijsh.2016.10.8.18
3. S. Jain, A. Vaibhav, L. Goyal, Raspberry Pi based interactive home automation system through E-mail (2014). https://doi.org/10.1109/ICROIT.2014.6798330.
4. A. Elshafee, K.A. Hamed, Design and implementation of a WiFi based home automation system. World Acad. Sci. Eng. Technol. (2012)
5. R. P. Saini, B. P. Singh, M. K. Sharma, N. Wattanawisuth, N. Leeprechanon, Designing of smart home automation system based on Raspberry Pi (2016). https://doi.org/10.1063/1.4942698
6. S. Kaur, R. Singh, N. Khairwal, P. Jain, Home automation and security system. Adv. Comput. Intell. An Int. J. (2016). https://doi.org/10.5121/acii.2016.3303
7. R. Iyer, A. Sharma, IOT based home automation system with pattern recognition. Int. J. Recent Technol. Eng. (2019). https://doi.org/10.35940/ijrte.B2060.078219
8. I. J. of A. E. and Research and Development, Research paper on android based home automation using Raspberry pi. Ms. Akanksha Rajendra Surve, Mr.Sampat Vaidya **4**(6), 806–808 (2017)
9. V. Lohan, R.P. Singh, Home automation using internet of things. *Lecture Notes in Networks and Systems* (2019)
10. B. Dave, S. Kubler, K. Främling, L. Koskela, Opportunities for enhanced lean construction management using Internet of Things standards. Autom. Constr. (2016). https://doi.org/10.1016/j.autcon.2015.10.009
11. D. Pavithra, R. Balakrishnan, IoT based monitoring and control system for home automation (2015). https://doi.org/10.1109/GCCT.2015.7342646

A Review on Impact of COVID-19 on E-Commerce

Madhav Singh Solanki(✉)

Department of Computer Science and Engineering, Sanskriti University, Mathura, Uttar Pradesh, India
swapnil.cse@sanskriti.edu.in

1 Introduction

Evolution of technology plays important role in our lives. One of the earliest triumphs in the communication field was achieved by Alexander Graham Bell by inventing the first telephone in the 19 century. In today's world everything is only one phone call away, from ordering food to other essentials and even keeping in touch with your near and dear ones. Also by each passing day newer technologies are taking the market by storm and people are getting used to them in no time and especially electronic media. Cyber communication provides valuable learning opportunities and independence, this has also dramatically expanded the social circle. With the increase in demand of technology even our phones change their states, a device merely less than the size of our palm has the potential to fulfil all our necessities. Technology acts as a catalyst to effectively enhance the applications for larger groups of society and has gained immense support from people in the past few years. With advancement in technology devices such as computers, laptops, tablets, mobile phones are used rapidly and with these devices some other technologies like conversing applications(What's app, Facebook etc.), e-commerce etc. E- Commerce has been the leading support system for the people in this time of pandemic when all over the world every person is homebound [1].

1. *Electronic Data Interchange (EDI)*

   For carrying business e-commerce involves use of internet and information technology like Electronic Data Interchange (EDI). EDI can be defined as technology to exchange of standard business dealings, invoice, shipping notes, inventory etc. [2]. Fig. 1 states all the benefits of EDI:
2. *Electronic-Commerce (E-Commerce)*

   E-commerce makes use of these technologies and offers all possible help in processing business transactions, inventory etc. to redefine value creation among different organizations as well as between organizations and individuals. There are following types of electronic- commerce:

   - Business-to-business (B2B)—commerce model that mainly focuses on selling goods and services to other companies

F. P. García Márquez (ed.), *International Conference on Intelligent Emerging Methods of Artificial Intelligence & Cloud Computing*, Smart Innovation, Systems and Technologies 273, https://doi.org/10.1007/978-3-030-92905-3_33

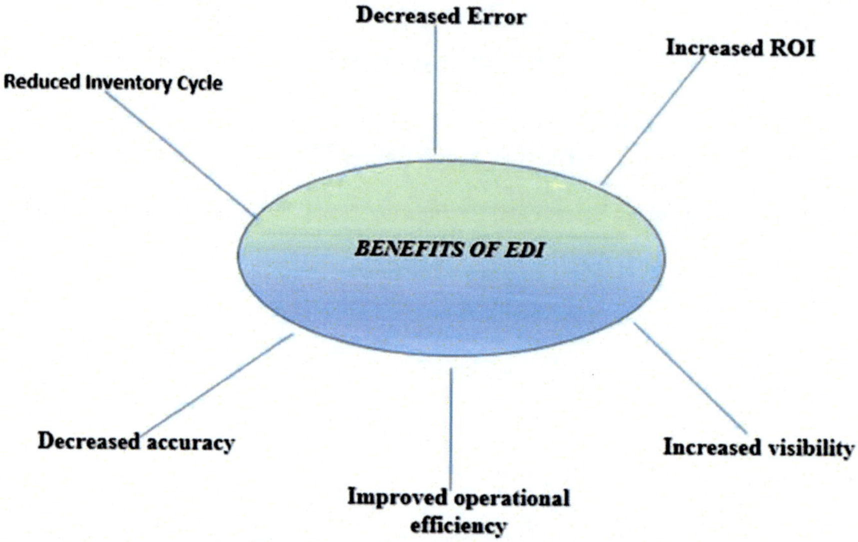

Fig. 1. The figure depicts all the benefits such as Increased Return of investment (ROI), Decreased Error, Reduce inventory cycle of Electronic Data Interchange (EDI)

- Business to- consumer (B2C)—commerce between business organizations and individuals.
- Business-to-government (B2G)—model that sells goods and services to federals, states or other government bodies.
- Consumer-to-consumer (C2C)—trading of customers with each other over an online environment [3].
- Mobile commerce (m-commerce)—Buying, selling of goods and via wireless portable devices like mobile phones, tablets etc. [4].

3. COVID -19 Outbreak

From bygone days e-commerce has been used by everyone in their day to day life but in this present scenario of pandemic online commerce has lended its hands and has been the lifeblood to the people living in cities by keeping people stay back at their homes. The factors which drove the growth of e-commerce business before COVID-19 outbreak were steady growth of internet users, low prices with certain offers, online launching of products, rising online shopping awareness and so on. Coronavirus is the large group of viruses that causes illness such as pneumonia, bronchitis and was first discovered in Wuhan city of China. People were forced into a lockdown because this disease had a contagious spread or outbreak. In the month of March the entire world was under the cover of this outrageous disease. Complete world was under shut down, people were not allowed to move out from their homes and in necessary conditions and wearing a mask became a compulsion for each person. Social distancing and staying home pushed the inclination of people towards e-commerce, as life has to go on and daily needs to be fulfilled. People in cities completely relied on e-commerce sites such as amazon, flipkart, big basket, and grofers to order their essentials from groceries to medicines and other goods. Due to transparency of the market, customer can compare the services of various e-commerce sites easily [5]. E-commerce acts a great help to customers from accessing the global market by just sitting back at their homes and ordering all the required stuff.

4. Global E-Commerce Scenario

   Globally e-commerce surveys are segmented or based on products such as electronics, personal care, healthcare beauty and others. The contagious outbreak of COVID-19 had a vast impact on these segments due to the uncertainty in the supply chain [6]. Due to an increased number of demands and factory closures worldwide, a huge shortage of products occurred. As whole electronics manufacturing is based in china due to shut down these products entered the circle of shortages. Going out for saloons, restaurants, shopping, theatres is not considered normal anymore in this time of pandemic. Now the normal is wearing a mask, maintaining social distance and wandering outside only when needed.

   In this situation of pandemic E-commerce has become a alternative source and considered, and e-retailers provide goods that usually consumers bought in superstore traditionally [7]. E-commerce has made life much easier as earlier people were inclined towards supermarket shopping in crowded places. Platforms such as swingy, zomato, Uber eats etc. which were solely the food ordering app have turned their way round and started delivering essentials (grocery, medicines, because sale of ordering food from restaurants became null and people were eating at their homes. The present study describes the impact and contribution of e-commerce during the epidemic of coronavirus.

2 Discussions

Before the pandemic, online shopping behaviour had clear distinctions between the people for whom it was a matter of comfort, or for whom it was imperative to shop from physical market and brick and mortar stores because think of it a more of an active process. Pandemic has clearly created a shift in the way people behave in their shopping habits, directly affecting the e-commerce industry.

Found in a market research before the pandemic outbreak, it was prophesied that the growth of e-commerce market in India would reach upto 200 billion US dollars by the year 2026. Post the outbreak current trends suggest that the goal of 200 US Billion dollar scan be concurred way before the predicted year because of the increasing number of people opting for online deliveries amid all the fear of infection while stepping out of the homes [6]. The E-commerce industry is affected from both supply as well as demand side, some companies are able to adapt and respond to the changes, others are struggling hard to cope up with the changes. Hospitality industry, marriage bureaus, tourism industry have faced a major backlash during the pandemic and their business saw the largest downfall in recent years.

It has been found that there is steady decline in sales of consumer packaged goods since the shutdown (Fig. 2). Survey conducted by Digital Commerce in the month of March 2020 came out with a study that among 304 e-commerce retailers over 36%

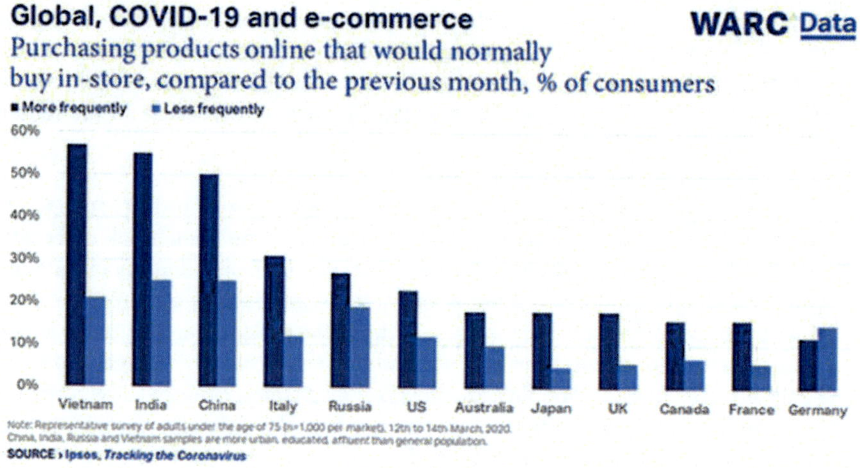

Fig. 2. Global, COVID-19 and e-commerce purchasing products and buying habits of customers after pandemic various countries. *Source* WARC Data

resolutely felt that their business will face a significant loss whereas 42% of them believed that there will be a shift in nature of consumer buying [6]. Another critical problem faced by companies was shortage of staff which led to non-fulfilment of orders placed by the consumers. A report suggested that online stores or supermarkets such as Grofers, Big Basket had access to only 60% to 50% of total staff respectively, due to COVID-19, hence were unable to meet with deliveries and frequency of filling up the customers demand reduced at a very high rate [6]. This whole situation has put immense pressure on the companies and as a result their business profit decreased. Similarly Amazon, Flipkart and other online shopping services suspended the deliveries of certain goods and restricted deliveries to only essentials such as grocery, medicines and other essentials in light of the reduced staff and logistical crisis as a consequence of lockdown. All the studies and report suggest the change in sales after the outbreak of the pandemic.

Globally, the online channels benefited from a demand boom (When almost every country worldwide entered into the lockdown) major supermarkets chains and online media, according to research from Stastia. Marketplaces and other tech retails saw a statistically low double-digit growth, while some other sectors declined or even collapsed as in the case of tourism, hospitality and event platforms. According to the study provided by COVID-19 Commerce Insight (Fig. 3), almost 25,000 brands across hundred countries, sterling e-commerce firms are holding up so well in countries like Europe and USA, Wherein Asian countries, Nigeria showed a steady decline in online sales. It seems that where e-commerce has established itself among consumers, there is a switch from supermarket shopping to online platforms, on the other hand in underdeveloped countries e-commerce markets were not sufficient enough to respond.

Some African entrepreneurs are finding small ways to overcome the crisis. Founder of some trading groups and investments have signalled an increase in demand for personal hygiene and basic foodstuffs: Items that are sold locally or in the market. According to study, e-commerce markets in African countries are now more focused on essential

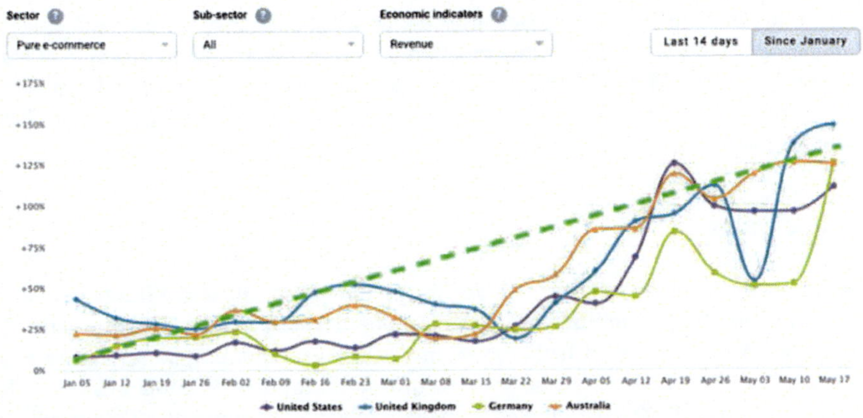

Fig. 3. Graph representing the data representing sales of various items in different countries during the pandemic of COVID-19. *Source* COVID-19 Commerce insight, https://ccinsight.org/ [8]

products. Products such as hand wash and sanitizers are two fast moving essential products. Even markets in such countries have planned to supply edibles (rice, sesame, etc.) to local or international vendors after the crisis [9]. African e-commerce market has gained a boom in pandemic and they have even tried their hands on exporting products in local as well as international market.

Trends and buying habits of customers has also changed during the pandemic. Browsing or fashion products and apparels have increased but there is a major decrease in sales of these products as they were of no use until needed in emergency situations. People during the pandemic have realised what are the necessary items to survive and have changed their buying or shopping habits. Most of the clothing brands have put their items on sale and also presented some other offers. Similarly Luggage and bags due to restriction of travelling have seen a decrease in sales. Due to shutdown people stayed home for almost four months due to this there was a major increase in sales of home goods and even sellers in the furniture industry have recorded 220% increased sales [10]. Now there is an increase in competition in several industries retailers even with shortage of staff are dealing with competition in the market.

The outbreak of COVID-19 has not only transformed the way consumers shop but has also seen a steady rise in online payments. Contactless payments received an unparalleled boost during the crisis, it was seen as a cleaner and hygienic way to pay in stores. Consumers are relying more on online payments even on online platforms so that there is contactless delivery. Customers are trying new payment methods while purchasing, and favouring those methods that have the stronger protection against fraud loses [11]. COVID-19 has changed everything from way consumers shop and the buying habits.

According to aforementioned findings, study shows that several industries have seen a major increase in sales while other industries are facing major backlash. The pandemic has changed the whole scenario in every aspect. People worldwide have shifted their interests and traditional habits to new emerging technologies and e-commerce is one of them. Although developed countries were making use of such platforms from way back in time, in developing countries after the crisis people have adapted the change. E-commerce market is dealing with a great amount of pressure due to shortage of staff and even goods. Supplies of products have also deteriorated due to shutdown of all transportation services and increase in demand of products. These online platforms have boosted themselves up for delivering products and the customers demand.

3 Conclusion

Information Technology (IT) has been playing an important role in the future growth of financial sectors and promoting new ideas for doing business in emerging economy like Bangladesh. Advancement of technology, increasing use of smartphones and internet has brought a lot of changes in all spheres of day to day life of human beings. This increase usage has shifted the interest of people towards e-commerce. This pandemic situation has increased the usage of e-commerce in every field from groceries to other essentials.

When the whole world was dealing with the outbreak of COVID-19 and people were forced to stay back at their homes the trend of mart or window shopping drastically decreased. When the news of this pandemic breaked worldwide panic buying started and people rushed to the market for stockpiling all the essentials and as shutdown came into picture the shift towards online shopping increased drastically. Various studies and researches have shown that e-commerce businesses saw immense growth in underdeveloped countries as well. Various industries had different impacts like tourism, hospitality, apparel came into great loss because sales and business of these industries became negligible. Contrary food industry, sanitizer and other essential items became the need of an hour and their business saw an immediate boom.

On the other hand small businesses and retails shops collapsed but some new opportunities in the field of e-commerce boosted during the time of pandemic, in India MukeshAmbani collaborated with Reliance retail and Jio platforms launched JIO MART. This venture came as a substitute for all online shopping platforms such as Amazon, Flipkart, Grofers, Big basket etc. Jio Mart delivers groceries and other essential items from the nearby store or small retailers providing them business and profit.

References

1. I. U. Dr. Shahid Amin Bhat, A Review Paper on E-Commerce
2. S. Narayanan, A.S. Marucheck, R.B. Handfield, Electronic data interchange: Research review and future directions. Decis. Sci. (2009). https://doi.org/10.1111/j.1540-5915.2008.00218.x
3. P.S. Aithal, A review on various E-Business and M-Business models & research opportunities. IJMIE **6** (2016)
4. Mohammad Wasiq Dr. Naseeb Ahmad Dr Mohd Tehseen and Burney, Future of M-Commerce services in India. Int. J. Mark. Financ. Manag.
5. A. G. Khan, Electronic commerce: A study on benefits and challenges in an emerging economy. Type Double Blind Peer Rev. Int. Res. J. Publ. Glob. Journals Inc (2016)
6. Impact of COVID 19 on the E-commerce Market.
7. A. U. K. Hamza Akram, E-commerce trends during COVID-19 Pandemic

8. INSIGHTS. https://insights.emarsys.com/observations/retaining-new-online-customers-is-key-for-e-commerce-brands/
9. J. Howe, The e-commerce response to COVID-19. https://www.intracen.org/covid19/Blog/The-e-commerce-response-to-COVID-19/.
10. V. Mejia, Understanding the COVID-19 effect on online shopping behavior. https://www.dmxmarketing.com/research-covid-effects-on-ecommerce/
11. GLOBE NEWSWIRE, COVID-19 Impact on e-Commerce & Online Payments, Worldwide, 2020 - Online Shopper Penetration Increases During the Pandemic. https://www.globenewswire.com/news-release/2020/05/29/2040716/0/en/COVID-19-Impact-on-e-Commerce-Online-Payments-Worldwide-2020-Online-Shopper-Penetration-Increases-During-the-Pandemic.html

Check for updates

Data Security Techniques in Cloud Computing

Pankaj Saraswat(✉)

Department of Computer Science and Engineering, Sanskriti University, Mathura, Uttar Pradesh, India

1 Introduction

Cloud computing refers to data centers providing on-demand computing services like servers, storage, databases, networking etc. available to users over the internet. In cloud computing, "cloud" means collection of networks Cloud uses multiple servers to store data. As per the demand and requirement, cloud computing modalities can be used by the user [1]. The user prefers any provider mediator for internet service in cloud computing. For the online services for services, one must criteria is security for privacy prevention and avoiding of leakage of information but it must be enough strong for avoiding various attack of breaking [2]. Cloud computing provides the users different services as service models. In cloud computing various devices such as mobiles, laptops, computer systems and other devices are interconnected by the internet to cloud. Cloud has of 3 major components storage devices, framework and applications [3]. These three are the services that are provided by cloud service providers to the users. Cloud computing made life easier and less expensive as the user only have to pay for the service being used. Cloud computing made possible for accessing the infrastructure (computing ram, storages etc.) online without installing any hardware and software. Cloud computing provides security and accessibility. Cloud computing revolves around three functional components that are cloud service provider, client/owner and user [4]. In terms of security means, cloud stores data over multiple servers and secure from data loss and data breaching attacks and in terms of accessibility means, cloud provides accessibility to the data anytime from anywhere. As given in Fig. 1, key features for cloud computing can be Elasticity, Resource Pooling, broad network access, measured service and on self-service on demand.

1. *Services of cloud computing*

A number of services are provided through cloud computing by cloud service providers. As mentioned above, there are three main components of cloud that are Storage, infrastructure and applications. These services are defined as three services namely Infrastructure as a service (IaaS), Software as a service (SaaS) and Platform as a service (PaaS) [5]. These 3 service provide user storage space for their information, applications and software and users can also host their own applications and software on cloud through a web browser and an internet connection.

F. P. García Márquez (ed.), *International Conference on Intelligent Emerging Methods of Artificial Intelligence & Cloud Computing*, Smart Innovation, Systems and Technologies 273, https://doi.org/10.1007/978-3-030-92905-3_34

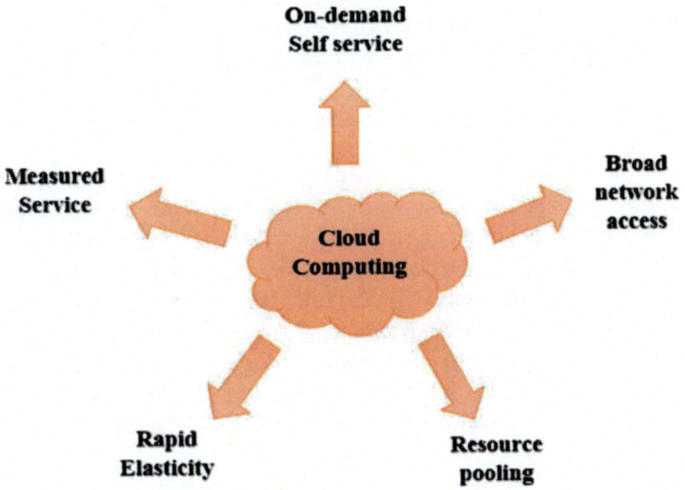

Fig. 1. There are 5 key features of cloud computing provided by every cloud service provider. These features are essential to maintain security and accessibility of data

1.1 *SaaS (Software as a service)*

Software as a service is used by end users and clients who do not have any prior knowledge of backend components. As the user does not know where the data will be physically stored. A third party provider hosts applications which are made available to the customers over the internet. The user utilizes the third party software or application and pays on the basis of the time software is being used. User or client needs to have an internet connection and web browser. In this user is not provided with control or access over Operating system, middleware, networks, virtual machines etc. SaaS applications are accessed by users through web browsers, without downloading and software and installing any hardware. With SaaS, it is easy for enterprises in streamlining thesupport and maintenance, since everything needs to be manage through vendors: runtime, applications, middleware, data, Operating system, virtualization, server, networking and storage [6].

1.2 *PaaS (Platform as a service)*

Platform as a service (PaaS) provides environment for runtime of applications, application deployment and application development. The runtime environment is formed with facilities provided by Infrastructure of IaaS (Infrastructure as a service). This service supports overall cycle of life for delivering and building the applications of the web and service completely through the internet. In cloud, an overall deployment and development whose resources enable the user to deliver everything. With infrastructure provided by IaaS With PaaS, the cost and complex nature of buying as well as maintaining licenses of software is reduced. PaaS supported an overall web application lifecycle: building, testing, deploying, managing and updating. The user is provided with the facility to develop their own software applications with the provided platform from PaaS.

1.3 *IaaS (Infrastructure as a service)*

IaaS provides resources in form of network, storage, hardware, OS and on demand storage device. Various benefits of using IaaS (Infrastructure as a Service) includes smooth workflow, reduced costs of hardware infrastructure. IaaS offers secure and automated backup of data. The user is provided with the full control over database, virtual machine, middleware, networks etc.another advantage of using IaaS is disastrous recovery and continuity. Mostly enterprises and businesses use IaaS to perform tests, developments, backups and recovery faster. As IaaS also provides the facility of processing power to more economically mining of large data sets may be defined as big data analysis.

2. *Type of clouds*

There are majority three type of clouds being used and provided by cloud service providers namely: Private Cloud, Public Cloud and Hybrid Cloud. First one is open and software deployed can be used by users whereas the private cloud is expensive that provides private clouds to users for storing sensitive information and Last one is amalgamation of public cloud and private cloud.

Public Cloud: Public clouds are multi tenant environments that provide multiple users to store information on a same server. Public cloud is pay-per-use model. User have to pay every time for the time and data user consumes. Public cloud is used by multiple organization as all the organizations store the information on same server. There are more chances of Data loss and data leakage in Public cloud. Public cloud allows enterprise to compute, store and host any application and software.

Private Cloud: In this type the one business organizations stores data in one cloud. Private clouds offer high security and secure environment for the sensitive information to store on the cloud. It is also called the corporate cloud. This cloud uses internal hosting and firewalls to make data more secure and prevent from data breaches attacks.

Hybrid Cloud: It is amalgamation of private and private cloud. Business uses hybrid because it provides the business organization the security and safety with low expenses. Hybrid cloud comprises of a IaaS model with private cloud. Hybrid cloud is capable of running any critical level of application and even more dynamic level of application [1].

3. *Cloud Storage architecture*

There are two important components of cloud storage namely: back end and front end. Latter is component which is shown to the user of cloud service whereas backend comprises a number of computers, devices for storing data and servers. This works- the user accesses the service through the front end component of cloud on the web browser and the cloud provider provides the services through the client-chosen data model when requested. One data server needs to connect to internet. The client sends data to cloud server user needs to store and the user can access, delete and update the data when required. The front end will export API for accessing storage databases. In conventional storage devices the application programming interface used are Small Computer System Interface protocol. For cloud computing the protocols have been developing. Cloud has services such as storage, interfaces, networks, hardware which enabled delivery of service of computing [7]. The backend of the cloud is the physical storage databases. The cloud architecture consists of many cloud components. All of the components of clouds are loosely coupled. Mainly there are two main components front end and back end. According to Fig. 2 Front end of the cloud is client or user part which is accessed by the client. Front end of cloud consists of applications and interfaces. These applications and interfaces will be used to access the platform of cloud. On the other hand, back end of the cloud comprises of databases, servers, storage devices, virtual machines and deployment models etc. John Harauz, Lori M. Kaufman Bruce Potter did research and named it "Storage" Data Security in Cloud Computing's world" and mentioned that ensuring integrity, confidentiality as well as availability of data is job of provider of cloud service and provider must provide mainly of three capabilities an encryption schema for securing the data, access control module for maintaining authorized access, data backup [8].

Security of Data in cloud computing: Protection ofdata and security are most important factors for making the cloud technology successfully used. Security of data in cloud computing involved a number of encryption of data. Requirement of security of data depended on three model of service SaaS (Software as a service), PaaS (Platform as a service), IaaS (Infrastructure as a service).

Two states of data exist which could be a threat to security:

- Data which is at rest
- Data in transit.

Data at rest: Data at rest implies that data which is stored in cloud or which is accessible by internet, it can be backup data or data which is live. Many a times becomes difficult organizations in protecting the rest data if a private cloud has not been maintained by them since no physical control of data can be maintained [9]. Data at rest also means the data inside a storage device as floppy disk, hard disk, Disks etc. It is advisable to store more sensitive data in data in rest mode as the attacker find the data in rest is less valuable than data in transmit.

Data in transit: Data in transit means data in a form file or a database is transmitting in and out of the cloudthrough a network and can be accessed from anywhere. Data in transmit is also known asdata in motion and can be transmitted from local device to cloud server. Whenever the data is uploading on cloud, data at time of uploadation data in transmit. The data under transmit can be very sensitive data like usernames, passwords, bank account details and needs to encrypted a number of time. The best practice to get secure from attack is not to rely on reactive security and use proactive security. To protect data in transmit user can also use robust network security controls (Fig. 3).

Protecting sensitive information like usernames, passwords, bank details is all important for organizations from hackers. As hackers have increasingly advanced ways of

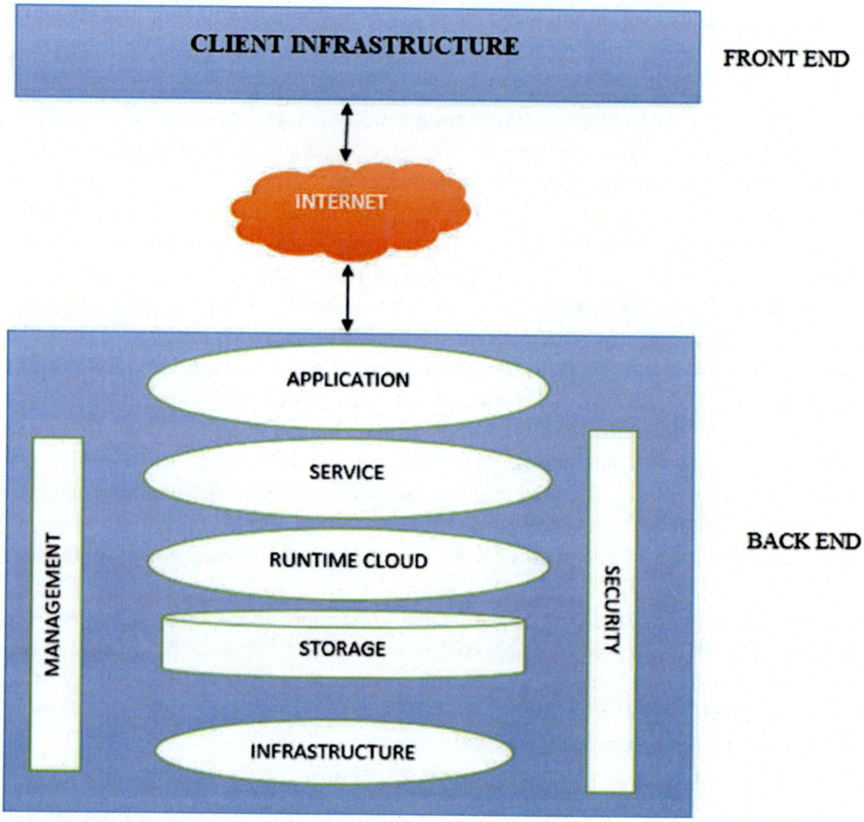

Fig. 2. Cloud storage architecture- which comprises of mainly frontend (operated by user) and backend (all the databases, services and storage etc.)

stealing data and the security techniques used by cloud service providers are vulnerable enough to get attacked.

Cloud computing saves money and time of an organization but having truston system of provider is important since real benefit of an organization is data they save in cloud [10].

2 Discussions

As businesses, organizations, enterprises and individuals are all facing challenges in managing their data. Whether an individual or an organization, almost every person stores data over the cloud which saves a lot of time and money. The requirements security has dissimilarities from any one type to another type like from data in transmit, in rest and data in use. The data security techniques use the concept of CIA trio: Confidentiality, Integrity and Availability, but applicable to virtualized, dynamic, distributed architecture.

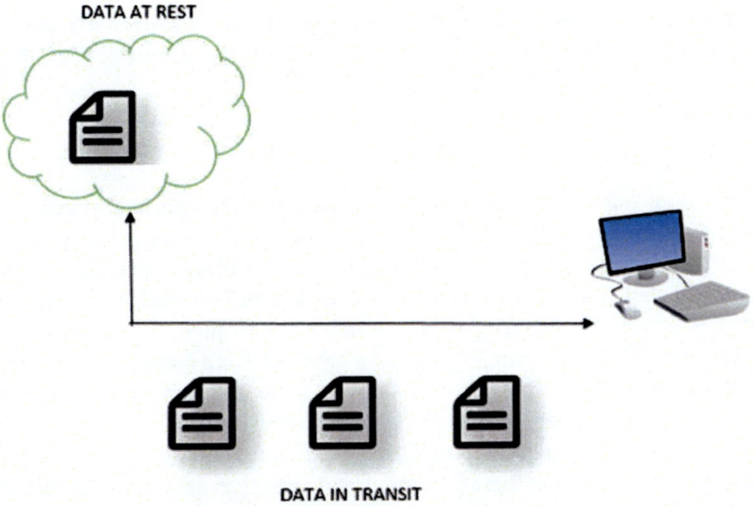

Fig. 3. The data at rest in cloud and data in transit. The rest cloud data is stored by the user whereas data in transit is data transmitting in the air.

1. *Confidentiality*

Multiple users store data over cloud remote servers and the data may be accessed by anyone. For protecting the data from unauthorized access confidentiality is introduced. Confidentiality means protection of the sensitive data from people who are not authorized and making sure that data is accessed by only authorized person. In a context of decentralized computing the problems of confidentiality are a lot and important because server which is hosting data need not belong to user necessarily [11]. There are two basic techniques to achieve confidentiality that are cryptography and physical isolation. Yusuf Haider and Siva SelvanfromManipal Universityconducted a survey and discussed about various techniques that can be used for enhancing confidentiality in cloud servers such as Secret Sharing Scheme, Biometric Encryption, HPI_Secure, K-NN classifier and preserving confidentiality through obfuscation and encryption etc.

2. *Integrity*

Data Integrity refers to making sure that the data stored on the cloud by a user must be updated, accessed or modified by only authorized party. Integrity of data also comprises of continuing keep on the data accuracy, data consistency and data trustworthiness. Integrity means maintaining the correctness, and completeness of data. Integrity can be maintained by conducting redundancy checks regularly and digital certifications along with another features like authentication, authorization and identification [12]. In centralized system, it is simple in performing, but in distributed environment it is a complex task as cloud computing. Providers use Keyed hash function for maintain the data integrity.

Keyed hash function is easy, simple and easily applied. Various other schemes are also used such as Provable Data Possession (PDP) and Proof of Retrievability (PoR).

3. *Availability*

Availability means making the data available to authorized people at anytime, anywhere when needed by them and availability of data is the utmost concern of cloud service providers. If due to any problem or issue the provider is failing at providing data, many users will be affected. Service providers ensure that data will be available in any situations from normal to disastrous. Ensuring the data availability includes the time for processing the data. To decrease the downtime two nodes are used Active and passive nodes. This works as if one node is down the other node will come into play and support the load. This process is repeated every time at the time of execution.

As data security is the most important factor in cloud computing. No organization, business, enterprise will save their important and sensitive data on cloud provided by a service provider. There are many cloud service providers and every service provider uses multiple techniques for ensuring security of data of clients from various data attacks implemented by hackers in order to steal sensitive information. As discussed, cloud computing provides various services: PaaS, IaaS and SaaS. These are the service models where security is needed while communicating and transmitting the data between browser and service provider cloud. All of this is based on degree for abstraction. SaaS model has a high degree of functionality which is integrated having a minimum control of customer and extensibility [13]. In contrast PaaS model has offered larger extensibility and bigger control to customer, mainly of comparatively lesser abstraction degree, IaaS offered larger tenant or control of customer on security than PaaS or SaaS [13]. The IaaS (Infrastructure as a service) is more secure as compare to PaaS (Platform as a Service) and SaaS (Software as a Service) also offers more features and control as compared to PaaS (Platform as a Service) and SaaS (Software as a Service).

SaaS provide application services when demanded by user of cloud. As user has less control over security between three models. There are various security concerns that arise when SaaS is accessed such as Application security, multi-tenancy, data security and accessibility. If there is any weakness or limitation in application, then it will create vulnerabilities which will help the hacker to steal or exploit the data and multi-tenancy is added to the model for efficiently using the resource but with less scalability. The risk of data-leakage is high between tenants because the data of more than two tenants is stored in the same database.

IaaS provides services in the form of resources like servers, networks, storage etc. in virtualized form. As the services are provided in virtualized form there is exposure of a new layer that needs to be secured. With that the hackers have more new opportunities of accessing the system unethically and stealing the data. The hacker can migrate the Virtual machine on a malicious server which will compromise its confidentiality and data integrity.

There are many researches relevant to security of data in cloud computing. R. Sridaran, Dean of faculty of computer applications, Ms. Disha H. Parekh, Assistant professor, discussed challenges of security in cloud computing in a paper named "An analysis of security challenges in cloud computing". The study had analyzed every threat of security found over both models of cloud network and models, has revealed solution to a few of them [14]. To ensure security there are a number of techniques for security for protecting the uploaded over the clouds. All of the techniques provide confidentiality, access control, authorization and authentication.

Authentication is necessary to ensure the services provided to the user is authorized for that service. Ensuring the identity of the user means the user can now access the saved files on the cloud. Every cloud service provider implement various authentication methods for increasing the security and those methods confirm the identity of a user when a service is requested from the user end. Various authentication methods are used depending upon the cloud service provider such as usernames and passwords, biometrics, multifactor authentication, single sign on, Mobile trusted module and many more. Private and Public type of cloud use a number of designs to authenticate with RSA. This cryptosystem has accept various models to authenticate such as knowledge based authentication two-factor authentication, and adaptive authentication [15].

A data protection confidentiality security mechanism is used. Encryption technique is used for ensuring the safety in which the plain text which is transmitted from one place to another is encrypted into cipher text first and then transmitted and after receiving the cipher text is decrypted again into plain text. Mathematical algorithms are used by encryption techniques for transforming the data from plain text to encrypted text. Encryption keys are used while the data is encrypted and stored on cloud and further used when the data is accessed by the user. The encryption keys are then again used to decrypt the data into plain text. Access control is also a very important factor in data security. This ensures only authorizable person are receiving requested information saved over cloud. Access control is very important phase for business organizations for protecting their data from reading, writing or updating the data stored by them on cloud. Access control is of 2 types namely logical access control and physical access control. But today numerous cyber-attacks (DoS attacks, side channel attacks, authentication attacks etc.) are executed and it is the job of access control of cloud computing for protecting the information by any access which is unauthorized. A various number of techniques maintaining the access control like detection system for intrusion, firewall and segregation both of obligation can be implementation on various networks and layers of cloud [15]. A method foraccess control is FADE proposed by Y. Tang and team. FADE provided access control which is fine-grained and deletion assured for any data which is outsourced on cloud [16].

Authorization is the phase employed after authentication in cloud computing security techniques to enable proof of their identity. In cloud Authorization methods include using MAC, RBAC, DAC and ABAC. Since the authorization and identity verification is the most important step in cloud computing and as these techniques have evolved and as taking weak points in consideration there are concerns affecting the security and needs to be looked at such as API's, Continuous authorization, zero trust and authorization services.

3 Conclusion

Cloud computing is technology that has provided IT and every sector an environment of storing, building, hosting and consuming applications or any information without paying the whole expenses of the hardware and software. Cloud computing offers the users the pay-per-use method as the user needs to pay only for the service that is being used. But as the business organization and individuals store sensitive data over the cloud but as with the increasing cyber attacks the data stored is not secured. Cloud service providers use multiple security techniques to secure the data servers from being attacked. The purpose of this paper is to discuss about various security techniques and methods being used by various cloud service provider and are these techniques and methods efficient enough.

The present study discussed that as the hackers are innovating new techniques and measures to execute the attacks and the techniques and methods being used in security are not hack-free. So there is a need of introducing some new techniques to enhance the security of a cloud. As the availability, authorization, access control and encryptions techniques needs to be evolve so that the user can trust the cloud provider on their service because every business organization and enterprises are moving towards cloud for storing their data, infrastructure, software and platforms without paying the whole expense of hardware installation.

In future if the cloud security techniques tend to evolve and become more efficient and provides every feature with low cost expenses and secure data with backup and recovery options. Every individual, enterprise, business, organization of every sector will shift towards cloud for storing their information, hosting applications, developing and testing.

References

1. P. Srivastava, R. Khan, A review paper on cloud computing. Int. J. Adv. Res. Comput. Sci. Softw. Eng. (2018). https://doi.org/10.23956/ijarcsse.v8i6.711
2. D.G. Pradnya, B. Godhankar, Review of cloud storage security and cloud computing challenges. Int. J. Comput. Sci. Inf. Technol. **5** (2014)
3. S. Aldossary, W. Allen, Data security, privacy, availability and integrity in cloud computing: issues and current solutions. Int. J. Adv. Comput. Sci. Appl. (2016). https://doi.org/10.14569/ijacsa.2016.070464
4. S.K. Sood, A combined approach to ensure data security in cloud computing. J. Netw. Comput. Appl. (2012). https://doi.org/10.1016/j.jnca.2012.07.007
5. S.-E. M. Mohamed, E. Hatem S. Abdelkader, Enhanced data security model for cloud computing
6. B.S.K.V. Spoorthy, M. Mamatha, A survey on data storage and security in cloud computing. Int. J. Comput. Sci. Mob. Comput. **3**(6), 306–313 (2014)
7. R. Piyare et al., Integrating Wireless Sensor Network into Cloud services for real-time data collection (2013). https://doi.org/10.1109/ICTC.2013.6675470.
8. L.M. Kaufman, Data security in the world of cloud computing. IEEE Secur. Priv. (2009). https://doi.org/10.1109/MSP.2009.87
9. G. W. Albugmi, Ahmed, Madini O. Alassafi, R. Walters, Data security in cloud computing
10. Y. Sun, J. Zhang, Y. Xiong, G. Zhu, Data security and privacy in cloud computing. Int. J. Distrib. Sens. Netw. (2014). https://doi.org/10.1155/2014/190903

11. A.Z. Lynda Kacha, An overview on data security in cloud computing
12. V. Chang, M. Ramachandran, Towards achieving data security with the cloud computing adoption framework. IEEE Trans. Serv. Comput. (2016). https://doi.org/10.1109/TSC.2015.2491281
13. K. Hashizume, D.G. Rosado, E. Fernández-Medina, E.B. Fernandez, An analysis of security issues for cloud computing. J. Internet Serv. Appl. (2013). https://doi.org/10.1186/1869-0238-4-5
14. M.D.H.P.R. Sridaran, An analysis of security challenges in cloud computing. Int. J. Adv. Comput. Sci. Appl. **4** (2013)
15. K. Jakimoski, Security techniques for data protection in cloud computing. Int. J. Grid Distrib. Comput. **9** (2016)
16. Y. Tang, P.P.C. Lee, J.C.S. Lui, R. Perlman, Secure overlay cloud storage with access control and assured deletion. IEEE Trans. Dependable Secur. Comput. (2012). https://doi.org/10.1109/TDSC.2012.49

Radio Frequency Identification Technology Used to Monitor the Use of Water Point for Grazing Cattle

Parmeshwar Kumawat(✉)

Department of Electronics Engineering, Vivekananda Global University, Jaipur, India
Kumawat.permeshwar@vgu.ac.in

1 Introduction

Artificial water points in northern Australia (e.g. lakes and bores) often providing beef cattle with some sources of drinking water. Grazing animals have a high water turnover rate and regular drinking water access is important. It is currently recommended to have at least 1 water points per 30 km 2 within a maximum separate of 6 km in between the water points. The current recommendation takes into account cattle grazing spread from water points, the effect of grazing across water points and grazing evenness. It does not consider the effects on living organism development, replication and longevity of water points use by the cattle (example: regularity of the visits) and water availability (example: distances in between the water points). In the recent study, the current paper suggested some practical knowledge about how cattle utilize water sources. However, it is difficult to determine how much water points to be install and how far the alone they will be to encounter the needs of cattle water. Several experiments have tried to understand the watering actions of beef cattle grazing. Therefore, basic facts about how much water bovine consume and how often bovine drinks are not well understood by grazers or scientists under different conditions of climate, pasture and water availability. This paper has shown in a recent review that the level of cattle drinking affects the amount of water bovine intake and may impact certain efficiency attributes. The review reported that ad libitum-accessed dairy cows drank 11–13 percent more than a cows with limited access to the water (1 or 2 a day) and had the higher milk surrender and milk fat. Once a day, beef bovines with enter to the water consumed the 16–25% higher than bovines with enter to the water once in the every 2nd or 3rd day and had more intakes of the feed. In the study further notes that in reaction to voluntary drinking regimes, water intake and grazing efficiency of beef bovines were not observed. Scientific analysis of the behavior of beef bovines watering under usually grazing conditions, including difficult interactions between living organism, management and their climate is important to advise grazer decisions for water point allocation and to improve the efficiency of cattle.

Remote weighing technologies combined with automated Radio Frequency Identification (ID) logging could be used to research the use of cattle grazing water level. In remote consider of grazing bovines was firstly introduced in a 1960s in order to invalidate the disadvantages of standard weighing application such as the stress and

F. P. García Márquez (ed.), *International Conference on Intelligent Emerging Methods of Artificial Intelligence & Cloud Computing*, Smart Innovation, Systems and Technologies 273, https://doi.org/10.1007/978-3-030-92905-3_35

costs connected with collection and design. At the entry of a surround water point, the technology is strategically installed to attract cattle to walk through the system. The RFID-equipped ear tag is checked every time an animal accesses water as its walks previously the RFID reader and the date and time is registered. The weights of the object are also calculated as goes through the automated weighing of the platform. However, as a different to the traditional of time-consuming and observation expensive methods, the RFID recording constituent must be used for the autonomously collect behavioral information. In a recent study, it reveals that RFID data from remote weighing devices are used effectively to assess maternal parentage of the calf. Over 90 percent of maternal cow-calf pairs were correctly identified by the numbers of the time a cow and her calf passed via the remote weighing technologies within the predefined times period. Due to the implementation of remote weighing technologies at water points, the RFID tracking portion basically tracks the dates and time of the cattle visiting to the water points and will be used to research the use of water point by the grazing cattle. A purpose of the studies was to enter whether remote weighing technology RFID reader information will be used for the examine the use of bovine's water point. So the idea was that the device might track behavioral differences depending on the availability of atmosphere and water and thus be a practical tool for researching the use of cattle water level.

2 Methods

The study was accompanying as a retrospective reviews of RFID information from 3 different studies that has remote weighing equipment built to track the live weight of cattle. Every station was situated in a different area of farming, reflecting different climates and conditions of water availability. A selection of spring–summer RFID data from each project was used (October to February). The data subsets were generated in such a manner that the time period across the three stations was accurate.

1. *Remote weighing technology*:
   At each trial location, a cattle yard was designed to enclose an indefinite water points so that the cattle access through the one-way puncture gate into the water point and escaped through a different spear gate. At the entrance to the water point, the remote weighing technology was set up and equipped with an electronic RFID panel reader. Each cattle were wearing a RFID fitted ear tag to the right ear and had at every times ad to the libitum enter to a water sources. Every time an object comes within the amount of the antenna of the RFID user, i.e. the RFID number and live weight were reported with the dates and time on each visit to the water site.
2. *Study Sites and Animals*:
   The attribute of the investigation paddock and cattle at every station are summarized The paddock at cattle center was 6600 ha in the size and was contain the 66% Barkly1 of (black soil) and 34% of Wonorah in (red soil) lands types. A red soil containing areas of a paddock had the scattered trees so that will have provided the shade. A paddock comprising 2 water points are located approximately in the 6 km apart. The 1 water points (No. 19 is drill) was located closed to the main center of a paddock and the different water points (Stud bore) are saturated in the northwestern corners of

the paddock. So at both the water sites or remote weighing apparatus was mounted. A maximum interval from a water points to the paddock's farthest point was about 6.5 km. The paddock was grazed by a group of 80 weanling steers and 544 young, non-lactating of cows who calved from the October to April. A steers have a total live weights of the 326 kg (August) and were breeding from Brahman (Bosindicus) and Charbray (B. indicus as well as Bos Taurus). A cows had an average age of the 7 years and an average living weight of the 475 kg (August) and were raising Brahman, Santa Gertrudis and Charbray (B. indicus B. Taurus). The Lansdown grazing area consisted of a three-ha block of loamy alluvial paddocks used in a rotation. There was shelter in the sparse leaves. There was one water source in the main paddock. For cattle to reach the water point from both sides of the paddocks, a 20 m long alleyway was built. Beginning mid-November, gates between the paddocks were opened to enable cattle to graze two or three paddocks at the same time. The maximum distance from the water point to the paddock block's furthest point was 0.75 km. The paddocks were grazed by a party of 20 steers.

3. *Weather*:
   The collected daily weather changing were minimum temp. (T min, °C), the average ambient temp. (Tavg, °C), the maximum temperature (T max, °C), the average relative of humidity (RH, percent), the average wind of speed (WS, km/h), average of cloud cover (CC, eight), average of solar exposure (MJ, SO m^2) and rainfall (mm, RF). The weather stations were 1 km and 45 km respectively from the investigation paddock. Both weather a stations were situated at the same climate zone of the respective sites of the project. The weather station was used to measure regular sunrise and sunset times.
4. *Data Processing*:
   Three columns of data were acquired as text files for the RFID reader [1–4] data: RFID, date and time. Each data row represented a single animal's recorded visit to a water point. Microsoft Excel downloaded the details and stored them using Microsoft Excel. The data is cleaned by deleting rows without electronic recognition [5–7] and eliminating data beyond the time periods picked. Rows of information with the following characteristics were the removed: (i) An incorrect RFID (i.e. then RFID amount was incompatible within the animals incorporate in the studies); (ii) A date relating to animal's care, as human intrusion could have influenced cattle behavior. Time periods between bovines visiting to the water points were determined per animal that are subtracting by the date and time of each event from the actual record date and time. During missed data times (due to animal care or equipment failure) the first reported time was discarded as the missing information may skew the results. Records within the 30 min of the previous records were also detach because cattle loitering around the water point were more likely to generate them than bovines making separate visits to the water points to drink.
5. *Data Analysis*:
   Daily weather changeable were compared in between the 3 stations. Every data materializes to be commonly distributed in the quantile–quantile (QQ) plots excepting for RF, in which they had a (+vely) skewed distribution. A weather changeable with common distributions of (T min, T avg, T max, RH, SO THI, WS and CC) were analyzed to using the Welch's analyzing of variance (ANOVA) due to the heterogeneity

of the variances according of a Levine's test of ($P < 0.05$). The non-parametric of bootstrap methods was used for analyzing the information from the given RFID scanner. The non-parametric of boot strapping is a method of examine used to solved the simulation issues such as simultaneity, interdependence, nonlinearity, volatility, heteroscedasticity, non-normality, limited datasets and missing info. The bootstrap was used in this situation to handle uneven sample sizes created at each station by the varying number of the animals.

The measures in a bootstrap process were follows: (i) An even amount of the records each month were inconstantly selected with replacement from every station to the create a balancing bootstraps sample (ii) The linear mixed-effect models was applied to a bootstrap representative using a' lme4′ R kit and summary statistics was measured (3) Measures 1 and 2 were replicated 100 times (4) Summary statistics were average. Two facets of the use of cattle water points were examined: bovine visits to the water points and time intervals in between bovine visits to the water points. Visiting bovine times at the water points was the time of a day (minute and hour) when bovine reached a points of water. The time periods (hours) in between bovine visits to the water points were the length of time in between consecutive water point visits. To analyze each facet of water point use, separate models were used.

3 Result

Table 1 provides a description of the details of the RFID app [8–11]. Stud bore data included 2152 records. The apparatus failures happen during 60 the days (45% of the study period) due to the some or every of an older system constituent breaking down to the grating environments. The information from the Stud bore will be omitted due to the higher frequency of the equipment failure. So the overall information set for the Belmont containing of 136 days and 5073 documents of data. Due to the animal handling, 5 days were detaching (215 records). The Over of 10 days, apparatus failures happen (7% of the study period).

1. *Time intervals between cattle visits to water points*:
   Information from the RFID revealed that most bovines visits to the water points took place within the 48 h of the previous visit (Fig. 1). Approximately 71% of the visits took place at Brunchilly within 24 h of a past visit and 85% within 48 h of a past visit. The Approximately of a 60% visits took place at Belmont within 24 h of a past visit and 96% of visits took place within 48 h of a past visit. The approximately 95 percent of the water level visits took place at Lansdowne within 24 h of a past visit and 98 percent of the visits took place within 48 h of a past visit. There were significant the differences between the stations in the time intervals of bovines visits to the water sources (Table 2).

   a. Means with different superscript letters in the same column differed between experimental sites ($p < 0.05$).
   b. Average daily RF is shown. Total RF during the study period was 496 mm at Brunchilly, 275 mm at Belmont and 222 mm at Lansdowne.

Table 1. Description of the details of the RFID app

	Station			
	Brunchilly (No. 19 bore)	Brunchilly (Stud bore)	Belmont	Lansdown
Days				
All data	135	135	151	136
No. animal handling days	0	0	5	3
No. equipment failure days	4	60	10	0
Total	131	75	136	133
Records				
All data	59,473	2152	5536	5388
No. with erroneous RFID	38,661	2152	171	0
No. on animal handling days	0	0	215	163
No. on equipment failure days	0	0	1	0
No. within 30 min of previous visit	2573	0	76	90
Total	18,239	0	5073	5135

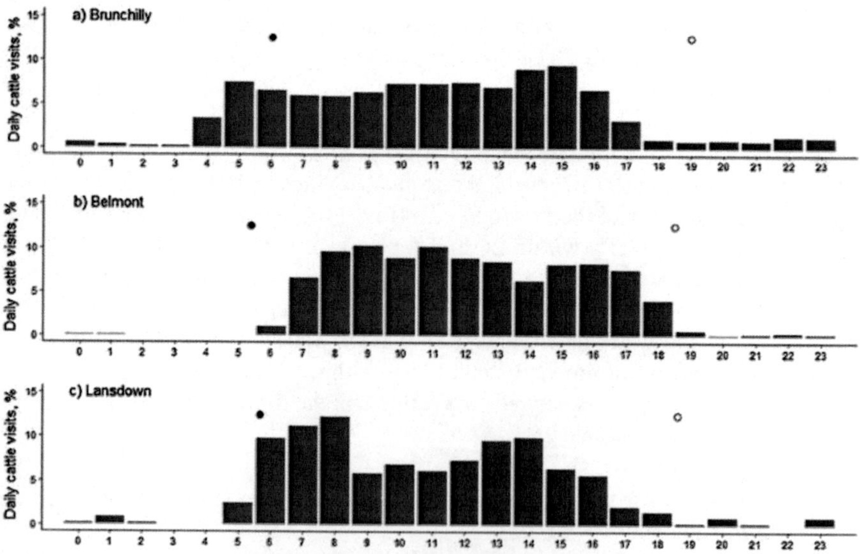

Fig. 1. Hour of the day

Table 2. Summary of RFID data from remote weighing technology at three stations

	Tmin(°C)	Tmax(°C)	Tav(°C)	THI(%)	RH(%)	WS(km/h)	CC(eights)	SO(MJ m^2)	RF(mm)	Sunrise (h)	Sunset (h)
Brunchilly	23.7 ± 2.2^{a}	35.8 ± 3.2^{a}	29.3 ± 2.5^{a}	76.2 ± 2.8^{a}	44.3 ± 18.7^{b}	14.0 ± 3.3^{ab}	3.7 ± 2.1^{ab}	25.1 ± 5.9	3.9 ± 2.5^{a}	$6{:}02 \pm 0.25^{a}$	$19{:}00 \pm 0.22^{a}$
Belmont	21.4 ± 2.4^{b}	32.9 ± 2.7^{b}	26.2 ± 2.1^{c}	75.1 ± 3.1^{b}	66.0 ± 7.2^{a}	13.2 ± 3.4^{b}	3.6 ± 2.0^{b}	23.9 ± 5.3	2.0 ± 2.1^{b}	$5{:}22 \pm 0.29^{c}$	$18{:}30 \pm 0.27^{c}$
Lansdown	23.6 ± 2.1^{a}	31.4 ± 1.6^{c}	27.0 ± 1.4^{b}	76.6 ± 2.8^{a}	67.8 ± 6.7^{a}	18.9 ± 4.7^{a}	4.2 ± 1.8^{a}	23.8 ± 5.3	1.7 ± 1.4^{b}	$5{:}39 \pm 0.20^{b}$	$18{:}36 \pm 0.27^{b}$
P value	<0.001	<0.0001	<0.0001	<0.0001	<0.001	<0.001	<0.05	<0.051	<0.001	<0.001	<0.001

Means ± SD of daily weather conditions observed at the three stations. Daily rainfall data were obtained from weather stations located at each station and the other variables from the nearest official weather station.1 T min, minimum temperature; T max, maximum temperature; Tav, average temperature; THI, temperature, humidity index; RH, relative humidity; WS, wind speed; CC, cloud cover; SO, solar exposure; RF, rainfall

The median time in interval between Belmont visiting was 24.5 h, corresponding to about one visiting per day. The mean period in time in between the Lansdown visiting was 12.5 h, equating to around 2 visiting per day. Differences in the time intervals in between the bovines visits to the water points were also observed by day, months, THI and CC. A negative relationship in between the THI and visit time intervals indicates that the cattle accessed a water points much frequently as the THI rose. The time periods during trips to the water point at Brunchilly ($P > 0.05$) were no different between the cows and steers.

4 Conclusion

RFID reader data from remote weighing devices is considered an appropriate method for tracking cattle visiting times and time intervals during cattle visits to the water points autonomously. A technology's practical environment makes it satisfactory for the field experiments and application by the industry. Potential studies on the cattle water's wants, cattle watering actions and the impact of water availability on the growth, reproduction and survival of cattle needed to improving grazers 'of decision making. Watering behavior of the cattle is complicated and many other factors are likely to interact and influence how water points are used by cattle. For individual situations, personalized water infrastructure installation guidelines may be needed. In addition to the current focus on grazing distribution, the future experiments on the cattle water at a point using will sanction the development of a water structure with much direct benefits for the cattle production.

References

1. Kamesh, N. Sakthi Priya, Security enhancement of authenticated RFID generation. Int. J. Appl. Eng. Res. (2014)
2. I. Erguler, A potential weakness in RFID-based Internet-of-things systems. Pervasive Mob. Comput. (2015)
3. C. Rong, G. Zhao, L. Yan, E. Cayirci, H. Cheng, RFID Security, in *Network and System Security: Second Edition* (2013)
4. D.M. Segura Velandia, N. Kaur, W.G. Whittow, P.P. Conway, A.A. West, Towards industrial internet of things: Crankshaft monitoring, traceability and tracking using RFID. Robot. Comput. Integr. Manuf. (2016)
5. E.A. Baldwin, J. Bai, A. Plotto, S. Dea, Electronic noses and tongues: Applications for the food and pharmaceutical industries. *Sensors* (2011)
6. S. Park et al., Electronic system with memristive synapses for pattern recognition. Sci. Rep. (2015)
7. J.K. Aggarwal, M.S. Ryoo, Human activity analysis: A review, *ACM Computing Surveys* (2011)
8. E. Valero, A. Adán, C. Cerrada, Evolution of RFID applications in construction: A literature review. *Sensors (Switzerland)* (2015)
9. J. Gubbi, R. Buyya, S. Marusic, M. Palaniswami, Internet of Things (IoT): A vision, architectural elements, and future directions. Futur. Gener. Comput. Syst. (2013)
10. S. Amendola, R. Lodato, S. Manzari, C. Occhiuzzi, G. Marrocco, RFID technology for IoT-based personal healthcare in smart spaces. IEEE Internet Things J. (2014)
11. L. Kumari, K. Narsaiah, M.K. Grewal, R.K. Anurag, Application of RFID in agri-food sector. *Trends in Food Science and Technology* (2015)

A Study of Continuous Variable Transmission

Vijay Kumar Pandey(✉)

Department of Mechanical Engineering, Vivekananda Global University, Jaipur, India
vijay_pandey@vgu.ac.in

1 Introduction

A "Continuous Variable Transmission (CVT)" [1, 2] in an unpredictable kind of transmission system which can change steplessly through persistent and interminable no. of powerful chain proportions between max and min esteem. This appears differently in relation to other mechanical transmission system that just permit a couple of various particular chain proportions. The adaptability of the "Continuous Variable Transmission" grants the driving shaft to keep up a specific rakish speed over a predefined go of opposite speeds. This can give preferred efficiency over other transmission systems by empowering the engine to run at its most proficient RPM for a scope of vehicle speeds. With a particular ultimate objective to endure new guidelines for auto effectiveness and transmission, the "Continuous Variable Transmission" continuously creating as a key development for upgrading the fuel adequacy of cars with Internal Combustion (IC) engines. "Continuous Variable Transmission" use boundlessly adaptable drive extents as opposed to discrete chain to achieve perfect engine execution. Since the engine constantly runs at the most capable RPM for a given vehicle speed, "Continuous Variable Transmission"-arranged vehicles accomplish liked gas mileage and animation over cars with regular transmissions.

As can be found in Fig. 1: the "Continuous Variable Transmission" is a smaller system as will be depicted, it doesn't require the utilization of massive chain sets or the same number of segments as in the traditional transmission. A "Continuous Variable Transmission" system is involved two funnel each other, their tapered shape parcels of every pulley draw nearer or more remote away from each other, their tapered shape makes the belt rise and fall between parcels of every pulley. Contingent on the condition of the belt, the dynamic chain proportion [3] is changed. Rather than exchanging between cumbersome fixed apparatus which just supply a predetermined number of apparatus proportions, the "Continuous Variable Transmission" pulley make a ceaseless trade of chain proportions by always adjusting the conditions of the belt between the pulleys scarcely any instruments that permit the control of the pulley breath incorporate engine speed, flyweights, 3 springs and a torque incline. At the point when these components work at the same time, they act to speed up easily while keeping up the engine speed at a specific worth. This highlights of engine speed upkeep is conceivable because of the coherence of chain proportions. "Continuous Variable Transmissions" are not all that new to the car world, however their effectiveness and steady quality have been restricted all through. Most recent advancements in gear reduction and assembling have

F. P. García Márquez (ed.), *International Conference on Intelligent Emerging Methods of Artificial Intelligence & Cloud Computing*, Smart Innovation, Systems and Technologies 273, https://doi.org/10.1007/978-3-030-92905-3_36

prompted even increasingly vigorous “Continuous Variable Transmissions” which thusly makes it conceivable to be use them in progressively differing car applications. As “Continuous Variable Transmission” advancement proceeds, expenses would go down and execution would increment further, which would make further improvement and use of “Continuous Variable Transmission” innovation bodacious.

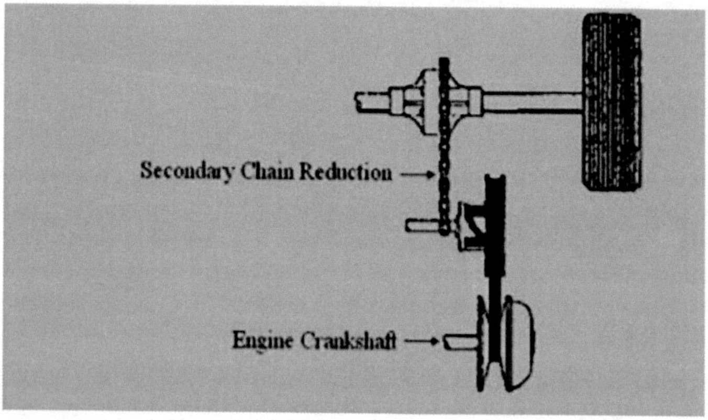

Fig. 1. CVT system in which the 2 belt pulleys represent the CVT’s primary and secondary gear reduction

2 Literature Review

Leonardo technologist, conceptualised a stage less factor transmission. The 1st patent for a contact based belt was recovered by Benz, and a patent on strong “Continuous Variable Transmission” by a North American nation was conceded.

- A V2-motor bicycle was structured by Celestial Point Motorcycles with the Gradua-Gear that is likewise a “Continuous Variable Transmission”. This Zenith-Gradua was subsequently thundering in slope climbing, that it completely banished, so elective produces got an opportunity to exceed expectations.
- A people bicycle brand Rudge-Whitworth planned the Rudge multi-gear. The multi was significantly better contrasted with Zenith’s Gradua-Gear. Browne offered a motorbike with variable-stroke ratchet drive.
- Early hardly, any employments of “Continuous Variable Transmission”, was compelled to just Clyno car (British), presented. A “Continuous Variable Transmission” suggested to as Aromatic, was structured and brought to use by center point van Doorne, fellow benefactor of DAF, around nineteen fifties, particularly to give control transmission for a little and sensible car. The essential DAF car exploitation van Doorne’s “Continuous Variable Transmission”, the DAF six hundred was made. His licenses were later moved to an association suggested to as VDT (Van Doorne Transmission B.V.).

- Many snowmobiles make usage of versatile belt "Continuous Variable Transmission". Rokon offered use of "Continuous Variable Transmissions" in some "Automatic Transmission Vehicles". Polaris director transformed into the first to be furnished with CVT. Subaru moved the Justy in national capital with "electronically controlled variable transmission (ECVT)" [4] developed by Fuji industries. The Justy transformed into an essential age association in U.S. to supply "Continuous Variable Transmission" development while Justy could simply observe limited accomplishment, Subaru is up till now using "Continuous Variable Transmission" as a piece of its cars till now. It gives "Continuous Variable Transmission" on the gift besides outback (Linear-electronic) [5].
- In the spring, Ford started to furnish the European autos with steel belt "Continuous Variable Transmission" (as a trade for less solid elastic belt model) [6]. Ford and Fiat together build up a "Continuous Variable Transmission" innovation, named the Ford CTX, which was then utilized in the transmission systems.
- Nisan March helped the Fuji genuine industries "electronically controlled variable transmission ECVT" and N-CVT [7]. Around late nineteen nineties, Nisan structured its very own CVT permitting more noteworthy power and encased a convertor. This shell utilizing a roller-type CVT back at that point. Its strong CVT, Extroid was reachable with at that point. Its strong CVT, Extroid, was possible inside the neighbourhood nation showcase in Y34 Nisan Gloria and furthermore in V35 skyline GT-8. Be that as it may, the shell wasn't utilized again once the Cedric/Gloria was snubbed by Nisan Fuga. The Nisan Murano what's more in this manner the Nisan Scalawag furthermore uses "Continuous Variable Transmission" in their transmissions. In the midst of Nisan affirmation, Nisan announced a tremendous development to "Continuous Variable Transmission" systems after they doled out their Xtronic (CVT) for all models of Versa, Cube and Maxima vehicles in North America. One essential typical method of reasoning fit as a fiddle a change to "Continuous Variable Transmissions" was an under territory of their 'green program 2010' furnished towards reducing nursery gas release by 2010. The "Continuous Variable Transmission" found in Nisan's Maxima, Murano and as such the V6 adjustment of the Altima is reflected over to be the world's beginning "3.5 class" belt-sort CVT and may hold a significant proportion of progressively conspicuous power masses appeared differently in relation to alternative belt-sort CVTs.
- After making sense of pulley-sort CVT [8] for significant long time, Honda likewise exhibited their claim specific structure on Honda Civic VT-I. Named Honda Multi Matic, this CVT shell recognised higher power than out of date CVTs, and furthermore consolidates a convertor for "snap" action. The CVT is immediately utilised inside the Honda Town ZX that is mechanical office made in republic of India and Honda Vario plant made in West Pakistan.
- Toyota had usage of an "Impact Split Transmission (IST)" [9], furthermore, every one future Toyota and Lexus hybrids sold-out all around still use of system. The HSD is in addition talked as "Electronically Controlled Variable Transmission (ECVT)". The basic time permits either the electrical motor or the "internal combustion engine (ICE)" [10] to incite the vehicle. In ICE-just mode, a bit of the engine's energy is normally coupled to the drivetrain, with the opposite half inspecting a generator and an engine. The measure of power occupied through the electrical way certify the effective gear

quantitative association. Toyota also offers a non-crossover CVT insinuated as Multi-drive for models like Avensis, Audi has, offered a chain-sort CVT (multi- electronic) as an option on a no. of its greater engine models. For example the A4 3.0 L V6 Fiat offered a cone-build CVT as choice as for one of its productive model Punto (16 V).

- Ford presented a chain-type CVT called the CFT30, in their Ford race, Ford 500. The transmission was planned in participation with German producer ZF Friedrichshafen and was made in Batavia, Ohio at Batavia transmission LLC. The Batavia plant also made the belt-type CVT that was then utilized in the Ford Focus C-MAX. portage also sold out Escort and Orion models in Europe with "Continuous Variable Transmission"
- Agreements were marked between MTD items and Torotrak for the essential full system to be plant made for outside power instrumentally like fly skis and skimobiles. The Dodge Caliber and along these lines the associated machine Compass and machine national utilize a CVT as their Discretionary transmission system.
- Mitsubishi trooper model was used with CVT. American state models and EST models got a normal CVT with drive and chain; the GTS model was given with a normal drive and furthermore a sport-electronic mode that empowers the thought process power to utilize six totally unique fixed apparatus proportions. The SEAT Exeo was introduced with CVT (multi-electronic) as a decision for the 2.0 TSI 200 HP (149 kW) fuel engine, with selectable "six-speeds".
- A North American brand gave territory 7, 647, 768 B1 for a progression of pressure driven power converters as a CVT. It'll be mated to a couple of four chamber fighter engine.

3 Principle

In any application, regardless of whether it be on-or-off-roadway, a consistently factor transmission CVT. A programmed transmission [11] that replaces regular apparatus proportions with a chain or belt and a system that modifies breadths on either side of two pulleys to produce consistent advances of speed and torque. "Continuous Variable Transmission" can be utilized as a component to give the vehicle the exact spread of proportions it needs. With this kind of transmission, the objective is to attempt to keep the engine speed at a particular point or sweet spot, regardless of whether the need is to enhance efficiency or to boost the power or torque conveyed to the drive wheels or the execute. To outline what a CVT is, and how it works, Lubrizol has made two recordings-one itemizing a chain CVT, the other clarifying a push-belt CVT, both traveller vehicle CVT models.

As a general guideline, the CVTs superseding advantage is its capacity to keep engine speed as close to steady as could be allowed. Most engines favour relentless state running. Superfluous additional fuel is scorched during homeless people and when engines accelerate and back off during gear changes; valuable rotational and longitudinal force is likewise lost during these procedures. Having a "Continuous Variable Transmission" instead of a traditional manual gearbox can help wipe out those issues. A 3D view of CVT is mentioned in Fig. 2 below for the better understanding.

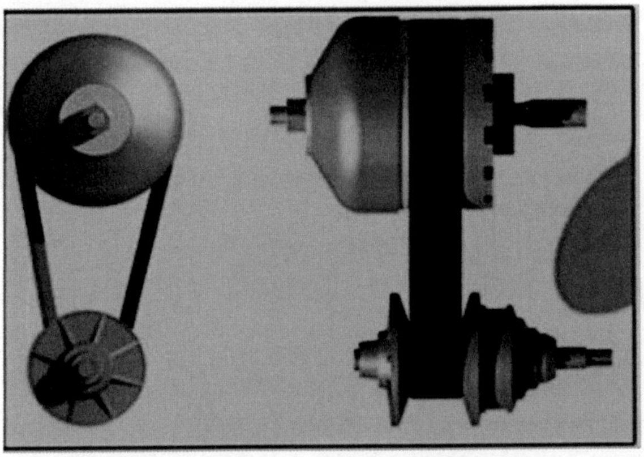

Fig. 2. 3D view of CVT

4 Working

Regular transmission systems utilize a chain set that furnishes us with an assortment of chain proportions. The transmission system (or the driver) changes gears to supply the principle appropriate extent connection for a given condition: lower gears for starting out, halfway ones for increasing speed and bigger ones for eco-friendly cruising. At present, there are a huge assortment of "Continuous variable transmission" systems yet the greater part of the vehicles utilize a mixture of variable-dia pulleys, each framed kind of joining 2 contradicting cones, with a metal belt/chain running between them. One of the cones is associated with the engine (Input shaft) and the inverse to the wheels (Output shaft).

The two parts of the machine are movement allowed; on the grounds that the machine parts will in general come back closer along the belt is compelled to ride higher along the stature of the cones, form this time forward bringing about biggerdia of the pulley. Change in dia of the pulley shifts the connection appropriately. For example, the chain of a 10-speed bicycle courses over bigger/litter apparatus to fluctuate the connection. Keeping the input cone litter and the output cone bigger offers us a strong connection for higher low-speed increasing speed. Since the car gathering quickens, the pulleys change their dia from higher to lower as speed rises. The abovementioned marvel can be steady factor a conventional programmed/manual transmission will be that as it may while customary transmission system changes the size connection a little by moving of chain, "Continuous Variable Transmission" changes the extent connection consistently-below a Fig. 3 is mentioned showing the connection of pulleys.

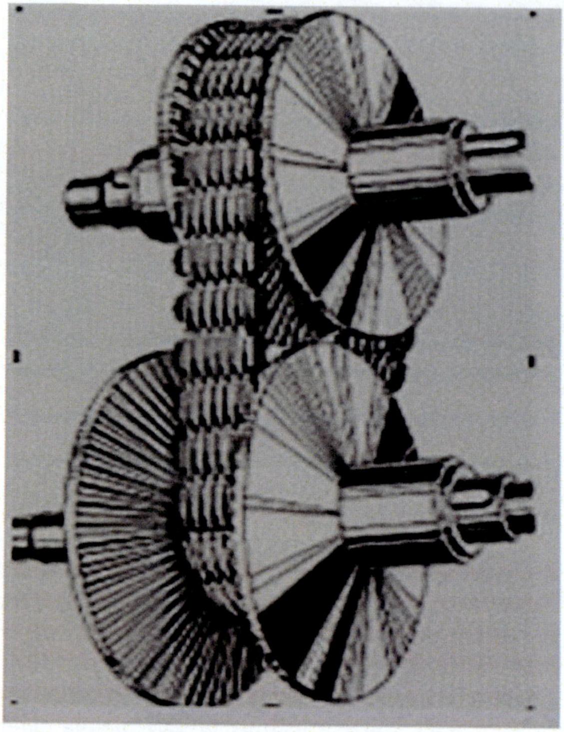

Fig. 3. Chain driven CVT

5 Mechanism of Pulleys

In the instrument structure of the pulleys, the followings ought to be noted:

The pole like area associated with the steady half pulleys ought to be manufactured punctured to fulfil the following needs:

Oil of the tank of the mobile pulley and oil of mobile pieces of the pulley Lower weight, making lower rotational idleness what's more, devouring less crude materials in assembling Associating the related shafts to the pulleys' system Smooth and frictionless development of the versatile half pulleys while the proportion changes Foreseeing the heading's area and making a stage what's more, decreasing the measurement in various pieces of the pulleys' pole for a superior get together and forestalling the development of the direction and controlling the range of the development of various parts. Anticipating the essential instrument for the pulleys to come back to their beginning state Predicting the portion area of the yield or decrease outfits on the auxiliary pulley's shaft Designing the type of the pulleys' tank's shell.

Structuring a few teeth on the optional pulley for the park gear system To delineate how pulleys' system works, two circumstances ought to be thought of: one when the pulleys' proportion changes from low proportion to overdrive and the other when the pulleys' proportion changes from overdrive to low proportion. The significant point in comprehension the system is that there is a belt between the pulleys which has a fixed length and is feeling the squeeze. In the first circumstance, when the weight of the grease behind the mobile half pulley of the auxiliary pulley diminishes, the mobile half pulley of the auxiliary pulley is pushed back by the applied power of belt, which is along the pole, on a large portion of pulleys' surface and subsequently the revolution breadth of the auxiliary pulley begins to diminish. Due to the fixed length of the belt and applied power from the belt's obstructs on one another, the belt's distance across in the essential pulley alongside the volume of the ointment behind the portable half pulley of the essential pulley increments and in this way the versatile half pulley of the essential pulley begins its development and makes the important strain to give the typical power of the surface of the half pulleys to be applied on the belt in request to fulfil required grating.

In the second circumstance, everything is equivalent to the first, but a spring is utilized for creating the typical power to give the contact of the optional pulley. Since every pulley ought to have the ability to make the vital power to push ahead or in reverse, in the essential pulley, the progress ahead of the mobile half pulley is done through expanding the volume of the grease in its tank however its retrogressive development is done through the pivotal power of the belt on its surface. The spring gives the progress ahead of the mobile half pulley of the auxiliary pulley and the regressive development of the auxiliary pulley is finished by expanding the volume of the ointment in its tank, in spite of essential pulley.

The making of the opposite development of the optional pulley in correlation with essential pulley regardless of expanding the oil's volume in both the pulleys is because of the way that there are two shells in the mobile half pulley of the auxiliary pulley in which the inward shell is fixed to the pole and with the expanding of the grease's volume the made power is applied on the external moving shell which is fixed to the mobile half pulley and makes the regressive development. Be that as it may, in the versatile half pulley of the essential pulley, there is just one fixed shell and the progress ahead is made through the expansion of the grease's volume in the tank.

6 Conclusion

Further, new research wildernesses ought to be examined with regards to CVT plan and design. Certain new setups of "Continuous Variable Transmission" structure have been answered to accomplish persistent varieties in transmission proportion with lower misfortunes, in any case the scope of appropriateness of such "Continuous Variable Transmissions" for high torque applications is yet to be dissected/confirmed. A constantly factor transmission is a promising car transmission innovation that can give a higher efficiency, reduced outflows, and better vehicle execution. The flow paper not just tends to the best in class investigate achieved towards comprehension "Continuous Variable Transmission" elements and control yet in addition ideally features the difficulties/bearings for future research which could cultivate better understanding and structuring of such system and their controllers. A "continuously variable transmission" or CVT mixes the simplicity of a programmed transmission with the productivity of a manual transmission. "Continuous Variable Transmission" is certainly an innovation of things to come with its higher eco-friendliness, interminable gear proportions, lower producing costs, consistent cruising speeds and better speeding up capacities. This innovation has discovered such wide applications recently. Therefore, the greater part of us need to become adapted to the elements of a "Continuous Variable Transmission"-prepared vehicle for its better rise.

References

1. K.S. Ivanov, Synthesis of toothed continuously variable transmission (CVT). Mech. Machine Sci. **3**, 265–272 (2012)
2. Y. Kwak, C. Cleveland, Continuously Variable Transmission (CVT) Fuel Economy, in *SAE Technical Papers*, 2017, vol. 2017, October
3. D.W. Ge, S. Ariyono, D.T.T. Mon, A review on continuously variable transmissions control, in *National Conference in Mechanical Engineering Research*, vol. 9501, May, pp. 543–554 (2010)
4. I.I. Mazali, K.B. Tawi, B. Supriyo, N.A. Husain, M.S.C. Kob, M.S.C. Kob, Transmission ratio calibration for electro-mechanically actuated continuously variable transmission. Int. J. Adv. Mechatron. Syst. **7**(3), 127–133 (2017)
5. 雅裕丹羽et al., The Japan Institute of Energy NII-Electronic Library Service. バイオマス科学会議発表論文集 **86**(5), 86–87 (2010)
6. W.M. Lai, D. Rubin, E. Krempl, The elastic solid, in *Introduction to Continuum Mechanics* (2010)
7. B. Heim et al., Atmospheric continuous-variable quantum communication. New J. Phys. (2014)
8. V.D. Kulakovskii, S.S. Gavrilov, S.G. Tikhodeev, N.A. Gippius, Polariton nonlinear dynamics: theory and experiments, 43–65 (2012)
9. Y.X. Zhao, H. Xiao, Y.Q. Huang, Dynamic split tensile test of Brazilian disc of coal with split Hopkinson pressure bar loading. Meitan Xuebao/J. China Coal Soc. **39**(2), 286–291 (2014)
10. D.E. Klett, E.M. Afify, K.K. Srinivasan, T.J. Jacobs, Internal combustion engines, in *Energy Conversion, Second Edition* (2017)
11. A.J. Drake, L. Liu, Intergenerational transmission of programmed effects: Public health consequences, *Trends in Endocrinology and Metabolism* (2010)

Review on Color Image Processing Techniques

Anil Bagaria(✉)

Department of Civil Engineering, Vivekananda Global University, Jaipur, India
anil.bagaria@vgu.ac.in

1 Introduction

The premature trade in digital imaging color-processing was in the publishing industry, and undersea cables respectively London and New York distributed pictures. An encrypted tape was created with a daily express printer in the early development of the digital image. Within the color image analysis processing method which enhances the Bartlane system, the photographs and photography process are of greater quality. Digital image processing employed in medical, terrestrial remote and astronomical imaging. The digital image processing business today has expanded steadily. It is employed today in the geographical, biological, nuclear, archeological, legal and civil law sectors.

Processing the composite image is been using for the production of a digital image and in a 2-dimensional format. Digital photo production defines the processing by modular synthesizer of images. For each pixel, the Color image gives a picture detail. A pixel is a small unit with the color picture configurable color. Disk color pictures are stored as just a raster or raster map. Raster graphics are contained in image files in different formats of computer graphics. A raster image is shown via the monitor, document or other media. The data form of a dot diagram is a rectangular pixel map.

The processing of photographs includes three simple color templates named RGB (red, green, blue). They are the recommended practice and are suitable for all implementations of digital design systems. These elements of RGB color are strongly correlated with a variety of processing techniques. For digital screens, the RGB color space has been used. The frequency and ceremony of the light are measured by color images. There are three-pixel values. An image is a lot of additional details and is used to render image analysis simpler [1–4].

Image processing offers a variety of features such as photo narrowing, image scaling and compression. The compression of photographs is dealing with grayscale pictures in BMP format. Rotating the image by a defined angle is employed to transform the photograph. Likewise, the scale of each pixel is decreased by a common element. Diminishing photograph is employed to decrease the image and to save space on the disk. The scrolling and widening of a picture require the pixel size. The color model employs HSI (Hue, Saturation and Intensity) as a standard application. The color of the domineering part of the spectrum is calculated. The colors on the ultraviolet spectrum vary from about 400 nm (violet) to 700 nm (rot) (Fig. 1).

F. P. García Márquez (ed.), *International Conference on Intelligent Emerging Methods of Artificial Intelligence & Cloud Computing*, Smart Innovation, Systems and Technologies 273, https://doi.org/10.1007/978-3-030-92905-3_37

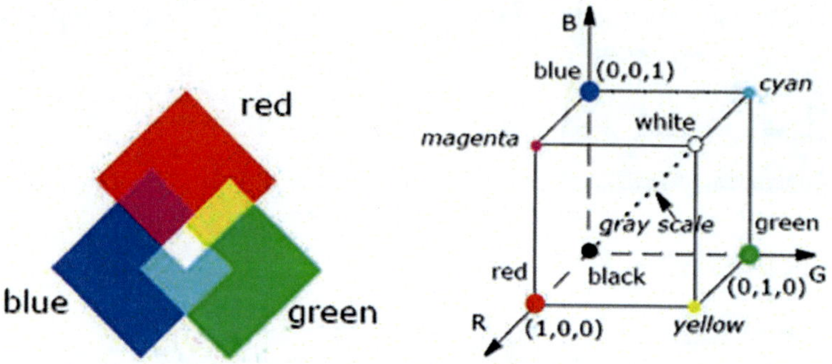

Fig. 1. RGB color model used in image processing

2 Review on Image Processing

Most vision-based systems rely on strongly contrasting and vivid photographs, which provide sufficient detail for the accurate description of the captured objects in a frame.

Dangerous conditions are sometimes undesirable but can be minimized for taking pictures. Intensity histogram equalization methods are common to maximize the quality of a diagram but over-improvement also contributes to unnecessary artifacts. Additionally, inadequate color richness and transparency are often found when limited to an intensity-based amplification. A pipeline strategy incorporating a color channel extending process, histogram equalization step, an amplitude compression treatment and saturation prioritization stage is suggested to resolve these constraints at the same time. Theoretical and practical results acquired from simulations on a wide range of natural scene images show the efficiency of the proposed strategy over other techniques to reduce artifacts while enhancing color and juxtaposition [5–10]. The proposed work in 2016 is illustrated in Fig. 2.

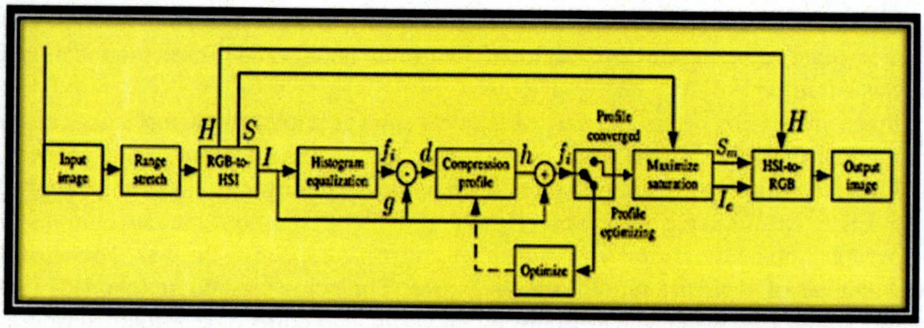

Fig. 2. Structure of the proposed color image enhancement approach

This paper evolved and portrayed a framework to increase the quality of colored images. A number of experiments, which use qualitative evaluation and quantitative measures including entropy, gradient, color, and saturation, have shown that the pipeline technique is designed to raise image information, enhance juxtaposition and sharpen color pictures without undermining colorfulness and brightness. The implementation of dynamically chosen hyperbolic tangent Profiles suppressed unwanted artifacts frequently used throughout histogram equalization. These results indicate that the monochromatic color improvement technique is an adequate choice if deliberate color picture preprocessing is a primary concern. The below pics shows the result of the proposed work (Fig. 3).

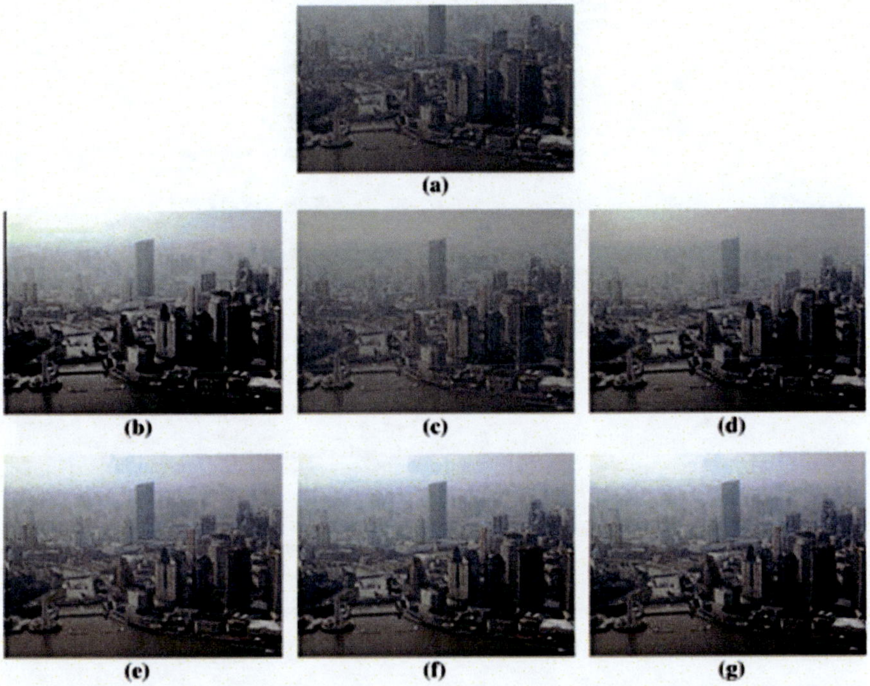

Fig. 3. Test image. **a** Input image, **b** uniform histogram equalization, **c** adaptive contrast equalization, **d** smoothed histogram equalization, **e** color stretching, **f** color stretching with compressed, and **g** color stretching, compressed equalization and saturation maximization

In 2019, a new method was introduced by a researcher to evaluate the movement of soil, especially at the time tillage operations. Tillage activities constitute an essential part of the development of farm crops. Farmers must manage soil laying using optimal tool parameters in order to obtain the beneficial soil situation on economically and environmentally grounds. A major phenomenon deserves to be acknowledged as a result of the soil perturbation caused by soil engagement devices, particularly soil layer mixing. Modern soil layer blending methods are restricted, particularly if mixing of various layers needs to be researched. Many effective methods is the use of biological lasers. Though this technique can continue providing some good information, as it does not adequately represent the shifting or combining of soil layers, it cannot be mistakenly estimated roughly. In this study, the blending of ground layers with multicolored sands placed at varying distances depths in shades was examined (to continually study the mixing of soil layers). A new color multivariate regression was evolved using a PCA (principal component analysis) approach to analyze this multi-colored layer mixing efficiency as shown in Fig. 4.

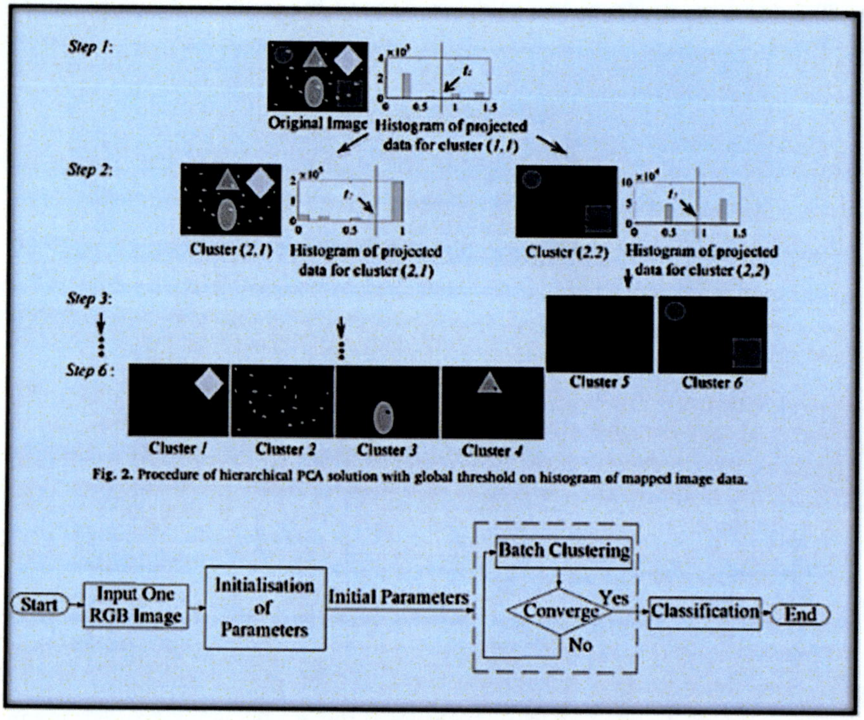

Fig. 4. Entire algorithm for clustering of soil color image

A bright color-clustering algorithm has also been evolved for analyzing the blending of the sediment coating of the crop rotation in this study that can sequence and group different colored sand particulates mixed with one another and into natural soil. A new method of error handling using a hierarchical PCA application was used for the neural network developed by batch categorization and K-means. A weighted average test was performed with reference to two current approaches in order to substantiate the implemented algorithm. The soil bin check was then conducted out with multi-colored layers to verify the suggested method in a practical application of soil mixing. The outcomes of examinations of both soil and benchmarks show that the neural network formulated can recognize highly effective and accurate different color categories. In addition, the suggested algorithm will also be helpful to evaluate soil and tillage device blending and simulator outcomes of the DEC process which is a computational discontinuous approach used for soil-model interaction. Future research will concentrate on the analysis of the computer program and on assessing various agricultural techniques even under weather conditions with separate soils (Fig. 5).

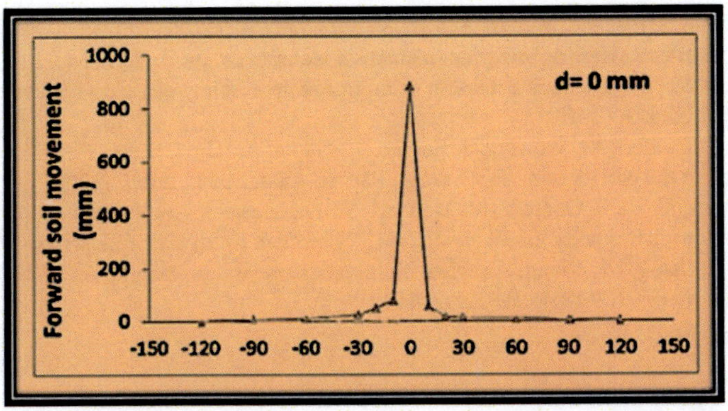

Fig. 5. Forward soil movement of the tracker (located at the soil surface) in the direction of travel for a vertical opener()

3 Conclusion

In the profession of computer research, the usage of digital file image processing means the processing of digital images by algorithms. Digital image analysis has many financial benefits over electronic image processing in a forum-category or ground of digital signals process. It helps the compiled code to be exposed to a broader range of techniques and avoids noise built-up and degradation issues throughout analysis. Virtual imagery can be represented in the context of multifaceted structures when images are represented in two directions (perhaps more).

Photo correction is the especially post-processing of the spatial and radiometric relations via satellite data. Improved picture processing is implemented in order for the resulting object representation to be accurately reflected. Digital categorization is the basis for knowledge revocation and is used to produce analog thematic maps. Numerous image processing strategies are presented in this review paper.

To scholars and educators this appears to be a difficult task. It is difficult to compare the efficiency of the production method without using similar data sets and methodologies. Research are performed on various types of data sources in different papers pertaining to certain image processing technologies.

References

1. D. Kaur, Y. Kaur, Various image segmentation techniques: a review. Int. J. Comput. Sci. Mob. Comput. (2014)
2. A. Chourasiya, N. Khare, A comprehensive review of image enhancement techniques. Int. J. Innov. Res. Growth (2019)
3. G. Kumar , P.K. Bhatia, A detailed review of feature extraction in image processing systems, in *International Conference on Advanced Computing and Communication Technologies, ACCT*, 2014
4. R. Besenczi, J. Tóth, A. Hajdu, A review on automatic analysis techniques for color fundus photographs. Comput. Struct. Biotechnol. J. (2016)

5. M. Nixon, A.S. Aguado, *Feature Extraction & Image Processing for Computer Vision* (2012)
6. H.P. Narkhede, Review of image segmentation techniques. Int. J. Sci. Mod. Eng. (2013)
7. A. Mohan, S. Poobal, Crack detection using image processing: a critical review and analysis. *Alexandria* Eng. J. (2018)
8. A.W.M. Smeulders, M. Worring, S. Santini, A. Gupta, R. Jain, Content-based image retrieval at the end of the early years. IEEE Trans. Pattern Anal. Mach. Intell. (2000)
9. M. Dowlati, M. de la Guardia, M. Dowlati, S.S. Mohtasebi, Application of machine-vision techniques to fish-quality assessment, *TrAC - Trends in Analytical Chemistry* (2012)
10. V. Dey, Y. Zhang, M. Zhong, A review on image segmentation techniques with, in *ISPRS TC VII Symposium – 100 Years ISPRS* (2010)

Building Demolition Technique

Harshil Bhatt(✉)

Department of Civil Engineering, Vivekananda Global University, Jaipur, India
harshil.bhatt@vgu.ac.in

1 Introduction

Relocation is the removal, relocation or fall of big structures during their useful existence. The demolition procedure is performed using certain tools or other legitimate means, accompanied by the local authorities 'permission. We are mindful that each system is usually planned for 100 years with a particular lifespan. The system becomes very unsafe for its inhabitants and the adjacent structures after the operation life has finished. The demolition is aimed at avoiding the unintended collapse of some portion of the structure and maintaining the health of staff, the public and neighboring property. Detailed building evaluation utilizing surveys and related tests, including a building survey and a framework survey are mandatory prior to any building demolition. Further environmental conservation issues have prompted states to enact laws to allow the usage of destroyed waste after recycling. The sites of destruction are the origins of significant amounts of hazardous waste, often known as a waste disposal. Design activities, on the other hand, often combine recycled products with waste, stone and dirt, growing their worth and rendering recycling hard or uneconomic. A new solution for eliminating environmental emissions by rising waste supplies is a construction waste recovery cycle as aggregates. Given the large volume of building waste, the processing of construction waste is a fairly recent challenge for the planet. Demolition is one of the most significant structural engineering activities. If demolition is not performed properly, severe repercussions can occur. The foregoing are the various factors of the effects of building destruction [1–6].

- Human disabled people are hurt when they are not allowed to join or work in a reconstruction house.
- Drop from the collapse building with tiny items or rubble.
- The destruction of the building in section.
- Unstable structural instability due to destruction of the initial system.
- Inadequate techniques of destruction are used.
- Collision with large excavation vehicles as the partly destroyed building was not protected properly.
- The partly destroyed building failed when vast quantities of unknown contaminants needed to be accommodated.
- The site atmosphere is congested that quickly hurts staff or third parties in the proximity of the demo1ition facility.

F. P. García Márquez (ed.), *International Conference on Intelligent Emerging Methods of Artificial Intelligence & Cloud Computing*, Smart Innovation, Systems and Technologies 273, https://doi.org/10.1007/978-3-030-92905-3_38

- Difficult entry for staff who reach a renovation building
- Strong demolition equipment may be at risk of failure due to inadequate assistance.

It is important to recognize the full cycle of building collapse to eliminate such effects from demolition research.

1. *Safety measures*

1.1 *Training and Communication*
Proper workplace protection instruction is given to demolition staff and factory or machinery operators and the possible risks must be alerted by completing all training courses and on the workplace. The Construction Industry Management Agency already conducts suitable site supervisor/foremen and plant or machinery technicians training courses.

1.2 *Equipment Maintenance*
Until usage, all equipment must be reviewed and verified. This is handled and managed properly. Equipment is checked regularly and inspection reports are reported appropriately. When required, a comprehensive protection guideline may be provided to comply with different project circumstances.

1.3 *Electrical Safety*
For excavation sites the properly linked electricity supply from a central power provider or handheld power generator is used. The health criteria set out in the Regulations on Electricity (Factories and Manufacturing Enterprises) are adhered to.

1.4 *Fire*
Both inflammable materials shall be excluded from the premises, until the job concerned needs them. All such flammable goods are kept in safe packaging. Before welding, all furniture, timber, locks, etc. must be extracted. Under operating environments, firefighting devices shall be equipped and preserved. The construction site rules (safety) allow the contractor to maintain all firefighting equipment on such a construction site in good repair and free from defects.

1.5 *Occupational Health*
In association with the applicable subordinate legislation of the Plant and Industrial Undertaking Law and the Workplace Protection and Health Law, the welfare of employees on site shall be adequately maintained with special regard to the subsequent extents:

- Dust Exposures;
- Exposure to Chemicals;
- Ventilations;
- Exposures to Noises;
- Facilities for medical;
- Cleanliness; and
- Work-relatedIllnesses.

2. *Demolition of buildings*

The teardown of houses and other objects is destruction. Demolition compares with deconstruction, which entails removing a structure while retaining precious objects for reutilization. The methods of demolition will rely on the location, time, construction materials, the intent of destruction, and how the wreckages are settled. Tools for reducing time are more effective than quicker models.

2.1 *Already planned Demolition*— the multiple phases until the initiation of the phase of demolition are-
2.2 *Location analysis*— In the sample, numerous criteria of specific structural views and their surroundings are studied. The two major types of surveys are house surveys and systemic surveys.
2.3 *Removal of dangerous materials*—If any, including asbestos bearing products, chemical waste and nuclear contaminants, etc. occur, before beginning renovation, hazardous materials are eliminated at the site. Further analysis and disposal by specialists of these toxic content shall be undertaken.
2.4 *The Demolition Plan and Policy*—which involves the following, is designed to schedule for demolition activity. The location of the house to be demolished. a. Site topography with contours of the land, portions of the slopes and of the landscape protected, if necessary, by the structure. c. Field deletion and/or backfilling information.
2.5 *The Stability Report*—This shall contain the following sections-

- Necessary Stability report. A study on the condition of the house to be demolished in the renovation process.
- In the case of driven hydraulic structures or machinery, design measurements are extended on both external bracings and supports.
- A Stability Statement on surrounding property and existing house.

2 Methodology

1. *Methods of Demolition*

The sequence of destruction depends on such aspects as the design, position and destruction methods selected. In addition, houses and infrastructure will be replaced by 'sequential destruction' in exchange for their development. The different demolition approaches are as follows:

a. *Destruction of structure by Top Down- Manual Technique*

Manually operated procedures are performed from top to bottom, typically from roof to floor. Based on location circumstances and building element to be destroyed, the precise demolition sequence can differ.

b. *Demolition of Cantilevered Structures and Balconies*

The building can be constructed on a pedestrian footpath or in certain situations over part of the traffic lane by cantilevered walls, balconies or canopies. The general sequence of removal is defined in the following—Layers and Beams [7–9].

- The outside wall will first be dismantled.
- Until removal of the cantilevered framework, any device or dead loadsmaintained by the supported framework must be eliminated.
- The concrete is torn down from the outside edge, operating inward, to its pillars, slowly from the cantilevered base.

c. *Demolition of Floor Slabs*

The concrete floor slab of reinforced concrete is removed by taking the concrete off slowly. When the mortar is damaged the support stays and is cut off.

d. *Collapsing Structures by Top-Down by Machineries*

The destruction process by computer is usually similarto the manual top-down system, although most demolitions are performed by mechanical engineering. The dismantling commences when the hydraulic system is relocated to the first floor of the house. The staff will be covered or held farto the region in the range of the cable or tie wires while the cord or tie wire is being pulled. The concrete will first be destroyed before the reinforcing is cut off. The reinforced concrete surface may even be removed by removing the saw. As seen in the following estimates, the destruction series will be.

e. *Demolition of building by Wrecking Ball*

A Machine fitted with balls of stells composes the wrecking ball demand. The house is demolished by the collusion energies of a stainless-steel balls that is hanging from the machine. The methods suggested for the wrecking with the help of balls involves.

- Vertical decline—free decline on a frame of the wrecking balls and
- Swinging in lines—balls fall in lines with the jibs. Figure 1 demonstrates the Wrecking Balls activity.

f. *Implosion of a Structure*

Implosion is the precise position and pacing of the destructive content to allow a building to crumble in a matter of seconds to mitigate internal harm to the immediate area. The procedure weakens or eliminates essential reinforcement to keep the system from fighting

Fig. 1. Wrecking balls demolishing a building

the gravitational force and from collapsing under own weight. Implosion is discussed below [10].

- Systemic pre-weakening is intended to provide systemic stabilization prior to disintegration.
- A channel or package wall is built externallyto the house to hold the rubble away, whether there is a shelter, to reduce the dispersion of construction debris onto adjacent land after implosion.
- The foundation may collapse into the middle of the house and/or the covered region from the successful design.
- A successful design should provide ample and adequate time to only 1 or 2 stories of the structure residual to collapse at one instance to the earth, so that the severity of the land effect may be reducesd.
- Throughout the implosion the plan will also create an evacuation zone to evacuate all occupants or visitors.
- In all points before implosion, the structure's integrity of the structure to be collapsed must be tested also approved to be solid as well as secure.

2. Safety measures for destruction

Health measures should be adopted to secure the workers on the locations and the local community and to safeguard items expected to be demolition-affected.

a. Precautions regarding Machinery/Equipment

For machinery/appliances, the following steps should be taken:

- All dismantling equipment with appropriate preparation shall be run by trained individuals.
- Such instruments will, as specified by the producer or provider of the hardware, be utilized and maintained.
- A trained inspector will closely check the lifting equipment for such items or products at least once every twelve months or once in a half year for such lifting workers.

b. Safeguards concerning Scaffolding

Safeguards given below should be adopted for scaffolding:

- The tower or triple steam shell should be designed, assembled, replaced, changed, or removed at a height greater than 4 m (excluding the tower and triple scaffold).
- Platforms for research would be clear of waste.
- At least one entry point will be accessible for each scaffold.
- Be vigilant to prevent injury to the elements of scaffolding from dropping waste.

c. Safety Measures Concerning the Safety of Community

Safety measures given below must be adopted for the safety of community:

- For acceptable alert signals, the demolition site will appropriately be barricaded.
- In the demolition zone, no unauthorized access should be permitted.
- If the outside walls or roofs are removed, capture platforms should be built.
- The demolition techniques must take into consideration the passage of equipment from floor to floor.
- The use of deconstructed waste to shape ramps for machinery may result in floor overload.

d. Precautions regarding Worker Safety

The measures for the health of workers should be taken:

- Appropriate preparation and guidance to carry out demolition operations must be given for staff engaged in demolition work safely.
- Oversight is necessary to insure that the job is only approved for certain workers with experience and guidance.
- Safety boots and goggles will be issued to all workers inspecting the demolition site.

- Staff will have appropriate personal safety devices such as lids, hearing protectors, secure harnesses, boots, protective suits, etc., issued if necessary.
- Suitable ventilation equipment can also be equipped for particles or chemicals.

3. Difficulties in Destruction of Structures

Difficulties which might occur during the destruction of structures are given below:

- Accident triggered by people from small, insecure workplaces and openings.
- Accidents triggered by individuals struck by items that fell.
- The rapid and unforeseen collapse of the building will contribute to workers 'deaths.
- Insecure building structures.
- The health of staff is compromised by proximity to pollution, contaminants and noise. Building destruction requires multiple destruction operations that impact the atmosphere and potentially trigger pollution.

3 Conclusion

This is agreed that before the renovation is carried out, the construction assessments must be carried out cautiously in order that the structure of the house to be demolished may not inflict significant health, public and neighboring harm. Every building style to be demolished depends on the technique, including its structural integrity and environment, on various factors including the ground state, size of buildings, building age, development standard and economic conditions.

To order to insure the health of both staff and the public, the regulated destruction of structures is required in order to minimize the amount of damage and disaster involved. The current favored approach for the effective and secure destruction of the massive buildings involving extremely high accuracy is explosive or implosive destruction. The demolition process should be performed in order to reduce the risks of harming the community and infrastructure, to harm the health and welfare of site workers and to threaten the atmosphere in the city.

References

1. G.A. Blengini, Life cycle of buildings, demolition and recycling potential: a case study in Turin, Italy. Build. Environ. (2009)
2. A. Coelho, J. De Brito, Influence of construction and demolition waste management on the environmental impact of buildings. Waste Manag. (2012)
3. A. Power, Housing and sustainability: demolition or refurbishment?, Proc. Inst. Civ. Eng. Urban Des. Plan. (2010)
4. M. Aksözen, U. Hassler, N. Kohler, Reconstitution of the dynamics of an urban building stock. Build. Res. Inf. (2017)
5. M. Mália, J. De Brito, M.D. Pinheiro, M. Bravo, Construction and demolition waste indicators. Waste Manag. Res. (2013)

6. J. C.P. Cheng, L.Y.H. Ma, A BIM-based system for demolition and renovation waste estimation and planning. Waste Manag. (2013)
7. K.P. Verian, W. Ashraf, Y. Cao, Properties of recycled concrete aggregate and their influence in new concrete production. Resources, Conservation and Recycling (2018)
8. D. Veeraselvam K, Utilization of demolished concrete waste for new construction. Int. Res. J. Eng. Technol. (2017)
9. J. Bolden, T. Abu-Lebdeh, E. Fini, Utilization of recycled and waste materials in various construction applications. Am. J. Environ. Sci. (2013)
10. M. Missimer, K.H. Robèrt, G. Broman, A strategic approach to social sustainability - Part 2: a principle-based definition. J. Clean. Prod. (2017)

Method of Noise Control for Building

Ravikant Pareek(✉)

Department of Civil Engineering, Vivekananda Global University, Jaipur, India
ravikant_pareek@vgu.ac.in

1 Introduction

This study is an overview to building noise controls. It is not a comprehensive method, but it does put designers into some core concepts and terminology. The knowledge given here should provide designers a strong start in basic implementations on actual projects to tackle the issues of acoustic performance. Designers would need to incorporate more comprehensive guidelines on more acoustically specific designs. Commercially and with government agencies, there are excellent acoustic engineering treatises which give practitioners the guidelines required to use such more stringent methods [1].

1. *Sound Absorption*

There are sound absorption properties for all materials. The not consumed sound energy should be mirrored or distributed. The sound absorbing property of a material is typically defined in a given frequency range as a sound absorption coefficient. The noise mitigation coefficient, the typical sound absorption coefficient of 250 Hz to 2 kHz, the key frequency of the speech spectrum, is used to test sound absorbing products used in houses. The potential spectrum of the NRC varies between completely absorbent (NRC = 1.0) and opaque (NRC = 0.0).

The introduction of sound absorption materials to a space also is a question for the library's inner architecture. A variety of solutions are available to provide appealing and maintainable sound amplification on walls and ceilings. Absorbent fabrics also have acoustically translucent textures including textiles, perforated metal and wood slats. This surface areas permit the propagation and absorption of sound energy by the substrate behind it. Perforated metallic panel are typically utilized to achieve a particular coating. The substance with the smaller holeradius and the broader open region (amount of holes) would be as thin as possible to produce better performance. Any absorption fabrics, such as regular suspended ceiling tiles, are enticing to see. Thicker porous materials typically enhance the absorption of air. Whenever mounted in a lay-in grid ceiling, the dense 5/8″ ceiling tiles have an NRC of 0.50. An NRC value of 0.80 or higher may be a 1-inch thick glass fibre ceiling stand.

F. P. García Márquez (ed.), *International Conference on Intelligent Emerging Methods of Artificial Intelligence & Cloud Computing*, Smart Innovation, Systems and Technologies 273, https://doi.org/10.1007/978-3-030-92905-3_39

2. *Principles of Sound Insulation*

   - By removing it or reflecting it with an intermediary solid material panel sound energy is minimized from one region in a building to another, sound transmission loss (TL) is name. Built products usually attenuate higher frequencies noises than lower frequencies noises. The advanced a wall masses or weights, the extra vibrating it requires energy. This is why a large wall has a higher LT than a smaller panel at all frequencies [2].

Such reduction of sound partitions are required for specific acoustic spaces or acoustic privacy criteria. In certain situations, they are often required as an exterior building cover, where a specific issue is ambient noise at the location. Walls, floors and ceilings that include places where excessive noise, such as mechanical rooms, are created, typically need a strong sound reduction level [3].

3. *Noise criteria*
   This segment provides details and discussions on commonly accepted standards for indoor noise in suitable living/work conditions. These requirements may be used to determine the appropriateness of current indoor and construction areas.

   - *In homes, noise standards*: Room Standards (RC) and Noises Requirements (NC) reflect 2 generally known principles utilized when determining the appropriateness of noise to indoor inhabited spaces by noisy mechanical machinery. In order to determine the detrimental impact of disturbance on speech expression the speaker intrusion standard (SIL) is used.
   - *Noise Criterion (NC) Curves*: Reasonable indoor sound values is established or measured by utilizing NC curves as a consequence of building mechanical equipment operations. Such curves *Send the SPLs to the octave frequency ranges*: This results in sound pressure rates. The lowest NC curves specifies quiet rates for rest and sleep, while the upper NC curves describe very loud working areas that are complicated and restricting except for speech communication. Under this whole spectrum the curves will be used in almost all usual indoor functions to determine the target noise rates goals. The sound rates of the hydraulic or ventilation systems under installation will be in a specific definition equivalent to or below those of the NC aim curve chosen for every octave band to achieve the designs objective. In action, though, NC requirement can be deemed satisfied if the sounds frequency not crosses the NC curves by 1 or 3 dB in one or two octave sections.
   - *Rooms Criterion (RC) Curve*: RCs curve, such as NC curve, are usually utilizeded for the estimation or measurement of interior sounds thresholds arising from mechanical machinery activity. In three essential ways, the RC curves vary from the NC curves. Initially, the low-frequency spectrum of 16 and 31.5 Hz octave band was expanded. Further, the high-frequency spectrum at 2 thousand and 4

thousand Hertz is considerably lower and the octave bands at 8,000 Hz has been excluded, since there is so little noise generated by most mechanical devices [4].

4. *Distribution of Sounds in Indoors*

4.1 *Room's Sound Pressure Levels*:

Equipment suppliers may also achieve the sound's pressure rates at specified distances or sounds strength rates for each apparatus element. The sound levels of whichever position inside a surrounded space might be measured after the properties of the sound sources have been established. The Sounds pressure frequency (SPL) increases at aamount of 6 dB per for the repetition of the gap from the source on the outside 'open area' (no reflective surfaces but ground). But all the containment surfaces of a space in an internal circumstance hold the sound energy such that it cannot extend endlessly and be scattered with time. When sound waves echo around the house, a sound frequency rises, as the sound energy becomes "catched" inside the space and gradually exits.

4.2 *Effect of distance and absorption*:

When you are going around the space away from the sound source, the change of sound intensity level inside depends on the environment in which the sound amplification is situated, the distance from the region and the source. The curves in Fig. 1 reflect all this quantitatively. In the case of a piece of mechanical equipment, Fig. 1 offers a means of calculating a reduction in SPL (or another form of sounds origin in a house) as an individual move from fairly near distances to some other distances in a space, given the sounds-absorbing characteristics of a house are established (Room Constant). Figure 1 offers a way of measuring the acoustic reduction from a given source in a space, providing that the distance is compatible and that the volume of absorption therapy is increased. For a special case of distance and space unchanged, the table 1 reflects a simplification of Fig. 1.

4.3 *Sound absorption coefficients*:

The coefficients on sound absorption differ with frequency for certain surfaces and materials; henceforth the Room Constantsshould be determined for all value frequency. But surfaces not known as absorbent usually have low absorption quantities. In general, 31, 63 and 8,000 Hz frequencies do not determine the coefficients of absorption of air. If the data in this range is not usable using 40% of the 125 Hz values in 31 Hz, 70% of 125 Hz in the 63 Hz band and 80 percent of the 4 thousand Hz in the 8 thousand Hertz octave bands. The supplier will acquire values of sound absorption coefficients for different acoustic products [5–7].

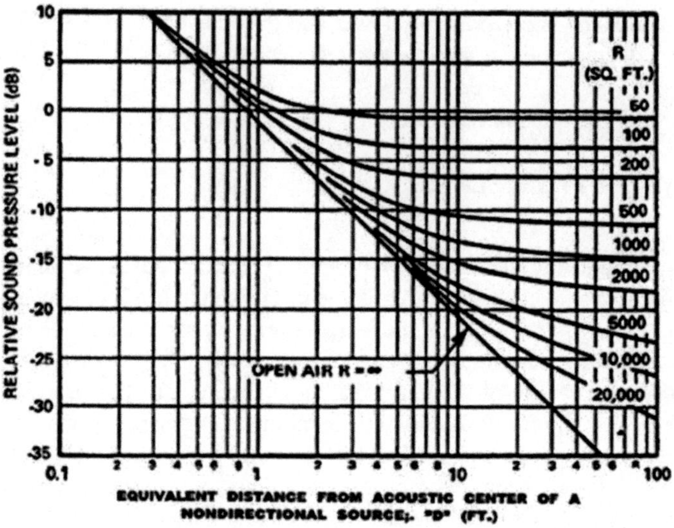

Fig. 1 A graph on relative sound pressure level vs equivalent distance from acoustic center of a nondirectional source

Table 1. 1. Decrease of SPL in dB from standardized distance of 3 feet to 800 sf room constants and room constant to every other range. 2. The negative decrease meaning means the sound volumes are increased

Room constant "R" (ft.^{2})	Distance "D" (in ft.) from equipment)								
	3	5	10	15	20	30	40	60	80
100	−5	−4	−4	−4	−4	−4	−4	−4	−4
200	−3	−2	−1	−1	−1	−1	−1	−1	−1
320	−2	0	0	0	0	0	0	0	0
500	−1	1	2	3	3	3	4	4	4
700	0	2	4	4	5	5	6	6	6
1000	1	3	5	6	7	7	8	0	0
2000	1	4	7	0	9	9	10	10	10
3200	2	5	0	9	10	11	12	12	12
5000	2	6	9	11	12	13	14	14	15
7000	2	6	10	12	13	14	15	15	16
10,000	2	7	11	13	14	15	16	17	10
20,000	2	7	12	14	16	18	19	21	22
Infinite	2	7	13	16	19	22	25	20	31

4.4 *Estimation of Room Constant*:

Some of the details in a room cannot be decided at the early stages of a design eventually, but certain aspects of the plan will also have to be carried out. Fundamental space measurements are expected, but all decisions on side walls, floors and decking materials are not appropriate. This simplification provides a less accurate measurement than the more comprehensive method, but makes quick estimation of Room Constants with gross yet unspecific improvements in room material and requirements for sound absorptions.

Then thorough calculations for the equation not provided here may be rendered if a desired state is detected.

5. *Sound isolation between rooms*

Data and methods to quantify sound level adjustments are addressed here as one follows the direction of "energetic surge" within a soundorigin to the receiver via components like wall, floor, door, etc. Second, when you step away from the source, sound intensity levels in the space surrounding the source are reduced. Some noises are captured on the walls of the room, some are mirrored back into the room and others are transmitted to the surrounding spaces through the walls (which often occurs on the surfaces of the floor and of the ceiling). This topic is dealing with the cumulative consequences of this absorption, reflection and transmission [8–10].

5.1 *Transmission Loss (TL)of Sound, Noises Reductions (NR) And Sounds Transmission Classes (STC)*: In order to monitor the noise interference from surrounding industrial machinery rooms, materials and architecture are selectable at appropriate rates, depending on the experience of acoustic insulating walls and floors. The propagation of sound is impaired by the noises insulation features of the decay structure, the environment of the decay walls, ground or roof and the acoustic features of the silent space.
5.2 *Walls Transmission loss (TL)*: The wall TL is the ratio of sounds strength emitted by the walls to the sounds strength of the airborne incident on a building, measured in decibels. The LT of a wall is therefore the feature of efficiency that is solely based on its weight, structure and structures, and the acoustic condition on either side of the wall or on the wall region does not influence its numerical meaning. Procedures are set out in ASTM E 90 to assess the lack of communication in the laboratory. These details are usually contained in literature and acoustic guides from most manufacturers.
5.3 *In field systems, laboratory scores are seldom obtained*: The laboratory losses of transmission are typically four to five dB greater than that which may be obtained in the fields even though reasonable building approaches are found. ASTM-E336 is a typical form of evaluating sound insulation at (in situ) structures. There are also references to the efficiency of building material transmitting failure. Many manufacturers still sell their goods transmission failure.
5.4 *"Low noise of a curtain" (NR)*: It is the transferred sound strength that is of concern as the signal is emitted from one space (the "source space") into a neighbouring

room (the "receiving room"). The lack of wall transmission is characteristic of the wall structure's efficiency, but even its area depends on the overall sound intensity emitted by the wall (i.e. the greater the region, the more sound energy transferred).

5.5 *The space constant of the hall affects the sound efficiency loss (SPL) in the building*: A wide Space Constant decreases the sound volume in the space at a reasonable distance from the wall. Therefore, in the reception area three things affect the SPL: the wall TL and the wall area among the spring and the reception areas, and the Continuous R2 of the reception office. Such 3 elements are coupled in an extraordinarily wide-ranging way.

5.6 *"Sound transmission class" (STC)*: The latest literature applies to the word "Speech Transmitting Class" (STC). For certain frequencies there is a range of measuring loses. In its connection to the protection of speech intruders or related forms of vibrations, this classification is used for calculating the performance of walls, doors and windows and other acoustic dividers. In the 500–2000-Hz frequency range this one-number rate structure is heavily weighted. The use of mechanical system noise is not advised, the key frequencies being less than the 500- to 2000 Hz range. But vendors who cite STC scores should have 1/3 octave bands TL information within which STCs' value are extracted, so that when certain forms of partitions are required for the insulation of noise in mechanical equipment, TL data may be demanded. ASTM E 413 norm defines the protocol to assess an STC ranking.

2 Conclusion

Sound absorption is not the same as sound separation that avoids the movement from the vibration into different places, for instance walls, ceilings or ground. Vibration absorption Sound-absorbing materials can turn some of the sound energy absorbed into heat rather than transfer it, enhancing sound isolation, but should not be used as a substitute for adequate sound isolation. In addition, sound amplification is added to ground, walls, doors, partition surfaces and items like chairs or bookshelves in the process of therapy. The usage of sound absorbers is also growing.

References

1. A. Y. Elghazouli, R.J. Peppin, Seismic design of buildings to eurocode 8. Noise Control Eng. J. (2011)
2. J. Sikora, J. Turkiewicz, Sound absorption coefficients of granular materials. Mech. Control **29**(3), 149–157 (2010)
3. X. Qiu, Principles of Sound Absorbers (2016)
4. N.B. Roozen, L. Labelle, C. Glorieux, M. Rychtáriková, H. Mullner, On the use of laser doppler vibrometry in building acoustics. Euronoise **2020**, 551–553 (2015)
5. V.G. Babashov, A.S. Bespalov, A.V. Istomin, N.M. Varrik, Heat and sound insulation material prepared using plant raw material. Refract. Ind. Ceram. (2017)
6. E. Gerretsen, European variety of descriptors for building acoustic performance and translation into proposed harmonized descriptors, in *42nd International Congress and Exposition on Noise Control Engineering 2013, INTER-NOISE 2013: Noise Control for Quality of Life* (2013)

7. M. Luykx, M. Lautenbach, M. Vercammen, Pop venues in living areas, in *Euronoise 2015* (2020)
8. M. Kleiner, J. Tichy, *Acoustics of small rooms* (2014)
9. V. Hongisto, J. Varjo, H. Leppämäki, D. Oliva, J. Hyönä, Work performance in private office rooms: the effects of sound insulation and sound masking. Build. Environ. (2016)
10. B. Rakerd, W.M. Hartmann, Localization of sound in rooms. V. Binaural coherence and human sensitivity to interaural time differences in noise. J. Acoust. Soc. Am. (2010)

Fish Tank Monitoring System Using IoT

T. P. Deepa(✉), Basana Khadka, Nirlipta Chatterjee, S. N. Rahul, and Sherwin Kopparam Sridhar

Department of Computer Science and Engineering, Jain (Deemed to be University), Bengaluru, Karnataka, India
tp.deepa@jainuniversity.ac.in

1 Introduction

Fishkeeping is a common thing in houses. For an indoor fish, a person has to take care of its food, check the water levels and many other things. Usually, the caretakers face various problems in the maintenance of the health of the fishes [1]. Most of the health issues faced by the fishes are because of the failure to provide a suitable environment for their survival. The challenges of maintaining the fish tank are the temperature and pH value of the water. It is also important to provide the right amount of water in the fish tank. Constant monitoring of all these parameters is of great difficulty [2]. Therefore, there is a need for a cost-effective and efficient system for monitoring these arguments. The method suggested is intended to ensure that essential characteristics like pH, temperature and water level are controlled and measured in time. The device was developed to calculate and warn automatically if any of these variables were found to be dangerous to minimise the risk of killing fish [3].

The device uses electronic sensor for measuring real-time parameters, processing the controller, data storage and instantly alerting the consumer by SMS. To build this method, elements discovered in previous studies and guidelines were incorporated. The methods used to measure pH and temperature are restricted to the manual use of a test kit of chemicals [4]. The presented method is based on physically monitoring consistency of the water so that whenever fishes are in the risks of adverse circumstances ensuing from insecure pH and temperatures levels. This project is aimed to crack this problematic issue by developing the device which monitors as well as regulate the pH, temperatures, automatically [5]. Fish is a prevalent option of among all the pets because of the following advantages:

(i) Health Benefits

- Reducing Stress
- Reducing Heart Ratesand Bloods' Pressures
- Alzheimer Disease

(ii) Silent Animals
(iii) Education Prospect for Kids

F. P. García Márquez (ed.), *International Conference on Intelligent Emerging Methods of Artificial Intelligence & Cloud Computing*, Smart Innovation, Systems and Technologies 273, https://doi.org/10.1007/978-3-030-92905-3_40

(iv) Small Creatures, take less space
(v) Cost effective as compared others
(vi) Less Upkeep
(vii) Satisfying Relaxation
(viii) Appealingly Attractive.

This project work is based on the automated aquarium technology Internet of Things (IoT), which is used to update the aquarium module function. On a large scale, aquarium monitoring is done manually. A person may do mistakes due to human nature. It is difficult to maintain an aquarium for working people. Unacceptable changes in the levels of water parameter values affect the life of aquatic animals [6]. This motivated us to build a system that automates the manual work to maintain the aquarium with minimum persons required using internet technology. So, we design a system that helps to maintain fish's needs like food, temperature, pH, water level, and cleaning of an aquarium.

This manual checking is time consuming in everyday life and often produces inadequate as parameters to constantly measure adjustments. It is easier if you can automatically track and send alerts somehow to the customer [7]. To solve this challenge, advanced innovation must therefore be introduced in aquaculture. Thus, an individual must choose the right technology to make this type of progress more selective. A SIM card is inserted on the GSM module where the updates are being sent to the phone. This makes it easier for the user to know the updates on the fish. All this monitoring is done by this IoT-based fish tank. The main objectives of this research are given below.

(i) To check if the pH of a fish tank is in range (6.5–9).
(ii) To make sure that the water temperature in the fish tank is within the range (25–27 degrees Celsius).
(iii) To screen the water levels inside tank.
(iv) To provide enough light source to the fish tank.
(v) To make sure that the user gets an SMS if any of the above parameters are not in range.
(vi) To make it easy for common people to maintain their fish tanks without professional support.

2 Literature Review

Sabari et al. researched the topic of water quality checking and monitoring via IoT in their research paper. The US Clean Water Foundations founded Water Quality monitoring Day in 2003 as an international education initiative designed to raise awareness of water supplies throughout the globe and to contribute to it. On September 18, World Water Monitoring Day is held. The control of tank levels of water is utilized to prevent flooding and sensitive levels of water inside tank. Application of the water management system allows possible in-home systems of importance. The present automatic level detection system is defined and could be used for setting up/off a computer. In addition, the traditional level management system for domestic apparatus is generally to begin the feed pumps at a lower level to facilitate it to operate till the water tank reaches a greater degree. It is not assisted adequately with a good control scheme. In addition, liquid level tracking devices for tracking level of water, rivers, silos and lakes, etc. are commonly utilized. The Water Level Tracking Microcontroller (ARM 7) is used to determine the tank level of water of the operator. The Water pollution monitoring on sensor is determined legitimately by the detectors and is tracked by an operator, using all these variables like water levels, turbidity, gas, and temperatures. This paper is motivated by the use of technologies and the system as an everyday living gadget like a computer or cell phone to avoid waste water [8].

Ramya et al. carried another research on the topic of IoT-based smart monitoring systems for fish farming. The sensors are preserved in this planned system of a fish pond with a microprocessor. Throughout the pond structure the ultrasonic and PH detectors are fastened to measure the water depth and consistency. The ultrasonic sensors are utilized to sensitive the object of the fishes throughout the pond as well as the pulse for transmitting and receiving the sound echo in the pond environment. The PH sensor is utilized inside the pond environment for sensing water levels. The microprocessor is utilized to monitor a complete device circuit and the LCD is attached to the microprocessor with errors. The approach addresses but water streams to fish dams are monitored carefully and accuracy ensures by checking variables including the fixing of hydrogen particles, turbidity, humidity, measuring compounds and much more. If someone talks about the listings of good water quality and also the influences on and ways to enhance the bad quality of the water in rivers. It implies that ensuring good water quality is a necessary prerequisite for increased viability and otherwise direct need should be met. And therefore it is essential to every person that the health of living creatures and cultures evolve in adapted water quality index [9].

Saha et al. conducted another research on the topic of Automated Fish Farm based on the IoT technology and aquaculture monitoring system. The DFRobot analogue pH Sensor is used to calculate the pH of the water at Arduino (SEN0161). This pH sensor has a simple communication and functionality. It is specifically designed for Arduino. The sensor must be connected to Arduino by a BNC connector. This sensor's range is about 0–14 pH. At a normal temperature of 25 °C, its precision is ± 0.1 pH, with a working range of temperatures of 0–60 °C. There are just only several portions of the detector throughout the liquid. Whenever the liquid is cleaned and one month for water of high alkalinity, the efficiency of this pH sensor will last for 6 months. Sensors for pH, conductivity and temperature are attached to Arduino and Arduino to AB USB cable is attached to Raspberry pi for serial communications. The Arduino Software accomplishes the retrieval of data sources. In the serial line, the Arduino programme outputs the results. There's also a Raspberry pi application which gathers sensor information from the range and saves sensor dataset to the Raspberry pi-installed MySQL database [10].

Limitations of the current work:

(i) The water has to be changed manually by the user if the values are not in range.
(ii) The sensors are delicate, might break if not handled properly.
(iii) The instruments might get corroded.
(iv) The wires might get disconnected if not handled properly.

3 Methodology

3.1 *Design*

The different steps involved in the algorithm of the developed system are as follows and shown in Fig. 1:

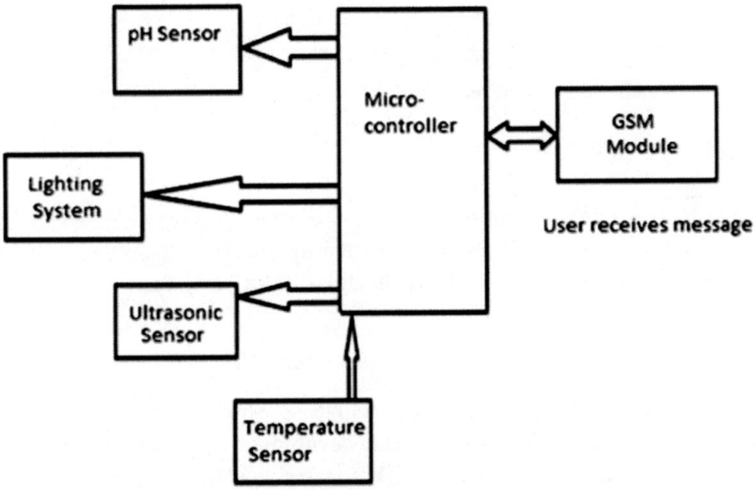

Fig. 1. Illustrates the system architecture for the proposed method

(i) To check if the pH of a fish tank is in range (6.5–9).
(ii) To make sure that the water temperature in the fish tank is within the range (25–27 degrees Celsius).
(iii) To screen the water levels inside tank.
(iv) To provide enough light source to the fish tank.
(v) To make sure that the user gets an SMS if any of the above parameters are not in range.
(vi) To make it easy for common people to maintain their fish tanks without professional support.

3.2 *Instrument*

3.2.1 *Hardware Requirements*:

(i) Arduino UNO
(ii) GSM SIM 900A
(iii) pH sensor
(iv) Temperature Sensor DS18B20
(v) LDR

(vi) LED
(vii) Ultrasonic Sensor HCSR04
(viii) Jumper Wire
(ix) Breadboard
(x) Mobile Phone
(xi) SIM.

3.2.2 *Software Requirement*:

(i) Operating systems:Windows10
(ii) Coding Language:Embedded C
(iii) Software used:Arduino IDE 1.8.9.

3.3 *Arduino IDE 1.8.9*:

The Arduino Integrated Development Environment (IDE) is theappwhich is in C++ and C function. IDE utilized to transcribe and upload programs to compatible Arduino board, but also several vendorsgrowth boards by the aid of third party core.

3.4 *Arduino UNO*:

The Arduino Uno is a microcontroller board with free software. The Arduino Uno board is an ATmega328-based microprocessor. Our development plan framework, similar to every one of the board, Arduino Software (IDE), is designed for UNO. The USB interface comprises of six analogue input pins and 14 digital I/O pins, which enable users to connect many extension boards. The UNO is indeed a simple USB platform which enables USB as a serial device interface. It supports a wide range such as timers, alerts from outer and internals, PWM pins and various sleep mode. The Arduino UNO consists of various classifications of pins. They are:

(i) Power Pins
(ii) Analog Pins
(iii) Digital/PWM Pins
(iv) Communication Pins.

Since Arduino UNO is a free software, it is used also trouble shooting by a wide group of people, which makes it simple to assist in debugging projects. Six analog and thirteen numeral pins are available. Such form of pin helps you to externally link the hardwares' to Uno Arduino boards. Such pins are utilized as the essential component for expanding Arduino Uno's computing capabilities into the real world. Only plug in the socket which corresponds to every such pin with your electronic devices and sensors and you ready to go. The ICSP connector enables users to bypass the USB port and communicate directly with the Arduino in a serial manner. This port should be rebooted to load your chip if it destroys and could no more be used by your device. The code is stored by a 32 KB Flash Drive. A digital pin 13 is connected to an on-board LED for fast code debugging and facilitating the debugging process. The software on the chip has a key to restart.

4 Results and Discussion

For the testing or testing of this theory a breadboard has been used. There is no need to weld because connectors can easily be changed and parts replaced. Parts will not be destroyed and will then be reused. There are also few small sockets on breadboard called holes organized on a series of 0.1" grids. The interfacing of Arduino Uno with GSM Module SIM900A was done to establish a communication link efficiently and pragmatically. There are taken multifarious components such as ultrasonic sensor, temperature sensor, pH sensor, light-dependent register (LDR) and light-emitting diode (LED) bulb. Figure 2 represents the picture of the components' connections during test case 1. Figure 3 represents the image of the components' connections during test case 2. Figure 4 shows the mobile screen where the user gets notified. Table 1 shows the test case, expected output, actual output, error description and their outcomes. We calculate these parameters—pH (6 to 7.5), the intensity of light, depth, and temperature of water (21 °C to 33 °C) and have a particular range for each of the parameters. So, when the readings are not in range, we use the GSM module so that the user gets an SMS and they can change the water or increase or decrease the level of water. When there is no proper source of light, the bulb turns on and when there is enough light the bulb turnsoff.

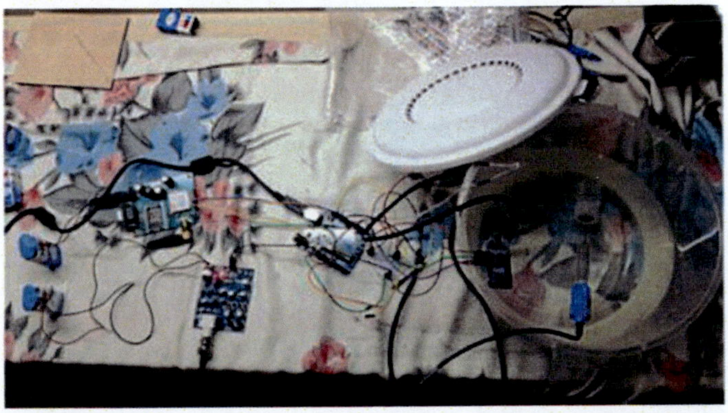

Fig. 2. Illustrates the image of the components' connections during test case 1

5 Conclusion and Implication

The project is an IOT based model that improvises the monitoring and maintenance of aquariums. There are various factors we need to keep in mind when it comes to maintaining a fish tank. If any of these factors are out of range, it creates an unfavorable environment for the fishes, and they eventually die. Thus, people find it hard to maintain the fish tank. Instead of using different tools and getting external help which can cost them a lot, we have come up with a project that integrated essential sensors to monitor the readings and notifies the user. The sensors used include pH sensors, Lightning Sensor, Ultrasonic Sensor, Temperature Sensor, and GSM Module. We calculate these

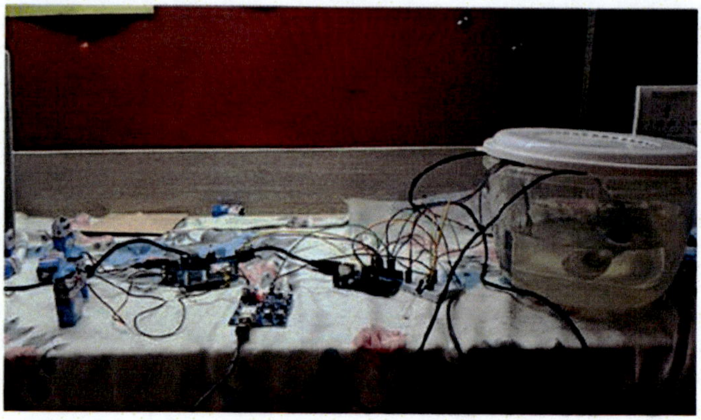

Fig. 3. Illustrates the image of the components' connections during test case 2

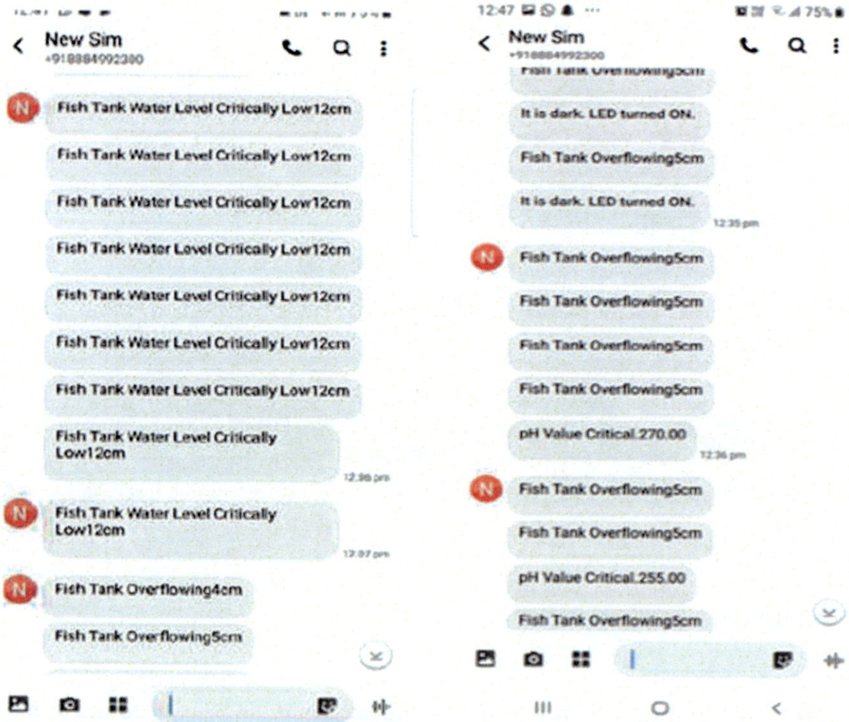

Fig. 4. Illustrates the mobile screen where the user gets notified

Table 1. Illustrates the test case, expected output, actual output, error description and their outcomes

Test case	Expected output	Acutal output	Error description	Action taken
Ultrasonic sensor	Less than 6 cm	5 cm	Overflow condition	User notified
	Greater than 11 cm	12 cm	Underflow condition	
Ultrasonic sensor	6 cm–11 cm	9 cm	None	None
Temperature sensor	Below 20 °C	19 °C	Temperature critical	User notified
	Greater than 23 °C	26 °C		None
Temperature sensor	Between 20 °C–23 °C	21 °C	None	None
pH sensor	Below 7.2	6.9	pH critical	User notified
	Above 7.6	8.0		
pH sensor	Between 7.2–7.6	7.4	None	None
LDR	1.0 or higher	5.0	Darkness, Bulb ON	User notified
LDR	0.0	0.0	None	None

parameters—pH (6–7.5), the intensity of light, depth, and temperature of water (21 °C to 33 °C) and have a particular range for each of the parameters. So, when the readings are not in range, we use the GSM module so that the user gets an SMS and they can change the water or increase or decrease the level of water. When there is no proper source of light, the bulb turns on and when there is enough light the bulb turnsoff.

The future scope of this project is to add more sensors to make the fish tank a more ideal environment by checking on other parameters. And also, in the proposed system we can monitor the fish tank. In the future, we can develop the system to automate all the controls and handle the parameters by the system itself, instead of notifying the user. By using this monitoring system, we can easily prevent the extended use of time and labor, along with saving the lives of the fishes in the aquarium.

References

1. Ss. Rani, IOT based fishery management system. Int. J. Ocean. Oceanogr. (2019)
2. S. Nocheski, A. Naumoski, Water monitoring IoTsystem for fish farming ponds. Int. Sci. J. "Industry 4.0 (2018)
3. C. Lee, Y.J. Wang, Development of a cloud-based IoT monitoring system for fish metabolism and activity in aquaponics. Aquac. Eng. (2020). https://doi.org/10.1016/j.aquaeng.2020.102067

4. Y. Kim, N. Lee, B. Kim, K. Shin, Realization of IoT based fish farm control using mobile app (2019). https://doi.org/10.1109/IS3C.2018.00055
5. S.U. Kiruthika, S.K.S. Raja, R. Jaichandran, IOT based automation of fish farming. J. Adv. Res. Dyn. Control Syst. (2017)
6. H.N. Binti Hasim, M. Ramalingam, F. Ernawan, R. Puviarasi, Developing fish feeder system using Raspberry Pi (2017). https://doi.org/10.1109/AEEICB.2017.7972422
7. J. Lee, A. Angani, T. Thalluri, K.J. Shin, Realization of water process control for smart fish farm (2020). https://doi.org/10.1109/ICEIC49074.2020.9051285
8. M. Sabari, P. Aswinth, T. Karthik, C. Bharath Kumar, Water quality monitoring system based on IoT (2020). https://doi.org/10.1109/ICDCS48716.2020.243598
9. A. Ramya, R. Rohini, S. Ravi, Iot based smart monitoring system for fish farming. Int. J. Eng. Adv. Technol. (2019). https://doi.org/10.35940/ijeat.F1089.0886S19
10. S. Saha, R.H. Rajib, S. Kabir, IoT based automated fish farm aquaculture monitoring system (2018). https://doi.org/10.1109/ICISET.2018.8745543

Blockchain Based Freelancing System

K. S. Shilpa(✉), Brahadeesh Kishore, P. Neil, Nilesh Jain, and Jay Jain

Department of Computer Science and Engineering, Jain (Deemed to be University), Bengaluru, Karnataka, India
shilpa@jainuniversity.ac.in

1 Introduction

A Blockchain is essentially an immutable, publicly visible distributed database that was first implemented by the cryptocurrency bitcoin. Its working is explained in Satoshi Nakamoto's paper "Bitcoin: A Peer-to-Peer Electronic Cash System". A copy of the Blockchain is maintained by many computers in a network [1]. Each computer is known as a node. Unlike a centralized system such as a traditional relational database which has centralized authority, there is no single ownership of the Blockchain by any of the participating nodes [2]. This results in a highly secure and fault-tolerant system. If one of the nodes goes down, every other node will have a copy of the Blockchain. Secondly, if a hacker manages to corrupt the Blockchain on one node, the network will be able to find anomalous copies of Blockchain and will successfully declare the copy as invalid [3]. Blockchains enforce immutability through the computation of hash values [4]. These hash values are computed in real-time, and the next block in the chain will contain the hash value of the previous block. Hence changing internal values of a block will change its hash, causing all blocks after it to become invalidated [5].

Aside from Blockchain, the concept of smart contracts has also been introduced. Smart contracts are self-executing, programmatic contracts. They are executed when the predetermined conditions stated by the contract are all met [6]. These smart contracts are stored on the Blockchain and can be used to make certain actions or conditions occur without the need for human or third-party intervention. Although the Blockchain cryptocurrency was the intended use of Blockchain, its features have proved to be useful beyond just cryptocurrency for a variety of use cases that require features such as immutability and irreversibility. Some of these applications include electronic voting, supply chain management, and crowdfunding. Our project explores the use of Blockchain and smart contracts in implementing a platform for freelancing. Freelancers are individuals who are self-employed and are not committed to any particular employer or organization in the long term. They are hired on a need basis by companies which are then provided with their services for a fee. Both part-time and full-time freelancing has become increasingly popular in the last several years [7]. This growing popularity can be attributed to several factors, such as the desire for more flexible work hours, the option to work from home, and an additional side-income for part-timers. We hence aim to create a platform that could further benefit these individuals.

F. P. García Márquez (ed.), *International Conference on Intelligent Emerging Methods of Artificial Intelligence & Cloud Computing*, Smart Innovation, Systems and Technologies 273, https://doi.org/10.1007/978-3-030-92905-3_41

2 Literature Review

To have a deeper understanding and broader perspective on this project, we selected several research papers relevant to the topic at hand to assist us. The section below aims to summarize our understandings of them.

Nakamoto et al. researched the topic of electronic cash systems. This research paper was authored by the group responsible for the creation of bitcoin, the first and most successful use case of Blockchain. Several shortcomings were observed with the traditional system of transactions in which a financial institution serves as a trusted third party. These include mediating disputes between the parties engaging in the transaction, which cannot be avoided by the financial institution. These shortcomings prompted the creation of a system that allowed for irreversible electronic transactions without the need of a trusted third party, a system that this research paper by Satoshi Nakamoto introduces. This system or structure is known as a Blockchain. The Blockchain uses cryptographic proof instead of trust, hence eliminating the need for a third party by making it computationally very expensive to reverse a transaction. By making it computationally impractical to reverse transactions and for attackers to modify the network, the double-spending problem is also resolved. A Blockchain is represented as a chain of transactions each linked together by cryptography. It is also a distributed and decentralized system, as there are a large number of computers participating in the hosting of the Blockchain called nodes, and none of these nodes have central authority over the system. A copy of each Blockchain is kept on each node participating in the Blockchain. Every time a new transaction occurs, itisbroadcast through all the other nodes. For an attacker to gain control of the Blockchain, the amount of CPU power possessed by them must exceed that of the nodes used to host it. That is, the attacker needs at least 51% of the CPU power to sabotage the chain, making the Blockchain highly secure [8].

This paper was written by Dr. Gavin Wood, the founder of Ethereum, who explains its design, implementation details, and the opportunities it provides. Mindful of the great success behind the Blockchain structure as well as its usefulness outside of just Bitcoin, Wood sought to create Ethereum, a generalized implementation of a Blockchain. The main goal of this project was to facilitate transactions and autonomous agreements between different entities thatwould otherwise have no means to trust each other. Wood describes Ethereum as a transaction-based state machine, in which valid state transitions are brought about by successfully executing transactions. These transactions are placed in structures known as blocks which are cryptographically hashed and chained together, forming the Blockchain. The blocks act as a permanent record or journal of what has occurred in the past. To avoid network abuse issues, Ethereum requires fees in the form of gas every time programmable computation is to be executed by the Blockchain. Gas is a sub-denomination of Ether and is used only to facilitate the use of the network's computational power. Gas and Ether are both exchangeable. The amount of gas required depends on the complexity of the computation to be performed [9].

This research paper examines how smart contracts are created in Ethereum, how users and contracts interact with each other, as well as how Ethereum and smart contracts are being used in general. The biggest difference between Ethereum and other Blockchain is that Ethereum acts as both a digital currency as well as a distributed virtual machine. This virtual machine is responsible for executing the code specified by the smart contracts. Ethereum has two types of accounts: external accounts which are controlled by users and a contract account, controlled by code. Hence a smart contract can be viewed as an automated user. The Ethereum virtual machine (EVM) supports over a hundred different basic instructions known as opcodes, with the CREATE and self-destruct primitives being used to create and end the contracts respectively. A combination of these instructions is what leads to more complex logic. The case study in the paper observes how Ethereum is being used, noting that its usage has increased drastically along with its increase in value. The value of 1 Ether rose from 10 USD in 2017 to over 1,400 USD a year later, and then later fell to around 500 USD, although very volatile. In the year 2017 alone, the number of transactions per day increased from 40 K to 1 M and remains stable despite the drop in the value of Ether. Additionally, it had also been observed that the number of transactions involving smart contracts constitutes 1/3rd of all transactions, hence confirming their importance [10].

3 Methodology

3.1 Design

The general methodology or workflow of the freelancing platform can be described as follows. Figure 1 shows the methodology from the client's point of view. Figure 2 shows the methodology from the freelancer's point of view.

- The user registers their account on the platform by using their public Ethereum wallet address. After registering, they may then modify their user profiles.
- The user then creates an account on the smart contract. Using this, they may accept or create jobs.
- When requesting for a job to be done, the client sends a request to the smart contract. The Ether they are willing to pay for the job to be done is temporarily locked into the smart contract. This is done to make sure the client indeed does have the funds to pay the freelancer.
- Freelancers search for jobs on the platform and get in touch with those who have posted jobs.

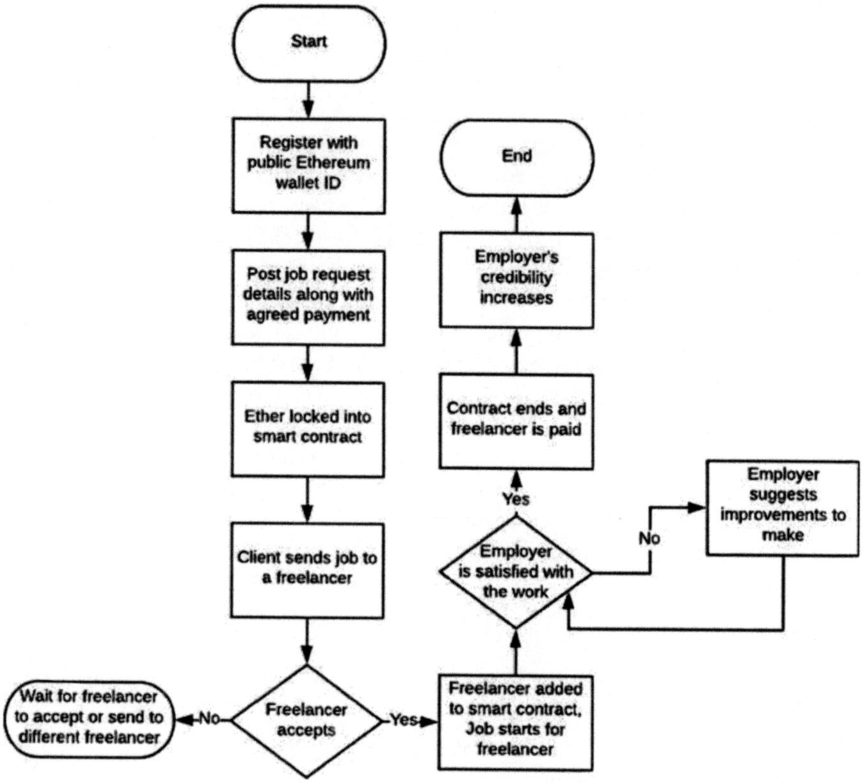

Fig. 1. Illustrates the methodology from the client's point of view.

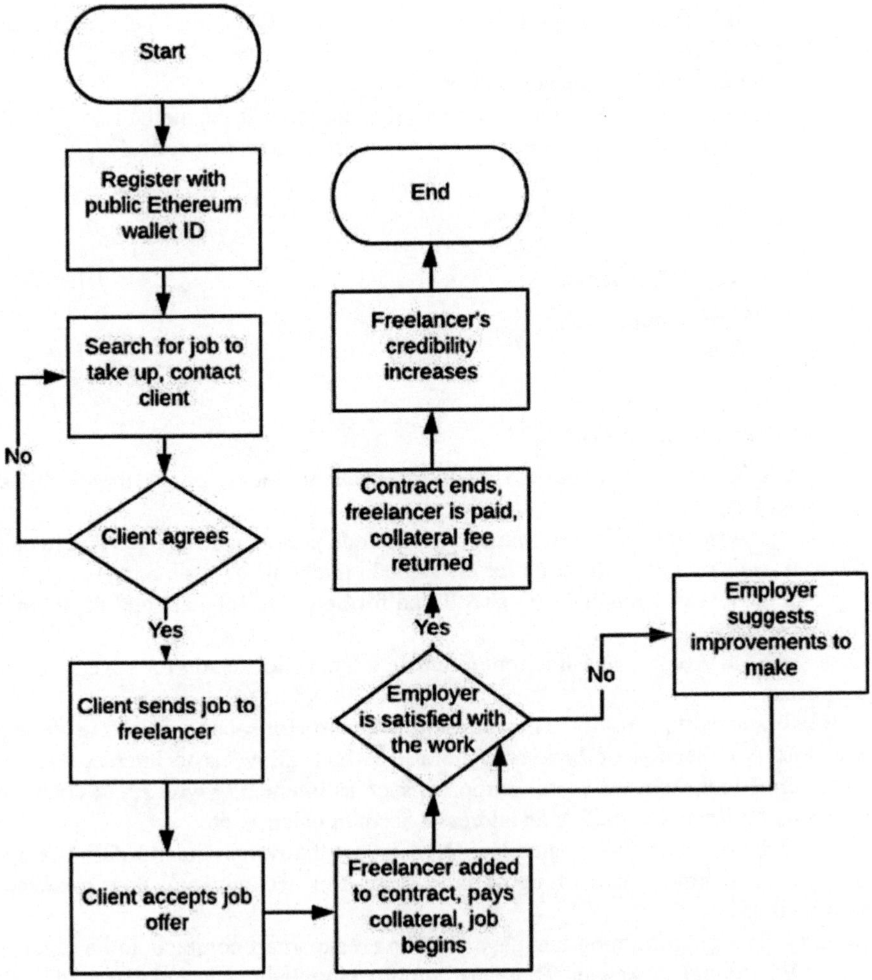

Fig. 2. Illustrates the methodology from the freelancer's point of view.

- The client then sends the job request to the freelancer. Once the freelancer accepts this job offer, they will be made to pay a small amount of Ether as collateral. This will be refunded to them after the completion of the job.
- Given below are two flow charts that describe the high-level methodology of the system from the perspective of the client and the freelancer, respectively.

3.2 Instrument

3.2.1 Hardware Requirements

- RAM: 4 GB, minimum 1 GB
- Storage: 25 GB.

3.2.2 Software Requirements

- Node.js/JavaScript: Language used to implement client-side scripts and the server-side of our project.
- Express.js: A minimalistic web framework for Node.js which provides an API to build the back-ends of web apps. Used for the same in this project.
- Operating System: Ubuntu 18.04 LTS. Used for both development and deployment for our project.
- Ganache CLI: A command-line tool to create a local Blockchain for developmental purposes.
- Truffle Framework: A utility to compile and write tests for smart contracts in Node.js.
- Web3.js: A collection of JavaScript libraries which allow us to interact with the Ethereum Blockchain for useful purposes, such as interacting with smart contracts, initiating transactions, finding an address's account balance, etc.
- Mongo DB: A document-oriented NoSQL database thatstores data in JSON-like documents. This project uses Mongo DB to store user accounts and their associated information.
- Solidity: The programming language used to create smart contracts to be executed on the Ethereum Blockchain. These are compiled into byte code and executed on the Ethereum virtual machine.
- Any web browser that supports HTML5, CSS3, and JavaScript. These will be used in the implementation of the user interface.
- Visual Studio Code: Text editor used to build the project.

4 Results and Discussion

Smart contracts are the heart of this project and implement the most important functionality. The contracts we use primarily act as escrow for funds that are to be transferred from the client to the freelancer after the freelancer has completed their work. The freelancer is paid the full amount after the client approves of their work. To de-incentivize unethical behavior on the platform, our smart contract takes advantage of the fact that transactions performed are publicly visible on the Blockchain, hence alleviating any suspicions of hiding information.

In addition to the client's deposit of ether into the escrow contract, a small fee will also be taken from the freelancer as a form of collateral, which will then be returned to them after the completion of the task. For now, this collateral is set to a minimum of 1/10th of the amount that the job pays. This is done to de-incentivize negligence from the freelancer, since using a small amount of Ether as a form of collateral means that they will have something to lose by being negligent. The collateral fee ensures that only freelancers serious about the task will apply for the task. The collateral fee is a fraction of the amount which will be returned to the freelancer after work completion.

Through the smart contract, users may perform the following actions:

- Create a user account
- Create a job with the necessary funds
- Delete a job
- Send a job to a freelancer
- Accept a job
- Mark a job as completed and pay freelancer.

The actions the user performs depend on whether the user plays the role of client or the role of the freelancer. The smart contract code written in solidity is given below.

```
library SafeMath
{
    function add(uint a, uint b) internal pure returns (uint)
    {
        uint c = a + b;
        require(c >= a, "SafeMath: addition overflow");
        return c;
    }
    function div(uint a, uint b) internal pure returns (uint)
    {
        return div(a, b, "SafeMath: division by zero");
    }
    function div(uint a, uint b, string memory errorMessage) internal pure returns (uint)
    {
        require(b > 0, errorMessage);
        uint256 c = a / b;
        return c;
    }
}
contract Freelancing
{
    event CreateAccount(address user);
    event CreateJob(address client, uint pay);
    event DeleteJob(address client);
    event AcceptJob(address freelancer, address client);
    event SendJob(address client, address freelancer);
    event MarkJobComplete(address client, address freelancer, uint pay);
    using SafeMath for uint;
    struct Job
    {
```

```
    string name;
    string desc;
    uint pay;
    uint collateral;
    address creator;
    address payable freelancer;
    bool deleted;
    bool finished;
}
struct User
{
    bool exists;
    uint[] created_job_indices;
    uint[] job_request_indices;
}
mapping(uint => mapping (address => bool)) past_freelancers;
mapping(address => User) public registered_users;
Job[] public jobs;

function get_user (address user_address) public view returns (User memory)
{
    return registered_users[user_address];
}
function get_jobs () public view returns (Job[] memory)
{
    return jobs;
}
function get_job (uint index) public view returns (Job memory)
{
    require(index < jobs.length, "Array out of bounds");
    return jobs[index];
```

```
}
function create_account () public
{
    require(!(registered_users[msg.sender].exists), "Account already registered"); // User
    must not already be registered
    registered_users[msg.sender].exists = true;
    emit CreateAccount(msg.sender);
}
function create_job (string memory name, string memory desc) public payable
{
    require(msg.value > 0, "Fee must be nonzero");
    require(registered_users[msg.sender].exists, "Account not registered");
    Job memory job;
    job.name = name;
    job.desc = desc;
    job.creator = msg.sender;
    job.pay = msg.value;
    registered_users[msg.sender].created_job_indices.push(jobs.length);
    jobs.push(job);
    emit CreateJob(msg.sender, msg.value);
}
function delete_job (uint index) public
{
    require(registered_users[msg.sender].exists, "Account not registered");
    require(index < jobs.length, "Index out of bounds");
    Job storage job = jobs[index];
    require(job.creator == msg.sender, "Sender must be the job creator");
    require(!job.finished, "Job cannot be finished");
    require(job.collateral == 0, "Job cannot be underway");
    require(!job.deleted, "Job cannot be already marked as deleted");
    uint pay = job.pay;
```

```
    job.pay = 0;
    job.deleted = true;
    msg.sender.transfer(pay);
    emit DeleteJob(msg.sender);
}
function send_job (address payable freelancer, uint job_index) public
{
    require(registered_users[msg.sender].exists, "Account not registered");
    require(msg.sender != freelancer, "Cannot send job to oneself");
    require(registered_users[freelancer].exists, "Freelancer not registered");
    require(job_index < jobs.length, "Index out of bounds");
    require(jobs[job_index].creator == msg.sender, "Must be the job creator to send the job");
    require(jobs[job_index].collateral == 0 && !jobs[job_index].deleted, "Job has already
    started or has been deleted.");
    jobs[job_index].freelancer = freelancer;
    if (!past_freelancers[job_index][freelancer])
    registered_users[freelancer].job_request_indices.push(job_index);
    past_freelancers[job_index][freelancer] = true;
    emit SendJob(msg.sender, freelancer);
}
function accept_job (uint job_index) public payable
{
    require(registered_users[msg.sender].exists, "Account not registered");
    require(job_index < jobs.length, "Index out of bounds");
    require(jobs[job_index].freelancer == msg.sender, "Job was not sent");
    require(jobs[job_index].collateral == 0 && !jobs[job_index].deleted, "Job has already
    started or has been deleted.");
    require(msg.value >= jobs[job_index].pay.div(10), "Collateral must be at least 1/10th of
    job's value");
    jobs[job_index].collateral = msg.value;
    emit AcceptJob(msg.sender, jobs[job_index].creator);
```

```
    }
    function mark_complete (uint job_index) public
    {
        require(registered_users[msg.sender].exists, "Account not registered");
        require(job_index < jobs.length, "Index out of bounds");
        require(jobs[job_index].creator == msg.sender, "Must be the job creator to mark job as
        complete");
        require(jobs[job_index].collateral > 0, "Job must have already started.");
        require(!jobs[job_index].deleted, "Job must not have been deleted.");
        Job storage job = jobs[job_index];
        job.finished = true;
        job.collateral = 0;
        job.freelancer.transfer(job.collateral.add(job.pay));
        emit MarkJobComplete(job.creator, job.freelancer, job.pay);
    }
}
```

Before creating the freelancing contract, we create a library contract called SafeMath which prevents potential integer overflows while performing math operations. It is used to add overflow-checking arithmetic member functions to the unit type (256-bit unsigned integer) in our main contract. A library contract is a special type of contract that cannot store any data or Ether. At the beginning of the contract, events are declared which may then be emitted in a function call. Emitting an event has no significance to the smart contract itself. It functions similar to a function that prints to standard output in other languages, except these events' arguments, are stored as logs along with the transaction block created on the Ethereum Blockchain when a function is executed. These emitted events may then be listened to externally and action may be taken based on them. For example, one may choose to write a script that notifies them every time a job is created or every time a job is deleted. Properties of Jobs and Users are encapsulated into their respective structures. An array of types of Jobs is used to hold all jobs, while mapping from contract address to the User type is used to hold corresponding user data of those who have registered on the smart contract.

The required function is used to ensure the validity of an expression and throws an exception with the specified message if the expression evaluates to false. Require statements are used in nearly every function to ensure that users do not perform malicious transactions or transactions they are not meant to perform, such as deleting another user's jobs or marking another freelancer's job as complete on behalf of another client. Creating transactions on the Blockchain through the smart contract comes with a fee known as gas. The amount of gas to be paid depends on the amount of data to be stored on the contract. Gas costs are usually of the order of several cents or rupees but must be paid for every transaction made. However, gas costs are negligible compared to the fraction of users' incomes taken by traditional freelancing platforms, so this can be considered a fair trade. The use of a smart contract comes with the benefit of using the Ethereum Blockchain to store everything vital to the project, which greatly decreases the need and dependence on a central server. Users may hence continue to interact with the smart contract even while the server is down.

5 Conclusion and Implication

One main limitation of this project is the inability to pay freelancers in currencies other than Ether. Exchanging currency for Ether is not a difficult process however and can be done through one of the various Ethereum exchanges. While it may be troublesome for a user to have to exchange centralized currency for Ether and then exchange Ether for currency when needed, we feel that this is an acceptable trade-off since the benefits that are provided by using Ethereum's smart contracts outweigh the downsides of having to exchange currency for Ether and vice versa. Payment also becomes less of a hassle due to the existence of only a single currency on the platform. Additionally, our project requires freelancers to accept working for a fixed amount of money paid at the end of completing a given task. Currently, there is no support for creating a task in which a freelancer is paid periodically in time intervals, or the form of invoices through a payroll system: All of which are commonly used, payment models. Creating additional ways to pay freelancers through the power of smart contracts is a useful feature that could be implemented in the future.

References

1. Z. Zheng, S. Xie, H. Dai, X. Chen, H. Wang, an overview of blockchain technology: architecture, consensus, and future trends (2017). https://doi.org/10.1109/BigDataCongress.2017.85
2. F.P. Hjalmarsson, G.K. Hreioarsson, M. Hamdaqa, G. Hjalmtysson, Blockchain-based e-voting system (2018). https://doi.org/10.1109/CLOUD.2018.00151
3. R. Azzi, R.K. Chamoun, M. Sokhn, The power of a blockchain-based supply chain. Comput. Ind. Eng. (2019). https://doi.org/10.1016/j.cie.2019.06.042
4. M. Turkanović, M. Hölbl, K. Košič, M. Heričko, A. Kamišalić, EduCTX: a blockchain-based higher education credit platform. IEEE Access (2018). https://doi.org/10.1109/ACCESS.2018.2789929
5. Q. Xia, E.B. Sifah, A. Smahi, S. Amofa, X. Zhang, BBDS: blockchain-based data sharing for electronic medical records in cloud environments. Inf. (2017). https://doi.org/10.3390/info8020044
6. J. Kishigami, S. Fujimura, H. Watanabe, A. Nakadaira, A. Akutsu, The blockchain-based digital content distribution system (2015). https://doi.org/10.1109/BDCloud.2015.60
7. R. Beck, J. Stenum Czepluch, N. Lollike, S. Malone, Blockchain—The gateway to trust-free cryptographic transactions (2016)
8. S. Nakamoto, A peer-to-peer electronic cash system. J. Gener. Philos. Sci. (2008)
9. G. Wood, A secure decentralized generalized distributed ledger. Ethereum (2018)
10. L. Kiffer, D. Levin, A. Mislove, Analyzing Ethereum's contract topology (2018). https://doi.org/10.1145/3278532.3278575

Achieving Efficient Data Deduplication and Key Aggregation Encryption System in Cloud

M. K. Jayanthi(✉), P. V. Naga Saithya, P. Sri Vaibhavi, and Y. Harshitha Reddy

Department of Computer Science and Engineering, Jain (Deemed to be University), Bengaluru, Karnataka, India
k.jayanthi@jainuniversity.ac.in

1 Introduction

Cloud storage has increased in popularity in recent years, not because of its cost-effective on-demand use, but because of the availability of vast volumes of storage on which consumers can conveniently outsource their records [1]. According to the majority of reports, redundant files or records occupy almost half of cloud computing capacity. The maintenance of the large, ever-increasing volume of data is a problem for today's cloud providers. De-duplication is used to make data storage more scalable. Protection and privacy are two of the most pressing issues in public cloud settings. Convergent encryption, rather than standard encryption, is used to guarantee data security for consumers [1]. The efficiency of storing data in a computer will improve day by day with the aid of cloud, and data access will be easy.

Data leakage, compromised apps, data misuse, privacy concerns, data location, protection at the device level, and data replication are some of the problems that users encounter when storing vast volumes of data in the cloud. When data was in reading or writing mode and was being transmitted from one network to another on the cloud, these problems interrupts the system's routine. Information is usually copied on different storage areas to increase accuracy and usability, as well as to provide disaster recovery. The bulk of this copied information contributes to the capacity framework's load in terms of additional storage and transmitting capacity used to transfer the copied data across the grid. Information storage reliability is important, and an information deduplication scheme is seen as a game-changing breakthrough for active storage of massive volumes of data. The deduplication approach is a one-of-a-kind information compression technique for extracting redundant data [2, 3].

The basic principle of cloud computing infrastructure is to pool various kinds of services in order to inspire multiple users or experts to utilize the same resource at the same time, raising privacy and confidentiality risks. Data storage is one of the many cloud technologies that has grown in acceptance in now days. This is attributed to the growing need for data subcontracting, as well as a number of other cloud advantages such as tremendously fault-tolerant storage of data through replication and delivery of data [4]. Since data isn't stored on the data owner's own hardware, cloud computing raises security concerns. While access control in the cloud offers a certain degree of

F. P. García Márquez (ed.), *International Conference on Intelligent Emerging Methods of Artificial Intelligence & Cloud Computing*, Smart Innovation, Systems and Technologies 273, https://doi.org/10.1007/978-3-030-92905-3_42

data protection. Any future customers are also reluctant to use cloud computing because of security worries. First and foremost, the data owner must have full confidence in third-party file sharing services; otherwise, they would feel they have no power of their data. This is also the case that the third-party data storage provider has all access control controls. Since the cloud can also be a multi-user network with services exchanged by many users, there's always the possibility of data leakage due to a number of reasons, like cooperatedaccounts of cloud user or mischievous cloud service hacks. Complete data stockpiled in cloud storage is circulated over the internet, that adds additional risk factor because, within thebest intentions of cloud storage company's, data is often stolen on route to its destination [5, 6].

2 Literature Review

Deduplication is a standard function of cloud storage providers that removes redundant data by maintaining only one copy of each file or block. Deduplication decreases the amount of space and capacity used for data storage facilities, and it functions well when used by many customers, as is the case for cloud storage services. The privacy effects of cross-user deduplication were studied. D. Harnik illustrate how deduplication is commonly used as a side channel to expose details about the contents of other users' folders. Deduplication is commonly used as a hidden medium by malicious software to connect with the core in a special situation, unaware of any firewall settings at the targeted. Cloud storage services are unlikely to leave cross-user deduplication due to the substantial cost savings it provides. As a result, we suggest simple protocols for cross-user deduplication that significantly minimise the possibility of data leakage [7].

Many existing systems, especially for cross-domain deduplication, have had to compromise productivity for the survival against brute-force attacks, that is invariably awkward in operation. Furthermore, only a few current schemes such as the message equality detail take into account (i.e., if two dissimilar cypher texts resemble to a reliable plaintext). In order to fix the above mentioned concerns, P. Christen recommended a multi-domain architecture along with an operative and privacy-preserving enormous data deduplication scheme. This scheme preserves data secrecy with the help of multi-domain deduplication on the other hand avoiding attacks from brute-force by producing a random tag as well asacontinuous number of random cypher-texts for all data. Moreover, this scheme preserves the message equality information to the fullest degree conceivable by allowing merely Cloud Service Provider (CSP) as well asthe agent to conduct intra-deduplication and inter-deduplication, respectively. Furthermore, comprehensive simulations illustrates that our scheme leave behind the existing competing schemes, particularly in terms of the cost of computation and thus duplicate search complexity of time [8].

Cloud storage should be able to store and exchange data with others in a safe, effective, and scalable manner. To help understand cloud data security, J. Wu et al. suggest a cryptosystem based on an unbroken size cypher text key and an attribute-based aggregate key. To access data stored in the cloud, this cryptosystem used an aggregate key that supported a user's attributes. Only those users whose characteristics fit the protection policy are given an aggregate key by the data owner [9].

The various cloud storage service providers like Mozy, Dropbox, and several others, use deduplication to avoid wasting space by storing only one copy of everyuploaded file. Clients who encode their files in the standard way, on the other hand, risk money. This tension is mitigated by message-locked encryption. It is, though, vulnerable to attacks of brute-force, which can recover files that belong to a recognised set. J. Baliga et al. suggest an architecture for safe deduplicated storage that is resistant to attacks of brute-force, and they impose it in the method of DupLESS. Clients encode with the help of message-based keys recovered from a key-server with the help of the unawarepulse repetition frequency (PRF) procedure in DupLESS. Further, it helps clients in storing encrypted data along with acurrent provider andon their behalfconduct the service conduct deduplication while retaining high degree of confidentiality. They illustrate that deduplicated storageencryption can save time as well as space as compared to the usage of the storage service for data of plaintext [10].

3 Methodology

When a user tries to upload a file to the cloud after registering and signing in, the cloud will search for replication, which means if the file has already been submitted, the cloud will refuse permission to upload a file since it is labelled as a duplicate file, which is known as deduplication. If there is a duplication, it will be blocked; otherwise, it will encourage you to upload a file, which will be added to the cloud. User: After registering and signing in, if a user wishes to open a file that has been transferred to the cloud, the user demands a key from the owner; if the owner wants to give the user permission to view the file, yes means the owner will transfer the key to the user and the user will be able to retrieve the file; if the owner does not allow the user to view the file, no means the owner will ignore the request.

3.1 Design

Our system is a multi-domain deduplication architecture that allows for both intra-deduplication as well as inter-deduplication. Basically the first level includes only those who help set up the structure. The Cloud Service Provider (CSP) and Agent (Ai), called honest-but-curious in the current literature on cloud storage security are inquisitive about the outsource decrypted datacontents. The CSP will work with Ai to estimate the content ofdirected cypher text with the help of an offline attack of brute force based on context information about the plain text space (Fig. 1).

Furthermore, owing to physical faults or monetary purposes, the cryptographic data contained in the CSP can be changed. Users are believed to be truthful but curious in the same manner as operators are expected to be truthful but interested. They cannot, for example, structure the CSP or the associated Ai by uploading null and void data. On the other hand, they might be involved in retrieving the encrypted data content without having to own it. Furthermore, if a mischievous user is conscious regarding the space of plaintext, he or she might effortattack of an online brute force by frequentlyimplementing the data upload technique in order to estimate the targeted cypher textcontent. Operators will not conspire with the Ai or CSP because of their own safety interests and the CSP

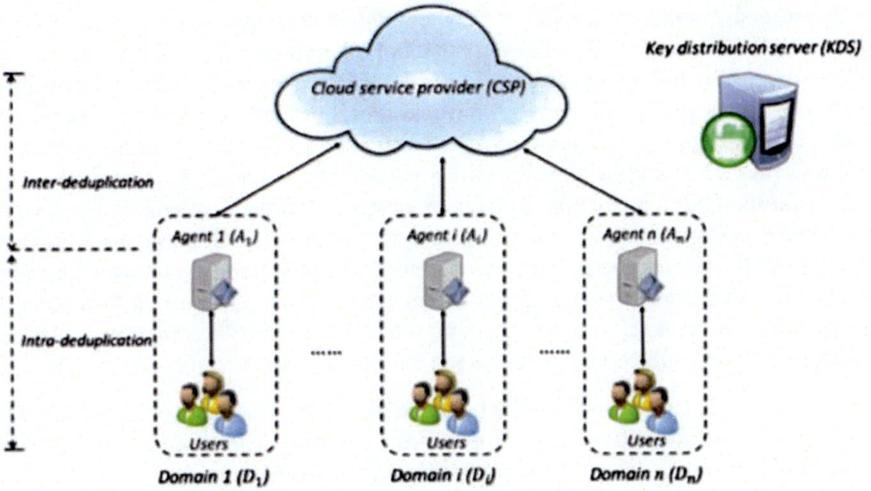

Fig. 1. The architecture of deduplication process. The CSP will work with Ai to estimate the content of directed cypher text with the help of an offline attack of brute force based on context information about the plain text space

or Ai's credibility. Furthermore, we believe that Domain (Di) users communicate with the appropriate agent Ai, and that all actions communicate with the CSP over secure channels. Figure 2 the Architecture of Key Aggregation Process. Di users communicate with the appropriate agent Ai, and that all actions communicate with the CSP over secure channels.

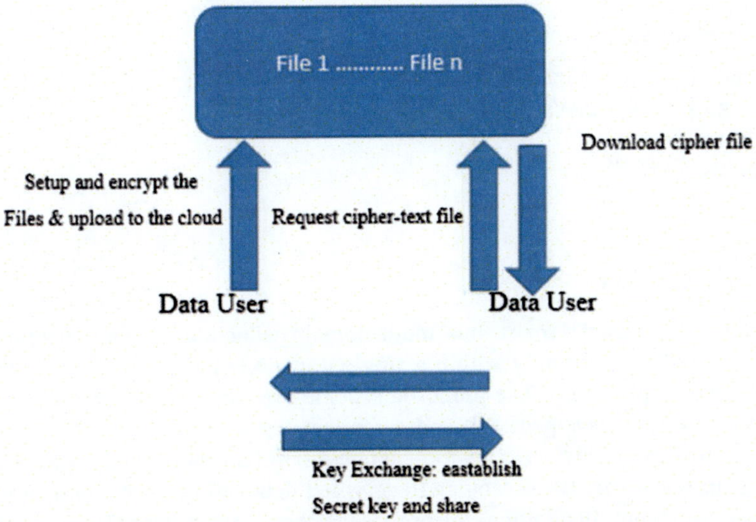

Fig. 2. The architecture of key aggregation process. Di users communicate with the appropriate agent Ai, and that all actions communicate with the CSP over secure channels

In our model of the Framework, we contemplate three components of a cloud-based system: CSP: This is typically a business that provides cloud-related software, and may fall into one of the following categories: Infrastructure as a Service (IaaS), Software as a Service (SaaS), or Application as a Service (PaaS) are all words for the same item (PaaS). In our scheme, a CSP is an organisation that offers cloud computing amenitiesto a Data Owner (DO) and handles the data's connection to Data Consumers. One may not rely on the CSP's protection services to ensure full privacy retention, because the Data Owner is responsible for enforcing security measures. DO is the one who wants to use the CSP's cloud computing facilities for the storage and exchange of personal information. The Data Owner imposes its own protection mechanisms to protect the confidentiality of its data, so that even though the CSP security mechanism is violated, unauthorised users will be unable to decode the private data.

Data User (DU): A DU is a person who demands access to data that the Data Owner has shared. Simply if it effectively completes the confirmation procedure with the Data Owner will it be given access to the private data.

3.2 Hardware and Software Tools Used

3.2.1 Hardware Requirements

a. System: Intel I3.
b. Monitor: 15″ LED.
c. Hard Disk: 120 GB.
d. Monitor: 15″ LED.
e. Ram: 4 GB.
f. Input Devices: Keyboard, Mouse.

3.2.2 Software Requirements

a. Operating system: Windows 7.
b. Coding Language: JAVA 1.8.
c. Tools: NetBeans.
d. Database: MySQL.
e. Cloud: Drivehq.

3.3 Implementation

For starters, the proposal is for a robust multi-domain deduction scheme that would allow deduplication of data not only within a single domain but also via multiple domains (called inter-deduplication). The proposed scheme uses the bilinear pairing method to generate a random convergent key and a random tag to protect data confidentiality. The proposed scheme just requires to generate a constant number of random cypring messages, to make sure the outsourced encrypted data can be correctly decrypted by users with ownership. In addition, the proposed scheme guarantees that only the cloud service provider will perform inter-duplication by matching random inter-tag derived from the Boneh-Goh-Nissim cryptosystem.

Second, a deduplication decision tree based on the B+ tree is built, which works well with massive data storage structures, to reduce the time complexity of duplicate search. The length of plaintext, in particular, can be used to represent the keyword in the B+ tree. Third, the suggested scheme's security is analysed to show that it can maintain high data confidentiality and integrity while avoiding brute-force attacks. Furthermore, comprehensive efficiency evaluations demonstrate the proposed scheme's efficacy in terms of computational, connectivity, and data overheads.

4 Results and Discussion

For cloud providers, data storage has been very popular. Cloud providers would rather rely on their main market than handle large volumes of data that are being introduced every day. This paper looked at different facets of data deduplication. The numerous cloud storage specifications, such as data management, data deduplication, and encryption, were investigated. The project created a situation in which the cloud server could deduplicate the data that was submitted. The required data is presented to the data consumers. By enforcing a proper access control system, the data was protected from unnecessary dissemination and unauthorised access. Allowed data deduplication helps to keep data private and stable by avoiding unauthorised access. Deduplication at the encryption stage removes a lot of memory, which can be put to greater use. Figure 3 shows the layout of home page. Figure 4 shows the layout ofregistration page. Figure 5 shows the layout oflogin page. Figure 6 shows the layout ofcloud server login. Figure 7 shows the layout of audit or login. Figure 8 shows the layout ofuploading a file. Figure 9 shows the layout of uploading to cloud. Figure 10 shows the layout ofdownload files. Figure 11 shows the layout ofactivation in cloud server page. Figure 12 shows the layout offile approval in admin page. Figure 13 shows the layout of updating page in admin. Figure 14 shows the layout ofdownload page in admin.

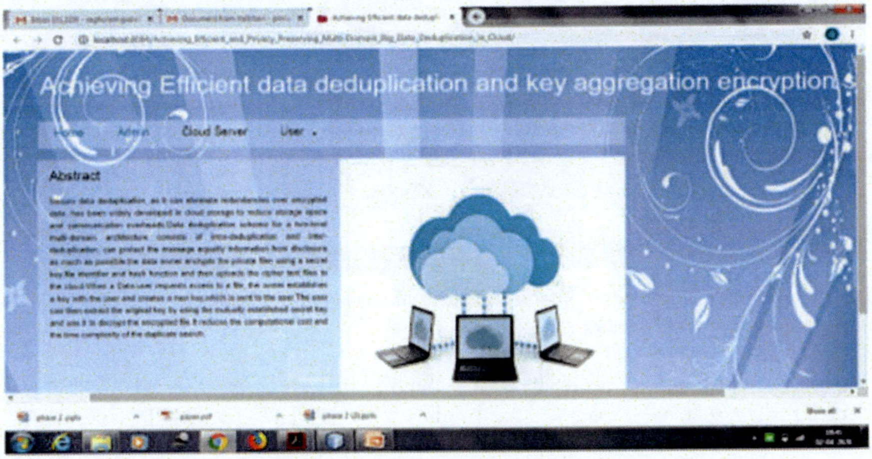

Fig. 3. The layout of home page. This provides interface to the end user

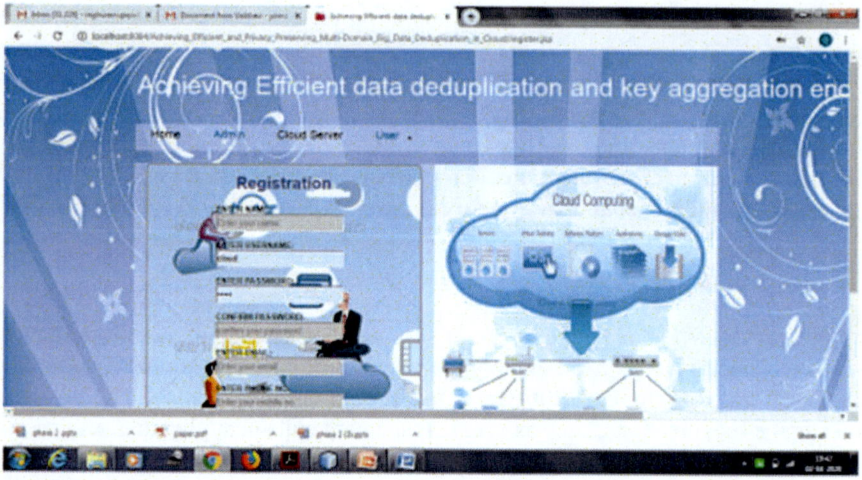

Fig. 4. The layout ofregistration page. The required information is feed in order to complete the registration process

Fig. 5. The layout of login page. The user name and password of cloud user are required to get authenticated and proceed further

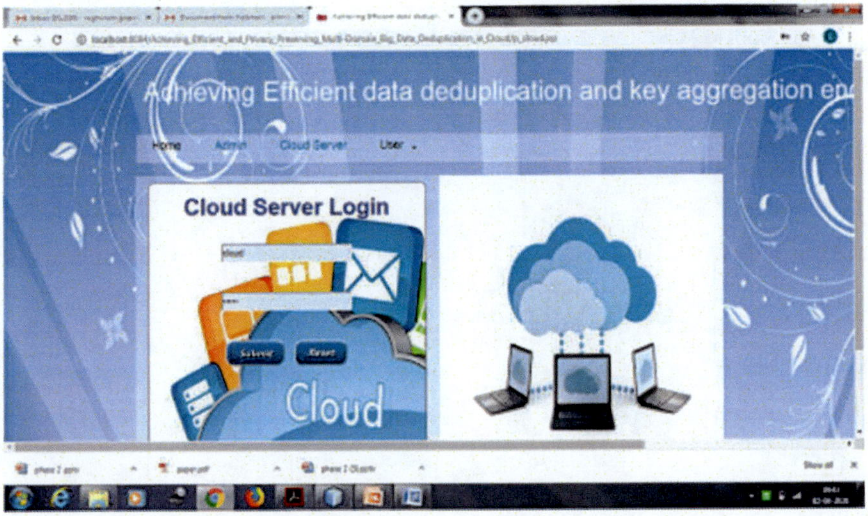

Fig. 6. The layout of cloud server login. The user name and password for cloud server are required to get authenticated and proceed further

Fig. 7. The layout ofaudit or login. The user name and password of auditor are required to get authenticated and proceed further

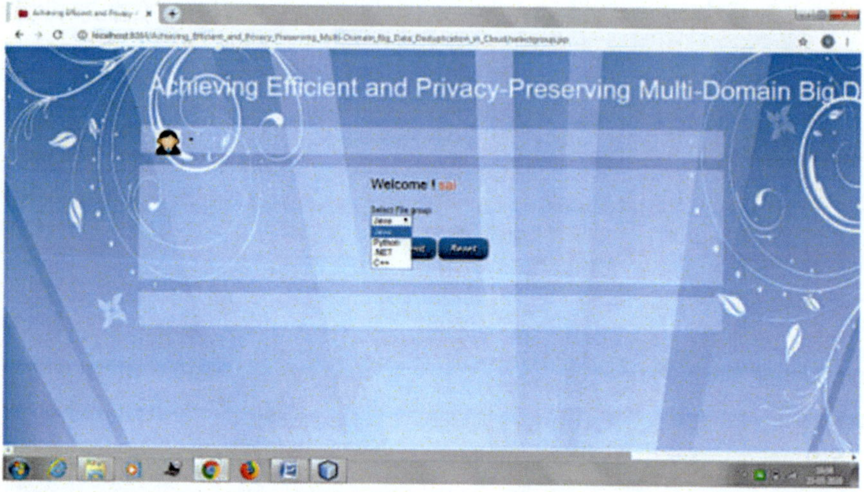

Fig. 8. The layout of uploading a file. A file to be uploaded is selected from the system

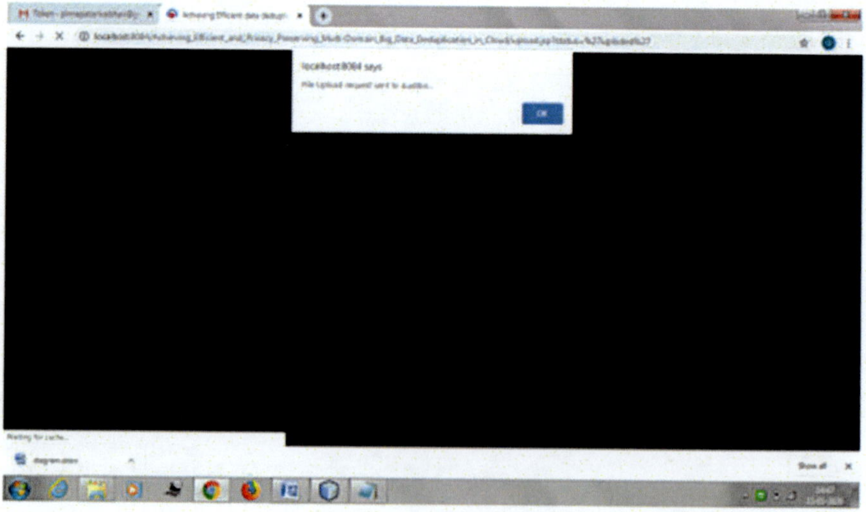

Fig. 9. The layout of uploading to cloud. The selected file is further uploaded to the cloud

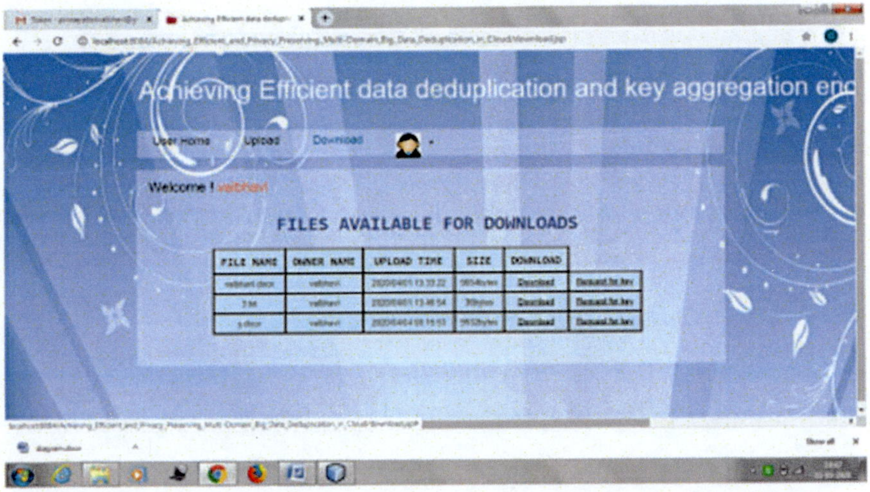

Fig. 10. The layout of download files. It shows the complete details of downloaded files

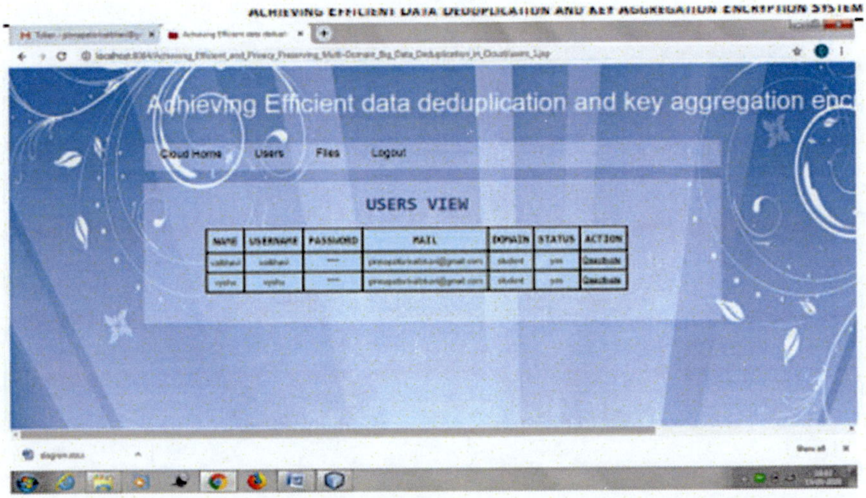

Fig. 11. The layout of activation in cloud server page. A user view is shown in this layout

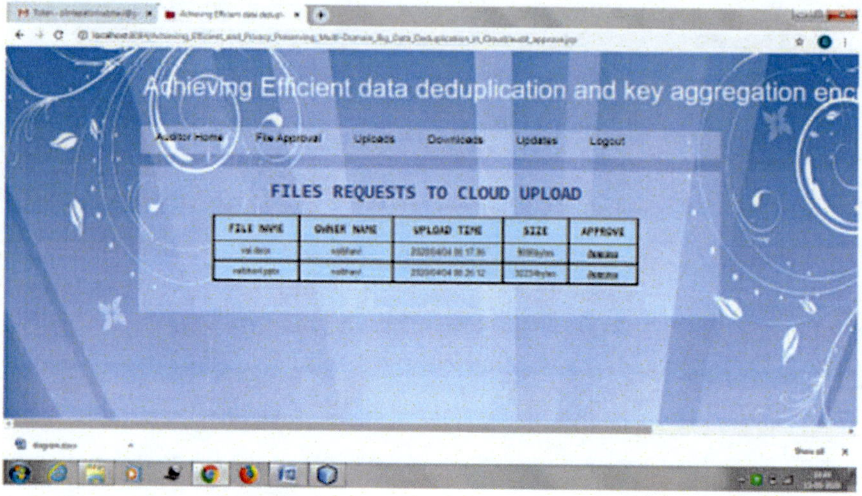

Fig. 12. The Layout offile approval in admin page. It shows the details of approved files

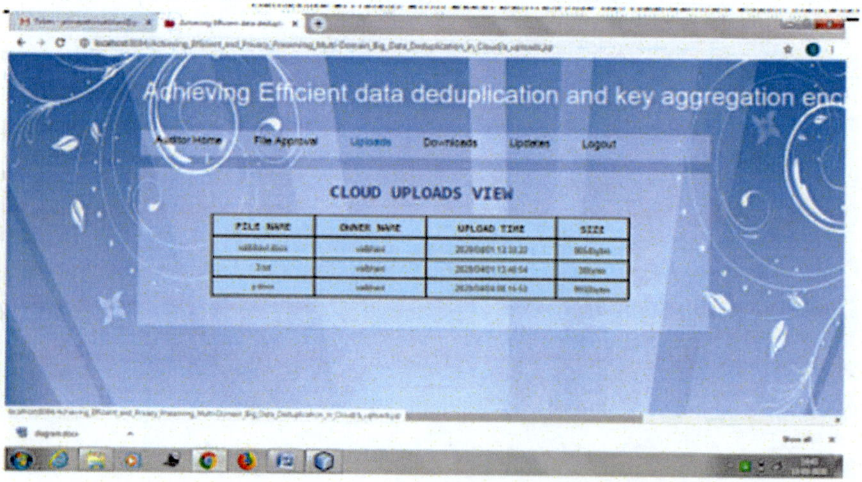

Fig. 13. The layout of updating page in admin. It shows the list of updated files in admin page

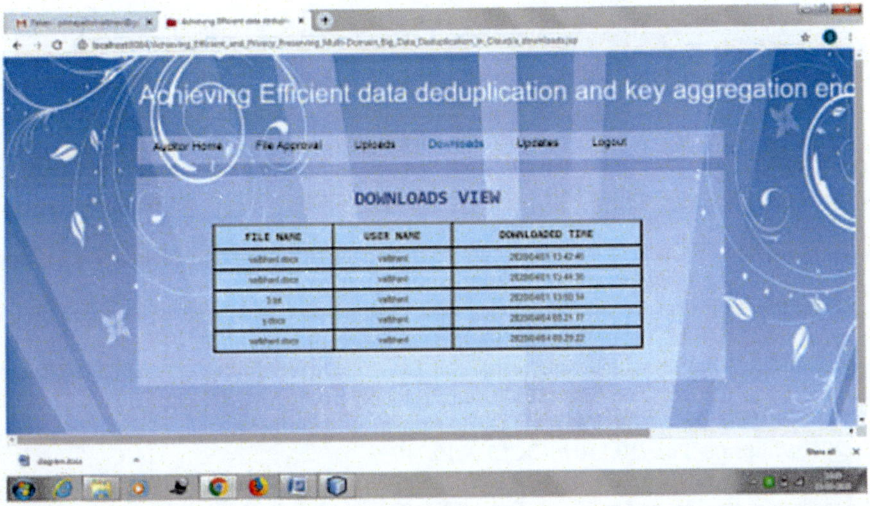

Fig. 14. The layout of download page in admin. It provides the access to download the files from the cloud server

5 Conclusion and Future Scope

We also implemented a replication scheme for two-story multi-Domain architecture that is open and privacy-preserving. The suggested strategy, by creating a constant number of cypher messages, ensures the semantic security of encrypted data and ensures the proper decryption of outsourced encrypted data by users. Moreover, only the CSP (agent) can evaluate whether two random inter-tags specified from different domains (the same domain intra-tags) relate to the same data or not, thus minimising any level of information on equality of message exposed. According to rigorous security analyses, the suggested method will achieve data secretiveness and confidentiality while preventing brutal attacks. Moreover, rigorous performance analyses demonstrate, in particular in computing costs and repeat search time complexities, that the proposed technology exceeds the current competition. Given that users often need data modification and deletion activities, future experiments would need an expansion of the scheme designed to achieve ownership control and revocation. Furthermore, when data deduplication methods have just one copy, outsourced information is susceptible to deduplication of data leakage or corruption.

References

1. J. Li, X. Chen, F. Xhafa, L. Barolli, Secure deduplication storage systems with keyword search (2014). https://doi.org/10.1109/AINA.2014.118
2. J. Li, X. Chen, M. Li, J. Li, P.P.C. Lee, W. Lou, Secure deduplication with efficient and reliable convergent key management. IEEE Trans. Parallel Distrib. Syst. (2014). https://doi.org/10.1109/TPDS.2013.284

3. L. Zhou, V. Varadharajan, M. Hitchens, Achieving secure role-based access control on encrypted data in cloud storage. IEEE Trans. Inf. Forensics Secur. (2013). https://doi.org/10.1109/TIFS.2013.2286456
4. M. Li, S. Yu, K. Ren, W. Lou, Y. Hou, Toward privacy-assured and searchable cloud data storage services. IEEE Netw. (2013). https://doi.org/10.1109/MNET.2013.6574666
5. M. Li, S. Yu, Y. Zheng, K. Ren, W. Lou, Scalable and secure sharing of personal health records in cloud computing using attribute-based encryption. IEEE Trans. Parallel Distrib. Syst. (2013). https://doi.org/10.1109/TPDS.2012.97
6. J. Li, X. Chen, F. Xhafa, L. Barolli, Secure deduplication storage systems supporting keyword search (2015). https://doi.org/10.1016/j.jcss.2014.12.026
7. D. Harnik, B. Pinkas, A. Shulman-Peleg, Side channels in cloud services: Deduplication in cloud storage, IEEE Secur. Priv. (2010). https://doi.org/10.1109/MSP.2010.187
8. P. Christen, A survey of indexing techniques for scalable record linkage and deduplication. IEEE Trans. Knowl. Data Eng. (2012). https://doi.org/10.1109/TKDE.2011.127
9. J. Wu, L. Ping, X. Ge, W. Ya, J. Fu, Cloud storage as the infrastructure of cloud computing (2010). https://doi.org/10.1109/ICICCI.2010.119
10. J. Baliga, R.W.A. Ayre, K. Hinton, R.S. Tucker, Green cloud computing: balancing energy in processing, storage, and transport. Proc. IEEE (2011). https://doi.org/10.1109/JPROC.2010.2060451

Book My Space: The Utilization of Empty Space

R. Kesavamoorthy(✉), M. Vijendrachar, Hites Chitalia, Ashrith V. Raghunanth, and Yash Kumar Jain

Department of Computer Science and Engineering, Jain (Deemed to be University), Bengaluru, Karnataka, India

1 Introduction

The aim of this project is to build a unified forum that enables warehouse owners or organisations (hereafter referred to as "Providers") and individuals or companies that need to use warehouses (hereafter referred to as "Clients") to communicate in a structured, streamlined, user-friendly, and coordinated manner. By introducing a framework that invokes continuity, predictability, uniformity, and, in certain respects, versatility of operations, this proposed platform is intended to support both clients and providers. Other benefits include cost savings, scalability, enhanced brand awareness, quality management, a broader customer base, and creativity that will improve over time. In short, by using a single platform such as "Book My Room," providers will not only improve efficiency and productivity, but will also provide clients with more varied and knowledgeable options [1].

There are currently no services that provide a unified platform for warehouses and potential clients to interact with one another in a simplified and standardised manner in the current business landscape. Individual warehouse providers currently interact with clients through separate platforms. This fragments the market and makes it difficult for customers to compare the services of various warehouse providers. Providers are also hampered in reaching a larger client base and expanding their business because of the existing fragmentation [2, 3].

Clients have the option of purchasing an existing warehouse or entering into a contractual agreement for a long-term lease of the required space under the established business model. This turns off clients who only need storage for a short period of time or for a limited amount of time, and it reduces the amount of business that providers would otherwise get. The current model places the entire burden of advertising and management on the providers, favouring those with the financial means to access platforms and services that would otherwise be out of reach for a small or underfunded provider. The proposed platform, which is the subject of this project, will establish a standardised and organised model that allows providers to reach a wider client base and clients to have access to diverse and better-informed choices by providing a unified platform that benefits both providers and clients in various ways [4, 5].

Progressive mobile apps mark an important improvement in the application development industry. Both mobile apps and websites play an important role in improving customer service. Progressive Web Application (PWA) applications are anadvanced step

F. P. García Márquez (ed.), *International Conference on Intelligent Emerging Methods of Artificial Intelligence & Cloud Computing*, Smart Innovation, Systems and Technologies 273, https://doi.org/10.1007/978-3-030-92905-3_43

in this direction. Itcombines the finest features of both the website and the mobile app. It's critical to identify that every corporation has their personal application. However, it is not possible to download and install apps for every product, service, or platform. PWAs offer a new perspective on the situation. Progressive web apps provide the benefits of website browsing without the need for installation by mimicking the look and feel of the website. As a result, they are expected to be the future of the app development industry [6, 7].

1.1 Ways to Improve Warehouse Space Utilization

When analysing multichannel warehouse operations, it was discovered that between 80 and 85% of space usage, productivity usually fell from a movement and storage standpoint. Pallet movement is severely limited, and each step and replenishment can necessitate several acts. Aisle movement can also be limited by temporary floor pallet positions. Having to make several steps raises costs and slows down the operation. Many companies need to increase warehouse space usage or relocate sooner than expected. There are relevant reports to bring up with management when addressing space usage in the warehouse:

- Use Vertical Space
- Analyze Department Space
- Consolidate Locations
- Right-Size Available Slots
- Use Off-Site Location for Overstock
- Drop Shipping as an Option
- Aisle Widths
- Cross Docking
- Best Utilization of the Building
- Depth of Storage
- Supply Storage
- Door Use
- Mezzanine Installation.

1.2 Steps to Calculate Space Utilization Rates

- ***Step 1***: Begin with evaluating the over-all dimension of the warehouse in terms of square feet.
- ***Step 2***: Remove the area for toilets, offices and so on. For example, if a warehouse has been rented that is about 80,000 m^2 but is just 10,000 m^2 in offices, then 70,000 m^2 of accessible warehouses will also be used.
- ***Step 3***: In order to calculate the stocking potential in terms of cubic feet, multiply 70,000 square feet with a warehouse simple height. For example, if the warehouse has a transparent height of 25 feet, it would be 1,750,000 cubic feet for storage or cube.

- ***Step 4***: Finally, measure the size of the stock cube by adding the amount of all merchandise in the store. Divide this by the overall size of the warehouse. For example, as 250,000 cubic feet of merchandise is stored in a warehouse of 1,750,000 cubic foot, 14.29% of the space available is for use.

2 Literature Review

Zeynep Tufekci investigates how a group of college students ($n = 704$), the vast majority of whom use Facebook and Myspace, negotiate boundaries between public and private. There is little correlation between online privacy concerns and information disclosure on online social networking sites, according to the findings. Students can control unwanted audience concerns by adjusting profile visibility and using nicknames, but not by limiting the information contained within the profile. Walls, locks, and doors, which are analogous to boundary regulation in physical space, are preferred; little adaptation is made to the Internet's key features of persistence, search-ability, and cross index-ability. There are also significant racial and gender differences, according to the author [8].

Cliff Lampe et al. investigates whether these students use Facebook to meet new people in their offline communities or to learn more about people they meet in person for the first time. Our findings suggest that Facebook is primarily used to learn more about people they meet offline, and that users are less likely to use the site to make new connections. Many first-year students at Michigan State University choose to join Facebook, according to this paper. Students who use the site expect their audience to be made up of their peers rather than other university members such as faculty and administration. Members of Facebook appear to be using it as a surveillance tool for maintaining previous relationships, as well as a "social search" tool to look into people they've met offline. There doesn't appear to be much "social browsing," or looking for people online with the intention of moving a relationship offline [9].

Aaron Norgrove et al. examines the Social Networking Site 'Facebook,' as well as the connections that can be drawn between those experiences and their individual research interests. This paper is split into three sections. Meredith Bean begins by outlining the connections between Facebook use and identity construction processes. Second, Aaron Norgrove establishes a link between the Facebook Terms of Agreement policy and information control politics. Following this theoretical context of their individual experiences, the authors conclude with a review of Facebook's broader importance for current and future social ties. The importance of the consensual dimension of hegemony and its relationship to online social networks, the role of computers and corporations in control societies, and the emergence of "digital individuals" through the process of online identity construction are all highlighted [10].

3 Methodology

3.1 Tool Description

Due to its iniquitousness over hardware and operating systems, the user-induced model of Security is a special forum for the development of applications and the fact that no one company controls its specifications and implementations. Thanks to its intrinsic liability, you will browse and share what you find with anyone else. It is all up to you when you visit a website and your involvement in it can be as short or as long as you want. Web apps can access anybody on any computer with a single codebase.

Native applications are well known for their high functionality and reliability. They're still there on home displays, docks, and taskbars. If you have a network link or not, you operate. As a standalone experience they are published. You can read and write data from a local filesystem, communicate with your system's data such as contacts and scheduling events, and use USB, Serial or Bluetooth hardware to access them. You can take pictures, display songs on the home screen and watch music replay with native applications in another app. Native applications have a feeling they belong to the computer on which they work. PWAs are used with modern APIs to provide any user with single codebase with native features, trustworthiness and install-ability.

3.1.1 PWA Description

A PWA is a web app that uses modern web capabilities to provide users with an interface that is very similar to (if not better than) a native app. PWA, unlike conventional apps, are a cross between standard web sites and smartphone applications. The word "progressive" refers to the fact that they add new functionality and, from the user's viewpoint, tend to be regular websites at first but eventually behave more like mobile apps, among other things multiplatform.

For most business and enterprise applications, PWA are a feasible alternative to native or hybrid software. These are seldom hardware-intensive applications, and they are often cost-conscious. In these cases, concentrating efforts on developing a single PWA rather than supporting multiple native platforms can save time and money during development. In general, it's a good idea to see if an app can be created as a PWA first before deciding to create a native app. The web platform is rapidly expanding its capabilities, allowing you to perform tasks ranging from 3D acceleration to hardware access.

3.1.2 Reasons for PWA

- Consumers fear mobile delays and unreliability: the stress level tempted by mobile delays is correspondent to that of viewing a horror film.
- Since they don't need to download an app, 50% smartphone users prefer to browse or shop on a company's mobile site.
- Insufficient storage is the top causes for uninstalling an app (although an installed PWA typically consumes size less than 1 MB).

According to our results, customers enjoy experiences that are easy, simple to instal, reliable and interactive. PWAs are a collection of best practises and modern web APIs to react quickly, securely and attractively to customers. Using a cloud provider to cache your services and to anticipate your web site more quickly and securely, for example. Through your site installable, you can quickly view your customers from a home screen or application launcher. New APIs including Web Push Notifications also facilitate the reengagement of consumers to personalised content to enhance their loyalty. Smartphone consumers are most likely to purchase products from websites with product ratings and 85% of smartphone users would find it helpful to have mobile updates.

3.2 Design

Progressive Mobile Apps are web applications that load in the same way that web pages or websites do in a web browser. It offers a rich mobile experience by integrating native-like features such as offline operating, push alerts, and application hardware usability. It has the same look and feel as a native app and has the same features. It is not necessary to purchase it from an app store. In a web browser, it loads, runs, and works. PWA bridges the gap between web and smartphone, offering users superior, native-like performance in their web browser. In the simplest terms, it is a type of web-based mobile application.

3.2.1 Features of PWAs

- *Progressive*: The term progressive applies to the fact that it works for all users, regardless of browser preference, since progressive improvement is a central tenet.
- *Responsive*: Adapts automatically to every form: desktop, web, tablet, and so on.
- *Load Time*: Progressive Mobile Applications are accessible right away.
- *App-like features*: Since it's based on the app shell model, it looks like a smartphone app with app-style interactions.
- *Fresh*: It's still up to date, so you don't have to keep refreshing it like other Android/iOS games.
- *Secure*: Content is delivered using HTTPS to ensure security.
- *Engaging*: It's very engaging thanks to features like push notifications and so on.
- *Installable*: Allows users to add the website as an app to their home screen without having to go to the app store.
- *Linkable*: It's easy to share via a URL and doesn't require any complicated setup.

3.2.2 Advantages of Making a PWAs Instead of Building a Fully Functional App

- *Cost-Effective*: The best profit for an app publisher is the cost savings in terms of app creation and maintenance. Since it is believed that designing a website is much simpler than creating an application.
- *Cross-Platform*: PWAs, unlike most apps, are not confined to a single platform. That means you won't have to build separate versions of the app for different platforms.

We can't assume that PWAs will totally replace Native Apps in the future. Nonetheless, there is an increasing interest in the community for this approach. PWAs are still in their infancy, and there are many obstacles to resolve. Nonetheless, they have the ability to change the way the internet works.

3.3 Implementation

HTML, CSS, and Bootstrap are used to build a PWA. The objects and elements are constructed on top of HTML, while the webpage is designed and styled using CSS and the Bootstrap framework. On the owner's side of the website, a database and a dashboard are installed. Firebase is used to build the database, which is connected to the website. The details entered on the site of the owner will be saved in the database. Since Firebase does not need data to be normalised, it is simple to use.

A service worker is a script that runs in the background of your browser, independent of a web page, allowing you to access features that don't need a web page or user interaction. You must register a service worker for your site, which you do in the JavaScript of your website. When you register a service worker, the browser will begin the installation process in the background. Usually, you'll want to cache some static assets during the installation process. The service worker is installed after all of the files have been successfully cached. The install phase will fail if any of the files fail to download and cache, and the service worker will not activate (i.e. will not be installed). Don't worry if this happens; it will try again the next time. However, if it does install, you'll know that the static assets are in the cache (Fig. 1). Figure 2 shows the layout of FireBase Dashboard. Figure 3 shows the layout of FireBase Authentication. Figure 4 shows the layout of FireBase FireStore. Figure 5 shows the layout of FireBase Hosting.

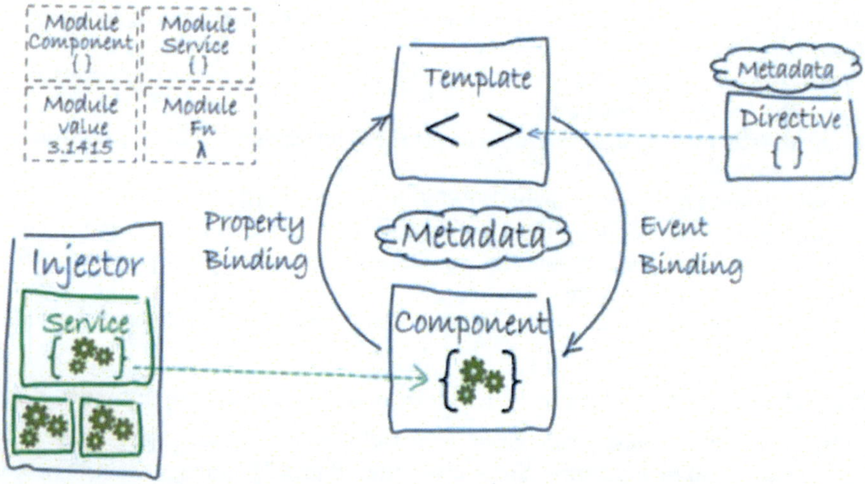

Fig. 1. Basic working of a service worker. A service worker is a script that runs in the background of the browser, independent of a web page.

For server interaction and database interaction, NodeJS is used. This helps the customer to see the details that the warehouse owner has updated. To build user-specific views, Angular is used to create various views of the information available in the database.

The overall cost payable per unit and the grand total for the booked slots are shown in the booking cart. The Authentication service checks to see if the user is already logged in. If the user is not signed in, it will ask them to sign in using Google, and the credentials will be saved in the database. The cart's checkout option confirms that the user is logged in and adds the order to the user's "my orders." The login part shows the user's name

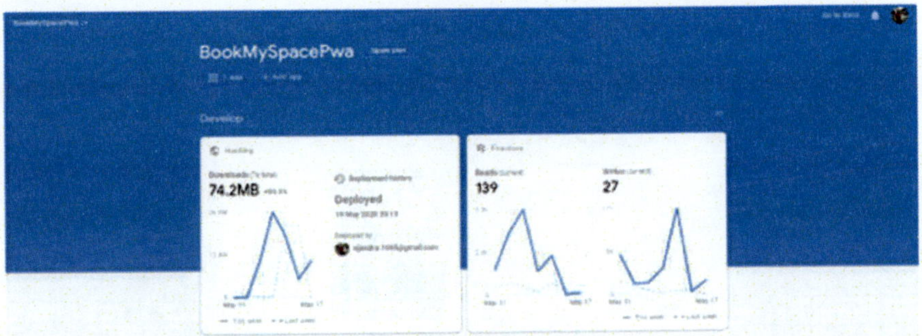

Fig. 2. The layout of FireBase dashboard. This helps the customer to see the details that the warehouse owner has updated.

Fig. 3. The layout of FireBase authentication. Here, the access is provided to the authenticated user.

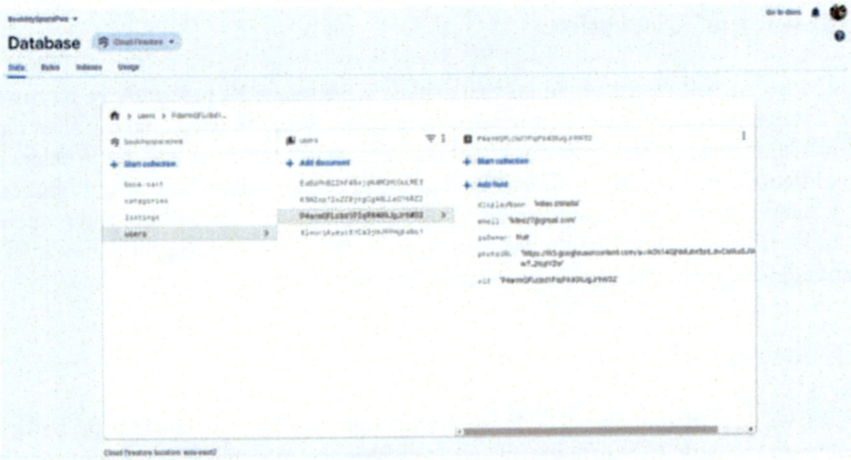

Fig. 4. The layout of FireBase FireStore. Here, the database of the FireStore is maintained.

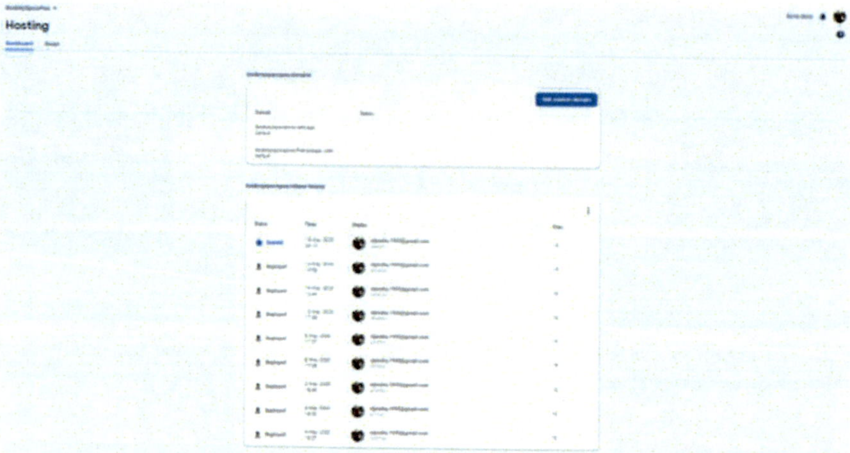

Fig. 5. The layout of FireBase hosting. This maintains the database of hosting details of FireStore.

and registered email address, as well as the option to add the user as an owner. By registering as an owner, a user can add, delete, or edit warehouse information, which are then modified for all users.

The navigation bar includes things such as the booking cart, news, contact us, and login button for users. If no users are logged in, a login button appears, which invokes the authentication service. If the user is already logged in, the login component retrieves the user's username and display image and replaces the login button with it. The navigation bar also shows the total number of items applied to the cart as well as the number of slots booked by the customer. The website's smooth operation is maintained by several components working together with services.

4 Result and Discussion

Lighthouse is a software programme that enhances the website's productivity. It answers with a series of suggestions for improving the usability of the page, making sites more accessible, following best practises and more. You may use this as a Chrome extension for continuous integration, or even a Node plugin. For quite some time now, Lighthouse offers many tips for optimising page loading speed, such as compression and reduction of scripts on render. The Lighthouse team aims to do new audits to give you ever more useful advice to improve the time available on your pages (Fig. 6).

5 Conclusion

Progressive smartphone applications represent a significant improvement in the software production industry. In enhancing customer experience, smartphone apps and websites play an important part. PWA applications are a measure in that direction. The best features both of the mobile app and the website are included. It is important to keep in

Fig. 6. Lighthouse is aprogrammed tool for enhancing the quality of the site. Chrome extension, or node module is used for continuous integration.

mind that all organisations have their own app. However, applications for each product, utility or network cannot be downloaded or installed. PWAs have a new perspective. The benefits of website browsing without installation by imitating the look and feel of our website are provided by progressive smartphone applications. The future of the software development industry is thus predicted.

References

1. S. R. Hiltz, K. Passerini, C. Dwyer, Trust and privacy: a comparison of Facebook and MySpace, in *AMCIS 2007 Proceedings* (2007)
2. C. Dwyer, S. R. Hiltz, K. Passerini, Trust and privacy: a comparison of Facebook and MySpace (2007)
3. C. Dwyer, S. R. Hiltz, K. Passerini, Trust and privacy concern within social networking sites: a comparison of Facebook and MySpace (2007)
4. C. Dwyer, S. R. Hiltz, K. Passerini, Trust and privacy concern within social networking sites: a comparison of Facebook and MySpace. Proceedings of AMCIS 2007. Assoc. Inf. Syst. AIS Electron. Libr. (2010)
5. R.C. Gomes, K.A.R. Ferreira, A.C.P. Oliveira, L. de O.M. Gomes, Determinant factors to local government performance: what is stakeholder's role in the process?. Encontro da Anpad (2009)
6. M. Friedewald, Privacy threats in the ubiquitous information society: an analysis of trends and drivers, in *Proceedings of the WebSci'09 Society On-Line, 18–20 March 2009, Athens, Greece*. In Press (2009)
7. Space mission analysis and design. Choice Rev. Online (1992). https://doi.org/10.5860/choice.29-5149

8. Z. Tufekci, Can you see me now? Audience and disclosure regulation in online social network sites. Bull. Sci. Technol. Soc. (2008). https://doi.org/10.1177/0270467607311484
9. C. Lampe, N. Ellison, C. Steinfield, A face(book) in the crowd: social searching versus social browsing (2006). https://doi.org/10.1145/1180875.1180901
10. A. Norgrove, M. Bean, Face(book)ing the future: identity, control and the formation of the 'digital dividual.'

Computer Vision Based Attendance Management System

Shree M. Rajani(✉), Mekala Ritheesh, VeeramAdiPranai Kumar Reddy, Pvs Harika, and Kurnool Jawaharlal Sai Sravan

Department of Computer Science and Engineering, Jain (Deemed to be University), Bengaluru, Karnataka, India
m.rajini@jainuniversity.ac.in

1 Introduction

One of the big obstacles in a smart learning setting is implementing an unobtrusive machine vision-based classroom attendance control framework. This paper describes an unobtrusive facial recognition-based smart classroom attendance control device that captures students' faces using a webcam [1]. For face recognition technique, the proposed system utilizes the Haar Cascade technique, and the model is trained using the Local Binary Pattern Histogram (LBPH) technique for student identification. In most cases, taking attendance in a classroom is a manual operation. In every period, the teachers must call the students by their roll number and record present and absent attendance on the attendance sheet [2]. This process takes up valuable class time for both students and teachers, and it can occasionally result in fake attendance. It's a time-consuming process for teachers, and it's difficult to process and maintain attendance sheets [3].

Face recognition is used in the proposed method to track attendance, with the face being the most significant factor. This approach overcomes the difficulties associated with conventional classroom attendance [4]. As a result, automating classroom attendance yields reliable outcomes and holds attendance in the database [5]. Every organisation has developed its own attendance management system. Some people still use the old-fashioned method of manually taking attendance, while others have switched to biometrics. In a wide classroom, the conventional approach makes it impossible to check students one by one [6]. Furthermore, the manual labour required to measure the attendance percentage becomes a significant task. Radio Frequency Identification (RFID) is a method of identifying large numbers of individuals using radio waves. It's quick and easy to use, and it lets you manage access without having to use your hands. It has been noted, though, that it can be exploited [7].

2 Literature Review

Samet et al. propose a mobile automatic classroom attendance management system based on face recognition that requires no additional equipment. Face recognition was designed using the euclidean-based filtering system measured with three face recognition

F. P. García Márquez (ed.), *International Conference on Intelligent Emerging Methods of Artificial Intelligence & Cloud Computing*, Smart Innovation, Systems and Technologies 273, https://doi.org/10.1007/978-3-030-92905-3_44

techniques. Three independent mobile applications are proposed for teachers, students and parents to operate on Smartphone have to monitor and execute the attendance process in real time. The proposed system was tested by students of Ankara University with excellent results. The system reduces the need for additional infrastructure and allows users to view data from anywhere and at any time. Smart devices for classroom attendance monitoring are very user-friendly. Teachers and parents can use the app in real time. Big and high-quality pictures can now be sent to the server with progressively increased Internet access speed. In addition, cloud processing capacity is constantly expanding. As a result of these technological advances, the proposed system's accuracy rating will increase [8].

Jayant et al. provide automatic attendance control framework based on facial identification and recognition techniques. Viola-Jones algorithms perform facial identification and facial recognition is done using a partial facial reconnaissance algorithm without synchronisation. Once a student has been correctly detected, it immediately changes the attendance in the excel sheet. The suggested scheme increases the effectiveness of current attendance control programmes by replacing the manual call, labelling and access to administrative websites [9].

Yu Chen et al. have built a coarse SR network for retrieving an HR image. The gross HR image is then submitted to two branches: a fine SR encoder and an approximation network that extracts image attributes and analyses hot images and mapping. In order to recover the HR file, both the image and previous data are sent to a good decoder. Furthermore, we propose to embed the generative opposing network of Face Super Resolution in FSRNet to generate realistic faces. The new appraisal parameters for face SR are two associated features, face orientation and parsing, which addresses the contradiction with conventional metrics in terms of visual perception. Extensive tests indicate that FSRNet and FSRGAN are both quantitatively and qualitatively advanced in the area of very LR face SR [10].

3 Methodology

3.1 Design

To satisfy particular specifications, the practise is known as device design to define the configuration, components, interfaces and data of a system. Systems philosophy is known as systems programming when applying to product creation (Fig. 1).A different solution would be to add a camera at the classroom's entrance that would detect each person's face when they entered. The face detector would have to do far less work in this situation, but there would only be one opportunity to catch a good picture. In the middle of the classroom, a frontal camera can take as many pictures as needed. The device is physically linked to the networks of South East European universities. Each classroom must have at least one internet-connected computer in order for the system to work. This computer connects to the LMS server, which is where the captured images are sent.

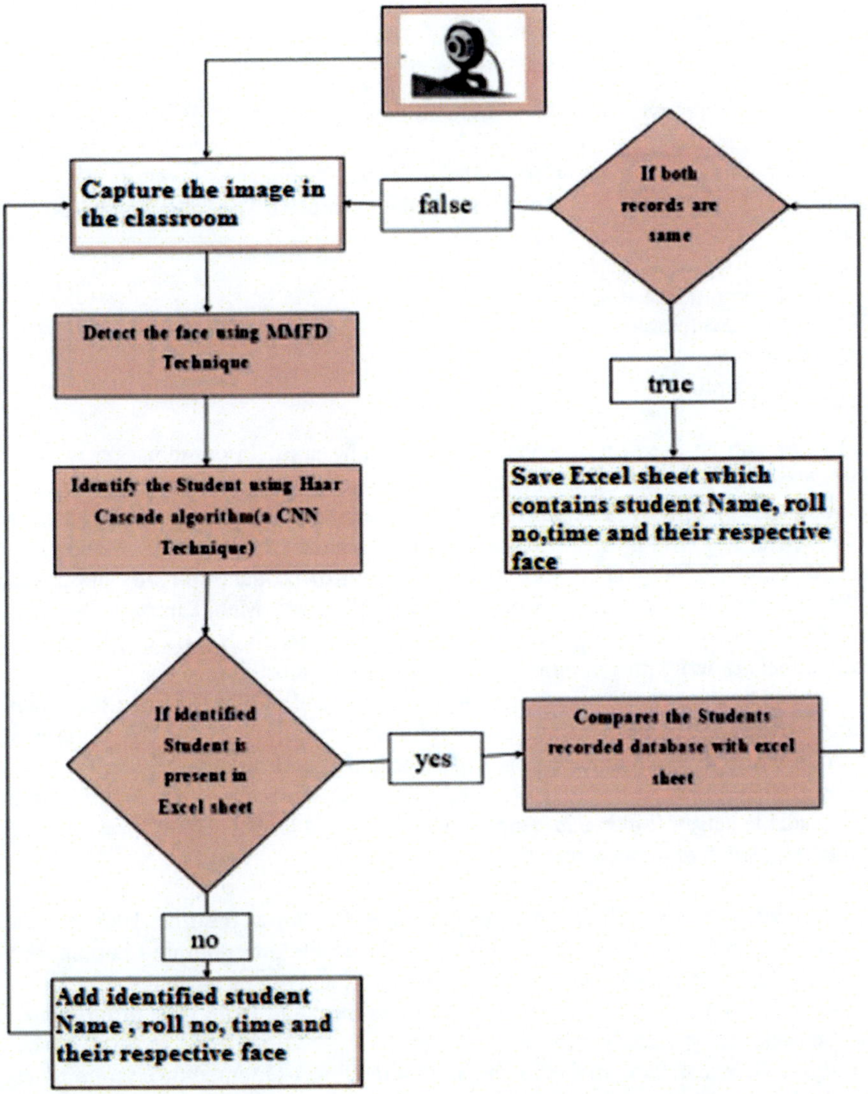

Fig. 1. The layout of system architecture. In the middle of the classroom, a frontal camera can take as many pictures as needed.

3.2 Tool Description

3.2.1 Python

Python is an interactive, object-oriented and general-purpose high-level programming language. Under the GNU General Public License Python source code is available (GPL). A comprehensive knowledge of the Python language is given in this tutorial. The programming language Python is one of the most used. The following purposes are used:

a. Web development (server-side),
b. Software development,
c. Mathematics,
d. System scripting.

Python can be used on a server to build mobile apps. In combination with other programmes, Python can be used to build workflows. Python has the ability to connect with database systems. It also has the ability to read and edit files. Python can be used to work with vast volumes of data and perform complex calculations. Python is also used for rapid prototyping as well as development of production-ready applications. Python can be used on a number of platforms (Windows, Mac, Linux, Raspberry Pi, etc.). The basic syntax of Python is similar to that of English. Syntax of Python allows programmers to write programmes with less lines than most languages.

Python is an interpreter language, which allows code to be executed immediately after it is written. The result is that prototyping is very simple to do. Three approaches are possible for Python: proceeding, object-oriented, and modular. Python is a high-level scripting language that interacts and focuses on objects. Python's aim is to be a truly readable script. It also uses English keywords and less syntactic forms than most languages instead of punctuation.

1. *Interpreted Python*: The interpreter processes Python at runtime. Before running software, one must compile it. This is similar to the programming languages like Hypertext Preprocessor and Practical Extraction and report language.
2. *Interactive Python*: You can sit at a Python prompt and write your programmes by communicating directly with the interpreter.
3. *Object-Oriented Python*: Python supports the Object-Oriented programming style, which encapsulates code in objects.
4. *Python as a Beginner's Language*: Python is an excellent language for beginning programmers, since it allows for the development of a broad variety of applications, ranging from basic text processing to web browsers and games.

3.3 Implementation

3.3.1 Image Capturing

The planned system involves a camera that will record all of the students. The Haar Cascade technique is used to detect the students' faces from these captured picture frames.

3.3.2 Face Recognition

This approach does not use sub-sampling and instead optimises through all sub-windows. This approach is also more effective at identifying all types of faces, including frontal, tilted up/right/left/down, and occluded faces. The figure below shows some examples of faces that were identified using the proposed process (Fig. 2).

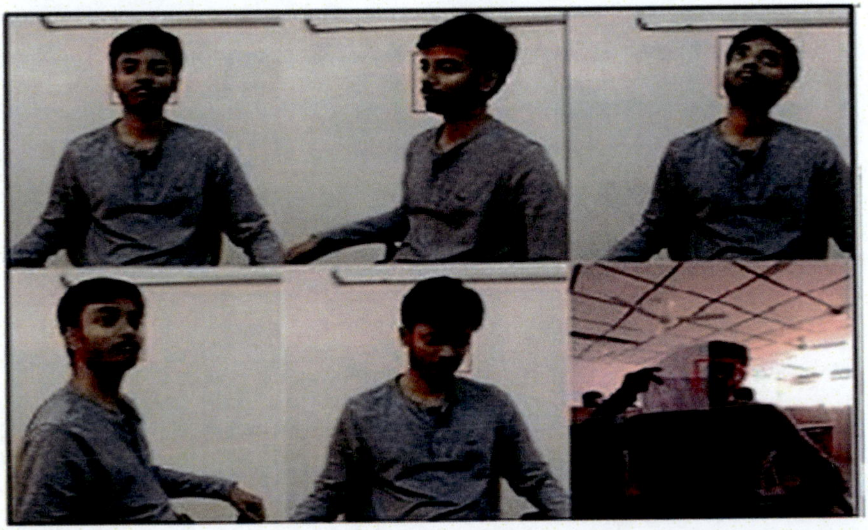

Fig. 2. Samples of detected faces in different angles. This technique is useful in the identification of all kinds of faces, including the dorsal, the right, the left and the overshadowed.

3.3.3 Training Phase

The number of students detected is determined by how the students are seated in the classroom. The proposed face detection technique has the highest accuracy rate for detecting all faces present. Using a face recognition algorithm, it was possible to recognise the students from the identified faces. Using more pictures of each student in the training process will increase the accuracy of student identification.

3.3.4 Algorithms

Haar Cascade Classifier In their paper 'Rapid Object Detection using boosted cascades of simple features' Paul Viola and Michael Jones presented an effective object detection method with Haar's feature-based cascade classification. It is a machine learning method that trains a cascade function with numerous positive and negative images.

Here we are going to deal with facial recognition. The algorithm involves a huge number of positive (faces) and negative images in order to train the classification system (images without faces). We ought to delete functionality from it afterwards. For this purpose, haar characteristics are used as seen in the picture below. They're very next to our fresh kernel. The pixel number below the white rectangle is removed from the total pixel of each function under the black rectangle (Fig. 3).

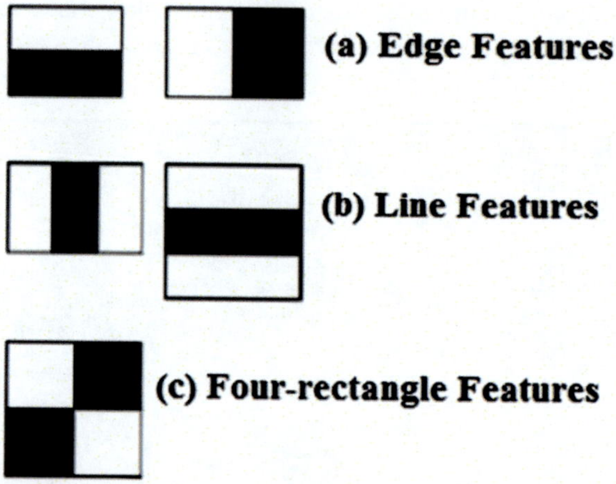

Fig. 3. Different features of Haar cascade. The integral image is used to solve the problem.

To measure a large number of features, all possible sizes and positions of each kernel are now used. For each feature measurement we must locate the number of pixels under black and white rectangles. To solve the dilemma, they used the whole image. It reduces calculations for a particular pixel to a four pixel process, regardless of the size of the image. The overwhelming majority of features, however, we estimate. In the below diagram, the top row shows two solid characteristics. The skin above the eye appears to be often darker than the nose and cheeks area. The second chosen feature is the black eyes than the nose's bridge. However, applying the same windows on the cheeks or somewhere else makes no sense (Fig. 4).

We do this by adding each function to all of the training pictures. It finds the best threshold for each function to identify the faces as positive or negative. Mistakes or errors will undoubtedly occur. We choose the lowest error rate functions that guarantee the reliable identification of images on the face and non-face. It isn't that easy technology. Any image at the beginning is assigned the same weight. The weights of incorrect images are increased after each grouping. Following this, it follows the same protocol. Error rates at new thresholds are calculated. New weights are also available. The procedure is replicated until the desired accuracy or error rates have been achieved or the appropriate number of functions have been found.

The final classification results in a weighted sum of these poor classifiers. It is considered weak, since it cannot recognize the image by itself, but it renders a strong classifier when combined with others. Also 200 features, according to the report, provide 95 percent accuracy in detection. About 6000 features were included in their final configuration. (Consider reducing the number of features from 160,000+ to 6000.) The non-face area of an image makes up the majority of the image.It is thus better to have a straightforward tool to see if a window is not a facet. Don't care about processing it, if

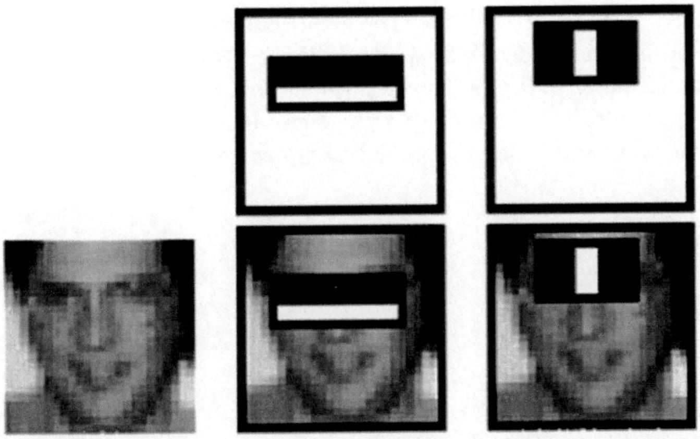

Fig. 4. Focusing the features. After each grouping, the weights of erroneous images are raised.

not, just drag it backwards. Focus on positions that can be a face, instead. This enables us to take longer to test potential areas. They developed the concept of a waterfall of classifiers. They are broken up into stages of classifiers and added one by one instead of adding all 6000 features in a single window. (In the majority of cases, even fewer features in the first few stages.) If the first inspection fails a slot, it should be removed. The other functions are not considered. Apply the second phase and resume the operation until the feature passes. A facial area is a window that crosses all layers. The detector was composed of over 6000 characteristics and 38 stages, with the first five stages including 1, 10, 25, 25 and 50 characteristics. (Adaboost has acquired the two best features in the picture above.) According to the reviewers, 10 out of more than 6,000 features of each sub window are evaluated on average.

4 Results and Discussion

This method is server-based due to the processor-intensive nature of the face detection algorithm. Detecting a face is basically an object detection operation, with the face as the object of interest. Face detection algorithms can be hindered by a number of factors, including face pose, size, location, rotation, light, image colours, and so on. When attempting to recognise (recognise) a face, the same issues arise, along with a few other stumbling blocks that will be addressed shortly. Face recognition from still images with multiple faces can be broken down into several stages. Many face recognition algorithms will identify a face precisely (or any other spec). There are various face detection algorithms that can detect a face (or some other unique object) in an image. Since most students in the framework presented here face the camera in the front, we decided to use the HAAR classifier for face detection. Intel's Free CV library is used to implement this classifier. The classifier works by using positive and negative face images to train a model. A positive image is one that includes the entity that needs to be detected, in this case a face. A negative image is one in which the desired object is not apparent. After the model has been conditioned, it can recognise facial features, which are then saved in an XML file. The architecture of Interface is shown in Fig. 5. The layout of the student details is shown in Fig. 6. The system capture and training configuration is shown in Fig. 7. The arrangement of saving educated pictures as seen in Fig. 8. The architecture of the student attendance live can be seen in Fig. 9. The state of the student attendance in the Excel sheet is shown in Fig. 10.

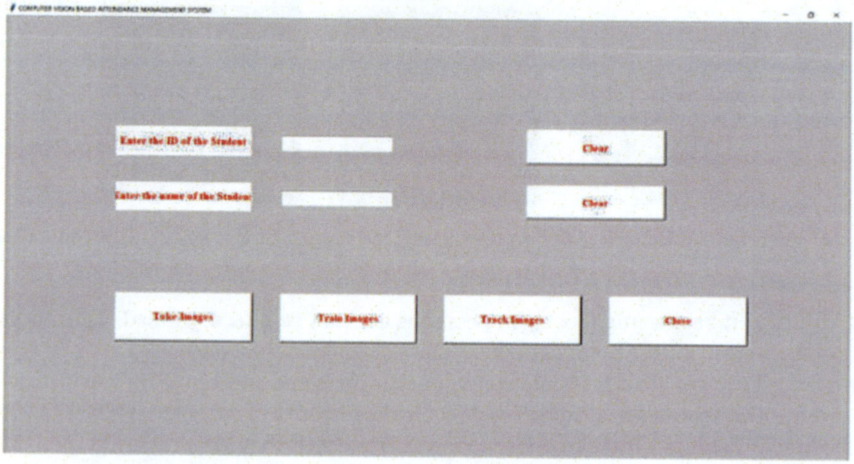

Fig. 5. The layout of Interface. This interface is used to fill student's detail.

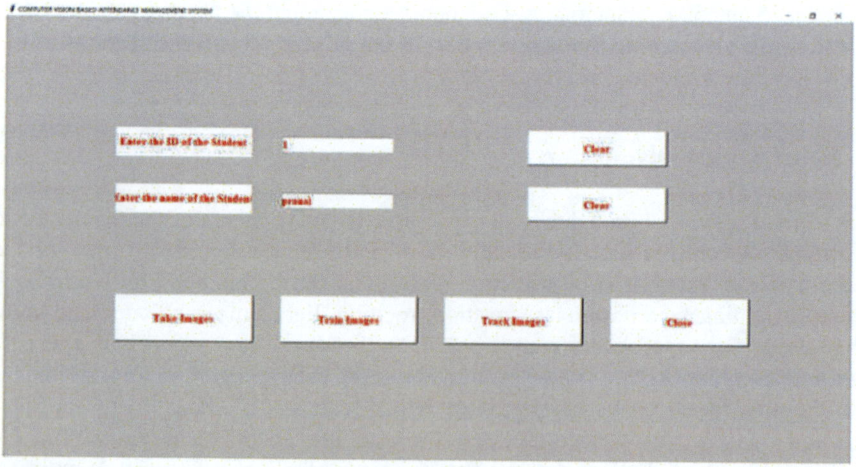

Fig. 6. The layout of entering student details. Student's details are entered in this page.

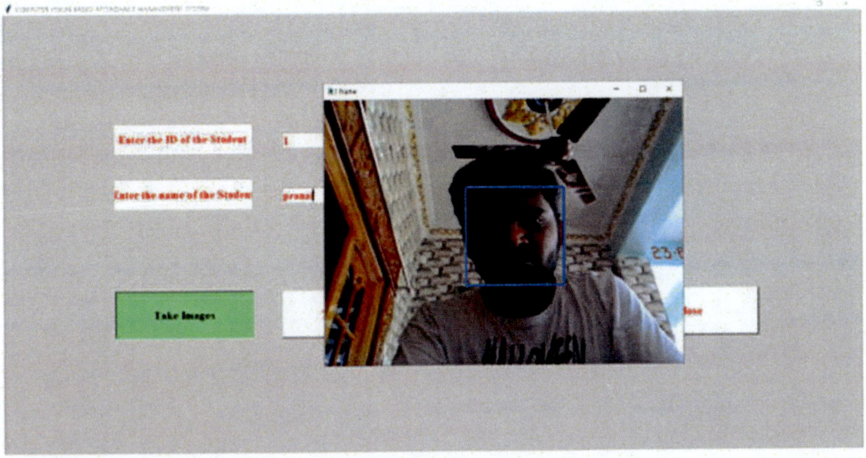

Fig. 7. The layout of capturing the video frame and training the system. A video of the student is captured for the system.

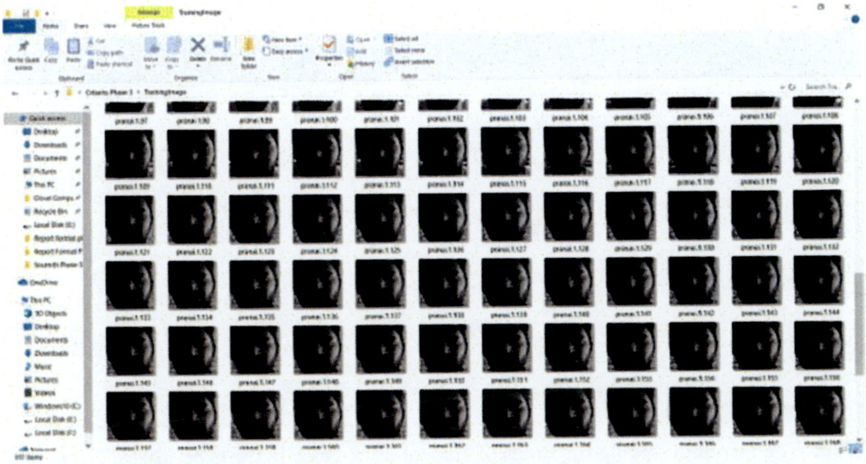

Fig. 8. The layout of saving the trained images. The trained images are saved in this page.

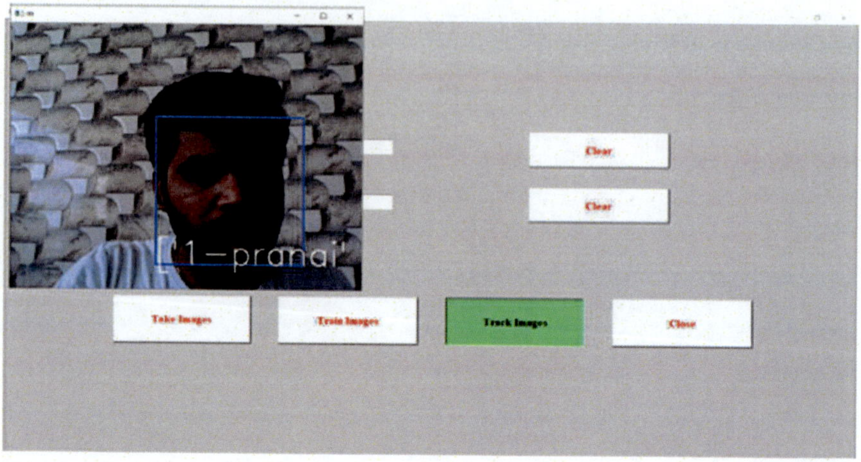

Fig. 9. The layout of tracking the live attendance of students. The continuous monitoring of student is performed.

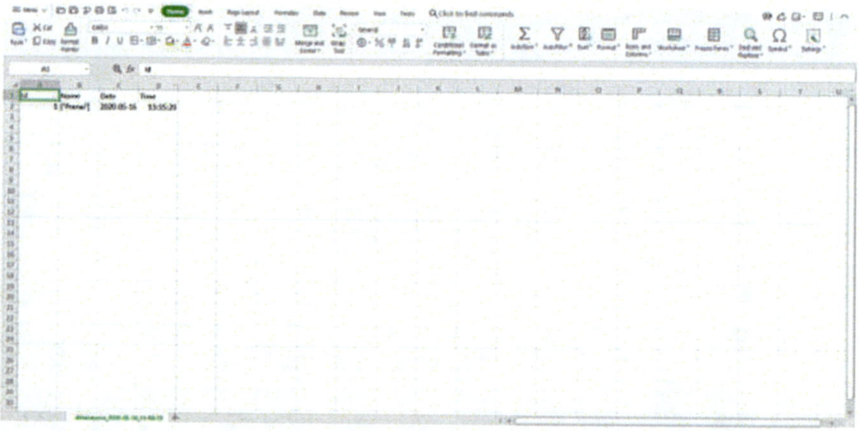

Fig. 10. The layout of storing the status of student attendance in the excel sheet. The complete details of student are managed in excel sheet.

5 Conclusion and Future Scope

In this paper, we suggest a deep automatic learning attendance framework which can be used with the aid of a camera in crowded classrooms. In our research, we used Haar Cascade technology and the LBPH algorithm, both focused on deep learning characteristics, and with just one enrollment face picture per subject we were able to achieve the highest accuracy. We assume that both teachers and students can save time by using our facial recognition automatic student attendance system while avoiding false participation. We can use more advanced algorithms based on reduced dimensionality to verify the spatial accuracy of the participants.

References

1. P.K.K. Pandey, RFID based smart attendance system. Int. J. Res. Appl. Sci. Eng. Technol. (2018). https://doi.org/10.22214/ijraset.2018.5080
2. P. Mehta, P. Tomar, An efficient attendance management system based on face recognition using matlab and raspberry Pi 2. Int. J. Eng. Technol. Sci. Res. (2016)
3. S.K. Gupta, T.S. Ashwin, R.M. Reddy Guddeti, CVUCAMS: computer vision based unobtrusive classroom attendance management system (2018). https://doi.org/10.1109/ICALT.2018.00131
4. D. Mery, I. Mackenney, E. Villalobos, Student attendance system in crowded classrooms using a smartphone camera (2019). https://doi.org/10.1109/WACV.2019.00096
5. U.A. Patel, Computer science and management studies development of a student attendance management system using RFID and face recognition: a review. Int. J. Adv. Res. (2014)
6. R. Joshi, V.V Shete, S.B. Somani, Android based smart learning and attendance management system. Int. J. Adv. Res. Comput. Commun. Eng. (2015)
7. E. Rekha, P. Ramaprasad, An efficient automated attendance management system based on Eigen face recognition (2017) https://doi.org/10.1109/CONFLUENCE.2017.7943223
8. R. Samet, M. Tanriverdi, Face recognition-based mobile automatic classroom attendance management system (2017). https://doi.org/10.1109/CW.2017.34
9. N.K. Jayant, S. Borra, Attendance management system using hybrid face recognition techniques (2016). https://doi.org/10.1109/CASP.2016.7746206
10. Y. Chen, Y. Tai, X. Liu, C. Shen, J. Yang, FSRNet: end-to-end learning face super-resolution with facial priors (2018). https://doi.org/10.1109/CVPR.2018.00264

Multidimensional Features Driven Phishing Detection Based on Deep Learning

P. Vigneshwaran(✉), A. Soumith Roy, B. Sunal Sathvik, D. Md. Nasirulla, and M. Lekha Chowdary

Department of Computer Science and Engineering, Jain (Deemed to be University), Bengaluru, Karnataka, India
vigneshwaran05@gmail.com

1 Introduction

The Internet has developed into a critical technology that provides enormous positive impact on human community [1]. Fortunately, the world wide web (WWW) is plagued with certain unavoidable protection issues, such as cyber—attacks, malicious malware, and data leakage, most of which have also posed significant risks to clients' finances [2]. Global APWG describes the phishing like a fraudulent scheme that involve online engineering and technological deception to capture customers' user details and income statement passwords [3]. Phishing is indeed a widely used network attack technique that results in data breaches, data theft, even harm to property [4]. Thus according Cyber Security figures, and over course of the last year, 29.4% of consumer machines was exposed to minimum one Malware-class network threat, and network virus protection modules identified 199, 455, 606 separate Hyperlinks as harmful. Furthermore, throughout 2017, monetary phishing accounted for nearly 54% among all malware incidences, up from 47.5% in 2016. Malware has evolved into the most serious online global problems [5].

Spam detection was no longer restricted to standard methods including e-mail, Text messaging, as well as rock windows. Although this growth in the mobile Internet as well as social networking sites has provided consumers with comfort, they are also used to broadcast malware, including certain QR code malware, spear-phishing, as well as spoof mobile apps, among other things. Furthermore, several clever malware assaults are hosted on pages with Http and SSL licenses, as many consumers perceive the HTTPS sites are probably genuine. Phishing is evolving in a variety of ways, posing strong authentication problems. Although scammers are cunning and sneaky, technology professionals as well as analysts has put a huge amount of work to identify phishing websites [6]. This URL's script series is a naturally created function which removes the subjective nature of arbitrarily feature set. However it does not necessitate the use of a third party or other previous experience of phishing. Even so, among the most complicated aspects of pattern decoding is efficiently extracting correlation as well as semantic information [7].

F. P. García Márquez (ed.), *International Conference on Intelligent Emerging Methods of Artificial Intelligence & Cloud Computing*, Smart Innovation, Systems and Technologies 273, https://doi.org/10.1007/978-3-030-92905-3_45

2 Literature Review

Alnajjar et al. researched the topic of phishing attacks on QR codes. Quick Response script is a vector serial number with graphical square blocks that enables simple connectivity among smartphone and sites either written content by eliminating the need to physically type a Link or contact info. QR code development is gradually being utilized in tablets, from either the magazine covers to the fronts of vehicles as well as billboards. Regrettably, scammers have begun to use Bar code for malware emails by exploiting a few of their functionality. The above authors propose "TrustQR," a novel method for detecting Link malware on Passcodes. It detects phishing URLs in Barcode material using QR code-specific functionality as well as Link characteristics. To produce the authentication password, many QR code-specific functions using Barcode information as well as its features including distance, size, and degree of error checking. The whole method employs the classification technique of artificial intelligence [8].

Canfield et al. researched the topic "Setting Priorities in Behavioral Interventions: An Application to Reducing Phishing Risk". Spam is now a bigger issue for businesses, states, and people. We illustrated a method for measuring the pros and cons of measures which address the most insecure people, based on facts which consumers differ greatly in the exposure to malware emails. The method employs Simulation to: (1) classify the applications as it's most susceptible throughout terms of sensitivity hypothesis; (2) evaluate the percentage of structure danger due to a very vulnerable persons; and (3) determine the percentage of threat assessment attributed to its most vulnerable individuals (3) calculate the material profit as well as expense of behavioral approaches addressing various degrees of security vulnerabilities; as well as (4) assess the effects' responsiveness to if the assaults are spontaneous or hammer [9]. Individuals discovered that even the most sensitive clients are far less vigilant and perhaps less willing to differentiate among malware and genuine communications, based on estimated coefficients from earlier studies.

Goel et al. researched the topic "Mobile phishing attacks and defense mechanisms: State of art and open research challenges". Phishers is also a digital money laundering scam where an intruder tries to obtain a recipient's private info, causing economic loss to both consumers and firms. Customers are increasingly using portable phones, particularly smartphones, considering the wide variety of features they offer. As a result of their small size and ability to perform functions close to that of desktops and laptops, hackers are increasingly targeting consumers of smart phones. Owing to the differences in implementations, detecting smartphone malware threats is a new challenge than detecting laptop phishing attacks. Furthermore, while little work was done in this area, detecting mobile malicious software with absolute speed is a significant research topic. While several anti-phishing methods for smart phones were developed, there would be a need for a complete solution. The purpose of the assignment is to conduct a thorough examination of smartphone phishing tactics as well as authentication methods. This article is categorized in four sections. First, mobile malware attacks are addressed in depth, including the past, hacker motivations, even mobile security risks [10].

3 Methodology

In the research publications and essential goods, there are many other protocols as well as data forms for spam detection. A malware Link and also its accompanying site have many characteristics that distinguish them from malicious Content. An attacker, for instance, might create a lengthy and misleading name to conceal its real web address.

The following is a list of features gathered from empirical research for phishing topic identification using machine learning approach.

1. The URL-rooted feature.
2. The domain-rooted feature.
3. The page-rooted feature.
4. The content-rooted feature.

Mainly there is the vastutilization of the NLP as well as the certain other machine learning methods.

Instrument:

Software Requirements.

- Operating system: Windows 7, Windows 10
- Coding Language: Python3.6
- Tools: Python IDLE.

Hardware Requirements.

- System: i3 and higher.
- Hard Disk: 120 GB.
- Monitor: 15″ LED.
- Input Devices: Keyboard, Mouse.
- Ram: 4 GB.

Scammers normally trick users by imitating the targeted site's learning materials. As a result, the text attributes of the website must be extracted. The extracting of successful website material as well as the remotely operated rendering of both the text are by far the most important aspects of such a step. We use pattern matching to delete unnecessary sections of the website, including Code generator, CSS, including mark symbols, in order to find accurate text content. Despite the fact that the multifaceted function approach is more accurate than CNN-LSTM, retrieval of WHOIS info including Alexa rating from either the Link, as well as the retrieval of site generate relevant including website textual content, all take time, which would be insufficient to true identification. As just a result, throughout this segment, we boost the CNN-LSTM method's softmax layer's classification performance.

Algorithm:

Multidimensional features technique.

Input: URL set to the U, U = {u1, u2,...,un}.

Output: The probability of phishing P(U).

(a): N = |U|,H = 0.

(b): for i in N do.

Fd = 0, Fu = 0, Fc = 0, Ft = 0, F = 0.

(c): Fd = CNN-LSTM(ui).

(d): Fu = extract_url(ui).

(e): Fc = extract_code(ui).

(f): Ft = extract_text(ui).

(g): F = Fd U Fu U Fc U Ft.

(h): H = H U F.

(i): end for.

(j): P(U) = XGBoost(H).

(k): return P(U).

4 Results and Discussion

The proposed methodology imports data-set of phishing as well as the legitimate links from database and then the imported data is pre-processed. Identifying phishing websites is performed based on the following category of URL features: domain, address, abnormal and HTML, JavaScript features. The URL features are extracted with processed data and values for each URL attribute. The analysis of URLs is performed using a machine learning technique that computes the range value and the threshold value for URL attributes. Then it is differentiated as phishing and legitimate URL. The attribute values are computed using feature extraction of phishing websites and it is used to obtain the range value and threshold value. The value for each phishing attribute is ranging from $\{-1, 0, 1\}$ these values are defined as low, medium and high. The classification of phishing and the legitimate website is based on the values of attributes extracted using different types of phishing categories and a machine learning approach. The attribute value of the URL is computed using the corresponding set of attribute values $\{-1, 0, 1\}$. The attributes URL_of_Anchor tag and Prefix_Suffix also have interlinked value and that needs to be computed for finding range and threshold value. The phishing features are extracted for each URL to find whether the website is phishing or legitimate. The URL_of_Anchor tag attribute is selected to identify the overlap values. The overlap value is the summation of the selected attribute value which is combined with other attributes. Table 1: Unit Testing Tables Fig. 1: Illustrates the screenshot 1 the Fig. 2

shows screenshot 2, Fig. 3 shows the screenshot 3 and Fig. 4 shows the screenshot of a graph, Fig. 5 shows thescreenshot of another graph.

Table 1. Unit testing tables.

Test case#	UTC01
Test name	File import format
Test description	To test whether an excel file with comma-delimited format is accepted or not
Input	A comma-separated excel file with a valid dataset
Expected output	The file should be read by the program and few lines of the file should be displayed
Actual output	The file is read and contents are displayed accordingly
Test result	Success

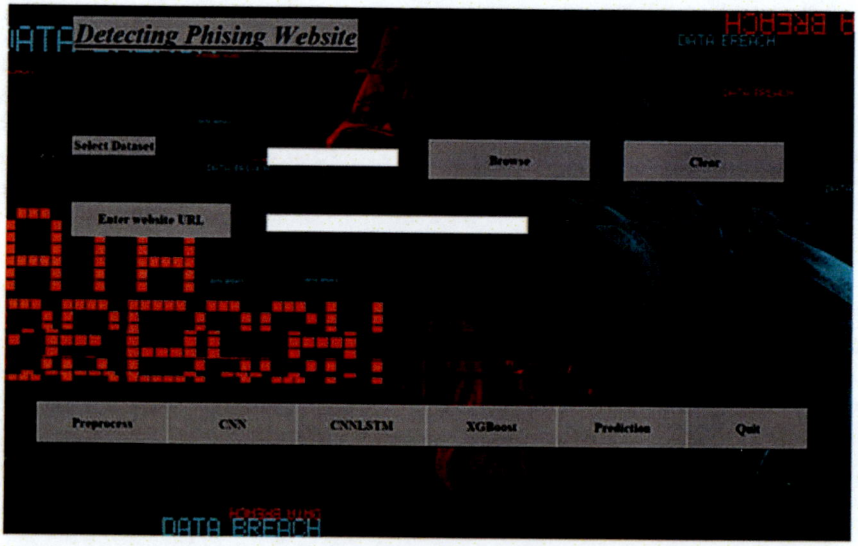

Fig. 1. Illustrates the screenshot 1.

5 Conclusion and Implications

A successful phishing site identification method has been well established for its successful positive changes, good precision, and a lower detection risk. This notion is supported by the suggested MFPD solution. The Link text series beyond phishing previous knowledge guarantees prediction performance underneath the guidance of a complex classification decision algorithm, as well as the multifaceted attribute identification guarantees

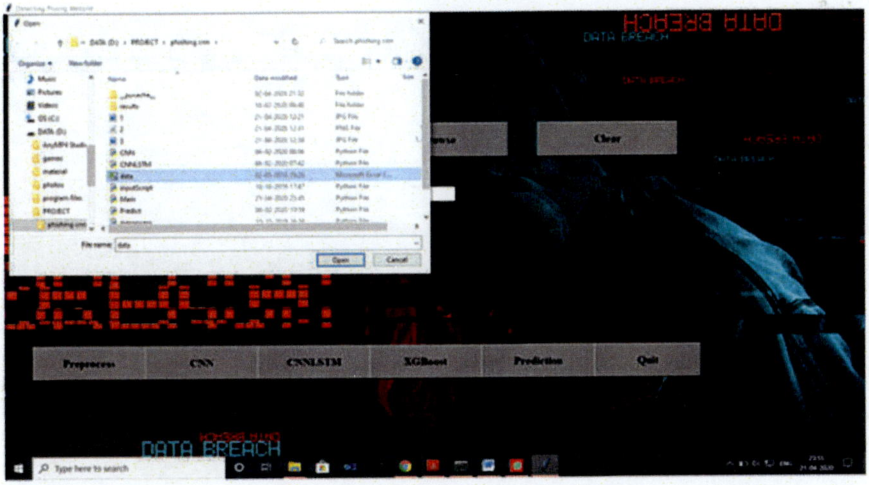

Fig. 2. Illustrates the screenshot 2.

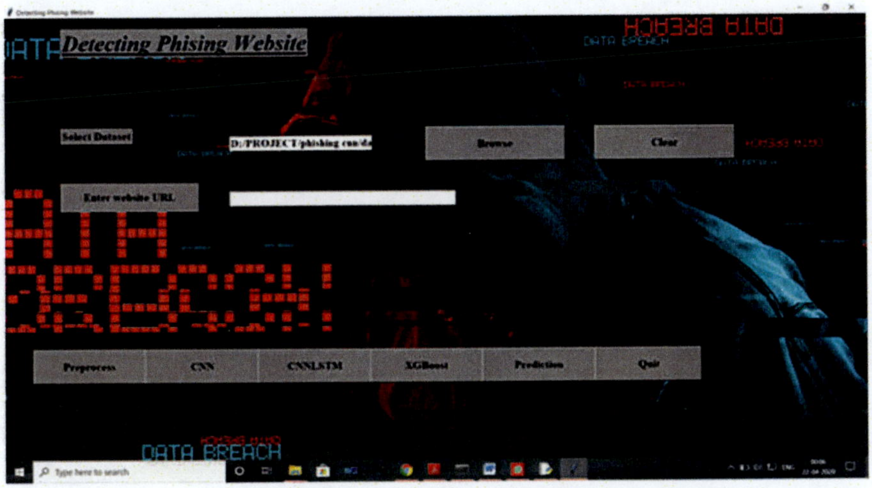

Fig. 3. Illustrates the screenshot 3.

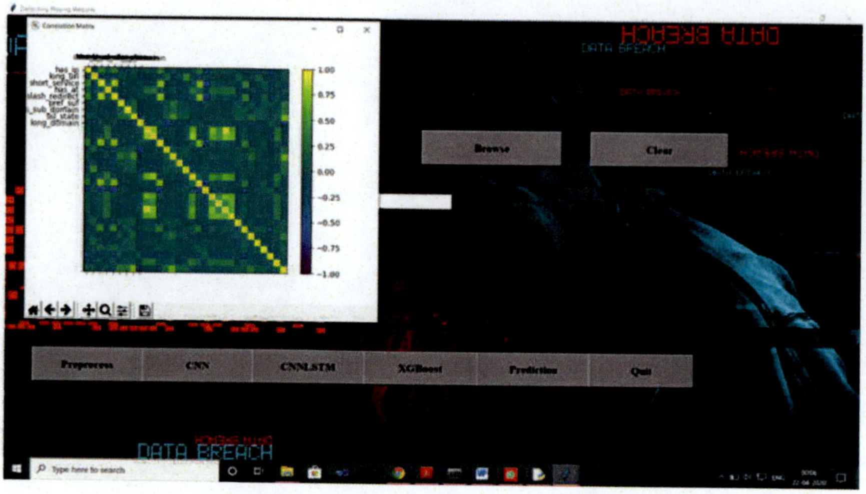

Fig. 4. Illustrates the screenshot of a graph.

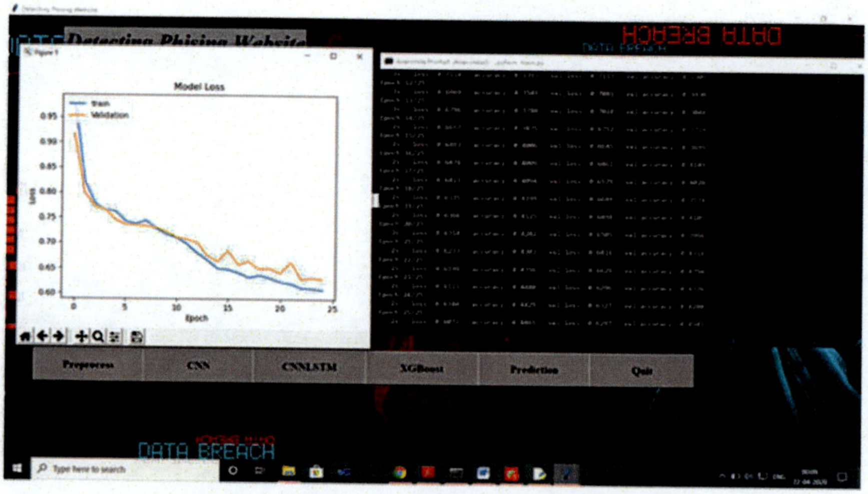

Fig. 5. Shows the screenshot of another graph.

prediction performance. On even a dataset containing billions of malware as well as legal URLs, we run a number of tests. According to the findings, using MFPD method is successful in terms of high precision, lower detection rate, including detection speed. This deep learning technique can be used to remove features from site code including script in the future growth of our methodology.

References

1. P. Yi, Y. Guan, F. Zou, Y. Yao, W. Wang, T. Zhu, Web phishing detection using a deep learning framework. Wirel. Commun. Mob. Comput. (2018). https://doi.org/10.1155/2018/4678746
2. P. Yang, G. Zhao, P. Zeng, Phishing website detection based on multidimensional features driven by deep learning. IEEE Access (2019). https://doi.org/10.1109/ACCESS.2019.2892066
3. B. Wei et al., A deep-learning-driven light-weight phishing detection sensor. Sensors (Switzerland) (2019). https://doi.org/10.3390/s19194258
4. X. Li et al., DeepSaliency: Multi-task deep neural network model for salient object detection. IEEE Trans. Image Process. (2016). https://doi.org/10.1109/TIP.2016.2579306
5. M. Moniruzzaman, S.M.S. Islam, M. Bennamoun, P. Lavery, Deep learning on underwater marine object detection: a survey (2017). https://doi.org/10.1007/978-3-319-70353-4_13
6. Z.Y. Wu, M. El-Maghraby, S. Pathak, Applications of deep learning for smart water networks (2015). https://doi.org/10.1016/j.proeng.2015.08.870
7. G. Liu, H. Bao, B. Han, A stacked autoencoder-based deep neural network for achieving gearbox fault diagnosis. Math. Probl. Eng. (2018). https://doi.org/10.1155/2018/5105709
8. A.Y. Alnajjar, S. Manickam, M. Anbar, S. Al-Saleem, O. Elejla, TrustQR: a new technique for the detection of phishing attacks on QR code. Adv. Sci. Lett. (2016). https://doi.org/10.1166/asl.2016.7102
9. C.I. Canfield, B. Fischhoff, Setting priorities in behavioral interventions: an application to reducing phishing risk. Risk Anal. (2018). https://doi.org/10.1111/risa.12917
10. D. Goel, A.K. Jain, Mobile phishing attacks and defence mechanisms: State of art and open research challenges. Comput. Secur. (2018). https://doi.org/10.1016/j.cose.2017.12.006

DNA Based Criminal Identification Using Blockchain

Narayana Swamy Ramaiah(✉), Abhishek Raj Dhungel, Daniel Thapa, Sonam Wangchuk Bhutia, and Bipul Giri

Department of Computer Science and Engineering, Jain (Deemed to be University), Bengaluru, Karnataka, India
r.narayanaswamy@jainuniversity.ac.in

1 Introduction

A Blockchain system is usually known as a 'Distributed Ledger Technology' and is an emerging technology in many different fields such as property, health, digital currencies and many more. As the name suggests, Blockchain holds a single public ledger that is viewable by everyone on the network but only the participants of the Blockchain can access and interact with the contents available inside. Blockchain technology has been prevailing in sectors that deal with digital transactions of physical objects such as properties, share markets, crypto-currencies and such. The main objective of introducing the system was to facilitate transactions between two people without the governance of a central controlling body to minimize time and cost [1].

A Blockchain is a connection of a huge number of computers (nodes) that are powerful enough to solve complex and resource-expensive mathematical problems. These computers are interconnected to each other and the users can traverse the whole chain through one of these nodes. The number of nodes that can be found in a fully developed and functional Blockchain can range between thousands to millions and forever increasing. In the present context, Bitcoin has over 10,000 nodes that are fully operational [2]. Each node holds a copy of the main ledger and any authorized changes that are committed during the processing of a transaction is also updated on every copy of the ledger present in every node. A node in a Blockchain typically holds three different crucial information:

- It holds information related to the transaction that the Blockchain verifies and is reflected upon the main ledger. The information can be present in forms like the date of the transaction, time of the transaction and in the case of crypto-currencies, nodes also hold dollar amounts.
- The node also carries the name of the user or the node itself. The name is not necessarily the real user name but usually, a name that is given to the node in the Blockchain. For example, how IP addresses work on the internet.
- A node also holds something called the hash value. A hash value is generated by a hashing function. The hashing function, depending on the data, produces a specific hash value. Any slight changes on the data will cause a new hash value to be produced

F. P. García Márquez (ed.), *International Conference on Intelligent Emerging Methods of Artificial Intelligence & Cloud Computing*, Smart Innovation, Systems and Technologies 273, https://doi.org/10.1007/978-3-030-92905-3_46

which is quite different from the previous one. This hash value is important as it helps to direct the user to the next connected node.

The Blockchain is arranged in such a way that the nodes point to the node that is present before them. A node will have the hash value of the node placed before them in position. A Blockchain starts with a 'Genesis Block'. The genesis block indicates the beginning position of a Blockchain hence the name [3]. It is different from other blocks as it does not point to a block because there are no blocks that are present in front of it. So, its previous block is set to 0 and the hash value for the genesis block is passed on to the next corresponding block so that it can point to the genesis block. This fashion is repeated until the end of the Blockchain. Any mismatch in the hash values will cause the whole chain to malfunction and corrupt it [4].

1.1 Block Chain Working

A series of steps happen consecutively for data to be stored inside a block and the block to be added to the Blockchain network. The steps have been mentioned below:

- A transaction needs to be initiated to record the details of the transaction into a block, which in turn will be added to the Blockchain ultimately.
- The transaction must be verified before it is committed to the main ledger and for the update to happen on all the present ledger copies. As Blockchain is a decentralized system that does not have a central controlling body, all the computers that are present in the chain take the responsibility of verifying the transaction. For a transaction to be verified, at least more than half of the active computers present in the Blockchain need to acknowledge the transaction that occurs and all of the details regarding it. After the confirmation from more than half of the participants present in the Blockchain then afterward the transaction is given the green light.
- After the transaction has been verified, a block is created that stores all the information related to the transaction. The information can be data like the time of the transaction, date of the transaction, dollar amount if you are dealing with crypto-currencies and other such data.
- Based on the data that the block holds, a hash value is generated from the hashing function that has been implemented. The hash value depends on the data present on the block. A slight change in the data can result in a very different hash value in comparison with the previous value. A hashing function will generate a certain hash value based on the data that has been passed to the hashing function but the data cannot be regenerated again based on the hash value.
- Ultimately, the block is added to the chain bypassing the hash value of the block before it so that it will be pointing to a block directly in front of itself.

1.2 Properties of a Blockchain System

The three main properties of Blockchain Technology which have helped it gain widespread acclaim are as follows:

- Decentralization
- Transparency
- Immutability.

1.3 Blockchain Security

Blockchains are visible to the general public as well. It is designed in such a way that the public can view the contents of the main ledger but cannot contribute to the Blockchain unless verified by the participants of the Blockchain itself [5]. The Blockchain uses two different keys to ensure that no outsider can invade the privacy of the chain as a whole. The two keys that are implemented to ensure integrity are:

- Public key
- Private key.

1.3.1 Public Key

This key denotes the address or name that is used by a block in the public view of the Blockchain. Other nodes will recognize the neighboring or other participating nodes through the public key. The main ledger is accessed and the contents are viewed by any block using the public key. The public key acts as the pass for accessing the main ledger and its contents [6].

1.3.2 Private Key

Private Key is a unique key that is different for every block and the key itself is only visible by the individual block itself. The key is used to access the contents of the block where the public key is used to access the contents of the main public ledger. Every block's content that is present in the Blockchain is only accessible through the private key. For example, every school has lockers that have been provided for the students to use. Nonetheless, other students and teachers can still slide some notes or letters inside the locker through some small openings on the sides. But the contents can only be viewed or accessed by the student who owns the locker as he will have the key. The key, in this case, will act as if the private key of a block in the Blockchain. Hence, to access the contents of the block a private key is used by the block and the key is only visible to the particular block itself.

The keys ensure the integrity of data and information throughout the Blockchain. One of the reasons for Blockchain becoming popular was because it provides a highly secure environment for transactions to take place that deals with highly valuable assets like properties, crypto-currencies and other valuables. The measures that the system has taken up to ensure such as secure environment has been discussed forward. The process of adding computers to the Blockchain is termed 'mining'. In a fully established Blockchain system, it can be seen that they are thousands to millions of nodes that are interconnected to each other. To ensure that all the nodes that participate in the Blockchain are trustworthy and eligible, the Blockchain demands every node a proof-of-work [7].

2 Literature Review

Before starting any experiment, it is usually a good idea to research the topic before beginning any concrete work on the topic. The mentioned papers below provided the team with insights and ideas about some of the Blockchain systems that were already developed and were being implemented. The papers also helped in finding their limitations which we, could overcome or avoid in our experiment. The list mentioned below contains the summary of some reputed journal papers and their proposed systems relating to the field of Blockchain development and their implementation in different fields like forensics, whichhave helped us to give an idea of the current systems in place for beginning our experiment altogether.

Hu et al. researched computer forensic systems in his research paper. The paper focuses mainly on gathering pieces of evidence dynamically and making it possible for the sharing of the pieces of evidence among the interconnected network. It also focuses on overcoming the drawbacks of the traditional forensic system such as the evidence loss chain [8]. Mansourvar et al. explored a novel computer-based system to support the smart forensic study. The experiment used CBIR (Content based Image Retrieval), which used visual contents to retrieve images from different accessible image databases. The experiment aimed to provide a smart way of figuring out important evidence for the crime using computer-aided technologies [9]. Rattanamuang et al. explored another paper on the topic of DNA forensic systems. This experiment created a distributed network that enabled the sharing of DNA records within Thailand. It was implemented as a web service i.e. it was accessible through the internet [10].

The limitations that the experiment faces are similar to the limitations that a Blockchain in the real world face. The limitations are listed below.

2.1 The Network Size

A Blockchain consists of a huge number of nodes that are connected and each node functions independently. A transaction is authorized based on the majority of votes that it gets. If the majority of the nodes do not authorize the transaction then the transaction is stopped altogether and not registered in the main ledger. There is a threat to the Blockchain network called the 51% attack. If the attacker can gain access to the majority of the nodes present then the hacker will be able to authorize the transactions for themselves and can change the Blockchain drastically for their advantage. So, the Blockchain that we have developed is quite small as compared to an actual Blockchain hence, the possibility of a 51% attack.

2.2 Network Speed

A Blockchain consists of a large number of nodes that make up the network which means that the number of transactions is also equally high and not to mention that the nodes may have a different amount of data and information to process respectively. This means that the main ledger has to be regularly updated and checked to ensure that the whole chain is valid at all times. For such distributed and a large number of transactions that each node performs needs a high-speed internet to ensure seamless transmission and update. This is one of the limitations that our experiments have faced but not to that extent as our network is quite small. But while considering an actual Blockchain then this is a serious limiting factor.

2.3 Human Error

The nodes present in the Blockchain are present to process the data that has been fed to it. It cannot differentiate or distinguish the data that is provided to it. So, it depends upon the data that is fed to it by a human. The data can carry human error which in turn will result in some inaccuracies in the processed result from the node. So this is one limitation that can be mitigated but can never be fully eradicated.

3 Methodology

3.1 Design

This section will demonstrate the methodology that has been implemented to create a Blockchain system that holds criminal data based on DNA. The methodology has been discussed below with a diagram to represent the flow (Fig. 1) of the overall experiment as a whole:

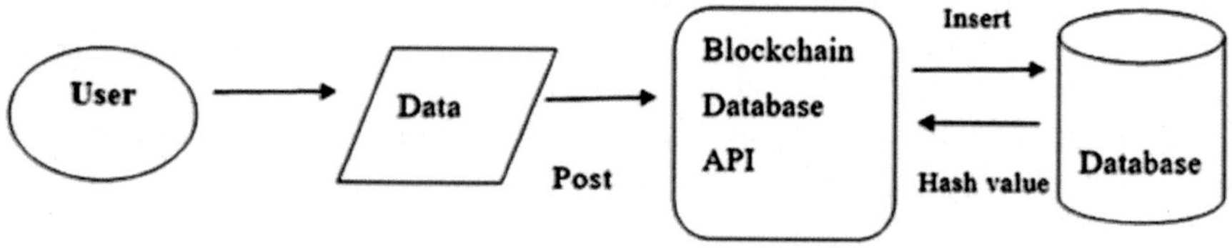

Fig. 1. Illustrates the database based on the blockchain data architecture.

- First of all, Node Js was installed onto the system. Node Js provides an efficient environment for JavaScript-related development hence it was selected as the preferred IDE. A check was also conducted to see whether Node Js was successfully installed onto the system through the terminal. Since the majority of the codes were UNIX-based, GitBash was installed and remained the primarily used terminal for the experiment.
- A check was conducted to see if NPM was installed properly along with Node Js through the terminal. The Jest library was also downloaded and installed through the NPM.
- A class was made which was named 'Block'. The purpose of this class was to generate blocks that can be added to the Blockchain. The class used a constructor so that every time the class was called upon the parameters like 'timestamp', 'data', 'latHash' and 'Hash' were also initialized for every block object that was formed.
- To create a Genesis block for the Blockchain, a function called 'genesis' was made within the Block class so that as soon as the Block class is called upon through an object, the Genesis block is already made before any other blocks are added to the Blockchain.

- Since Blockchain systems have the option of mining new blocks to the existing Blockchain, a mine function was also added to the experiment. The mine function has two parameters: 'last blockand' 'data' where the 'lastBlock' parameter is an object. This ensures that every new block which will be added to the Blockchain will receive all the data of the block present in front of them. A series of tests were also developed to ensure that the 'lastHash' received by the new block matches that of the block placed in front of it.
- A hashing function was required to generate the hash values for the blocks based on their data and other information that each block carried. SHA-256 was used by downloading the 'crypto hash library through the NPM'.
- The previous functions were developed to handle blocks. Now, a Blockchain class was created, which had the 'addBlock' function. The 'addBlock' function will in return call upon the 'mine' function and add the block to the existing Blockchain.
- For every block added, the Blockchain needs to check whether the chain formed after the addition of the new block is still valid. This is done by using the 'is ValidChain (chain)' where 'chain' is the object of the Blockchain class. This function will check if the newly formed chain is valid by tracing it till the Genesis block.
- A second validation to ensure that the added block is valid is done by the Blockchain by using a technique where, if the newly formed chain is no longer than the previous chain then it discards the newly added block. The experiment implemented this technique in the 'replaceChain()' function.
- To maintain the integrity of the Blockchain, nonce value and the mine rate were adjusted dynamically by adding the difficulty parameter in the block class.
- Downloaded and installed Express API to handle HTTP requests and for managing the port-related processing for the Pub/Sub method of communication.
- Downloaded and installed Redis to implement the Pub/Sub communication that would be required for broadcasting the changes to every other instance of the Blockchain.
- After installing and setting up Express API and Redis, the broadcasting of changes in any instance of the Blockchain was picked up by all the other instances so that all the Blockchains were synchronized.
- Finally, the Blockchain has been completely set up and all the instances of the main Blockchainare also synchronized so that any future changes will be reflected upon all the other instances of Blockchains.

3.2 Instrument

This section will list out the hardware and software tools that have been implemented in developing this experiment. A brief functionality of each of those mentioned tools will also accompany the list but the tools will be discussed in detail further in Chap. 3.

The only hardware tool that was used in this experiment was a Computer/Laptop.

As this experiment is mainly focused on cybersecurity, the experiment did not require a lot of hardware components. The software tools that were used are mentioned below:

- Node JS
- Express
- NPM

- React
- Redis
- GitBash/Linux Shell
- Jest.

3.3 Software Tools Description

Table 1 illustrates the software tools that are used in this experiment.

Table 1. Software tools used in this experiment.

Software tools	Description
1. Node Js	It is an open-source, cross-platform IDE for JavaScript that does not run on a web browser
2. Express	It is a web framework that is used along with Node Js to handle codes related to HTTP requests and is also efficient in port management
3. NPM (Node Packet Manager)	It helps in managing, installing and downloading packages in the Node Js environment
4. Redis	It helps in implementing the Pub/Sub concept as it can act as a message broker between the publisher and subscriber
5. React	React is a UI library that uses component logic for developing complex UIs
6. GitBash/linux shell	GitBash and Linux Shell are environments where we can run Git commands and Linux commands
7. Jest	Jest is a JavaScript testing framework which means that it provides the environment and tools to run tests on your JavaScript code

4 Results and Discussion

With the experiment's objectives been clear from the early sections, it has created a solid base for the experiment. The online research along with the literature survey has even strengthened the case for a need fora solution to the problems presented, in this case, to provide a high-security system for the use in DNA-based forensics and criminal identification, by making use of Blockchain technology. We were able to successfully create a robust system to tackle the limitations in the existing system. With the help of a user-friendly UI accessing data on the network by authorized individuals can be seamless and efficient. To begin the process of data recording and retrieval, the officials are presented with a front-page with clean and apprehensive options as shown in Fig. 2. The options presented are View, Search and Create. The view allows us to look at the

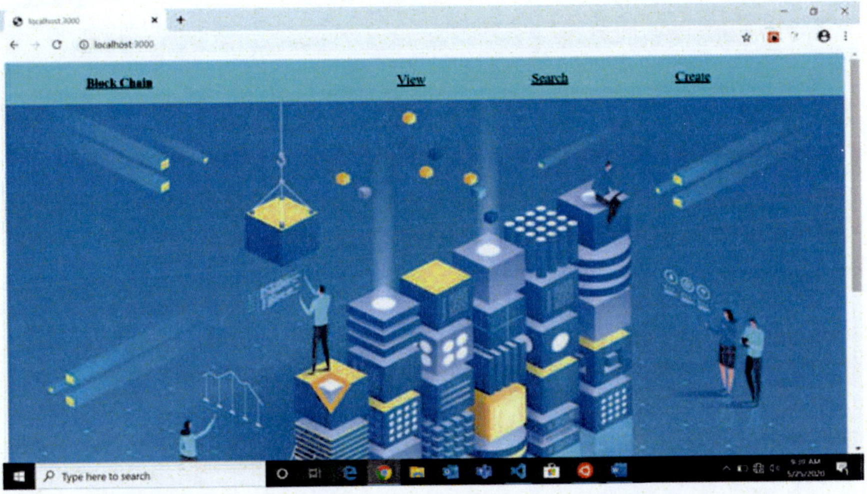

Fig. 2. Illustrates the image of the user interface (UI) of the home page.

blocks created on the Blockchain with their data. We can record the DNA data using the Create option. Identification can be performed on the Search page.

To initialize the Blockchain we begin with the main node, in this case, the localhost: 3000. Here we observe the initialization of the Blockchain which begins with the generic block. Every Blockchain requires a generic block to start a chain while other blocks are created gradually and added to the chain. We can also see the parameters of the generic block in Fig. 3. On entering the DNA sequence onto the search field, the results are instantly presented on the same page. Figure 4 shows the search results of the data entered in the search field. The page shows data like Name, DNA code, Address and the Crime committed for the DNA code that it matched with. Having an infrastructure where every step requires verification and caution does create a tough system but also loses out on a few limitations. Making transactions across the network will require a significant amount of network speed considering the number of transactions passing through numerous nodes simultaneously.

So, in a network with large number of nodes, every node will have to endure considerable amount of load. For a Block chain system to work normally the network size must be favorable and democratic. If enough nodes (51%) is controlled by a single entity then the block chain can be manipulated and the data or the system can be breached creating security issues. Since Block chain is a new concept and has just cracked the surface, not many people are familiar with its working which brings limitations to the people that can be hired to access the system. Proper training can be implemented to work around this issue. Even with few drawbacks this system can be further improved in the future with better internet speed to match the network load.

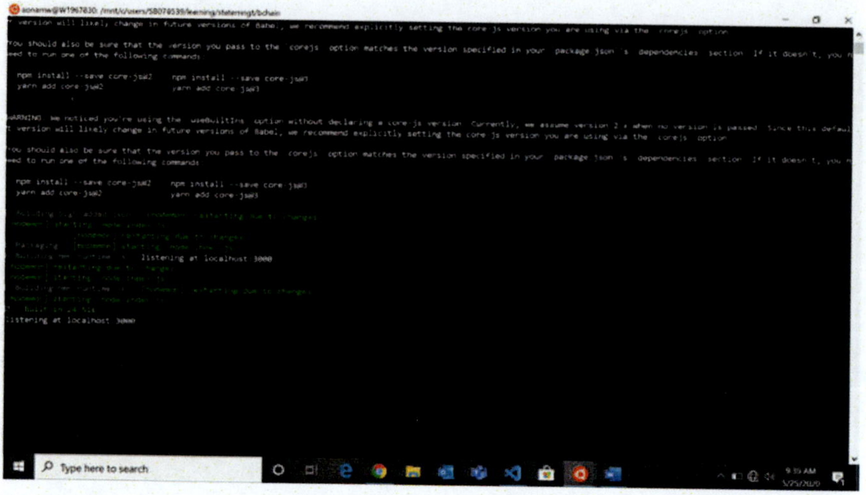

Fig. 3. Illustrates the back end showing the successful initialization of block-chain.

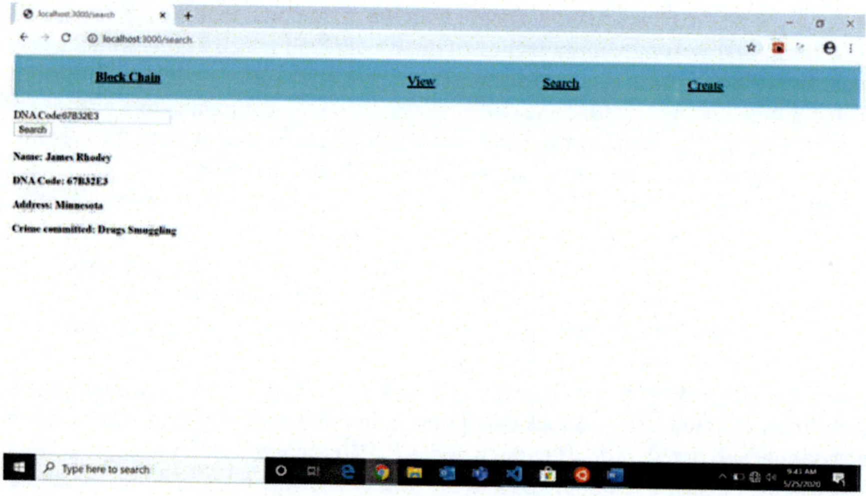

Fig. 4. Illustrates the search page showing search result.

5 Conclusion and Implication

This experiment as a whole reminds of the possibility of a forensic DNA identification system that can be a huge improvement over existing systems in terms of security risks with the help of block-chain technology. A decentralized system with a network of nodes spread around the world communicating with one another to maintain a single ledger or block-chain has great potential to transform the traditional method of data storage and retrieval. With the primary focus on creating a unreachable infrastructure to store sensitive data and match it with provided data for criminal identification, this experiment comes close to achieving the desired results with exceptions to few technological disadvantages. Even though this block-chain system has been specifically programmed for forensics purposes it has great potential for further use in various fields that require storage of sensitive information. For e.g., financial information like bank details or intellectual property or digital copy of a physical property which act as a proof of ownership are just few examples of such potential market for this technology. Block-chain has not been around for long and has just begun but it surely is here to stay and cause disruptions in numerous fields.

References

1. H. Nakamura, T. Muro, S. Imamura, I. Yuasa, Forensic species identification based on size variation of mitochondrial DNA hypervariable regions. Int. J. Legal Med. (2009). https://doi.org/10.1007/s00414-008-0306-7
2. V. K. Kashyap, T. Sitalaximi, P. Chattopadhyay, R. Trivedi, DNA profiling technologies in forensic analysis. Int. J. Hum. Genet. (2004). https://doi.org/10.1080/09723757.2004.11885864
3. M. Walport, Distributed ledger technology: beyond block chain. Gov. Off. Sci. (2015)
4. A. Dhungel, B. Giri, D. Thapa, S. Bhutia, N. Ramaiah, DNA based criminal identification using block chain. SSRN Electron. J. (2020). https://doi.org/10.2139/ssrn.3580116
5. BitcoinWiki, Block chain 2015 (2015)
6. T. Prathyusha, M. Kavya, L. Akshita, Block chain technology. Int. J. Comput. Math. Sci. (2018)
7. M. Nakasumi, Information sharing for supply chain management based on block chain technology (2017). https://doi.org/10.1109/CBI.2017.56
8. L. Hu, K. Tang, G. Shi, Nurbol, K. Zhao, DDCFS: a distributed dynamic computer forensic system based on network (2009). https://doi.org/10.1109/ICICTA.2009.729
9. M. Mansourvar, et al., A computer-based system to support intelligent forensic study (2012). https://doi.org/10.1109/CIMSim.2012.33
10. A. Rattanamuang, S. Chiewchanwattana, K. Sunat, B. Waikham, DNA forensic system for police forensic science center cooperation: architectural design and implementation (2017). https://doi.org/10.1109/ICSEC.2016.7859901

Check for updates

Driver Drowsiness Detection Alert System Using Haar Method

C. R. Manjunath(✉), K. Swathi Krishna, K. Bhavya Sree, and K. Naga Ramyatha

Department of Computer Science and Engineering, Jain (Deemed to be University), Bengaluru, Karnataka, India

cr.manjunath@jainuniversity.ac.in

1 Introduction

Due to driver drowsiness, most road collisions occur. In the latest report, it indicates that one accident out of five accidents is due to the driver's drowsiness, which is around 20% of road accidents, and it rises steadily every year. Nearly 100,000 crashes happened last year owing to operator sleepiness, including more than 1,500 fatalities and much more than 70,000 casualties as per the national highway road safety association [1]. Most road accidents occur between midnight and early morning. By knowing the health risks and ensuring measures conductors will mitigate hazards. The drowsiness of drivers, drink and drive are coming further major reasons for road accidents. The Traffic Safety Report finds that 42,4% of drivers in a typical week travel fewer than 6 h of sleep without at least 1–2 nights of rest [2].

Drowsiness is about being incapable to retain your eyes open or feeling exhausted or sleepy. Drowsiness, also known as tiredness in abundance. Since the drive is attempting to close its eyes, the drowsiness of the drive is also decided. The method of the behavior of the eyes calculates the frequency of blinking and therefore the interval of closing eyes to detect sleepiness. The Yawning approach is based on the frequency of the driver's yawn. The mouth opening of the drive is greater when yawning than when usually speaking [3].

This project begins with the capture of the video and the captured video is transformed using OpenCV into several image frames and the image frame is taken as an input and pre-processed to remove features such as eye and mouth for face detection from the input image. The method involves distinguishing characteristics such as eyes and mouth using the technique of LBPH. By using the Haar detection process, we detect the features of the eye (blink) and mouth (yawn) and find the parameter measurements such as Root Mean Square Error (RMSE), Eye Aspect Ratio (EAR) are measured.

1.1 Driver Drowsiness

Now-a-days Accidents are most common if the driving is insufficient [4]. The factors causing accidents are:

F. P. García Márquez (ed.), *International Conference on Intelligent Emerging Methods of Artificial Intelligence & Cloud Computing*, Smart Innovation, Systems and Technologies 273, https://doi.org/10.1007/978-3-030-92905-3_47

(1) Speeding: Majorly on highways drivers ignore the speed limits. Speeds kill and driving more than the regulation is a simple cause for an accident.
(2) Drunk Driving: Whenever the driver is intoxicated, he loses the capacity to concentrate and work properly, because it is risky when a car is operated. This is one of the most contributing factors of an accident [5].
(3) Reckless Driving: Improper driving as speeding up the vehicle or changing lanes too quickly or tailgating and many more can cause reckless driving. Reckless driving is when you're operating a vehicle without caring.
(4) Night Driving: It can be dangerous to drive during daytime, although at nighttime the chance of an accidents almost increases. If you cannot see just what is going on, you do not even know exactly what to expect when you go to it [6].
(5) Driving under the Influence of Drugs: Approved and illicit drugs will undermine the driving capacity in full. Mind clearance and control over the body are need else it's going to cause an accident.
(6) Drowsy Driving: Driver tiredness is not always stated, but how do we allow someone to operate if they find it difficult to remain fully conscious? The majority of traffic crashes resulting from drowsiness happen in the darkness [7].

Here we are concern about the factors which make the drivers fall asleep.

As the result's same for all the above-mentioned factors that are, Eyes become "heavy" and blinking time is higher. The driver tends to keep his eyes closed "at least for a few seconds and if the yawning exceeds the usual time [8].

Hence here we create a buzzer to warn the driver and the owner.

2 Literature Review

Lashkov et al. investigated a new method for driver dangerous state recognition by OpenCV technology. Real-time tracking of driving actions plays an important role in smart transportation networks by reducing and minimizing the possibility of possible traffic incidents, such monitoring improves traffic safety. The vision-based approach to hazardous circumstance identification like video cameras is certainly among the most visual and widely utilized in sensing drivers' atmosphere. In current case, a drivers' photographs, recorded with filmed camera, will define its facial characteristics, such as movements of head, eyes status, mouth status, and then classify the present fatigue state levels. In such post, we use the smartphone's already placed cameras that constantly monitors the facial characteristics as well as identify the drowsiness and distraction hazardous states of the driver at an early point. Recognition of dangerous states is divided into online and offline modes. Driving dangerous conditions on smart phones are measured in real time due to the reliability and accuracy of smartphones when traveling with the assistance of computer vision libraries Dlib and OpenCV [9].

Nahar et al. researched the topic of an Eye-tracking system for real-time driver drowsiness control using a mean change algorithm. The continuous occurrence of a car accident has become a familiar scene in our direction of travel. Many people are struck every year, many of them lose their lives, and due to vehicular misadventure, a massive amount of property extermination occurs. It is renowned that the main cause of road accidents resulting in deaths, serious injuries and monetary damages is due to a tired or a sleepy driver, according to analytical studies on road accidents of recent years. The drowsy condition can be caused for long periods by lack of sleep, medicine, medications or constant driving. But this sloppiness in the life of a victim will bring preponderant torments as it causes fatal harm to lives and property. It is, therefore, necessary to research an advanced technological method to prefigure the occurrence of an accident based on drivers' drowsiness and distraction to slim down vehicle accidents. The aim of this work is to identify a device that tracks the level of drowsiness and distraction of drivers based on eye-tracking. A distinction has been made between the approach suggested and another methodology. It has been shown that the approach proposed givesgood precision than others [10].

3 Methodology

3.1 Design

The main weakness of this eye-state information method is the variation in accuracy for particular characteristics of the vehicle and driving conditions. A drowsiness sensors system can be developed to sense drowsiness or fatigue in drivers to avoid crash and increase road safety. The closed or open driver eyes and baying can be reliably monitored by this method. The primary objectives are to capture the live video, to classify the driver's state of alertness as drowsy or awake and feature extraction using the HAAR algorithm.

We input the recorded video of the face of the driver which is transformed into many image frames, stores the image frames in a data set to perform image pre-processing. Preprocessing images is designed to increase image quality of the data where there are irregularities or boost those image characteristics by using geometric transformations like rotation, translation, scaling., to enhance the features of face detection. The characteristic of the mouth and eyes are extracted using the LBPH technique. We detect the characteristics of the mouth (yawn) and eye (blink) by using the Haar detection method and find the parameter measures like Root Mean Square Error (RMSE), Eye Aspect Ratio (EAR) are calculated. If the eyes are open or close, we check the blink rate of each eye and if the blink rate of the eye exceeds the normal criteria then it alerts the warning beep system showing the red signal and if not the steps continue executing showing the green signal as shown in Fig. 1.

3.2 Use Case Diagram

3.2.1 Use Case 1

The use case diagram represents the interaction of the user with the system. When Webcam takes the video in the form of the frames from the user. The system detects the Face and extracts the features and eye part is identified by using the point-based extraction method and the system tries to identify whether the user is drowsy or not by calculating the blink rate of the eye using EAR and RMSE. If the system detects the drowsy then the system immediately warns of a beep alarm. Figure 2 shows the use case diagram for eye detection.

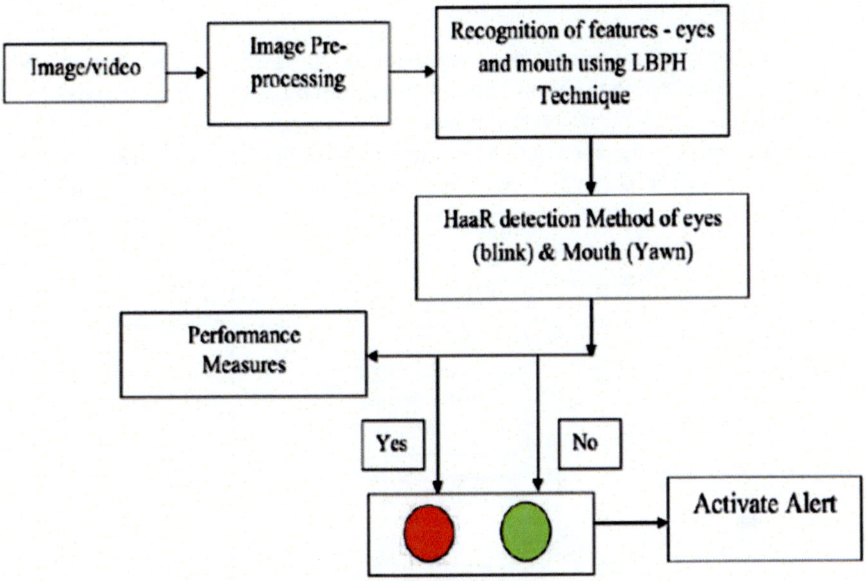

Fig. 1. Illustrating the block figure of the proposed methodology.

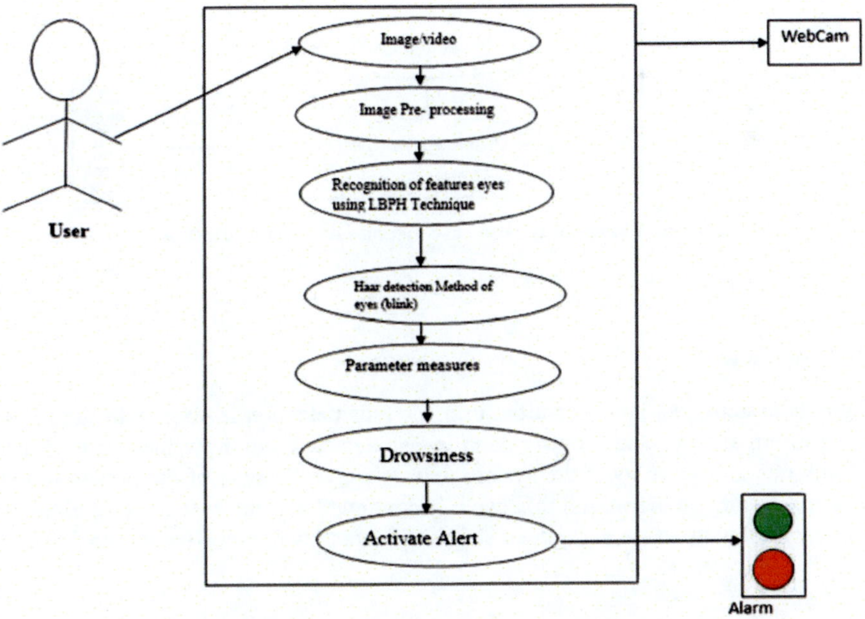

Fig. 2. Illustrates the use case diagram for eye detection.

3.2.2 Use Case 2

The use case diagram represents the interaction of the user with the system. When Webcam takes the video in the form of the frames from the user. The system detects the Face and extracts the features and the mouth part is identified by using the point-based extraction method and the system tries to identify whether the user is yawning or not using EAR and RMSE. If the system detects the drowsy then the system immediately warns of a beep alarm. Figure 3 shows the use case diagram for mouth detection.

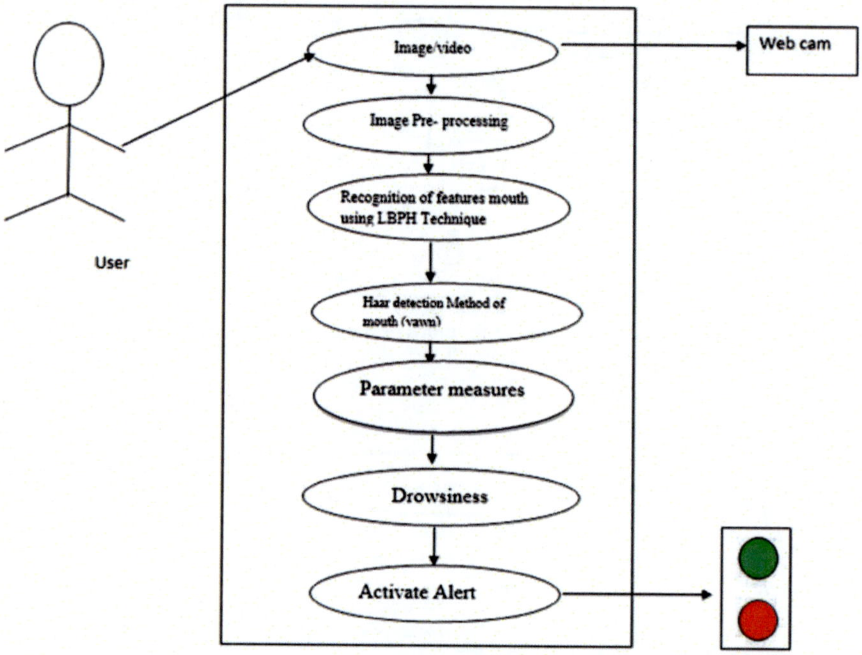

Fig. 3. Illustrates the use case diagram for mouth detection.

3.3 Data Flow

When the system gets input from the user the face detection is done with the help of the algorithm and eye and mouth region is also detected and drowsiness is calculated by blink rate and yawning if the system detects the drowsiness of the person it alerts the user by using an alarm and this cycle is continued and the system is continuously checking for the drowsiness. Figure 4 shows the data flow diagram for mouth detection.

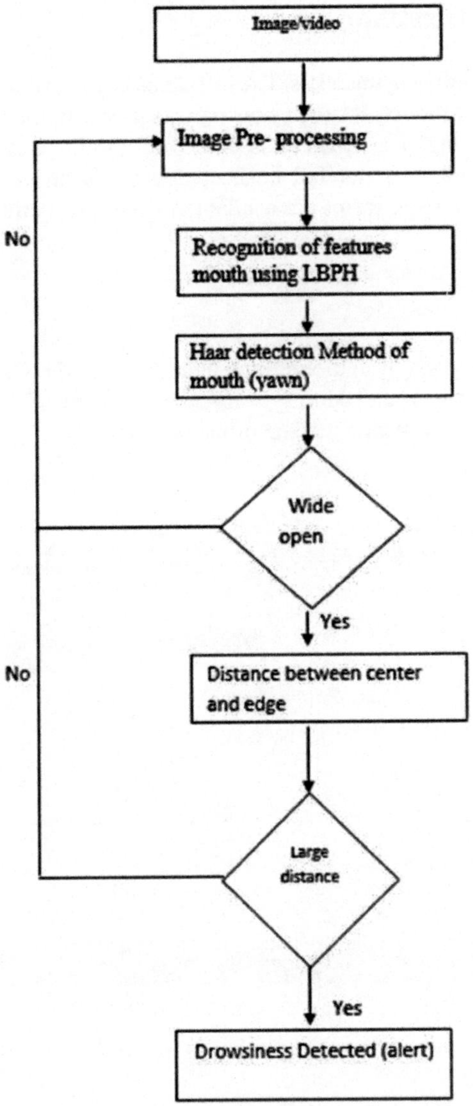

Fig. 4. Illustrates the data flow diagram for mouth detection.

4 Results and Discussion

Testing is aimed at identifying mistakes. The test is that any possible defect or flaw is tried throughout a working product. It shows how parts, sub-assemblies, assemblies and/or a final product are working. The application is the tool used to ensure that the Software framework satisfies and does not fail unacceptably to fulfil its needs and consumer preferences. Different testing forms are available. A basic test criterion is addressed by each test form.

4.1 Test Cases

The results in our proposed system are implemented using the OpenCV tool. Figure 5 shows the driver drowsiness detection system–menu login page. Table 1 shows various test cases considered for validation of the proposed method.

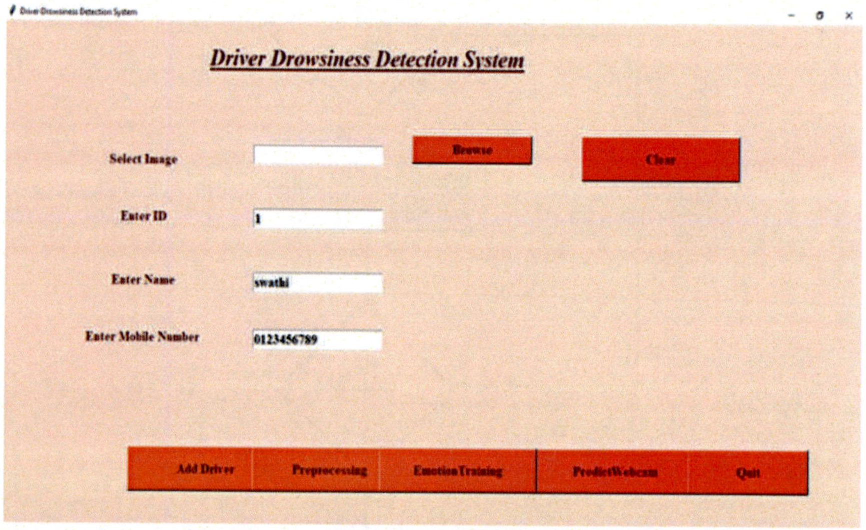

Fig. 5. Illustrates the menu page/add driver button (driver drowsiness detection system).

In Fig. 5, on the menu page we enter the driver details like ID, Name, Mobile number then the system starts recording. These credentials are used to identify the driver's face. When the 'Add Driver' button is clicked, the recorded video takes a one minute video of the users input face which stores as shown in below Fig. 6.

Images are stored in the form of frames which are converted into greyscale. On clicking the pre-processing button, we perform a geometric transformation (rotation) for the input image to enhance the image quality and once the features are detected using point-based, we display pre-processing is complete as shown in Fig. 7.

Table 1. Illustrates various test cases considered for validation of the proposed method.

Test case#	01
Test name	User input format
Test description	To test users' input as face
Input	Face
Expected output	The file would be read by the program and display on the monitor
Actual output	The file is read and display accordingly
Test result	Success
Test case#	02
Test name	User input format
Test description	To test users' input as face
Input	Not face
Expected output	Screen would not appear to wait for a face
Actual output	Idle state
Test result	Success
Test case#	03
Test name	Detection of face
Test description	To test whether detecting face or not?
Input	Face
Expected output	It Should detect and mark the face in a rectangle box
Actual output	detected and marked face in a rectangle box
Test result	Success

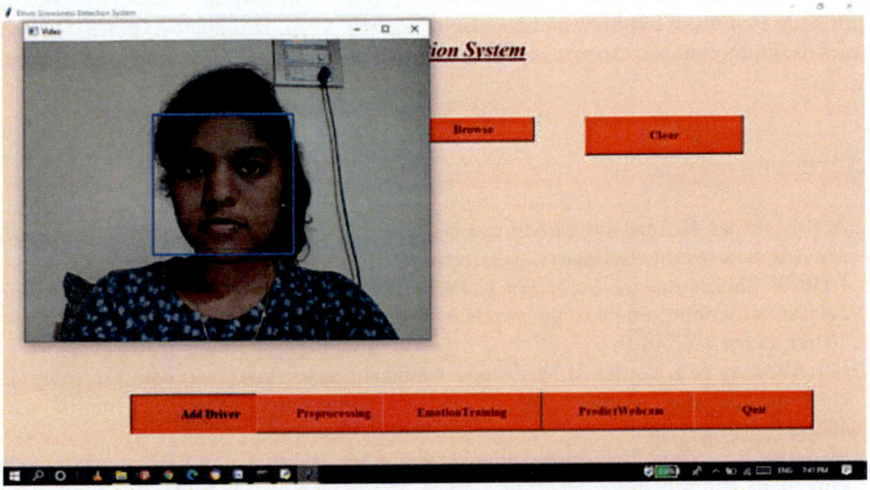

Fig. 6. Illustrates the menu page-add driver (next step).

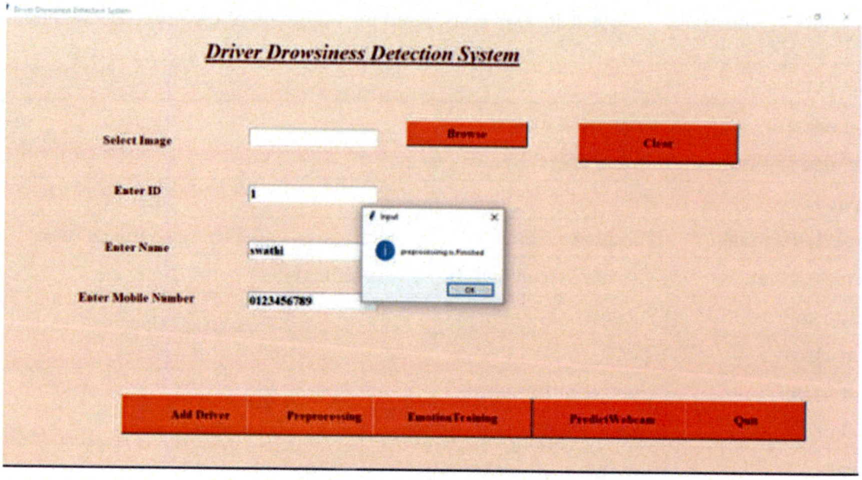

Fig. 7. Illustrates the menu page-processing button.

5 Conclusion and Implication

In conclusion, a mechanism to alert the driver drowsiness implementation of image processing is carried out to alert with the immediate buzzer in less than a second once the eyes blink and mouth yawn based on the parameters are detected to reduce road accidents. Afterimage acquisition, we apply geometric transformation—rotation for performing image pre-processing to improve the quality of an image. And we extract features like the eyes and mouth of the input image using the LBPH technique. Once the features like eyes and mouth are extracted, we use the Haar detection method to detect the eye (blink) and mouth (yawn). Based on the parameter measure like EAR results, we immediately get a warning alert that driver drowsiness is detected. We can further extend this work by identifying few more parameters like PSNR, Entropy, MI, and achieve better accuracy results to implement this driver drowsiness detection system using more real-time data sets.

References

1. A. Sahayadhas, K. Sundaraj, M. Murugappan, Detecting driver drowsiness based on sensors: a review. Sensors (Switzerland) (2012). https://doi.org/10.3390/s121216937
2. J. He, W. Choi, Y. Yang, J. Lu, X. Wu, K. Peng, Detection of driver drowsiness using wearable devices: a feasibility study of the proximity sensor. Appl. Ergon. (2017). https://doi.org/10.1016/j.apergo.2017.02.016
3. G.J. Al-Anizy, M.J. Nordin, M.M. Razooq, Automatic driver drowsiness detection using Haar algorithm and support vector machine techniques. Asian J. Appl. Sci. (2015). https://doi.org/10.3923/ajaps.2015.149.157
4. M. Sabet, R.A. Zoroofi, K. Sadeghniiat-Haghighi, M. Sabbaghian, A new system for driver drowsiness and distraction detection (2012). https://doi.org/10.1109/IranianCEE.2012.6292547

5. F.S.C. Clement, A. Vashistha, M.E. Rane, Driver fatigue detection system (2016). https://doi.org/10.1109/INFOP.2015.7489384
6. T. Vesselenyi, S. Moca, A. Rus, T. Mitran, B. Tătaru, Driver drowsiness detection using ANN image processing (2017). https://doi.org/10.1088/1757-899X/252/1/012097
7. R. Grace et al., A drowsy driver detection system for heavy vehicles (2002). https://doi.org/10.1109/dasc.1998.739878
8. L.F. Ibrahim et al., Using Haar classifiers to detect driver fatigue and provide alerts. Multimed. Tools Appl. (2014). https://doi.org/10.1007/s11042-012-1308-5
9. I. Lashkov, A. Kashevnik, N. Shilov, V. Parfenov, A. Shabaev, Driver dangerous state detection based on OpenCV & dlib libraries using mobile video processing (2019). https://doi.org/10.1109/CSE/EUC.2019.00024
10. L. Nahar, R. Palit, A. Kafil, Z. Sultana, S. Akter, Real time driver drowsiness monitoring system by eye tracking using mean shift algorithm (2019). https://doi.org/10.1109/ICAEE48663.2019.8975593

A Research Paper on Third Eye for Blind

Narayana Swamy Ramaiah(✉), Roshni Mishra, Anjali Sharma, and Timothy Iwoni

Department of Computer Science and Engineering, Jain (Deemed to be University), Bengaluru, Karnataka, India

r.narayanaswamy@jainuniversity.ac.in

1 Introduction

The proportion of persons with vision impairments is approximately 253, 36 million are blinded when 217 million suffering with mild to extreme vision disability, as seen by WHO survey and fact sheet revised on October 2017.Globally, the main cause of vision loss is chronic eye diseases while the top two causes of visual impairment are in-corrected refractive errors and un-operated cataracts. In this fast-moving world, visually impaired people are left behind and not treated equally. Several approaches and procedures were tested and implemented to support them and give them better some degree of comfort. Each of them is called flexibility and alignment. In this approach, a professional assists and guides the visually disabled and blind to lift themselves [1].

They are skilled to depend on their remaining senses to move autonomouslyand safely. Many techniques have been developed recently to enhance blind people's mobility. These techniques are developed based on signal processing and sensor technology. However, these tools tend to fail in a crowded are aorinan area where there are high volume so electronic waves. Current work provides a simplistic, customizable, and effective electronic guiding embedded vision system to help blind and visually disabled people's mobility within and outside. The machine contains three types of instruments, such as an IR, sonar and camera sensors. Micro-controller Raspberry Pi3 analyses the signal displayed from both instruments to identify an impediment [2].

The proposed guidance system can determine the obstacle distance, in addition to the material and shape characteristics of the obstacle. Furthermore, the system can name some of the detected objects and also it can read out loud eBooks for blind people who cannot read books. Moreover, neither cane nor another marked tool is needed to be carried by the user [3]. Like other systems, the proposed system can be fastened to a hat or a pen-sized hand mini stick. It has high immunity to both ambient light and the object's colour. This section of related work gives a general overview of Artificial Intelligence (AI), including its definition, history, advantages, disadvantages, complications, types and deployment methods. Artificial Intelligence is the science and engineering of making intelligent machines. In addition to computational intelligence, there have been other names that had been proposed for artificial intelligence as well, such as synthetic intelligence or computational rationality.

The Artificial Intelligence term also describes the property of machines and programs. We are surrounded by artificial intelligence and in some way or the other, we

F. P. García Márquez (ed.), *International Conference on Intelligent Emerging Methods of Artificial Intelligence & Cloud Computing*, Smart Innovation, Systems and Technologies 273, https://doi.org/10.1007/978-3-030-92905-3_48

happen to interact with it. These interactions with machines are limited by their ability to understand and process information. Here are four types of AI [4].

Type 1: Purely Reactive

This is the most basic type of AI in which the computer only processes one situation/environment and acts on it. It does not have any model, design or even the concept of the wider world. It specializes only in one area.

Type 2: Limited Memory

This AI helps the computers to take into account historical knowledge and adapt it to new solutions that are pre-programmed global representation. It only has sufficient knowledge to decide properly and to behave properly.

Type 3: Theory of Mind

This AI can understand thoughts and emotions. It can understand feelings, motives, intentions and expectations and has the capability to interact socially. It is yet to be built.

Type 4: Self-Aware

This kind of AI is self-conscious. It examines its inner state, forecasts other people's emotions and can produce abstract concepts and inferences. That is the machine's potential.

1.1 Complications in AI

When attempting to build a smart machine, there are several problems. An older AI or a basic AI is just a set of requirements where to act decisively on certain terms. This is not wisdom, and the development of real intelligence means an awareness of input and the relationship between this outputs and inputs. Fuzzy logic, which is designed after studying a human's excellent ability gives an approximate answer without any real values. However fuzzy logic has many complications. Conceptually, computer processes work in bits. Hence, processing words or concepts are considering a complicated task for the computer [5]. There are complications with image processing as well such as recognizing different photos locations from photos on the internet because they have a lot of variation. A computer cannot simply model the world from these photos because an average internet has immense variations. Image processing requires the data to be consistent. Rendering a 3D image is more complicated as the data has to be correct and more accurate with the least variations. The computer cannot simply detect an image. Finally, using all the set fields of AI to develop strong AI is very complicated. Designing and developing a system that has sentiments and conscience would require us to fully understand how the human brain works which we do not as of now [6].

1.2 Natural Language Processing

Human language is very complex and subjective at all times. The current form of communication with a machine is through input devices like a mouse, keyboard or verbal commands. This is totally different from how humans communicate with each other. This basic issue of accurately representing ideas with images or words is creating significant challenges for natural language processing. Implementations can manage natural language will have the capacity to articulate and decode languages at the level of human interpreters, understand the difference among blue water and blue sky and use computing orders such as giving me the purple stuff to regular exercise. Using current natural language processing techniques, the machine can recognize spoken words. Furthermore, it can even do a basic level of translation. However, compared to fully understand the complexities of human conversation, this is still primitive. Machines cannot summarize a novel as of now or process any data or conversation that requires moderate conceptual knowledge Devillers' research in the detection of emotions from human conversations gives much wider view and helps in understanding the meaning behind different pronunciation and words [7].

2 Literature Review

Varalakshmi et al. investigated a new navigating system for visually challenged persons by using the IoT. In everyday life, people who are visually impaired experience many challenges in life. Traveling from one venue to another is challenging for the visually impaired. Our primary objective is to build a smart stick for walking that uses an ultrasonic sensors as navigating tools to identify manholes and staircases in the surrounding. This paper reviews the efficiency of a number of approaches, strategies and algorithm (GCF, ML) used for the identification of manhole, pitfall and steps and its function reliability. IoT sensor (UV, Infrared, and Temperature) are utilized to identify and give alternative path through GSM for the visually impaired and also measure for their convenience the shortest distance to the journey's end. For image reorganization, different algorithms (three class and binary prediction) efficiency- and output measurements are tested and reliability is demonstrated [8].

Divija et al. explored integrated intelligent shoes for the blind person in their research. This study introduces a concept model and device architecture for smart electronic help supply people who are blind. This program is proposed to provide general object identification and assisting in actual environments using the GPS coordinates. The device consists of a microcontroller, an ultrasonic sensor, a smartphone, a vibratory circuit and a Bluetooth unit (GSM module).The goal of this project is to create an Electronic Travelling Aid (ETA) kit to help blind people find barrier-free routes. This ETA is permanently attached to the shoe. When the object is detected close to the shoe and if any person comes in front, it warns them with the aid of the vibratory circuit and also with the help of voice command speakers or headphones in advance. Here, the key criterion is a power supply, so the shoe is combined with a self-power generation unit to remove the need for a power backup [9].

Oladayo et al. investigated another technology that provides multidimensional walking assistance to visually challenged people. The application of engineering practices in medicine has made an important contribution to recent biomedical research findings. The creation of sophisticated aids for physically disabled persons is one of the results of this application. Visually impaired walking aids are developed and implemented in this paper using a network of ultrasonic sensor that can sense the directions and location of the obstacles. In comparison, an alert light and a voice guide signal related to a miniature headset increase reliability and flexibility. The voice recorded alerts the user of the location and position of the obstacle (s). With a voice alarm effective in the range from 0 to 1 m, the multidimensional walking support prototype identified obstructions on the left, the right and the front of your handle. The prototype test results have shown that the stick can effectively guide its user[10].

3 Methodology

3.1 Design

Figure 1 shows the flowchart for the Third Eye system. The flowchart shows the sequence of steps and decisions required to perform the Third Eye system process. The camera is considered the eye of this system. It continuously captures still images that are sent to the Raspberry Pi microcontroller. The raspberry pi microcontroller takes those images, processes them using artificial intelligence image processing algorithms and generates a matching response which is sent to the text to speech module. OpenCV framework is used to develop the AI model which gives Cafee model for the raspberry pi. Python program is used to get the input through a camera and handle the interaction between the model and the input. Then, it shows the output on the screen as text. Text to speech module takes the text as an input and converts it to an audio output which goes directly into the user's ears. Infrared sensors are placed in a way to match the angle of vision in a human. They are used to calculate the distance between the user and an object located at a far distance. Sonar sensors are placed at an angle of 30° from the vertical plane facing downwards. This inclination of 30 degrees helps sonar sensors to scan for the objects placed on the ground near the user. Hence, they can be used to calculate the distance from an object lying on the ground adjacent to the legs of the user. Both of these sensors send the data to the raspberry pi which in turn processes the data and sends the desired output to the text to speech module. This again takes text as an input and converts it to an audio response which reaches the user through a headset. This full process keeps going simultaneously and continuously until the user decides to switch off the device.

This project proposes the design of a portable AI based guidance system for blind Third Eye, which benefits the visually impaired community and also helps in their day to day mobility. The third eye provides the visually impaired community a new way to visualize the world by explaining to them about their surroundings. The whole system is controlled by a Raspberry Pi microcontroller. The third eye harnesses the maximum capabilities of the Raspberry Pi microcontroller which has enough potential to up hold the system with one advantage being the in built graphic card.

- The prototype uses various sensors such as IR Sensors, Sonar Sensors and a Camera module which helps the system to gather the required data. Additionally, the text to speech module is used to talk to the user.
- Python forms the heart of the system. It is used to program the whole system which helps the raspberry pi microcontroller to communicate with all the sensors. It then

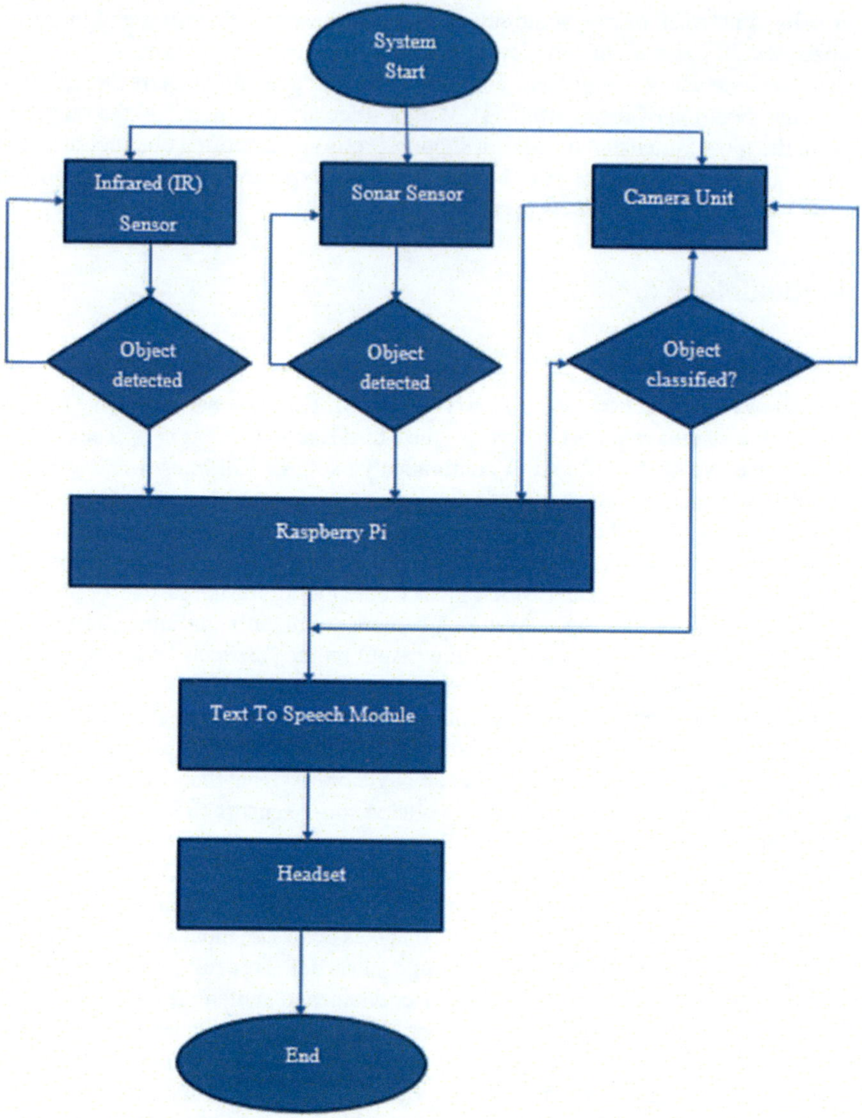

Fig. 1. Illustrates the flowchart of the proposed object detection system.

processes the collected data and converts it into information which is finally delivered to the end-user.

- The IR sensor is used to map the object's shape and size while the sonar sensors get the data about the distance of the object at regular intervals.
- The camera module plays an important role as it takes the pictures which are then processed using image processing technique to properly visualize the object.
- All the information is processed and converted to a text which is then fed into a text to speech module. The text to speech module delivers this information to the end-user in his/her ear using headphones.

3.2 Instruments

Hardware Used

- Raspberry Pi
- SD card
- GPS
- Ultrasonic Sensor
- Wireless transmitter
- Wireless receiver
- Button switches
- Wi-Fi
- Speaker or ear pod
- Camera.

Software Used

- Raspberry Jessie OS
- Python
- Open CV
- Yolo.

3.3 Objectives

- To learn machine learning and object detection.
- To learn raspberry pi and sensors.
- To develop a much faster response system for the blind using ultrasonic.
- To detect an obstacle and identify them using image processing.
- To integrate read eBooks features along with the improved blind stick.

4 Results and Discussion

The Third Eye AI based Navigation System for Blind is developed and integrated as per the design documented in this section. The systems are applied utilizing raspberry pi, sonar sensors, infrared sensors and a camera as the hardware base, python is utilized to program the hardware and OpenCV is utilized to setup and run the object detection model with python. The system has 3 separated components, a sonar sensor, infrared sensor and camera which work simultaneously. Python controls the sonar and infrared sensors. For object detection, OpenCV provides a framework for python to control the camera and perform object detection.

4.1 Implementation

The system is composed of three different components namely: sonar sensor, infrared sensor and camera. All the data from all three components are sent to the raspberry pi which processes it and gives the output to the user. The structure, implementation, and execution are similar for sonar and infrared sensors. The camera works on a different structure and algorithm. The features of the system include:

- To identify the distances of the objects (sonar and infrared sensors).
- To identify the directions of the objects based on the sensor that detects the object.
- To detect the type of the object (using object detection algorithm).

4.2 Distance of the Object (Sonar Sensor)

To get the distance of the object from different directions (left or right), two sonar sensors are placed in each direction. Each sonar sensor gets the distance using sound-waves after the wave hits an object and bounces back. It is capable of calculating distance every second. However, in this work, the system is configured to calculate the distance every three seconds, due to the simultaneous running of other components. The output of the sonar sensor can be seen in real-time on the screen with the actual distance value from the object. Figure 2 shows the sonar sensor output for the proposed method.

The output screen shows the distance of the user from the object every three seconds. At the same time, this information is being processed by the raspberry pi and being transmitted to the user through headphones. This guides the user about the position of the object. The above output screen shows the distance of the user from the object every three seconds. At the same time, this information is being processed by the raspberry pi and being transmitted to the user through headphones. This guides the user about the position of the object.

This is one of the most important parts of the system. To detect what is the object, the system needs an object detection algorithm to run on. This system uses OpenCV for the model and the backend and python to connect to the model and to run everything in the front end. The algorithm works on = caffemodel which is locally stored on the raspberry pi. When the program is running, python connects with the model and starts taking the video. Then, the objects captured in the video are compared with the model and the closest match is shown on the screen and sent to the headset (Figs. 3 and 4).

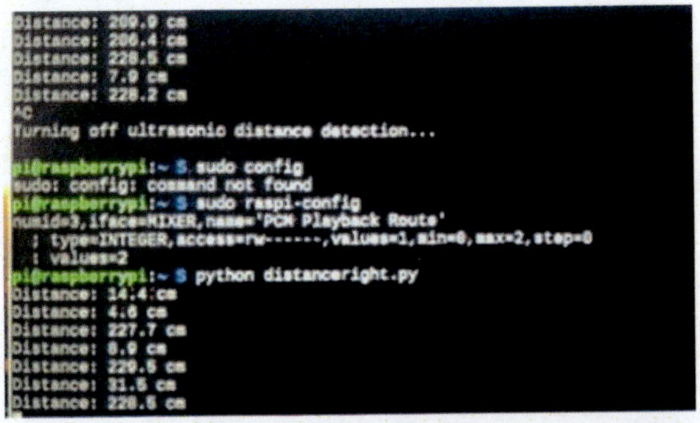

Fig. 2. Illustrates the sonar sensor output for the proposed method.

However, Raspberry Pi 3 does not have a high power GPU and CPU, thus the frame rate is not high. Currently, the system can detect up to nine types of objects which are sofa, table, chair, bottle, TV-monitor, aero-plane, train, person and dog. The system is compatible with the raspberry pi camera module as well as USB cameras.

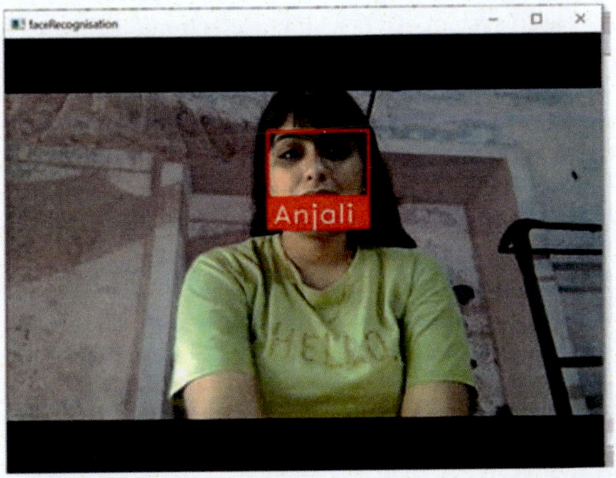

Fig. 3. Illustrates the image for the face recognition. The voice will come that the person is Anjali by mp3.

5 Conclusion and Implication

In the existing system, there is only technology to detect the obstacle and to notify the blind by having some warning sound. We have electronic components which make the stick detect an object but not recognize it. In case of emergency or theft, blind people will have no clue or description about the culprit. Visual abilities are still not given. In today 's world, disability of any kind for any person can be hard and it is the same case with

Fig. 4. Illustrates the image for object detection. The voice will come that the person is detected by mp3.

blindness. Blind people are generally left underprivileged. It is very difficult to give a vision to a blind person. In this paper, a new AI-based system called Navigation System for Blind Third Eye to control the navigation of a blind person has been proposed and developed. This AI-based system offers a simple electronic guidance embedded vision system which is configurable and efficient. The system helps blind and visually impaired people to be highly self-dependent by assisting their mobility regardless of where they are; outdoor or indoor. Results show that all the sensors work properly and give accurate readings, though the range of the prototype sensors is not high. Object detection algorithm utilizes 100% CPU which makes the raspberry pi hot and thus, in the future system, two raspberry pi are recommendedtobeused;oneforobjectdetectionandoneforallthesensors.

References

1. D.E. Gbenga, A.I. Shani, A.L. Adekunle, Smart walking stick for visually impaired people using ultrasonic sensors and Arduino. Int. J. Eng. Technol. (2017). https://doi.org/10.21817/ijet/2017/v9i5/170905302
2. Exertion pain received et al., Effects of single-task versus dual-task training on balance performance in older adults: a double-blind, randomized controlled trial. Arch. Phys. Med. Rehabil. (2009)
3. D.E. Gbenga, A.I. Shani, A.L. Adekunle, Smart assistance for visually impared person using Arduino, in *Smart Stick* (2017)
4. S. MacNamara, G. Lacey, Smart walker for the frail visually impaired (2000). https://doi.org/10.1109/robot.2000.844786
5. M. Bousbia-Salah, A. Redjati, M. Fezari, M. Bettayeb, An ultrasonic navigation system for blind people (2007). https://doi.org/10.1109/ICSPC.2007.4728491
6. N. Molton, S. Se, J. M. Brady, D. Lee, P. Probert, A stereo vision-based aid for the visually impaired. Image Vis. Comput. (1998). https://doi.org/10.1016/s0262-8856(97)00087-5
7. J. José, M. Farrajota, J.M. Rodrigues, J.M. Hans du Buf, The smart vision local navigation aid for blind and visually impaired persons. Int. J. Digit. Content Technol. Appl. (2011). https://doi.org/10.4156/jdcta.vol5.issue5.40

8. I. Varalakshmi, S. Kumarakrishnan, Navigation system for the visually challenged using internet of things (2019). https://doi.org/10.1109/ICSCAN.2019.8878758
9. M.V.S. Divija, M. Rohitha, Raveendra, Integrated smart shoe for blind people. Int. J. Manag. (2018)
10. O.O. Oladayo, A multidimensional walking aid for visually impaired using ultrasonic sensors network with voice guidance. Int. J. Intell. Syst. Appl. (2014). https://doi.org/10.5815/ijisa.2014.08.06

Robot Movement Based on Color Detection

C. R. Manjunath(✉), T. Prathyusha Reddy, Darla Gayathri, and Manaswini Yarka Reddy

Department of Computer Science and Engineering, Jain (Deemed to be University), Bengaluru, Karnataka, India
cr.manjunath@jainuniversity.ac.in

1 Introduction

Engineers in computer science construct programs and algorithm to monitor the behaviour of robots using external input. This place is rapidly expanding, & new robot designs to obey a variety of purposes are always emerging, with your imagination serving as the only limit. Earlier the robot will controlled by the wired networks, but at present time to make a robot more user-friendly, it is now made wireless [1]. The wireless robot is capable to operate for any application based on the requirement. It also can receive real-time data like video from a camera mounted on the robot. As the internet of things is Morden introduced in all fields. The concept is about handling objects with the used of the internet & the best models for these types of applications using raspberry Pi [2].

Monitoring is important when it come to protecting everything because it is a time-consuming process. Because it might be unsafe to constantly monitor anything, we are going to create a robot that can do so. This robot keeps an eye on things and sends a live stream of it to a designated human. As a result, tracking the work would be a little simpler, and because of technology, it will be more precise. When a live video is captured, the position of the required object is sensed or found by specifying the pixel range of the required object. And according to the pixel range, it is difficult to detect, using this technique very easy to find the position of the object. A different colour object contains different pixel values [3].

The aim of this project is to solve the problem of replacing humans with monitoring robots, thus reducing human resource damage. Robots are constructed in a miniature structure to allow them to mines, penetrate tunnels and small holes in buildings, as well as to continue in harsh and difficult climatic situations for long periods without damages. Military robots have been in development for many decades. Most of the system uses a robot with a camera for monitoring. Now the robot which has the camera mounted can move in any direction. These robots can prove to be more flexible than the standard cameras that are fixed to a single point. The surveillance robots are mostly built as a wheeled robots to give the advantage over the traditional cameras. Though these are robots can be more effective on a flat platform. The videos captured by a robot can be seen remotely over the device through wireless communication and the internet [3].

A robot, like a person, detects variations in the environment with sensors, contemplates about what procedure must be achieved with the logic circuits & electronic

F. P. García Márquez (ed.), *International Conference on Intelligent Emerging Methods of Artificial Intelligence & Cloud Computing*, Smart Innovation, Systems and Technologies 273, https://doi.org/10.1007/978-3-030-92905-3_49

components, and eventually performs necessary acts by controlling its working systems. For example, mechanical engineers deals with the designs construction and utilize of appropriate material for the structures of a robot [4]. The development of algorithms & writing programs to monitoring the actions of a robot depend on the external inputs is performed by computer science engineers. Robots are also utilized the automate industrial procedures like manufacturing assembly & quality checks. Now-a-days robots have become autonomous and are increasingly being utilized for the logistics purposes and shipping goods & also locating goods within a warehouse. These kinds of robots need a detection system to avoid obstacles in their way [5]. Hence this fields is fastly growing & new designs of a robot to serve different practical main are always coming up. There are different kinds of robots. They are utilized according to the situation and application required. These kinds of a robot are classified into three categories. They are:

- Manual robot which is controlled by humans for example cranes.
- Semi-autonomous robotics can be operated by humans or operate on their own.
- Autonomous Robotics can performing a task autonomously without any human's intervention.

Detectors are the modules that sense& detect variations in environment, then translate the data into analogue or digital data that the robot's internal circuitry can process. Autonomous or semi-autonomous robots include sensor-guided robots. The robot's sensors allow it to see, touch, hear, and shift in response to environmental feedbacks [6]. The sensors detects change in the environment and sends data to the robot's (CPU)central processing unit, which then controls the robot's actuators or perform a particular task using any parts of the robot [7]. These robots are utilized in several fields like industrial, medical & defence. Microcontrollers are intelligent electronic devices that are utilized inside of a robot [8]. They perform functions that are similar to those performed by a computer's microprocessor or CPU. Microcontrollers are programmed using a variety of programming languages. In this project, the microcontroller used receives an input signal from the camera connected to it. It then processes the signals received and sends the signals to the motors through the motor driver attached. Robots are driven by a variety of motors that allow them to move in a variety of programmable vibrations. The motor converts electrical energy to mechanical energy [8].

2 Literature Review

So et al. developed a LAN-dependent surveillance robot for the maintenance of the buildings. We plan to build a surveillance robot that allows for both active and passive surveillance, using an autonomous patrol mode and a manual mode that notifies the user when motion is detected. The robot will record video and stream it to an Android device over Wi-Fi. There are many disadvantages with surveillance robots currently on the market: they are quite costly, with prices ranging into the thousands of dollars; they require a large base station in order to communicate with the robot; and they have unnecessary features, such as being able to be thrown through a window without being damaged. Our robot will have several advantages over those already on the market. Is much cheaper and more convenient to use than many other surveillance robotscurrently on the market. Would be ideal for use in a secluded indoor area that no one should be able to access [9].

Hollinger et al. researched another robot that works based on the emotion-based technique. We define a robot that interact with human beings in the crowded conference setting in this paper. The robot recognizes the faces, decides the colors of onlookers' shirts, and responds with a mix of voice, musical, & also movement feedbacks. It updates the internal emotional states continuously, based on human psychology studies. The robot's emotional state is converted into a basic sets of sound & movement feedback using empirically defined mapping functions. At the AAAI '05 Open Interactions Case, we successfully demonstrated this system, demonstrating the potential for the emotional modeling for enhance human–robot interactions [10].

3 Methodology

3.1 Design

The block diagram depicts the basic operation of a Wi-Fi-enabled Raspberry Pi, with each block describing the operation. A USB webcam, a board with a motor-driven IC, and voltage regulator circuitry make up the robotic assembly. Within the Wi-Fi range, the user unit consists of a PC. DC motors are controlled using the keys in the keyboard. The sensed information from the keyboard is transmitted by the Wi-Fi modules which is related to the Raspberry pi. The images taken by the camera or the motor may be used as the output. It also displays the environment's live video streaming.

Object Tracking shows that the robot arm can be controlled by an object, which includes moving, placing, grasping, rotating & stopping the robot. At same time, the robot arm's movement data is being recorded. The main control module and the monitored module make up the teaching monitoring system. The main control module's functions primarily include gesture data collection, data optimization processing, object recognition, and data storage, with PC as the hardware. The main control module send the activation signals to the monitoring modules, can use any object to control the movements of the robot arms. In Fig. 1 shown the diagram of movements of the robot as per object placement. In Fig. 2 shown the diagram of proposed methodology.

The proposed method there are certain steps in order to detecting moving objects in videos surveillance.

Step 1: Read Video.

In this phase, it read videos from the camera that is mounted on the 3 wheeled robots.

Step 2: Foreground Extraction & Background Modelling (Detection).

it is divided into other part:,

(a) Background Removing using Gaussian Mixture model.
(b) For Motion Detection apply the appropriate threshold value for a complex image to a binary image.
(c) Detect real moving objects.

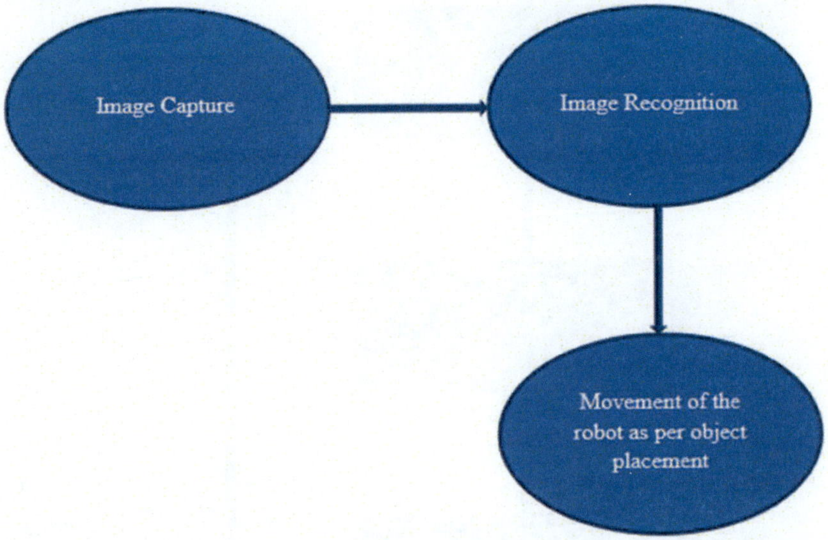

Fig. 1. Illustrates the diagram of the movement of the robot as per object placement.

3.2 Instrument

3.2.1 Software Requirements

- Operating system: Windows 10.
- Coding Languages: Python3, Linux (terminalcommands).
- Tools: Python IDLE, Raspbian OS.

3.2.2 Hardware Requirements

- Raspberry Pi B+ model
- 60 RPM Motors
- L239D Motor Driver
- Webcam
- Other miscellaneous items.

3.2.3 Hardware of Raspberry Pi

- 10/100 Base T Ethernet
- HDMI Socket
- USB 2.o
- RCA Video Socket SD Card Socket
- RCA Video Socket
- Powered from Micro USB Socket.

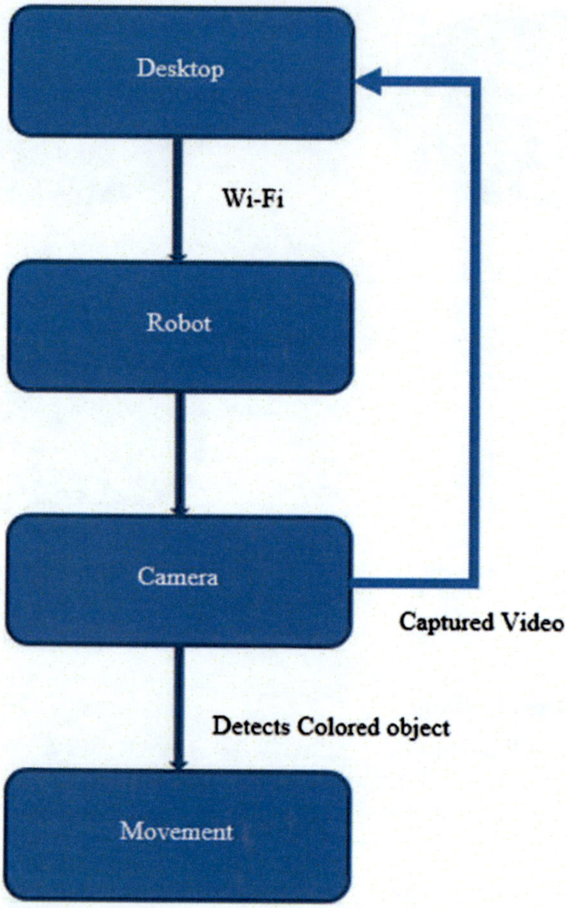

Fig. 2. Illustrates the flow diagram of proposed methodology.

3.2.4 Technology in Raspberry

- Powered by Broadcom BCM2835 System on chip
- Core frequency set to 700 Hz
- Video core 1 V GPU with HDMI and composite output
- Ability to boot from SD card
- GPIO pins can be used to run various electronic gadgets and appliances.

3.3 Concept of Surveillance

The surveillance concept is based on monitoring human behaviour, activities and other changes in the surroundings. The main purpose of surveillance is to gather intelligence and situation of the environment in which the camera is fixed, to prevent crimes and protect people and organizations. The mode of surveillance robot is through direct observation using electronic equipment and direct interception of information.

3.4 Obstacle Detection

If a robot is set to move autonomously, in real-time, then it will surely come across obstacles in its way. Obstacles can be anything that blocks the way of the robot from reaching its destination. Some of the common obstacles in the remote area are people, books, wooden tables and so on. As many obstacles obstruct the robot from moving further, so there are many sensors to detect those obstacles. In this project, the robot is designed to recognize the color of the object placed in front of the camera mounted on it. If the mobile robot recognizes the specific color, for example, red, then the robot moves away from that particular object without getting jammed. In this way, the path of the robot can also be drawn by arranging the colored objects in a particular route.

4 Results and Discussion

In this project, the microcontroller is the raspberry pi which acts as the brain of the system. Raspberry pi managed single-board computers with the capability of interfacing multiple peripherals. It is equipped with Ethernet, Wi-Fi and Bluetooth. It also has general-purpose input and output pins that can receive or transmit digital signals to other electronic components. It acts as the controller for the entire system. The Raspberry pi is the low-cost and card-sized device that can be plugged into a computers or a TV & can be operated by standard input devices such as a keyboard and mouse. It is a capable computing device to explore and program in python or scratch programming languages. It is capable of doing everything that is expected from the computer would do. Raspberry Pi has the ability to interact with the outside world through wired or wireless. The Raspberry Pi is used to monitor a robot from a distance using any computer. The time it takes to process the device's commands and react appropriately is negligible. The computer and laptop improve the system's stability and usability. Two gears motors are necessary to move the robot, & motor driver modules is utilized to provide enough present to drive 2 gear motors while shielding the Raspberry-pi modules. The key benefit of using a limited number of motors is that the robot's power consumption is minimized.

Types of models:

Model A has 1 USB, no Ethernet, SD card slot.
Model B has 2 USB ports, 10/100 Ethernet, SD card slot.
Model B+ has 4 USB ports, 10/100 Ethernet, Micro SD card slot.

Hardware of Raspberry pi:

- 10/100 Base T Ethernet
- HDMI Socket
- USB 2.0.

Automation of the Human Visual System-We humans, have a very robust visual system, which helps us to identify people and objects, play sports perform operations, drive vehicles, read, and so on. Although it might seem that we do not put any special effort to do most of these tasks, the human visual system is fairly complex to replicate and implement. Computer Vision, in the simplest terms, is the automation of such a visual system, so that computers or machines, in general, can obtain a high-level understanding of the environment from digital Images and videos.

Table 1 shows the computer vision application.

Table 1. Illustrates the applications of computer vision.

Area of application	Purpose
Industry automation	Label readings on different packages, sorting of objects and parts on circuits, assembly lines Inspection of circuit boards that are printed
Food markets	Identification of spoiled vegetables and fruits that are marketed in bulk
Medical treatments	Inspection in x-rays, counting of blood cells, identification of any tumors inside the body
Robotic engineering	Object recognition and avoidance of collision, monitoring of machines, and determination of hazards
Radar concepts	Providing guidance to helicopters, aircraft and also missiles in remote areas
Analyzing documents	Recognition of graphics, handwritten documents can be analyzed and characters are identified

4.1 Software Implementation

Raspberry pi is being used to analyze and control operations in the robot. It acts as the brain of the robot. There are various ways of connecting raspberry pi to the pc or laptop. The methods are both wireless and wired. In the case of a wired connection, the constraints in the movement of the robot increase. The wired connection will restrict the many attributes like distance, speed and efficiency of the robot which leads to less utilization in real-time. The wireless robots overcome these restrictions and provide better efficiency compared to wired robots. To connect the raspberry pi module to any device via Wi-Fi, there are certain protocols to be followed. Figure 3 shows the use case diagram of the proposed method.

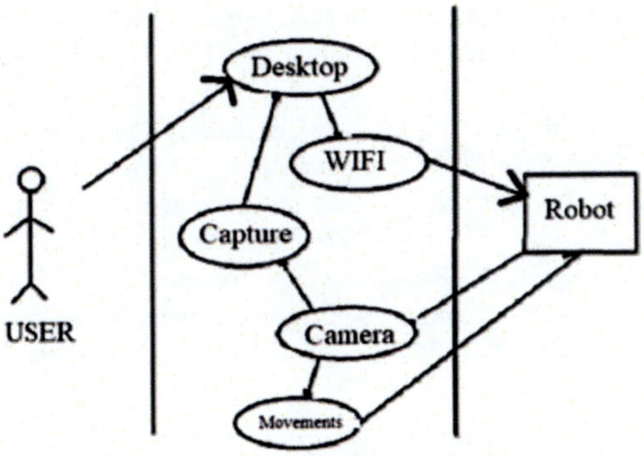

Fig. 3. Illustrates the use case diagram of the proposed method.

5 Conclusion and Implication

The object detection-controlled robot system gives an alternate way of controlling robots. It makes the user control a smart environment by object detection interface. It eliminates the use of external hardware supports like a remote control in the navigation of the robot in a different direction and reduces human efforts in risky environments. This robot is a stepping stone of such robots. This intelligent vehicle can both capture a video and transmit a live stream to the operators and also can move automatically. While in its motion, it can detect any obstacles in its path. One of the important qualities of this robot is that it can also detect the color of the object and mould its direction according to the command given in the program. This robot can be often referred to be an intelligent vehicle as it can analyze and act according to the situations.

References

1. T. Yoshimi et al., Development of a person following robot with vision-based target detection (2006). https://doi.org/10.1109/IROS.2006.282029
2. S. Bhattacharyya, A. Konar, D.N. Tibarewala, Motor imagery, P300 and error-related EEG-based robot arm movement control for rehabilitation purpose. Med. Biol. Eng. Comput. (2014). https://doi.org/10.1007/s11517-014-1204-4
3. A. Vourvopoulos, F. Liarokapis, Brain-controlled NXT robot: tele-operating a robot through brain electrical activity (2011). https://doi.org/10.1109/VS-GAMES.2011.27

4. M.B. Nugraha, P.R. Ardianto, D. Darlis, Design and implementation of RFID line-follower robot system with color detection capability using fuzzy logic (2015). https://doi.org/10.1109/ICCEREC.2015.7337058
5. D. Kim, D. Lee, H. Myung, H.T. Choi, Object detection and tracking for autonomous underwater robots using weighted template matching (2012). https://doi.org/10.1109/OCEANS-Yeosu.2012.6263501
6. S. Strauss, D. Heinke, A robotics-based approach to modeling of choice reaching experiments on visual attention. Front. Psychol. (2012). https://doi.org/10.3389/fpsyg.2012.00105
7. C.A. Cifuentes, A. Frizera, R. Carelli, T. Bastos, Human-robot interaction based on wearable IMU sensor and laser range finder. Rob. Auton. Syst. (2014). https://doi.org/10.1016/j.robot.2014.06.001
8. H.M. Do, M. Pham, W. Sheng, D. Yang, M. Liu, RiSH: a robot-integrated smart home for elderly care. Rob. Auton. Syst. (2018). https://doi.org/10.1016/j.robot.2017.12.008
9. A.T.P. So, W.L. Chan, LAN-based building maintenance and surveillance robot. Autom. Constr. (2002). https://doi.org/10.1016/S0926-5805(02)00003-1
10. G.A. Hollinger, Y. Georgiev, A. Manfredi, B.A. Maxwell, Z.A. Pezzementi, B. Mitchell, Design of a social mobile robot using emotion-based decision mechanisms (2006). https://doi.org/10.1109/IROS.2006.282327

Classification of Cervical Squamous Cells

T. P. Deepa(✉), Rajani Prajapati, Shubham Dubey, Manpreet Singh, and Dhiraj Bothra

Department of Computer Science Engineering, Jain (Deemed to be University), Bengaluru, Karnataka, India
tp.deepa@jainuniversity.ac.in

1 Introduction

Processing of digital images or Digital image processing is a phenomenon that uses algorithms of Computers to carry out the image segmentation. Image segmentation is a crucial and difficult image processing phase. The image segmentation technique divides an image into useful sections with identical characteristics and properties [1]. The chief goals of segmentation is to simplify an image by expressing it in a coherent and easy-to-understand way. The first step in image processing is image segmentation. Image segmentation is the process of separating an image into many segment/parts with similar features or attributes. The common applications of the image segmentations are Content depend image retrieval, Object detection, Medical imaging etc. Image segmentations can be divided into 2 basic stages: Local segmentation (concerned with a particular region or part of the images) & Global segmentation (concerned with the segmenting the complete images, consisting of the large amount of pixels) [2]. The Image segmentation can be approached from different perspectives such as Threshold approach, Region approach, Edge approach, etc. [3].

The histogram of the input image is typically used in threshold-based image segmentation approaches to detect single or multiple thresholds [4]. The aim of edge-based image segmentation is to discover the edges in an input image. As a result, segmentation is done by determining the input image's area boundaries [5]. Region-based picture segmentation techniques start by looking for some seed points in the input image, then use proper region-growing techniques to get to the object boundaries. Some medical image applications include image segmentation as well [6]. Highly qualified physicians spend hours determining certain regions of medical images to signify salient regions of medical image analysis. With the right image segmentation approach, this process can be completed in seconds [7].

2 Literature Review

Jie et al. researched the topic a two-level cascades classification structure was utilized to detect cervical cancerous cells automatically. By observing the contours of the cells, automated and semi-automatic approaches are utilized to distinguish irregular cell from the slides. The automated study methods of cervical cell pictures, which have been extensively studied and improved and is utilized to sense cervical cancers, was recently

F. P. García Márquez (ed.), *International Conference on Intelligent Emerging Methods of Artificial Intelligence & Cloud Computing*, Smart Innovation, Systems and Technologies 273, https://doi.org/10.1007/978-3-030-92905-3_50

developed. The cells are smeared on slides in this process, and images are collected using high-resolution cameras. After that, the photos are examined for irregular cells. This approach has the advantage of conserving vast amounts of human and material resources while significantly increasing screening effectiveness, reducing human errors, and improving screening accuracy. In this approach, the acquisition of cell characteristics, the designs of the cells organization systems, and the organization of cell all carry out important role. These three crucial aspects were explored in this report.

They suggested the subsequent innovative idea for automatic examination of cervical cellular image:

(1) Withdrawalof Features: gaining28-dimensional characteristics with contrasts as well as nuclear region, 8 texture, and 20 morphological characteristics, on the basis of distinction in cellular densitiesor morphology.
(2) Arrangement of Cells: Initially cells are separated in to neutrophils, lymphoid, junk and epithelial cell and then categorizing the epithelial cells into abnormal and normal cell.
(3) Method of classification: A two-level cascade is used to classify cervical cells. The cells are first categorized into abnormal and normal cells by a swift and rough classifiers, and then by a more precise LR classifier [8].

Ewert et al. researched the topic of screening for cervical cancers. Cervical cancers are among the most dangerous and widespread cancer in females, action is necessary, but it can be prevented with a basic monitoring tests so-called a PAP-smear. It is one of the most important cancer inhibition strategy that has been devised to date. However, visual inspection of the smear is time-overwhelming as well as costly, and attempts to automate the analysis have been made several times meanwhile the test was implemented before 6 decades ago. Around the turn of the millennium, the 1st profitable system for automatic cell sample examination emerged, but they had only a minor effect on screening costs. In this article, we look at the main issues that must be addressed when developing an automated research framework, as well as how these issues have been addressed in the past. The knowledge gained could be beneficial in exertions to develop a cost-efficient monitoring systems which would create reasonable cervical cancer monitoring accessible to all women worldwide, avoiding the majority of the one third a million premature demises triggered by this illness each year [9].

Habibzadeh et al. researched microscopic images in their research. Cytopathology is a branch of pathology that studies diseases at the cellular level. In today's world, a fatal illness like cancer has become a big challenge in medical science. If these diseases are identified in the early stages, then they are curable. There are many methods to ease, faster and produce accurate results of this analysis. One such approach is using image processing techniques. This paper applies segmentation techniques to find an important feature like cell count. This feature is further used to differentiate between normal and abnormal cells. Abnormal cells are the one which may turn into cancerous cells. Hence the early detection of such abnormal cells before they turn into cancerous cells help to diagnose patient early. The proposed method was evaluated using cervical cells [10].

3 Methodology

3.1 Design

Determining the pre-cancerous cells in the woman's cervix using watershed and fuzzy segmentation algorithms based on cell features thus overcoming the drawbacks of the traditional approach of doing it manually. The main reason for the death due to cervical cancer in women patients is that it cannot be detected at an earlier stage and the patients cannot receive any symptoms until they reach the final stage of cancer. The death ratio of the women patients can be reduced if it is detected at an earlier stage only. There are two types of Pap testing for these facilities i.e., Liquid-based Pap test and Conventional Pap test. The main problem with the above-mentioned systems is that they can be time-consuming and the result accuracy may be affected because it is performed manually.

A proposed solution to the aforementioned matter is a software-based classification of squamous cells, which classifies the cells as abnormal or normal depend on the features of the cells and also reduces the cost and time consumed by the existing method. The primary advantages of the suggested method are to identify the cervical cell in a given image, to obtain features of cervical squamous cells, to classify the cells into abnormal and normal. The main objectives are to identify the cervical cell in a given image, to obtain features of cervical squamous cells to categorize the cells into abnormal or normal. Figure 1 shows the diagram of the proposed methodology. Image segmentation is the significant aspects of image processing. So the main object of segmentation is to extract the ROI (Region of Interest) for the image analysis. Segmentation plays an important role as dividing a large images into different segments makes further processes simpler. It depends on other features like the color or texture of an images. The main object is to decrease the information and collect useful information for making easy analysis possible. There are much variety of the segmentation methods that has suggested till now such as Threshold-based segmentations, Region-based, Edge-based, Model-based, clustering techniques, etc.

3.2 Instrument

3.2.1 Software Requirements

- Operating system: Windows 7.
- Coding Language: Python3.6, MATLAB.
- Tools: Jupyter Notebook, MATLAB.
- Input: Microscopic cell images.

3.2.2 Hardware Requirements

- HardDisk: 120 GB.
- InputDevices: Mouse, Keyboard.
- Monitor:15″ LED.
- Ram: 8 GB.
- System: i5.

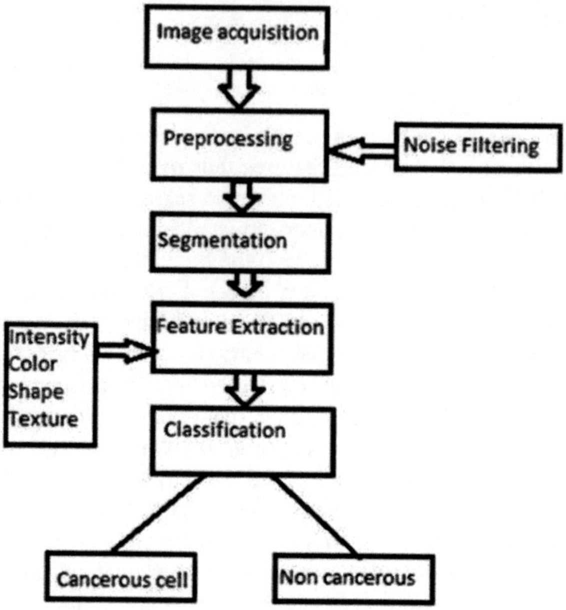

Fig. 1. Illustrates the diagram of the proposed methodology.

MATLAB has become one of the most useful software over the last decade due to many reasons. Certain useful tools within MATLAB offers pragmatic features to solve various problems. Various researchers prefer MATLAB in comparison to other software due to its simple user interface and less computational complexity. MATLAB is used for several applications such as deep learning, machine learning, neural networks, image processing and many more.

3.3 Implementation

3.3.1 Removal of Debris or Poison Noise from Image

Since the image taken as input is not ready to be processed. So to make it usable for further processing the cleaning of the image is done. To remove the debris (Unwanted cells) from the input image watershed algorithm have been used. Once the watershed has been applied to the image the output obtained is the cleared image i.e. the unwanted cells or the background objects of the cells which are of no use are removed from the image. The image obtained as output is now ready to be further processed.

3.3.2 Segmentation

The image obtained after applying the watershed algorithm is being processed for segmentation. Image segmentations is most important tasks in the image processing. The image is divided into different parts i.e. segments, and each of the segments is processed differently by the algorithm. Segmentation of an image reduces the time consumed to process an image, as processing one big image takes more time and is hard to execute. Therefore, segments of an image are easy to process and understand.

3.3.3 Feature Extraction

The feature are extracted from the cells of the images given as input. The data of the cells from each image is being collected in an excel sheet. The features are areas, Major Axis Length, Eccentricity, Minor Axis Length, Orientation, Circularity, Perimeter (mm), Class, etc. the excel sheet is given as an input to the classifier to train the software.

4 Results and Discussion

The whole process of removing Poisson noise is explained as:

- The colored image is given as input to the system and is converted into the greyscale image.
- Gradient Magnitude will be used as a segmentation function.
- The foreground object will be marked.
- Background Markers will be computed.
- Watershed transform of segmented function will be computed.
- Results visualized.

In above Fig. 3, the flowchart proposed system and its work is shown. First, the training dataset contains the input images of the raw data in the form of a microscopic image. The steps involved in the testing processare pre-processing of the image i.e. removing the Poisson noise and debris from the image. After pre-processing of an image, the segmentation is done i.e., dividing the image into different parts. Further step is to retrieve the features of cells in the image based on their characteristics like perimeter, area, size, texture, color, etc. The last step is to categorize the cells as normal or abnormal depend on the features extracted.

Figure 2 represents the (a) Original image (b) Grey scale image (c) Marked foreground object (d) Background object removed partially (e) De-noised image. Figure 3 shows the flowchart suggested system and its working is shown. Figure 4 shows the input image histogram. Figure 5 shows segments of the inputs image. Figure 6 shown the output of the trained classifier.

5 Conclusion and Implication

Cells are seen clearly as all the debris and noise are removed. The image is segmented and each segment is processed further for feature extraction. The datasheet is prepared based on the data collected of cells from input images. The software is trained by giving the data sheet as input. Currently, for this process we are getting accurate output, the cells can be identified. We have also compared the execution timing of serial and parallel code. We have worked on thin prep images. In the future, it can be done for conventional images also. The primary advantages of the suggested method are to identify the cervical

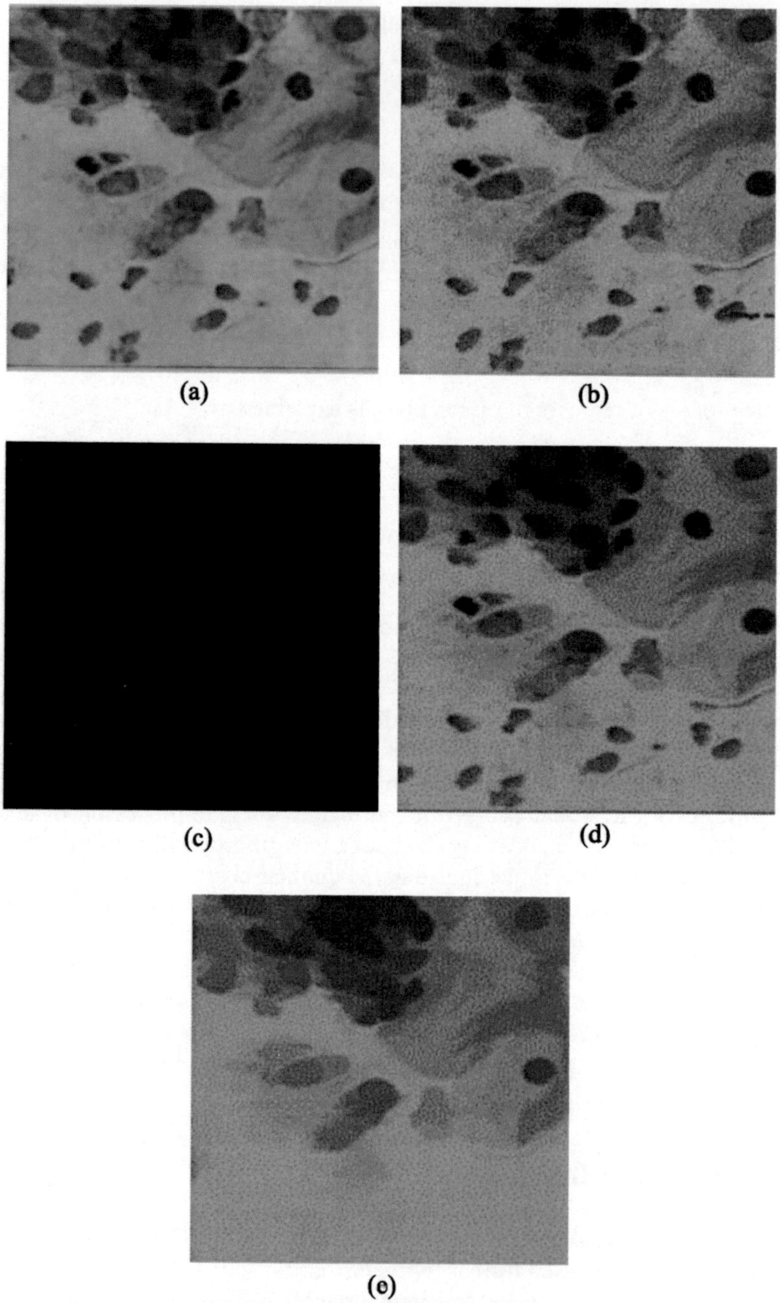

Fig. 2. Illustrates the images. **a** Original image. **b** Greyscale image. **c** Marked foreground object. **d** Background object removed partially. **e** De-noised image.

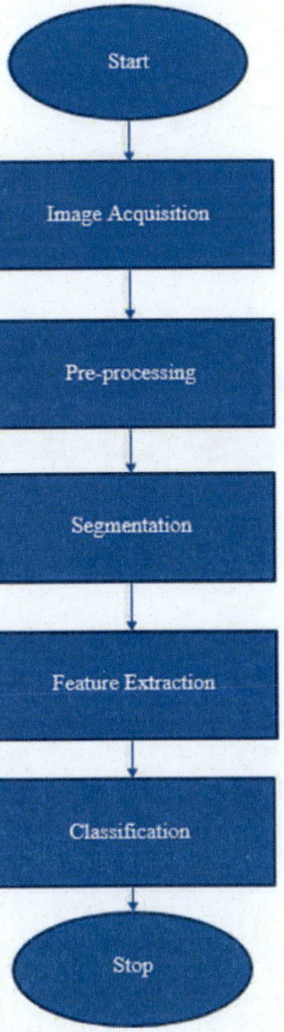

Fig. 3. Illustrating the flowchart of the projected system and its working is shown.

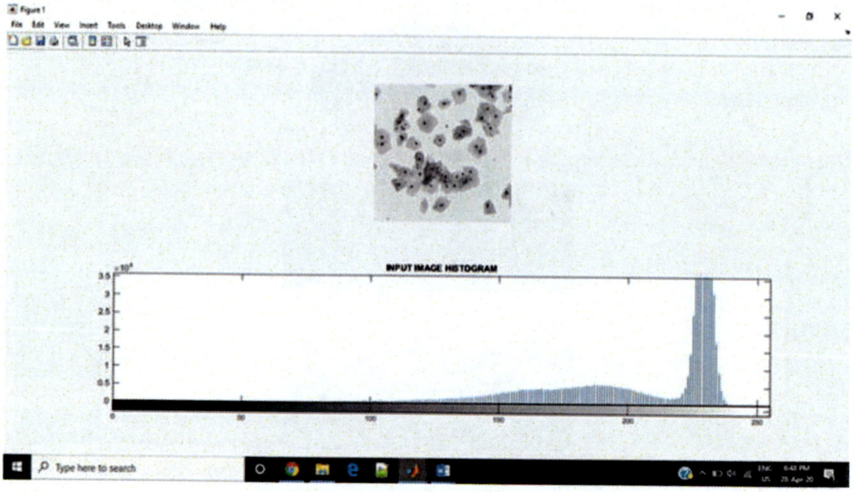

Fig. 4. Illustrating the input image histogram.

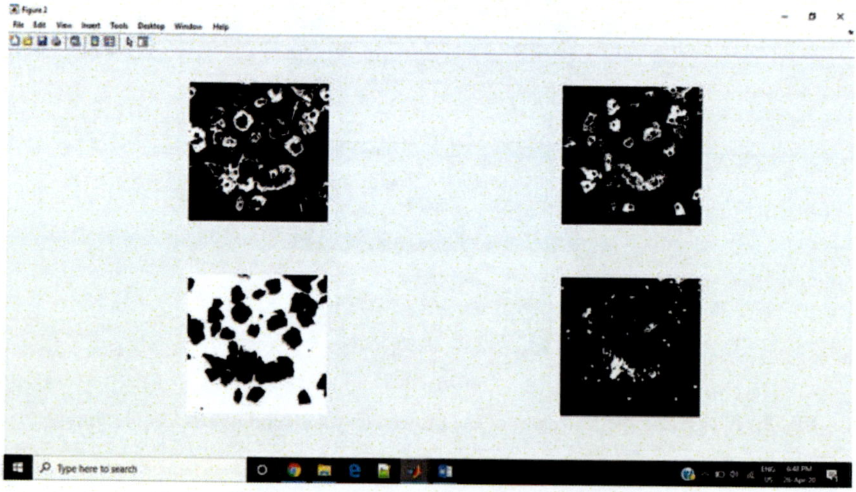

Fig. 5. Illustrates segments of the input image.

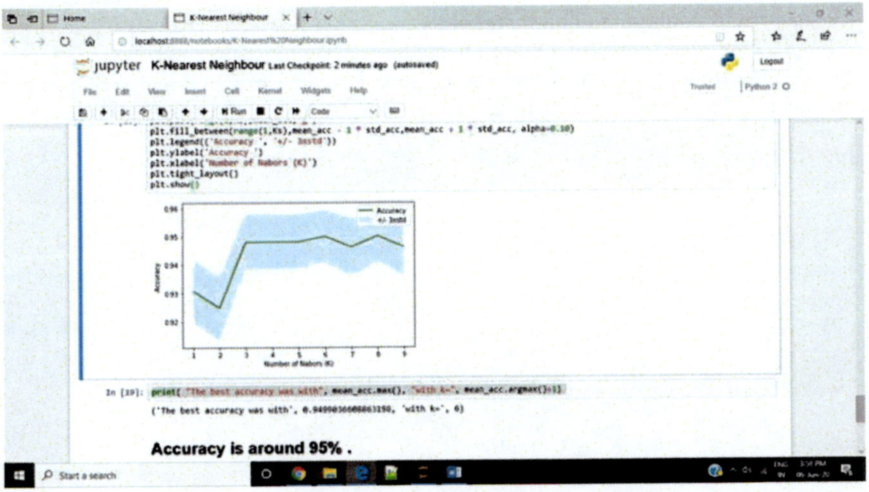

Fig. 6. Illustrates the output of the trained classifier.

cell in a given image, to obtain features of cervical squamous cells, to classify cells into abnormal and normal. The main objectives are to identify the cervical cell in a given image, to obtain features of cervical squamous cells to classify the cells into abnormal and normal.

References

1. A. Vyas, S. Yu, J. Paik, Fundamentals of digital image processing. in *Signals and Communication Technology* (2018)
2. L. Zhao, J. Yin, L. Yuan, Q. Liu, K. Li, M. Qiu, An efficient abnormal cervical cell detection system based on multi-instance extreme learning machine (2017). https://doi.org/10.1117/12.2281648
3. J. Martínez-Más et al., Classifying Papanicolaou cervical smears through a cell merger approach by deep learning technique. Expert Syst. Appl. (2020). https://doi.org/10.1016/j.eswa.2020.113707
4. L.H. Oliver, R.S. Poulsen, G.T. Toussaint, C. Louis, Classification of atypical cells in the automatic cytoscreening for cervical cancer. Pattern Recognit (1979). https://doi.org/10.1016/0031-3203(79)90007-4
5. Y. Xiang, W. Sun, C. Pan, M. Yan, Z. Yin, Y. Liang, A novel automation-assisted cervical cancer reading method based on convolutional neural network. Biocybern. Biomed. Eng. (2020). https://doi.org/10.1016/j.bbe.2020.01.016
6. E. Supriyanto, N.A.M. Pista, L.H. Ismail, B. Rosidi, T.L. Mengko, Automatic detection system of cervical cancer cells using color intensity classification (2011)
7. P. Sukumar, R.K. Gnanamurthy, Computer aided detection of cervical cancer using pap smear images based on adaptive neuro fuzzy inference system classifier. J. Med. Imaging Heal. Inform. (2016). https://doi.org/10.1166/jmihi.2016.1690
8. J. Su, X. Xu, Y. He, J. Song, Automatic detection of cervical cancer cells by a two-level cascade classification system. Anal. Cell. Pathol. (2016). https://doi.org/10.1155/2016/9535027

9. E. Bengtsson, P. Malm, Screening for cervical cancer using automated analysis of PAP-smears. Comput. Math. Methods Med. (2014). https://doi.org/10.1155/2014/842037
10. M. Habibzadeh, A. Krzyzak, T. Fevens, A. Sadr, Counting of RBCs and WBCs in noisy normal blood smear microscopic images (2011). https://doi.org/10.1117/12.878748

Automation of Animal Classification Using Deep Learning

G. Komarasamy(✉), Mapakshi Manish, Vadde Dheemanth, Devraj Dhar, and Manash Bhattacharjee

Department of Computing Science and Engineering, VIT Bhopal University, Bhopal, Madhya Pradhesh, India

1 Introduction

Classifying and recognizing animals is a key field not quickly explored. The challenge of separating photographs of various types of animals is an easy activity for humans. However, research shows that it is hard to differentiate them automatedly, even in minor systems including dogs and cats [1]. Animals have a fluid structure which can camouflage itself and are generally seen in a dynamic scene. They can also occur under various lighting conditions, perspectives and sizes, as with all things. Recognition techniques for animal pictures are attempted but there has lately been little concern for the special issue of animal categorization.

The complexity of animal groups which have complicated intra-class heterogeneity and inter-class similarities cannot adequately reflect numerous standard techniques which display impressive outcomes for human identification [2]. Many methods exist to resolve this issue, each of which has its benefits and drawbacks. The method suggested builds dynamic characteristics that better reflect and discrimination sample images, but it is difficult and problem-related. The detail is used for judgement rather than utilizing dynamic representation [3].

In this paper, we demonstrate that the approach suggested can be generalised to many other issues of recognition and classification. We assume that if data from two distinct forms of classification systems is combined, the success rate may be improved. We are building classification systems for this purpose on the basis of CNN functionality [4]. The interaction pattern among neurons represents an individual visual cortex, which is influenced by organic processes [5] by revolving networks. The signals were only answered by specific cortical neurons in a limited peripheral vision area known as the receptive field. The receptive fields of various neurons are partly overlapped to occupy the whole field of vision. Comparison to many other image recognition algorithms, CNNs use very little pre-processing. In other words, the network learns the filters in standard, handmade algorithms [6]. This freedom from previous experience and human initiative in designing functions is an obvious benefit in this respect. Its uses include image / video identification, device recommendations and natural language processing [7].

F. P. García Márquez (ed.), *International Conference on Intelligent Emerging Methods of Artificial Intelligence & Cloud Computing*, Smart Innovation, Systems and Technologies 273, https://doi.org/10.1007/978-3-030-92905-3_51

2 Literature Review

W. Tang et al. carried a research on the topic titled, "Enabling trusted and privacy-preserving healthcare services in social media health networks," [8] describes Social networking health channels as a promising paradigm for enticing patients in the digital healthcare to post and interact their overall health with the other consumers and social networking sites. Throughout this study, we recommend a customized and trustworthy strategy to the health sector in social media health systems that enable verified and private health systems to increase the trustworthiness of providers and patients by accurate reviews to providers and to ensure the privacy and confidentiality. In order to identify people with common medical problems privately, we specifically use our shared sorting framework to look for effective customized caring, flower filters to remove and chart personal medical problems and internal item. In addition, we create a system for the identification of attacks by Sybil that allows us to find false ratings and feedback of patients utilizing various pseudonyms to ensure credible scores for the care provider. Safety research reveals that our strategy will safeguard patient confidentiality and deter Sybil attacks.

Yick et al. researched the topic titled, "Wireless sensor network survey," [9]. This work addresses some safety problems through Wi-Fi networks (WSNs). A study is provided of recent developments in general defence, traditional safety risks, intrusion prevention system, main delivery systems and target localization. In order to promote applications requiring packet transmission through one or more transmitters to many receivers, safety is identified as a crucial and demanding objective in communication process. The challenge posed are critical for potential WSN delivery.

Sahi et al. researched the topic titled, "Privacy preservation in e-healthcare environments: State of the art and future directions," [10] describes The very next great wave of health is e-Healthcare. It gives both the patient and the customer all of the benefits and rewards that are possible. Present e-Health platforms are not yet completely integrated and mature, though, and there is no need to enforce broadly the levels of safety, honesty, confidentiality and users confidence. The consistency of healthcare performance and customer confidence in the healthcare company are two main features of any operating healthcare company. Trust is linked with topics such as privacy, honesty, transparency, authenticity, identification and information processing. Confidentiality considered one of the major barriers to the growth of eHealthcare approaches in the gain of confidence, as it addresses most safety issues implicitly. To fix privacy problems, compliance issues such as access management, authorization, disallowance, and transparency must be addressed, without the end-to-end confidentiality.

3 Methodology

3.1 Design

See Fig. 1.

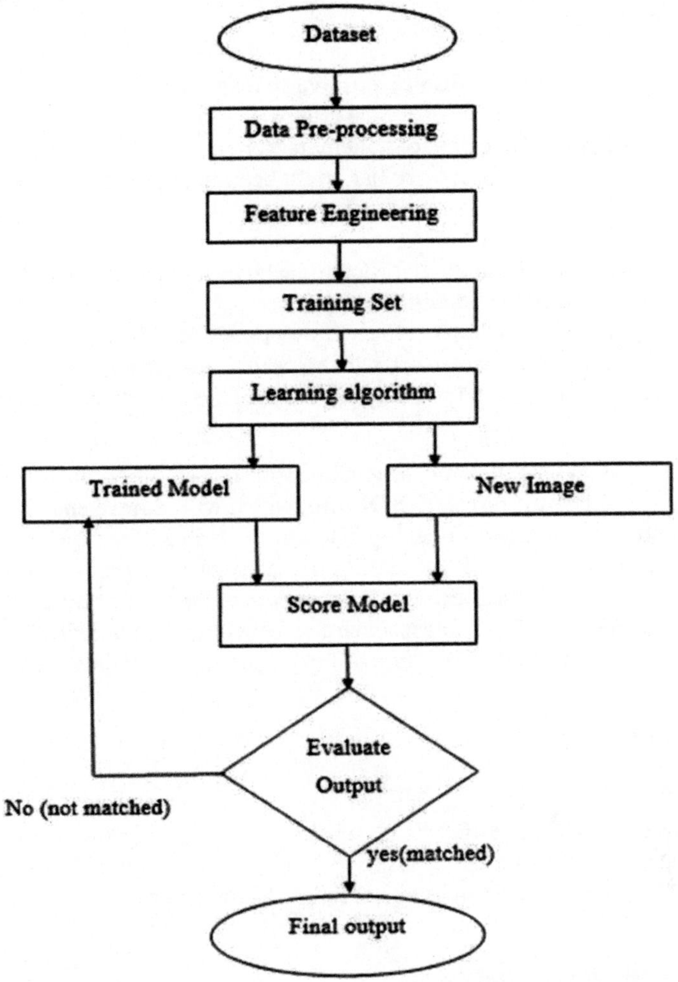

Fig. 1. Illustrates the system flow diagram of the process of the proposed methodology

4 Drawbacks

Many current approaches that display positive human identification outcomes do not adequately reflect the complexity of animal groups with complex intraclass variation and interclass resemblance. While previous attempts were made to implement classification methods in animal photographs, there has lately been little attention in the basic issue of animal classification.

- The animals with complex intra-class heterogeneity and inter-class similarities cannot adequately reflect variation of animal groups.
- Moderate contrasted with human efficacy for animal identification.
- In a multi-authority ABS framework with no centralized government, but the scheme requires several activities and significantly reduces the algorithm's performance.

For the animal facial classification, a novel approach focused on one of the common CNN features is suggested. We use CNN here, which will retrieve and understand and identify features automatically. In comparison to other image recognition algorithms, CNNs use comparatively less pre-processing. The network then discovers the filter which are handcrafted in standard algorithms. This freedom of previous experience and human initiative in the creation of features is a massive benefit in the classification of images. Figure 1 shows the system flow diagram of the process of the proposed methodology.

4.1 Instrument

4.1.1 Software Requirements

Operating system	Windows 10
Coding Language	Python 3.6
Tools	Anaconda, Jupyter notebook, Keras library

4.1.2 Hardware Requirements

System	i3
Hard Disk	120 GB
Ram	4 GB
Input devices	Keyboard, mouse

4.2 Implementation

When the dataset is used, primarily, data preprocessing takes place where the data is processed and cleaned as per the requirement. The fusion of both CNN and KFA is done to extract features accurately. A training set consisting of 30 images from eachclass is taken and CNN and KFA are used on it, which provides us with a trainingmodel. The new data and the training data are scored to match the image. The evaluated output is then obtained. As shown in Fig. 1, the following figure demonstrates the architecture

Fig. 2. Illustrates the overall procedure

of our model. The Scoring fusion approach makes the model more efficient with its accuracy compared to using a single algorithm. Figure 2 shows the overall procedure.

Level 0: Describes the overall process of this project. We are passing the animal dataset as an input and the system will classify the image of the particular animal using the CNN-KFA model.

Level 1: Describes the first stage process of this project. We are passing the animal dataset as an input the system will perform the pre-process and extract the important features.

Level 2: Describes the final stage process of this project. We are passing extracted features from level 1and trained data asan input the system will classify the given animal is matched or not using the CNN-KFA model.

4.3 Data Set

We are using the "LHI Animal Faces Dataset" as our base dataset in this model. This dataset consists of 20 classes. 19 classes of animal faces and 1 class of human faces. Each class consists of 20 images of a particular animal or human face in it and all images are collected based on different angles of the faces, different lighting conditions and different qualities and sizes of the faces. Using a lot of facial images in a particular class will help extract features more efficiently.

4.4 Data Pre-Processing

In the data pre-processing phase, cleaning of data takes place. This process is also called "Data cleaning". Here the obtained data or facial images are transformed as per our requirements. Some of them include converting the images from RGB to greyscale and vice versa, resizing the image as per our requirement and fabricating the data completely as per our requirement and our model's need.

4.5 Feature Engineering

This is an important phase in our model called "Feature engineering", in this phase, extraction of features takes place. Extracting features is an important concept of classification models in deep learning. We first use Convolutional Neural Networks (CNN) which automatically learns, extracts and classifies the features of the images. 90% of the work will be completed by using CNN. Then, Kernel Fisher Analysis (KFA) algorithm is added to it which is primarily appearance-based learning, which helps to classify the image more accurately.

4.6 Training Set

The next phase is taking the Training set, where it is extracted from the requirement. This training set is used to train the data as per our requirement from the new image. The trained set will be used as a trained model and a learning algorithm is applied to it to continue with the further process.

4.7 Learning Algorithm

The Learning algorithm used here is CNN. CNN plays a very crucial role here to match the trained model and new image and derive the required output. CNN automatically learns, extracts and classifies the image in our model. Using CNN in our model helps classification with more accuracy.

4.8 Trained Model and New Image

A trained model is the model derived from the Training set and which is ready to be applied to the input. A new image is aninput thathas to be applied. The learning algorithm is applied to both the parts here to continue the process of classification and will be scored using the Score level fusion method.

4.9 Score Model

In this phase, both the trained model and new image are combined or fusioned to classify the input. This process is called Score level fusion. This process is used to match the input.

4.10 Evaluate Output

In this phase, the output matched and scored in the previous phase using Score level fusion is evaluated. This process is used to check whether the output is deriving the right output or not. If it's theright output, it will be displayed as the right match and provides us with the animal. Else, it moves back to the learning algorithm and thus the process continues. This process acts as a decision-making statement.

4.11 Final Output

This is the final phase of our model. The output matched and proceeded from the evaluate model gives us the result of an accurate output. As we are using the CNN and KFA to extract features. We believe our output to be more accurate compared to the older models.

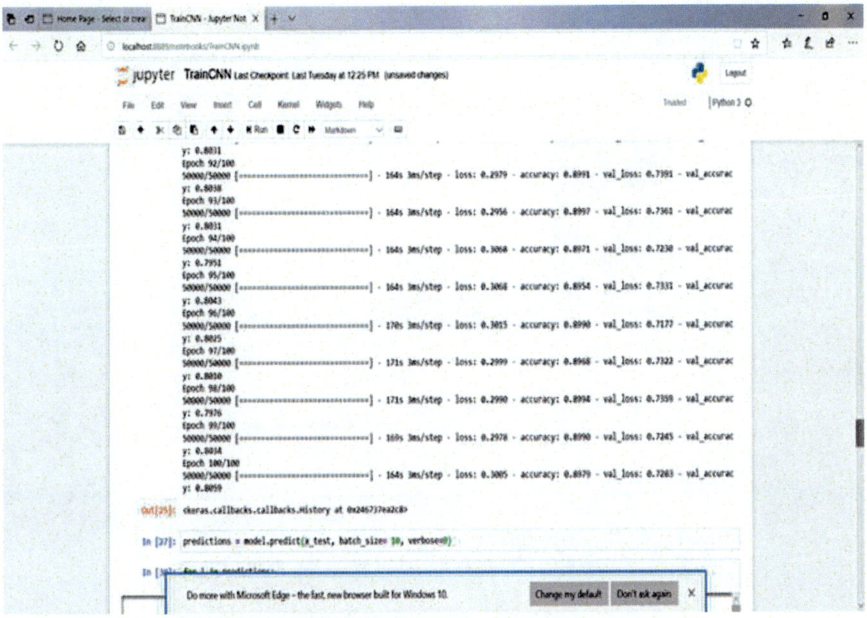

Fig. 3. Illustrates execution of code and accuracy

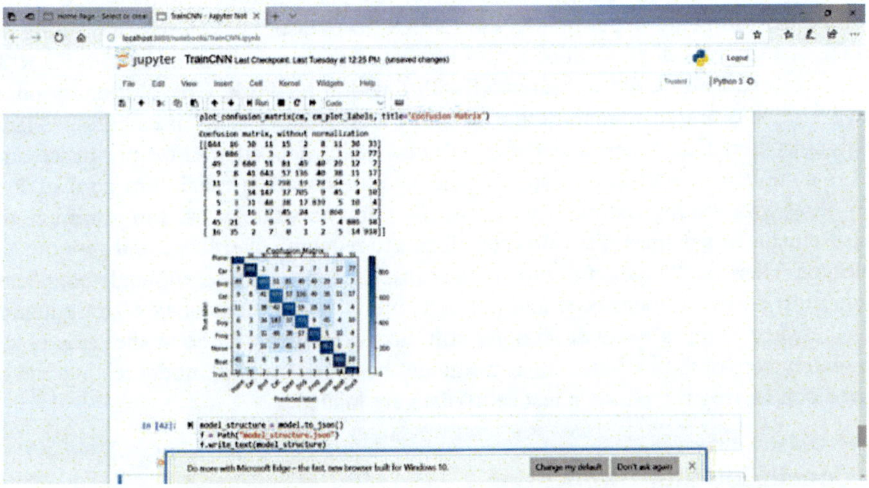

Fig. 4. Illustrates constructing, testing, evaluating of code and matrix

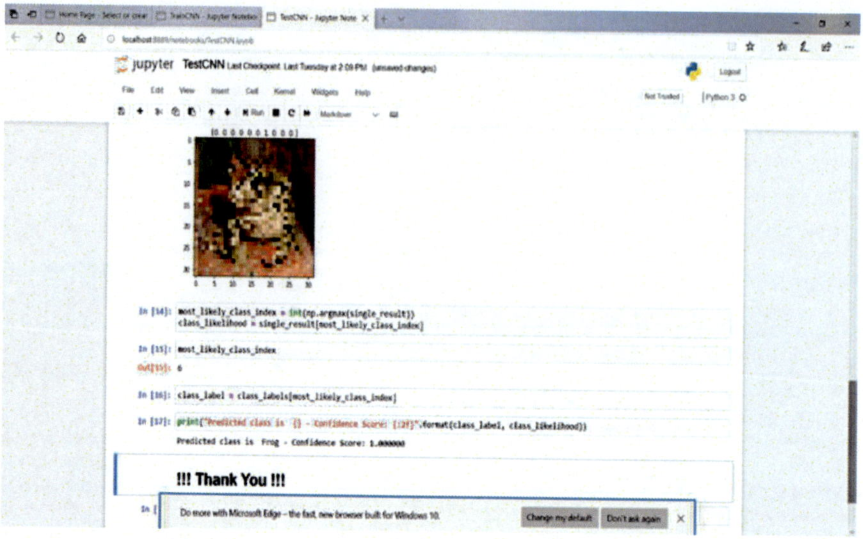

Fig. 5. Illustrates test phase output with the given image

5 Results and Discussion

Figure 3 shows the execution of code and accuracy. Figure 4 shows constructing, testing, evaluating code and matrix. Figure 5 shows the test phase output with the given image. In all of the following experiments, the CIFAR-10 dataset istaken into a training set and the test set istaken as a collection of images of the random animals with 32 × 32 size resolution. The model is generated with a neural network. Coevolutionary nerve network (CNN) are an unique neural network architecture. CNN uses certain visual cortex characteristics. A neural network will now be introduced after the pre-processing step is complete. It will have 3 layers of convolution with a maximum pool of 2 × 2. Pooling with the highest pixel value for the grid is a strategy used to minimize the measurements of a picture. This also contributes to reducing overfitting and generics the prototype. Then add 2 layers fully linked.Because the input of the completely connected layers must be two-dimensional and the outcome is four-dimensional, a straightening layer between them is needed. Another softmax layer seems to be at the ends of the completely convolution layers. Once the model has been trained, model testing can be carried out. During the phase, a test set of data are loaded.

6 Conclusion and Implication

The machine is a profound and advanced technology that is currently being studied. In particular in the image description, CNN is more appropriate in image analysis. We summarize with the experimental outcome, which is more than 90% accurate from the method built. The experiment results indicates that the method produced is relatively more reliable than certain previous procedures. Image recognition and prediction are a wide concept thatisvery deep. Deep learning concepts can be used in the subject to get efficient results. Concepts like Convolutional neural networks (CNN) can be used, which are effective in obtaining results for image identification and classification.

References

1. T.D. Pereira et al., Fast animal pose estimation using deep neural networks. Nat. Meth. (2019). https://doi.org/10.1038/s41592-018-0234-5
2. C. Tan, F. Sun, T. Kong, W. Zhang, C. Yang, C. Liu, A survey on deep transfer learning (2018). https://doi.org/10.1007/978-3-030-01424-7_27
3. R. Chen, R. Little, L. Mihaylova, R. Delahay, R. Cox, Wildlife surveillance using deep learning methods. Ecol. Evol. (2019). https://doi.org/10.1002/ece3.5410
4. H. Lee, L. Yan, P. Pham, A.Y. Ng, Unsupervised feature learning for audio classification using convolutional deep belief networks (2009)
5. H.T. Rauf, M.I.U. Lali, S. Zahoor, S.Z.H. Shah, A.U. Rehman, S.A.C. Bukhari, Visual features based automated identification of fish species using deep convolutional neural networks. Comput. Electron. Agric. (2019). https://doi.org/10.1016/j.compag.2019.105075
6. M.F. Hansen et al., Towards on-farm pig face recognition using convolutional neural networks. Comput. Ind. (2018). https://doi.org/10.1016/j.compind.2018.02.016
7. P. Popat, P. Sheth, S. Jain, Animal/object identification using deep learning on raspberry pi (2019). https://doi.org/10.1007/978-981-13-1742-2_31
8. W. Tang, J. Ren, Y. Zhang, Enabling trusted and privacy-preserving healthcare services in social media health networks. IEEE Trans. Multimed. (2019). https://doi.org/10.1109/TMM.2018.2889934
9. J. Yick, B. Mukherjee, D. Ghosal, Wireless sensor network survey. Comput. Netw. (2008). https://doi.org/10.1016/j.comnet.2008.04.002
10. M.A. Sahi et al., Privacy preservation in e-healthcare environments: State of the art and future directions. IEEE Access (2017). https://doi.org/10.1109/ACCESS.2017.2767561

Real-Time Eye Blinking for Password Authentication

T. R. Mahesh(✉), M. Sai Ram, N. Satya Sai Ram, Allu Gowtham, and T. V. Narayana Swamy

Department of Computer Science and Engineering, Jain (Deemed to be University), Bengaluru, Karnataka, India
t.mahesh@jainuniversity.ac.in

1 Introduction

People are constantly confronted with authentication mechanisms and must validate themselves using traditional knowledge-based methods such as passwords, As a result, one of the security criteria for general terminal authentication systems is that they must be easy, fast, and safe [1]. However, these methods are insecure because they can be observed malicious investigators who collect data encryption data using surveillance methods such as shoulder-surfing (observing a consumer while typing a code on the keyboard).There are also security issues as a result of weak communications between systems and users [2]. As a results, we suggested a 3-layered protection scheme for PIN numbers, in which users enter the password by the blinking their eyes at the required symbol in the correct order, making them immune to bear surfing [3]. Blinking is the normal form of interaction, and security system depend on blink monitoring provide a promising solution for device usability and security. The purpose of this project is to examine solutions or methods for coping with the eye- blink in a security system [4].

Personal identification numbers (PINs) are a popular user's authentication mechanism for a variety of application, including managing money in ATMs, authorizing electronic transactions, opening doors and unlocking digital devices [5]. PINs are a popular user authentication technique for a variety of application, such as handling money in ATM [6]. According to the European ATM Security, the number of ATM fraud attacks expand by 26% in 2016 as juxtapose to 2015. PIN entry is endangered to password attacks, like shoulder surfing and the thermal tracking, since an approved costumer must enter the codes in private or public locations [7].

2 Literature Review

Weaver et al. research on the topic of Gaze-based password authentication. Researchers have suggested schemes in which user entered passwords using an eye tracker by simply looking at the correct symbols a computer display in correct orders. This verification system is resistant to shoulder surfing, which involves secretly monitoring a legitimate participant's keystrokes while he or she types a code on a keyboard. We explain the Eye

F. P. García Márquez (ed.), *International Conference on Intelligent Emerging Methods of Artificial Intelligence & Cloud Computing*, Smart Innovation, Systems and Technologies 273, https://doi.org/10.1007/978-3-030-92905-3_52

Dent scheme in this document, in which user entered their password through looking at patterns on the on-screen of keyboard. When looking at each mark, existing eye-tracking depend authentication system enable the consumer to dwell or click a button. Instead, in Eye Dent, gaze targets automatically clustered to decide the consumer's selected symbols; this method has the advantage of enabling users to verification at their own pace other than a set dwell period. Furthermore, the lack of a clear trigger hides the amount of the symbols in password. Preliminary investigation show that using this scheme, fast authentication (3 s for the 4 digit PIN) are possible, but further work is required to justification for dynamically and calibration error adjust device attributes for individual users characteristics [8].

Mowery et al. researched the topic of the heat of the moment. In this paper, author look at how a thermal camera can used to retrieve codes typed into keypad in a number of situations. This attacks has the advantages over a traditional camera in that the codes do not have to be recorded when being typed and can instead be retrieved for a short time afterward. We consider several variables to get a wide sense of how successful such an attack could be: the keypad content, the user entering the code, the distance between the camera and the keypad, and the potential methods used to analyze the data. First, they present code proposed solution from our test data set's human review; this gives us a benchmark for the combat efficiency of thermal sensor attacks. Second, we automatically identify the software from raw different cameras using computer vision techniques, illustrating that this attack has the potential to spread well in operation [9].

3 Methodology

3.1 Design

3.1.1 Face Reorganization Using CNN

Let's take a look at how humans identify faces first. Face perception is difficult because it requires large and varied brain areas to recognize facial expressions. The fusiform gyrus, a region of the temporal lobe known to cause prosopagnosia when impaired, normally shows a lot of activity in brain imaging studies (particularly when harm present on both sides). The people learn to identify the faces from biological and can differentiate one individual from another by the age of four months. The eyes, nose, cheekbones, mouth, and brows, as well as the texture and skin colour, are the most noticeable features. At the same time, our brain procedure the whole face and can recognize an individual even though only half of it is visible. When the brain juxtapose the resulting image to the internal minimum pattern, it notices distinctive variations. What is the mechanism behind the facial recognition systems?

First and foremost, the face identification system must locate and illuminate a face in the image. The programme will do this using a number of algorithms, such as evaluating ratios and skin colour similarities, selecting contours in an image and comparing them to the shapes of faces, and selecting symmetries utilizing neural networks. The Viola-Jones system that can be utilized in real time is the most powerful. And when rotated 30 degrees, the device identifies faces with it. The system is depend on the Haar signs, which set of light and dark rectangle mask of various shapes. The filters are transposed on various areas of the images, and the algorithms measures the difference between the brightness of every the pixels in the images that are below the black and white parts of the cover. After determining the face in the picture, the machine juxtapose the results to the cumulative data and continues to monitor it to decide the best angle and image quality. The Motion vector optimistic about future or association algorithms can be utilized for this.

After selecting the most common images, the device performs face recognition and compares it to the current database. The software seeks the orientation point on the person's faces that make-up the single features in the same way that an artist paints portraits. The software usually allots about 100 points in this category. The space between the eyes, the length of the nostril, the width of the nose, the shape and height of the cheek-bones, and the width of the jaw, the height of the forehead, and other parameters are the most significant dimensions for face recognition systems. After that, obtained information is compared to those in the database, and the individual is defined if the parameters match. Figure 1 shows the overall process.

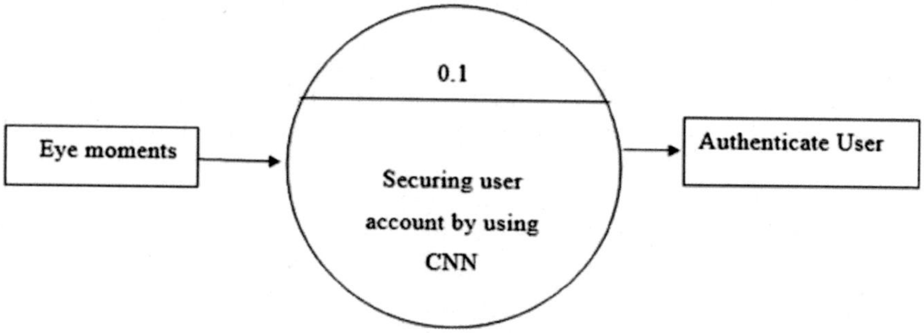

Fig. 1. Illustrates the overall process.

3.2 OTP Generation and Verification

This is the third level of security we are using the random numbers to generate the OTP and send to users email/mobile number using the OTP we can secure our accounts.

Level 0:

Level 0 describes the overall process of the project. We are using the user's eye moments as input. The system will use the conventional neural network to secure user account information from shoulder surfing attacks.

Level 1:

Level 1 describes the first stage process of this project (Fig. 2). We are user credentials as input the system will read the data and generate a password to store in the database. Figure 3 shows the second level of the overall process.

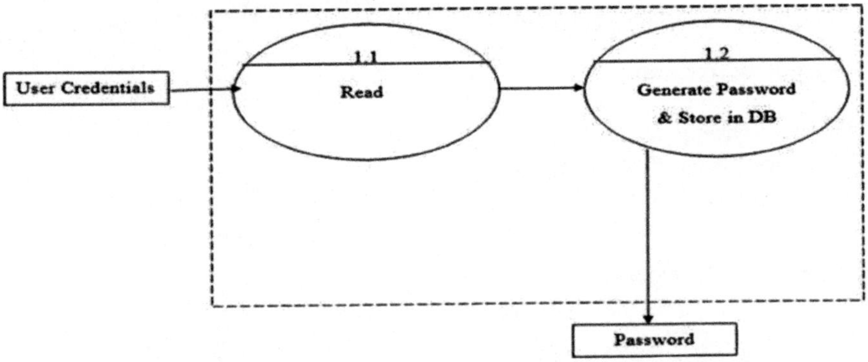

Fig. 2. Illustrates the describes first level of the overall process.

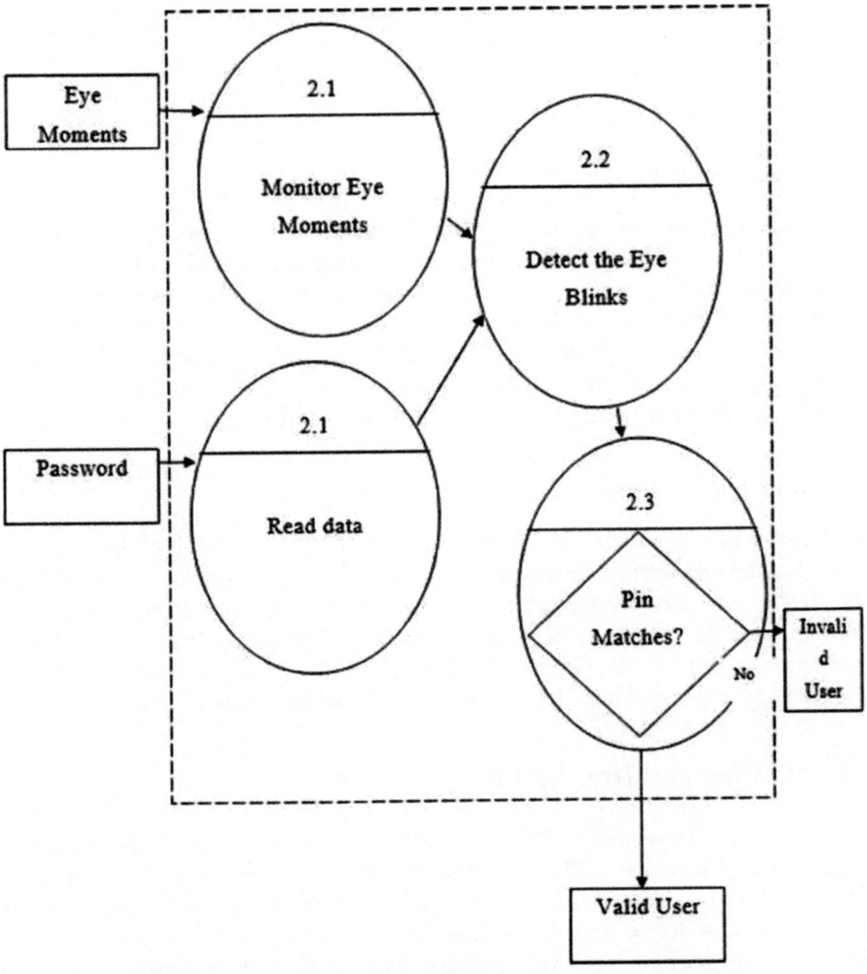

Fig. 3. Illustrates the second level of the overall process.

3.3 Hardware and Software Tools

Software Requirements

- Operating system: Windows 7
- Coding Language: Python3.6
- Tools: Python IDLE.

Hardware Requirements

- Hard Disk: 120 GB
- Monitor: 15″ LED
- Input Devices: Keyboard, Mouse
- Ram: 8 GB
- System: i5.

4 Results and Discussion

4.1 Localization of Face

Researcher use the symmetry-based method since the Face is symmetric. We discovered that using a sub-sampled gray-scale form of the images is sufficient. For each pixel-columns in the abridged images, a symmetry-values is then computed. If the images is defined as I(x, y), then S(x) = [abs I ((x, y − w) − (x, y + w)] is the symmetry values for a pixel-column. S(x) is calculated for X € [k, size-k], where k is the supreme detachment from the pixel-columns at which symmetry is restrained and x size is the images width. The face's centre is the x that resembles to the lower value of S(x).

4.2 Tracking of the Eyes

We follow the eyes in search of the dark pixels in the region predicted. Author make sure that none of the geometrical constraints are broken in sequence to recover from tracking errors. If they will, the eyes are re-located in the next picture. We focus the eye template at the dark pixel and then perform gradient descent to find the local minimum to find the best fit for the eyes template. Figure 4 illustrates the registration page. Figure 5 illustrates the login page. Figure 6 illustrates the client details on the database e login page.

5 Conclusion and Implication

A new applications for eyelid blink-based PIN recognition has been developed using a smart-camera-based eye-blinking framework. The device has been effectively tested with a 9-digit keypad and that can be extended to support characters and digit password combinations. The correctness of the detected pins is affected by the user's eye blink stability, which must be taken into account. PIN recognition is currently done after real-time eyes blinks and eye center calculations and recording have been completed.

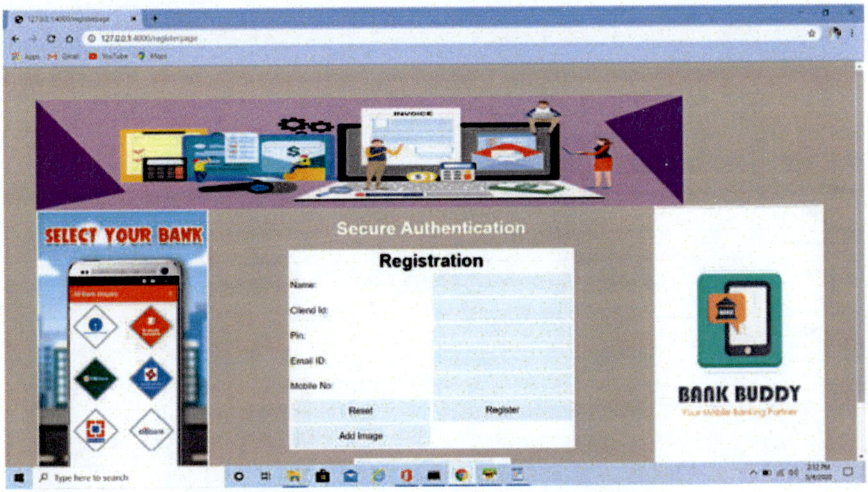

Fig. 4. Illustrates the registration page.

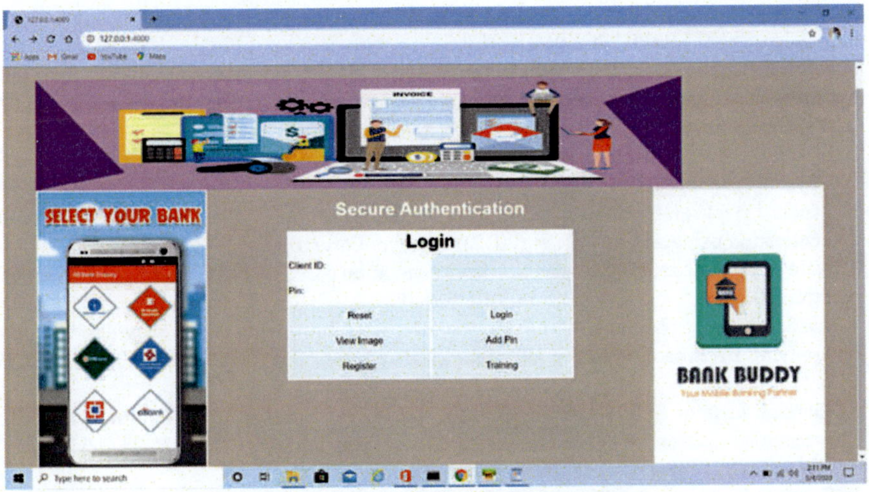

Fig. 5. Illustrates the login page.

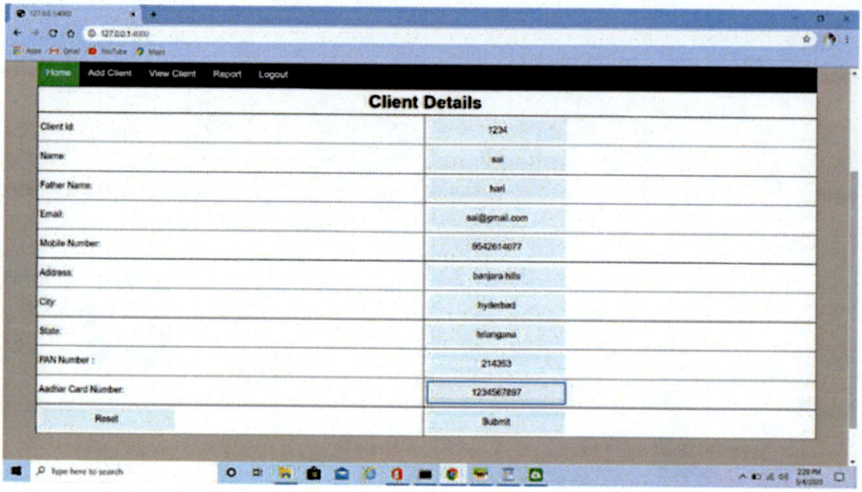

Fig. 6. Illustrates the client details on the database e login page.

References

1. M. Mehrubeoglu, V. Nguyen, Real-time eye tracking for password authentication (2018). https://doi.org/10.1109/ICCE.2018.8326302
2. S. Seha, G. Papangelakis, D. Hatzinakos, A.S. Zandi, F.J.E. Comeau, Improving eye movement biometrics using remote registration of eye blinking patterns (2019). https://doi.org/10.1109/ICASSP.2019.8683757
3. S. Ma et al., An empirical study of SMS one-time password authentication in android apps (2019). https://doi.org/10.1145/3359789.3359828
4. K. Mock, J. Weaver, B. Hoanca, M. Milton, Poster: Real-time continuous iris recognition for authentication using an eye tracker (2012). https://doi.org/10.1145/2382196.2382307
5. T. Trainys, A. Venčkauskas, Encryption keys generation based on bio-cryptography finger vein method (2018)
6. A. Asaduzzaman, A. Mummidi, M.F. Mridha, F.N. Sibai, Improving facial recognition accuracy by applying liveness monitoring technique (2016). https://doi.org/10.1109/ICAEE.2015.7506814
7. M. Gargi, J. Jasmine Sylvia Rani, M. Ramiah, N.T. Naresh Babu, A. Annis Fathima, V. Vaidehi, Mobile authentication using iris biometrics (2012). https://doi.org/10.1007/978-3-642-30567-2_28
8. J. Weaver, K. Mock, B. Hoanca, Gaze-based password authentication through automatic clustering of gaze points (2011). https://doi.org/10.1109/ICSMC.2011.6084072
9. K. Mowery, S. Meiklejohn, S. Savage, Heat of the moment: Characterizing the efficacy of thermal camera-based attacks (2011)

Emergency Location Sharing Using GPS Tracking

Harish Naik(✉), Nishant Kumar Yadav, and Shivam Mishra

Department of Computer Science and Engineering, Jain (Deemed to be University), Bengaluru, Karnataka, India
harish.naik@jainuniversity.ac.in

1 Introduction

In case of any kind of emergency, traffic signal changes instantly in green signal and further permits ambulance vehicle to cross road intersections immediately. The android app installed within a cellular-phone shares the global positioning system (GPS) position using interactive map API's (Application Programming Interface), may be shared ahead through the GSM cellular phone services [1]. By utilizing this collected data, ambulance vehicle may acquire the remote traffic aid from a nearby traffic control station immediately. The traffic control station and communication network, system administrators, trained professional persons primary objective is to work in a proper coordination in order to achieve the success [2]. The handling of the emergency situation is the primary concern and for this well planning of a system demands important work in order to safeguard the serving to public properly [3]. There are several administrative selections, plans for the physical frame, emergency vehicles such as ambulance con-statutes plays a critical role in the enhancement of system performance. There are multifarious criteria that are essential as well as decides the position of any emergency ambulance vehicle server positions [4]. In addition to this, detection of the accurate position of the emergency ambulance vehicles has become the most demanding now a days due to various reasons across the globe. This Android-based Emergency location sharing application will provide flexibility to the ambulances and other emergency vehicles to move faster when it stuck in traffic and also when any kind of accident happens it will help to arrange vehicle by sharing the location of the person and it will navigate ambulance from starting places to the nearby hospital. This information will be shared with the traffic police also to make the road clear for the ambulance [5].

In every intelligent traffic management scheme, traffic light control is important. In traffic light regulation, the sequence of green lights and the length of green lights are the two most significant factors to remember [6]. Most traffic lights in many countries have set sequences and light durations. Fixed control approaches, on the other side, are only sufficient for secure and daily traffic, not for dynamic traffic [7]. In the current position of operation, series of the green light is evaluated without considering the probability of existence of emergency ambulance vehicles. Thus the vehicles that are used in case of emergence or for certain other essential works such as police cars and fire brigades and

F. P. García Márquez (ed.), *International Conference on Intelligent Emerging Methods of Artificial Intelligence & Cloud Computing*, Smart Innovation, Systems and Technologies 273, https://doi.org/10.1007/978-3-030-92905-3_53

several others do not arrive at the destination and causes the loss of many people life globally now a days [8].

2 Literature Review

Igor et al. researched the topic of "Privacy in mobile computing for location-sharing-based services". The LSBS services are another category of the services that complements location-rooted facilities by utilizing the positions from a couple of the users as well as not only individual people. The data is becoming more important now a days from the security point of view worldwide and this is continuously becoming more challenging to reduce the chances of misuse of data by the third parties globally. The data security is a huge challenge in modern world and demands more attention and new technique in order to resolve the existing issues related to the data privacy. In this paper, authors addresses very relevant threat of the security in the LSBS that offers the practical and very much effective solutions of the existing problems. Moreover, in this paper authors, investigated two diverse privacy-preserving methods in order to resolve the existing issues of the FRVP as well as analytically determines the secrecy in both the scenarios namely the passive as well as active [9].

Selin et al. researched emergency exit planning in their research. In this research article, the authors explored multifarious methods in order to use the gamified data prototypes as well as the data analytics. This study provides comprehensive details about the emergency exist planning as well as the simulation platform that are implemented with the help of a novel gamified data prototype based on the commercial data center. In this research article, authors investigated and compared multifarious disaster exist location choices as well as searching of the crucial arena for the user evacuation. The customized consumer details was utilized in order to analyze movement capability as well as the situations of the old age or handicapped people worldwide. The validation of the proposed method were done with high degree of precision in order to validate the simulated results. The observed results provides the information that the gamification is one of the most important and best tool for the exit evacuation in case of high emergency situations [10].

3 Methodology

3.1 Design

This Android-based Emergency location sharing application will provide flexibility to the ambulances and other emergency vehicles to move faster when it stuck in traffic and also when any kind of accident happens it will help to arrange vehicle by sharing the location of the person and it will navigate ambulance from starting places to the nearby hospital. The proposed technique may offer the id to the ambulance for reaching to the hospitals in the very less time in comparison to the existing method. In the very much crucial time the proposed method holds good. This is one of the simplest method from the point of view of execution and utilizes less operating power comparatively, lowest cost as well as offer more advantages. The traffic problems are increasing day by day that demand more attention towards solution of such critical threats. In case of any kind of the emergency situations the traffic signal is controlled in most efficient and pragmatic manner as the traffic light spontaneously change in the green and permits an ambulance vehicle for passing via a road crossing immediately. The cellular-phone android application installed within this device shares the position of the GPS via interactive map. Furthermore, the APIs may be shared via GSM cellular-phone services and in the future a navigation sensor may be installed within existing system that would help of the ambulance vehicle in finding in the 360 degrees as well as for the controlling and handling four-way traffic in most efficient manner. Moreover, this would ensure that there is no death chances inside the ambulance and system would also respond in emergency situations faster.

3.2 Dynamic Traffic Control System

(1) How an Active Traffic Controlling System does looks?

- A Display Unit: This is one of the very basic as well as common traffic display unit that any of the ambulance driver or commuter may watch. This display unit may comprise incandescent lamp/LED arrangement.
- A Detector: This detector is a component which determines presence of ambulance vehicles as well as further transfers the acquired data to a controller for the processing.

(2) Detectors Types:

- Inductive Detector: This comprises of coil of the wires that is embedded inside a groove upon a road surface that is closed with the help of a rubber. This determines the variation in the frequency. This coil which is basically an inductor coil joint with a detector unit that determines any variation in resonance frequency of inductor coil as well as control and handle triggering of the another relay component accordingly that is utilized to trigger traffic light. This works on the basis of command signal provided in case when a moving car cross inductor coil, a property called inductance will decreases. Furthermore, the reduction in the inductance property causes in resonance or may be in the oscillation frequency increment as well as electronic circuitry transmits signal to the controlling unit for the switching operation as required. There are some drawbacks of such systems like the inductor coils causes the electromagnetic interference which is a huge problem for instance electromagnetic radiation of certain other devices create disturbance in the magnetic field.

On the other side to secure the lives of the people, the proposed approach is basically works on the queue length of emergency ambulance vehicles over traffic signals. There is presented simulation of four multifarious prototypes that are utilized in diverse parts of the world as well as showcase the results such as waiting time or total number of ambulance vehicles which are not served in earlier. Sharing of travel details from the API got me confused since there was a parameter required named subleg which was unknown for me and there is no clear description in the documentation too. Implementing the route match extension API was difficult since there was only a rest API option available in the documentation. The major problems in urban areas are traffic congestions and long queues at intersections during peak hours. Increasing traveling times because of the rising numbers of people that travels on the roads and minimal resources offered through the present infrastructures globally. There have been detected multifarious reasons why the traffic congestion is increasing day by day over the roads beyond the capacity that is the primary factor and demands more attention. Such types of the issues occurs due to the lower coordination.

3.3 Hardware and Software Tools

System Requirement:

- Windows 10 (and 64-bit operating system)
- 8 GB of the RAM
- Disk Space minimum 4 GB
- Android Studio
- JDK
- Here API's
- Routing API.
- Places API.
- Tracking API.
- Interactive map API's
- Positioning API's.

Technology:

- Java
- XML
- Firebase Database.

3.4 Proposed System Design

System architecture contains the three-module user, driver, and traffic police. The main purpose of the user is to access the interactive map by updating the current location and share location to the nearby ambulance driver using positioning API. The information gets stored in the firebase, current shared location is updated on the firebase database. Users will be able to see nearby hospitals using places API. Driver access share a map and share current location using positioning API. It will be updated on and the driver would able to a nearby Hospital. This location shared by a user would be able to see to the driver. Traffic police that can access an interactive map and vacancy driver seat location and they will update verification of passing of vehicle in firebase. Traffic police will track vehicle details and verify the details of the vehicle and trace the driver and weaving estimated time using routing API. This system architecture will be developed on Android studio using a software development kit (SDK). To run Android studio it will require Windows 10 (and 64-bit operating system) with 8 GB RAM minimal and JDK should be installed in the system. Figure 1 shows the relationship between modules.

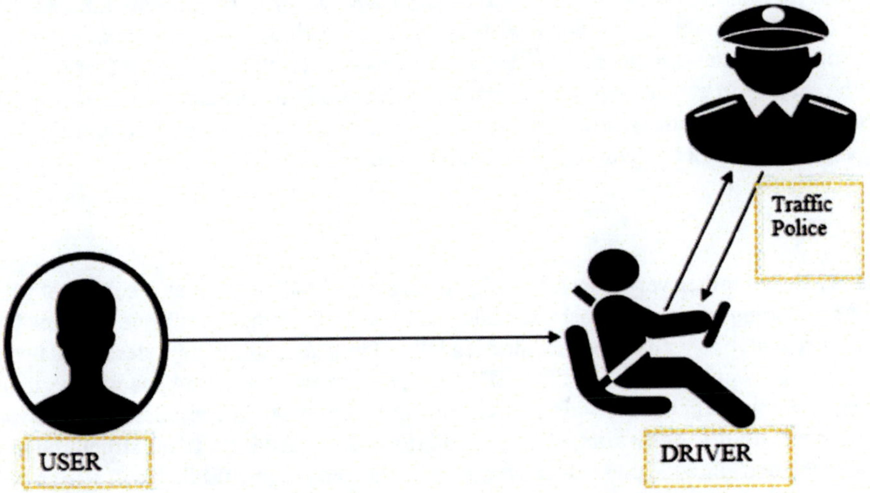

Fig. 1. Illustrates the relationship between modules.

- Our idea is to use a smartphone in emergency vehicles which can be used to track their location. So if we consider a case in which the vehicle is stuck in traffic or any issue, the location could be shared with the nearby traffic police for the essential arrival and effectual clearance of the roadways. Also provided with the information of the arrival of emergency vehicles to their respective locations.
- So this application should be installed by the emergency vehicle drivers (Maybe a local vehicle, ambulance, or any lifesaving vehicles) and the traffic police.
- If the driver needs the traffic police to clear the traffic, he can set his location in the app and send a request message to the traffic police. Once the request is sent to the traffic police, he will get a message alert and they can track the vehicle to help them.

3.5 Java Development Kit (JDK)

The JDK is a pragmatic tool which is intended for the use by the software developer and comprises of the various essential tools as well namely the Java compiler, a Jar, another a Javadoc and last but not the least a debugger. JDK is one of the most essential and the important software development framework that is utilized for the development phase of various kinds of the java applications as well as the applets. Moreover, this comprises JRE, another one a compiler, next one an archiver and certain other multifarious tools that are required for the java development phase.

The Java developers have been proposed two diverse JDK tools namely the Java and the Javac. These developed tools run via command prompt in the computer system. The source files of the Java primarily text files stored via an extension of the Java. At the end of the writing source code of java and further saving, Javac compiler invokes in order to make. Class files. After the process of class files creation 'java' command may be utilized for running the java code. For the coders who wants to code on the IDE, another JDK bundled through the Netbeans and may be download form Oracle site. The speed of this kind IDEs increase in the development phase via introducing multiple features such as the drag-drop etc. This platform that supports multiple things and tools comprises Window, Mac and many other operating system. The people who are using the Mac operating system need diverse software development tools.

3.6 XML

The XML is another very popular mark-up language which describes a couple of rules in the encoding of the document including a format which have multiple features like human-readable as well as the machine readable. The XML language is described by the World Wide Web (www) Board's XML 1.0 Specification from 1998, as well as many other similar specifications, which are all free open norms. XML's architecture priorities stress accessibility, generality, and cross-platform compatibility. It's a structured data type with good Unicode that aid for a range of human languages. While XML was created with the intent of representing files, it is now commonly used to display arbitrary data types including those found in internet services. There has been developed multifarious types of the schema systems in order to support the definition of the XML-rooted coding languages, but multifarious coders have explored multifarious APIs in order to assist the overall XML data processing. XML is one of the language that don't have any kind of the predefined tags for writing a code.

There are many basic tags for instance (namely <to> as well as the <from>) but not defined in all the existing XML language standards. The writer of the XML coding language "developed" such identifiers. HTML utilizes standardized labels such as <header> , <footer> , <section> and many other.

The creator of a XML file should specify that both labels as well as the format of the file. And if actual data is introduced, many XML programmes can continue to function normally (or removed). Consider a programme that displays the updated version of the note.xml file.

Consider an updated model of the note.xml only with <date> as well as the <hour> components introduced as well as the <heading> element omitted. Due to the extreme nature, XML is built, previous models of the programme may still function.

4 Results and Discussion

This Android-based Emergency location sharing application will provide flexibility to the ambulances and other emergency vehicles to move faster when it stuck in traffic and also when any kind of accident happens it will help to arrange vehicle by sharing the location of the person and it will navigate ambulance from starting places to the nearby hospital. This information will be shared with the traffic police also to make the road clear for the ambulance. Figure 2 illustrates home page diagram. Figure 3 illustrates police module diagram.

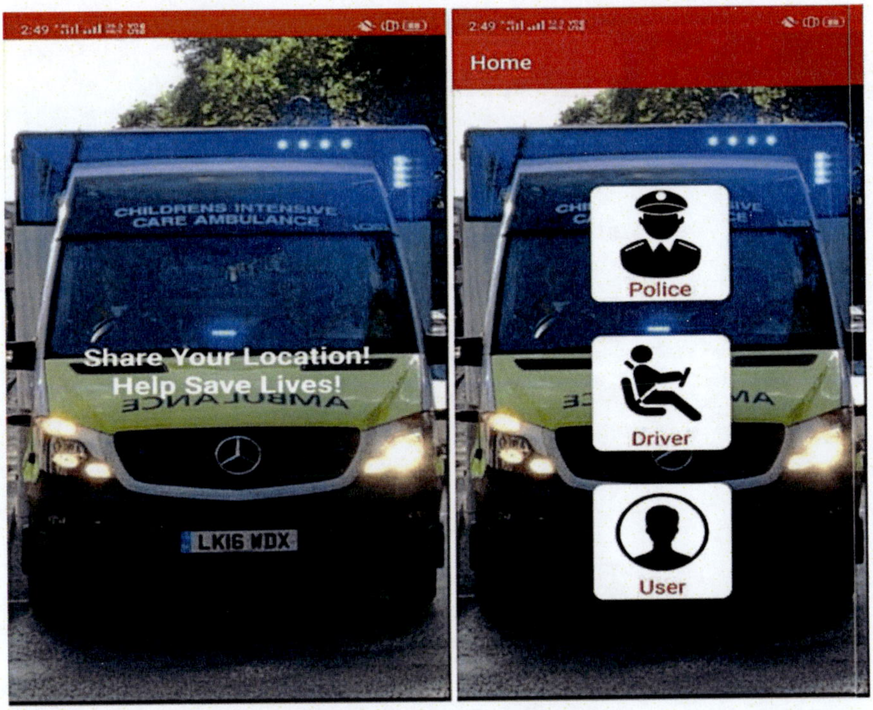

Fig. 2. Illustrates home page diagram.

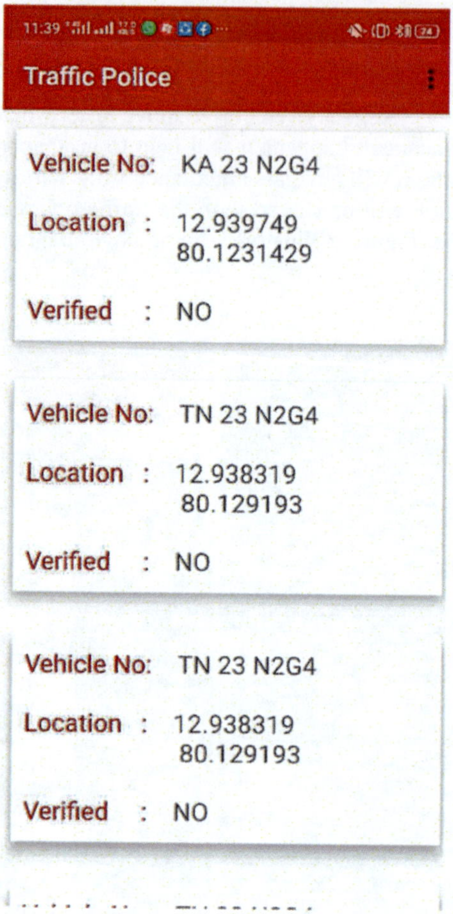

Fig. 3. Illustrates police module diagram.

5 Conclusion and Implication

Such approach can allow the emergency ambulance vehicle to arrive at the nursing home in much less time. Such principle remains true in sensitive cases. Our method is very easy to implement which requires relatively few processing power. It also has a minimal prices and other benefits. In an incident, the traffic signal goes green, allowing the vehicle to drive across traffic signals, as well as the Android device on the smartphone communicates GPS coordinates utilizing interactive map APIs that can then be transmitted further with GSM phone carriers. The emergency ambulance vehicle will get mobile traffic aid from either the emergency command center using that same information, decreasing the lag time. The outcome demonstrates Ambulance's effectiveness in reducing traffic problems. As comparison to many other methods, this is much more favorable. The mapping sensor will be integrated with both the current system at some point, making it easier to locate the emergency ambulance vehicle in 360° and monitor four-way traffic. This means that nobody ever dies in the emergency under any circumstances. The machine can answer even more quickly.

References

1. K.P. Tang, J. Lin, J.I. Hong, D.P. Siewiorek, N. Sadeh, Rethinking location sharing: Exploring the implications of social-driven vs. purpose-driven location sharing (2010). https://doi.org/10.1145/1864349.1864363
2. J. Fulk, Y.C. Yuan, Location, motivation, and social capitalization via enterprise social networking. J. Comput. Commun. (2013). https://doi.org/10.1111/jcc4.12033
3. A. Beldad, M. Citra Kusumadewi, Here's my location, for your information: The impact of trust, benefits, and social influence on location sharing application use among Indonesian university students. Comput. Human Behav. (2015). https://doi.org/10.1016/j.chb.2015.02.047
4. J. Frith, Communicating through location: The understood meaning of the Foursquare check-in. J. Comput. Commun. (2014). https://doi.org/10.1111/jcc4.12087
5. J. Freudiger, R. Shokri, J.P. Hubaux, Evaluating the privacy risk of location-based services (2012). https://doi.org/10.1007/978-3-642-27576-0_3
6. T. Kurashima, T. Iwata, G. Irie, K. Fujimura, Travel route recommendation using geotags in photo sharing sites (2010). https://doi.org/10.1145/1871437.1871513
7. S. Benford et al., Life on the edge: Supporting collaboration in location-based experiences (2005)
8. E. Novak, Q. Li, Near-pri: Private, proximity based location sharing (2014). https://doi.org/10.1109/INFOCOM.2014.6847922
9. I. Bilogrevic, M. Jadliwala, K. Kalkan, J.P. Hubaux, I. Aad, Privacy in mobile computing for location-sharing-based services (2011). https://doi.org/10.1007/978-3-642-22263-4_5
10. J. Selin, M. Letonsaari, M. Rossi, Emergency exit planning and simulation environment using gamification, artificial intelligence and data analytics (2019). https://doi.org/10.1016/j.procs.2019.08.204

Check for updates

Providing Voice to Susceptible Children: Depression and Anxiety Detected with the Help of Machine Learning

T. R. Mahesh(✉), G. Vamsi Krishna, P. Sathwik, V. Ajith Chowdary, and G. Hemchand

Department of Computer Science and Engineering, Jain (Deemed to be University), Bengaluru, Karnataka, India
t.mahesh@jainuniversity.ac.in

1 Introduction

Depression and Anxiety might appear in young children of measuring age of 4 years [1], However, because of the abstract feelings involved, signs are often ignored before children are able to more clearly articulate their distress and convey their impairment to help-seeking adults [2]. Directing a 50–80 min semistructure interviews with a professional doctor as well as their main caregiver is the existing gold standard medical assessment for young children [3]. Holding databases and expense premiums will retard the assessment process and insufficient internal childhood records can discourage many children from obtaining sufficient assessments and diagnosis [4]. Numerous early symptoms of or depression anxiety go unnoticed or are ignored as fleeting by well-intentioned yet uninformed parents [5]. We do know, however, that about 20% of children suffer from an adopting syndrome through their juvenile. Such psychopathy, if untreated, negatively affects the development and progress of an infant and forecasts serious health problems later in life the risk of suicide is greater [6]. As a consequence, the internalization circumstances entail substantial personal and society expenses, highlighting the value of timely identification [7]. New interventions to allow children, at routine paediatric appointments, to examine the extent and extent of their child's discomfort in a feasible and rational way, to providing an efficient understanding and relate children to initial development treatments, while neuroplasticity and symptomatic possible changes are at their height [8].

2 Literature Review

McGinnis et al. researched the topic of anxiety and depression. Anxiety and depression in children are often misdiagnosed. Such circumstances, mutually recognized as adopting illnesses, are linked to longterms harmful effects such as drug abuse and an increased risk of suicide if left untreated. Using a 3 min speech task, this paper provides a novel technique for recognizing young children with internalizing disorders. We demonstrate 80% precision for artificial intelligence to evaluate voice files from the mission in order

F. P. García Márquez (ed.), *International Conference on Intelligent Emerging Methods of Artificial Intelligence & Cloud Computing*, Smart Innovation, Systems and Technologies 273, https://doi.org/10.1007/978-3-030-92905-3_54

to determine children via an integral part (54% sensitivity, 93% specificity). Compared to controls, children impacted include close to zero sounds and recurrent sounds, a high-pitching reaction to unwanted stimuli. This new approach outpaced therapeutic standards on infant signs identified by parents, finding infants with a smaller value, comparable precision, and feedback. This results show that children should evaluate the whole technique for the internalizations of consequences in the forthcoming years, meaning that interventions can be carried out only when lengthy progress is at their peak [9].

Bufferd et al. researched mental health in preschoolers in their research. Many preschoolers, according to recent research, meet diagnostic requirements for mental disorders. However, there is a scarcity of data on the persistence of these diagnoses, especially from studies that looked at a wide range of disorders in community samples. Such study is required to determine the legitimacy and clinical relevance of psychiatric identifies in youths and children [10]. The researchers looked at the persistence of particular psychological conditions in a broad population sample of preschoolers from the preschool era (age 3) to the start of school (age 6).

3 Methodology

3.1 Design

Functionality Requirements:

This section outlines the system's functional specifications for those requirement that are articulated in natural language.

- Create a desktop application.
- Load the voice as input.
- The system should preprocess and extract the deep features from the voice frames.
- Applying the Logistic regression algorithm System should classify the given voice as depressed or normal.
- The application should provide high accuracy of classification of a child's voice.

Non Functionality Requirements.

There are non-functional conditions, i.e., restrictions that must be met in order for the device to operate (Fig. 1).

- The software will be self-contained in order to be transferred easily from one device to another. On the machine where the software is installed, it is presumed that a network link would be open.
- Capacity, availability and scalability: At all times, the device must be 100% accessible. The framework must be flexible in order to accommodate more clients and volunteers.
- Maintainability: As far as possible, the system should be configured for supportability or comfort of maintenance. This can be accomplished by using coding standards documentation, class libraries, naming conventions and abstraction.

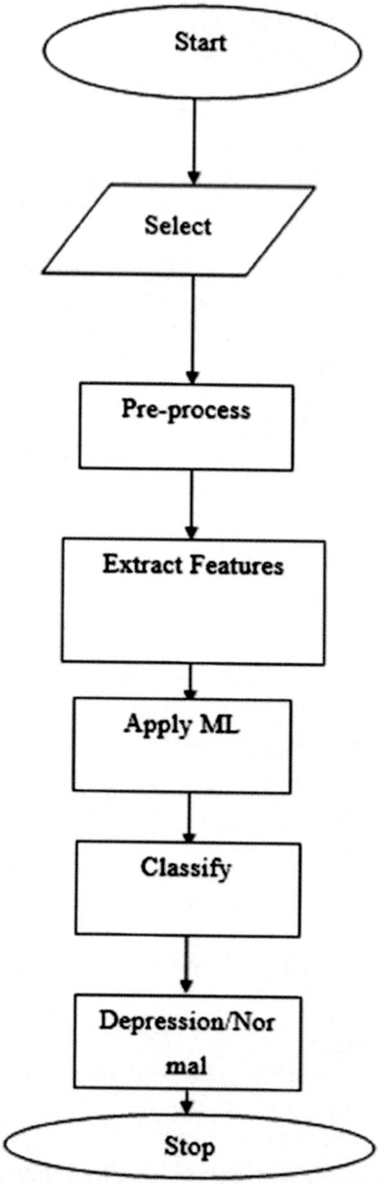

Fig. 1. Illustrates the flow diagram of the proposed method.

- Randomness, verifiability and load balancing: As far as possible, the systems should be configured for maintainable or reliability of maintenance. This can be accomplished by using coding standards documentation, class libraries, naming conventions and abstraction. It should be load balanced and have randomness to search the nodes.

3.2 Hardware and Software Tools

3.2.1 Requirement of Software

1. Operating systems: Window 7 or further Windows
2. Coding Language: Python 3.6
3. Tools: Python IDLE.

3.2.2 Hardware Requirements

1. System: Pentium i3
2. Hard Disk: 120 GB
3. Monitor: 15″ LED
4. Input Devices: Keyboard, Mouse
5. Ram: 4 GB.

4 Results and Discussion

Pseudocode

Procedure Feature Extraction(audio files):

Input: audio files.

Output: Extract the MFCC core features.

Begin

1. Step1: Read the dataset files
2. Step 2: for each file

    a. Step 2.1: Extract the core features from the file using the librosa library
    b. End for

3. Step 3: write the extracted features in the corresponding csv file.

End

Figure 2 illustrates the child depression prediction. Figure 3 illustrates the logistics regression metric value. Figure 4 illustrates the prediction depressed. Figure 5 illustrates select data set from the dictionary.

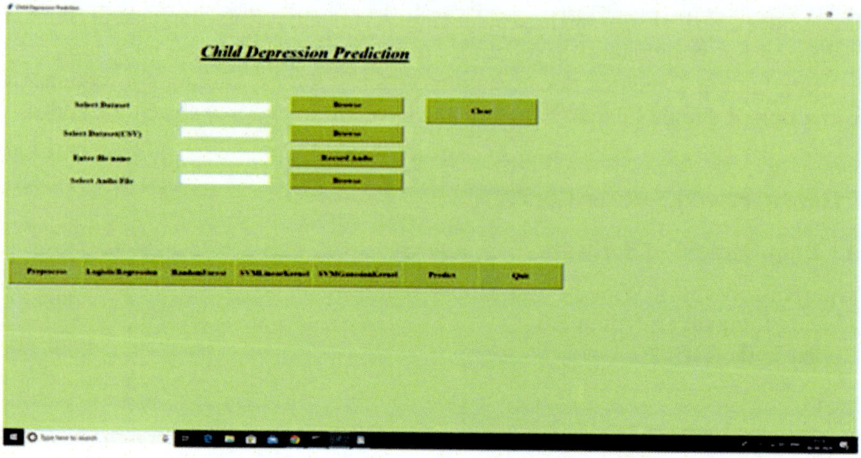

Fig. 2. Illustrates the child depression prediction.

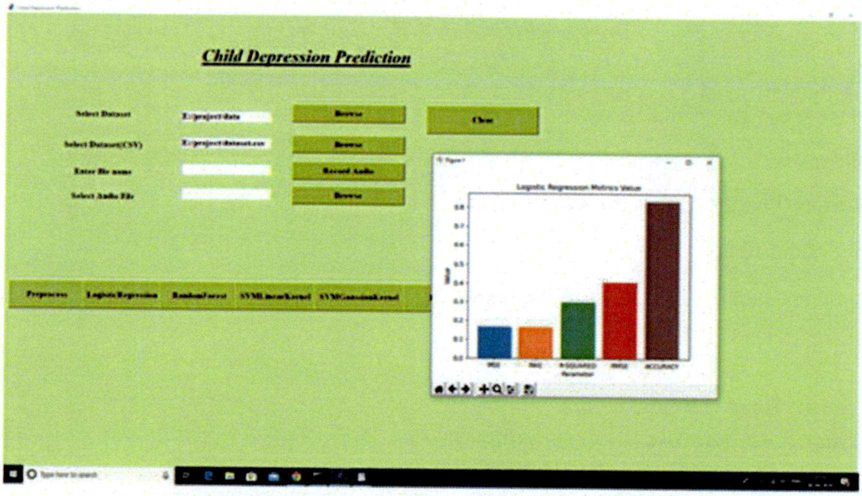

Fig. 3. Illustrates the logistics regression metric value.

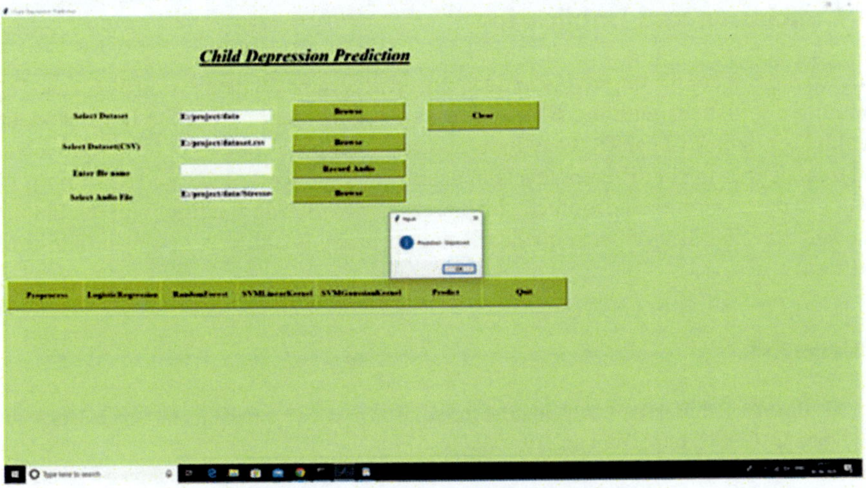

Fig. 4. Illustrates the prediction depressed.

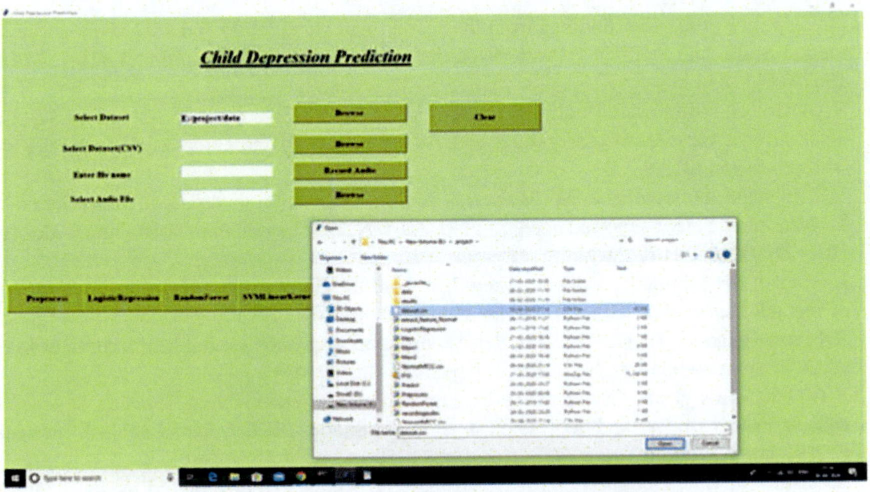

Fig. 5. Illustrates select data set from the dictionary.

5 Conclusion and Implication

There is more space for improving and scaling up programs for depression by focused and committed initiatives with the NMHP and its implementation arm, the District Mental Health Programme (DMHP). Easy tools and common procedures can be used to test for and treat depression at different levels of health care, the workplace, and education. The ability to depression medication can be increased by offering facilities and maintaining a constant availability of basic medications in primary health care settings (both urban and rural).

References

1. J.F. Epstein, P.R. Barker, L.A. Kroutil, Mode effects in self-reported mental health data. Public Opin. Q. (2001). https://doi.org/10.1086/323577
2. N. Stadnick, C. Chlebowski, M. Baker-, M. Dyson, A. Garland, L. Brookman-Frazee, Psychiatric comorbidity in autism spectrum disorder: Correspondence between mental health clinician report and structured parent interview. Autism (2017). https://doi.org/10.1177/1362361316654083
3. K.R. Merikangas et al., Prevalence and correlates of bipolar spectrum disorder in the world mental health survey initiative. Arch. Gen. Psychiatry (2011). https://doi.org/10.1001/archgenpsychiatry.2011.12
4. D. Lawrence et al., Key findings from the second Australian child and adolescent survey of mental health and wellbeing. Aust. N. Z. J. Psychiatry (2016). https://doi.org/10.1177/0004867415617836
5. R. McGee, S. Williams, R. Poulton, T. Moffitt, A longitudinal study of cannabis use and mental health from adolescence to early adulthood. Addiction (2000). https://doi.org/10.1046/j.1360-0443.2000.9544912.x
6. S.E. Johnson, D. Lawrence, M. Sawyer, S.R. Zubrick, Mental disorders in Australian 4- to 17- year olds: Parent-reported need for help. Aust. N. Z. J. Psychiatry (2018). https://doi.org/10.1177/0004867417706032
7. J. Alonso, J.P. Lépine, Overview of key data from the European study of the epidemiology of mental disorders (ESEMeD). J. Clin. Psychiatry (2007)
8. S. Mack, F. Jacobi, H.-U. Wittchen, Self-reported utilization of mental health services in the adult German population. Int. J. Meth. Psychiatr. Res. (2015)
9. E.W. McGinnis et al., Giving voice to vulnerable children: machine learning analysis of speech detects anxiety and depression in early childhood. IEEE J. Biomed. Heal. Informat. (2019). https://doi.org/10.1109/JBHI.2019.2913590
10. S.J. Bufferd, L.R. Dougherty, G.A. Carlson, D.N. Klein, Parent-reported mental health in preschoolers: Findings using a diagnostic interview. Compr. Psychiatry (2011). https://doi.org/10.1016/j.comppsych.2010.08.006

A State of Art Review on Blockchain Technology

R. Kesavamoorthy(✉), Animesh Guptha, Anmol Gupta, Anushka Gahlot, and Arpit Pandey

Department of Computer Science and Engineering, Jain (Deemed to be University), Bengaluru, Karnataka, India

1 Introduction

Blockchains are considered to include a number of closely linked blocks, public (and, recently, private) leads comprising transaction data inside their decentralized datatypes. In order to achieve leader durability, data confidentiality, accountability, repudiations and authentication as part of simple safety primitives, asymmetric cryptography and different consensus algorithm are deployed [1]. They are made irreversible by the distributed and egalitarian character of Blockchains so transfers cannot be manipulated after being formally authenticated by network members and recorded in the chain block. Around the same period, durability and strength are components that make Blockchains a trustworthy infrastructure that is built into transparent and reliable peer groups. We have been doing extensive investigations in recent years through the use of Blockchains in financial markets like digital content and intelligent transactions, IoT and credibility schemes. This is partly because Blockchains can enforce transfers and payments with no intermediate and therefore essentially threaten the manner in which conventional companies function to date [2].

Procurement and the management of the supply chain are seen to be areas where Blockchains are suited for a number of reasons. The data produced in each phase throuhout the lifecycle of a item can be recorded as a transaction that creates a lasting background of the product as it moves into the supply chain (from manufacturing to consumption) [3]. Blockchain technology, and others, will help effectively:

(1) Monitoring, along with physical properties (e.g., guarantees, certifications, copyrights, licenses, bar codes, unified documentation of every single asset (from commodity to container) via the chain supply nodes
(2) and tracking requests, refunds, receipts, transactions and all other authorized content. Besides that, via its transparent architecture, the Blockchain will easily contribute to exchanging knowledge among manufacturers and vendors on manufacturing processes, distribution, repair and wear off of goods, thus creating novel approaches of collaborating on diverse production lines.

F. P. García Márquez (ed.), *International Conference on Intelligent Emerging Methods of Artificial Intelligence & Cloud Computing*, Smart Innovation, Systems and Technologies 273, https://doi.org/10.1007/978-3-030-92905-3_55

1.1 Walmart

Walmart workers can trace those items practically with Blockchain. The Wall Street Journal reports that workers are able to see from mangoes or a few hundred other items via a supermarket app where all the product comes from and is kept in the back room. The technology is helping consumers understand the source of their foods and simplify the rebuilding process [4].

1.2 Maersk

In March 2017, the world's largest transportation firm passed its first Blockchain technology test, examining how its freight can be managed. The test enables anyone to view freight devices remotely from Maersk, the Netherlands Customs and the U.S. Homeland Security, while allowing technologies to streamline and protect foreign shipping.

1.3 Ups

In the hope of improving accountability across all the supply chain communities, the UPS shipping company entered the Blockchain in Trucking Alliance (BiTA) in November. Blockchain specifications for the cargo industry are being established by the company. "Blockchain principles and cooperation among companies help logistical strategies and approaches, enabling UPS consumers to be involved in international trade and finance," UPS said during a press statement. UPS is now examining how Blockchain could be used in its customs brokerage.

1.4 British Airways

The airline checked Blockchain in 2017, the Walloon Street Journal said, to handle flight information among London, Geneva and Miami. The airline designed to decrease overlapping flying details from portal displays, flight application and airline websites to use an unchanged background origin.

1.5 FedEx

In February, another logistics giant, FedEx, has supported BiTA to start up a Blockchain-powered consumer dispute-settlement pilot programme. The organization expects that the programme, in order to solve client problems, will explain what data can be placed on Blockchain. The Vice President of Tactical Strategy and Research Dale Chrystie told Coindesk that FedEx is now hoping to use this database to collect information.

1.6 Deloitte IoT Blockchain

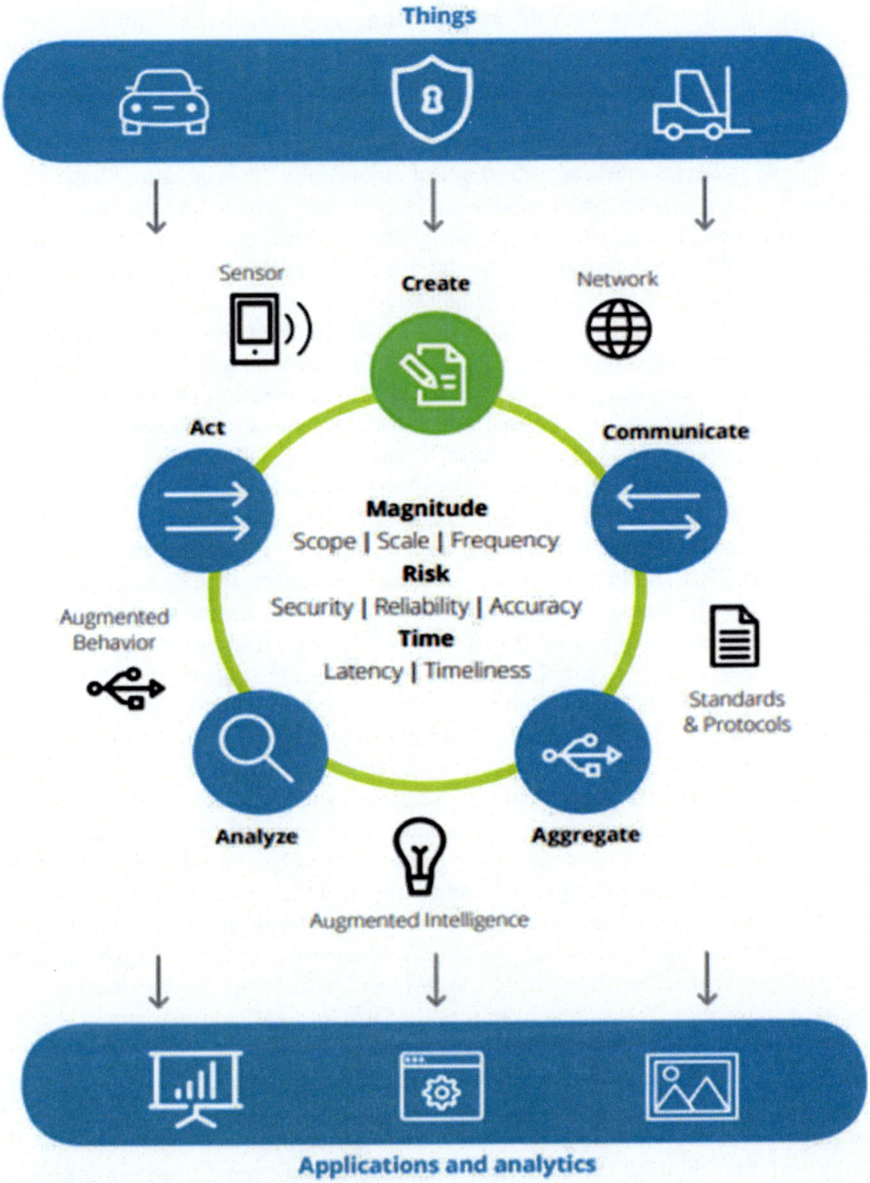

Fig. 1. Deloitte has combined Blockchain and IOT to define new rules to supply chain management in the financial supply chain [5].

2 Literature Review

In various fields, including electricity and financing, blockchains is effectively utilized. For the supply chain, ideas were theoretically suggested in real-world situations without substantial success. Recent published articles, in specific, have different facets of the specific field. In this context, researchers examine typical uses of supply chain and logistics administration for Blockchain on the basis of five technology principles: technical engagement; technological novelty; awareness of the local context; knowledge of new technologies; and expectation of technological innovations. Most situations have strong technological commitment, however when it refers to many other conscious usage values there are considerable variations. In the above, the researchers use the 'characteristics of the development system' approach to classify possible Blockchain implementations and to subequal categories the defined implementation areas as per the impact of four phases of change on the organizational systems and processes. The researchers defined potential adoption applications and offer a basis for identifying Blockchain prospects in the logistics sector through theoretical and empirical literature. These experiments may be especially helpful for business leaders to evaluate how to effectively implement Blockchain technologies in order to develop organizational capabilities [6].

If there is no general consensus on the meaning, a management structure cannot be used efficiently by professionals and scholars. This is the same in the case of the word "supply chain management," which uses so many meanings that there is not any common sense. This article is therefore intended to review current findings in order to explain the idea of "supply chain management." The SCM and the supply chain are checked, grouped and synthesized in different definitions. Description and structure of SCM enabling structures to develop a coherent way of conceptualizing SCM is then given. SCM has established the context and implications and SCM's limits as regards corporate functions and organizations. A conceptual model and unified SIM description indicating the origin, history and implications of phenomenon are then introduced [7].

Need for integration of supply chain management science and practice with ecologically responsible decisions is rising. A literature review revealed that a wide framework framework is not sufficiently defined for green supply chain management (GRCM). There is also a shortage of regulatory authorities to draught legislation that address social and environmental issues that support market development and the economies. A brief classification is needed to help academics, scientists and doctors grasp structured GrSCM from a broader perspective. In addition, ample literature to guarantee this designation is accessible. This article examines the GrSCM field in an integrated and new way. The GrSCM literature is extensively addressed by its conceptualisation, mostly in the context of 'reverse logistics' aspects. The information on GrSCM is categorised according to the issue sense in the main supply chain fields that use the rich amount of knowledge, particularly previous research with very little prospects. The technique and strategy are both classified. Different mathematical techniques employed in publications are marked against the GrSCM frameworks. A schedule is also given which indicates the relevant documents [8].

3 Discussion

3.1 Existing System and Target Industries Benefits

For several problems in the supply chain business, Blockchain could be implemented like complex records management and stock monitoring. As an alternative to centralized databases, less inherently corrupt and more automated. This is how Blockchain in supply chain business can be helpful.

3.2 Provenance Tracking

There are also components of their supply chains for big corporations and organisations. As a result, even multinational companies almost cannot keep track of any record. The lack of clarity contributes to cost problems and client relations that dilute brand name at the end of the day. With the aid of sensing devices and RFID tags, recording and provenance monitoring in Blockchain-based supply chains are easier to access because product information can. Blockchain is able to track down a product's background from its roots to where it is now. In addition, precise provenance monitoring may in any stage of the supply chain be used to prevent attacks [9].

3.3 Cost Reduction

Efficient inventory monitoring with the help of This system decreases the total expense of shipping products across a supply chain. And over one third of the people identified cost savings as the greatest advantage of Blockchain's use in supply chain management, this is from an APQC and Digital Supply Chain Institute (DSCI) Supply Chain Survey.

As Blockchain is used to accelerate administration of supply chains, additional costs arising in the system are immediately minimized and transaction safety is still guaranteed. The removal of mediators in the supply chain prevents theft costs, duplication of products, and also saves money. Suppliers and customers in the supply chain will transfer funds by cryptocurrency instead of using EDI. instead of consumers and partners. Performance can also be increased, and with precise records management the chance of missing goods is minimized.

3.4 Establishing Trust

For safe operation, confidence is important in complex supply chains with multiple participants. For e.g., a producer must be likely to agree on its products with the following factory security requirements while sharing their products with suppliers. Furthermore, assurance plays an important role in regulatory enforcement, for example with customer enforcers. Blockchain's unchangeable essence is well conceived to discourage disturbance and trusting.

3.5 Benefits of Supply Chain with Blockchain

One of Blockchain's most attractive advantages is that it makes more interoperable with the info. As a result, knowledge and information sharing with distributors, retailers and providers become simpler for businesses. Openness of Blockchain needs to decrease discrepancies and differences, while keeping products in the supply chain. Although the chances of error are rare since each material could be monitored in real time [10].

Blockchain has scalability that makes any massive database available from many places all over the globe. It also offers better security practices and the flexibility to adapt to the database server. In addition, Blockchains may also be generated secretly to enable direct accessibility between the parties authorized for the use of the information [11].

The utility of Blockchain technology is derived from having the ability to bind separate headers and data sets, whilst preserving various participants' data integrity. The characteristics of Blockchain's simplicity and untrace ability are helpful in the elimination and maintenance of corruption in the supply chain. Figure 1 shows combined Blockchain and IoT and Fig. 2 illustrates the system architecture.

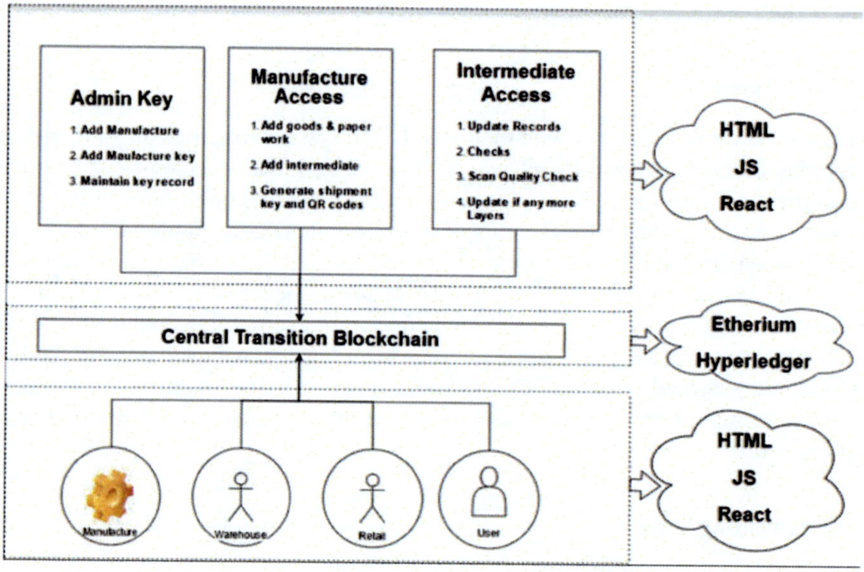

Fig. 2. Illustrating the systems' architecture.

4 Conclusion and Implication

One of Blockchain's most attractive advantages is that it makes more interoperable with the info. As a result, knowledge and data sharing with distributors, retailers and providers become simpler for businesses. Openness of Blockchain intends to minimize discrepancies and differences, while keeping products in the supply chain. Because any product could be monitored in real time, it is unlikely that mistakes will occur. Blockchain has robustness that makes any massive database available from many places all over the globe. It also offers better security practices and the flexibility to adapt to the database server. In addition, Blockchains can also be privately generated, allowing transparent access to this information among parties with authorization.

References

1. M. Andoni et al., Blockchain technology in the energy sector: A systematic review of challenges and opportunities. Renew. Sustain. Energy Rev. (2019). https://doi.org/10.1016/j.rser.2018.10.014
2. M.A. Khan, K. Salah, IoT security: Review, blockchain solutions, and open challenges. Futur. Gener. Comput. Syst. (2018). https://doi.org/10.1016/j.future.2017.11.022
3. F. Casino, T.K. Dasaklis, C. Patsakis, A systematic literature review of blockchain-based applications: Current status, classification and open issues. Telematics Inform. (2019). https://doi.org/10.1016/j.tele.2018.11.006
4. R. Kamath, Food traceability on blockchain: Walmart's Pork and mango pilots with IBM. J. Br. Blockchain Assoc. (2018). https://doi.org/10.31585/jbba-1-1-(10)2018
5. IoT for all, Article (2021). https://www.iotforall.com/human-rights-iot-tracking-violations-improving-protections
6. J. Frizzo-Barker, P.A. Chow-White, P.R. Adams, J. Mentanko, D. Ha, S. Green, Blockchain as a disruptive technology for business: A systematic review. Int. J. Inf. Manage. (2020). https://doi.org/10.1016/j.ijinfomgt.2019.10.014
7. J.T. Mentzer et al., Defining supply chain management. J. Bus. Logist. (2001). https://doi.org/10.1002/j.2158-1592.2001.tb00001.x
8. S.K. Srivastava, Green supply-chain management: A state-of-the-art literature review. Int. J. Manage. Rev. (2007). https://doi.org/10.1111/j.1468-2370.2007.00202.x
9. S.A. Rehman Khan, Z. Yu, Introduction to supply chain management, in *EAI/Springer Innovations in Communication and Computing* (2019)
10. S.E. Fawcett, G.M. Magnan, M.W. McCarter, Benefits, barriers, and bridges to effective supply chain management. Supply Chain Manage. (2008). https://doi.org/10.1108/13598540810850300
11. Z. Yu, H. Yan, T.C.E. Cheng, Benefits of information sharing with supply chain partnerships. Ind. Manage. Data Syst. (2001). https://doi.org/10.1108/02635570110386625

Check for updates

Review of Data Storage and Security in Cloud Computing

Durgesh Wadhwa(✉)

SBAS, Sanskriti University, Mathura, India

1 Introduction

Cloud computing is characterized as a parallel and distributed computing platform consisting of a pool of interconnected and virtualized computers dynamically supplied and provided to users a single computing resource based on pre-agreed Service Level Agreements (SLA) [1]. This helps users to run their applications remotely as well as store data with the advantage of a highly available on-demand service; without the issue of local hardware and software management. Data is stored on several third-party servers with cloud storage, rather than on the dedicated server used in standard networked data storage [2]. Third-party service providers are entrusted with user data and most users do not know exactly where these data are kept for security purposes. Cloud computing uses the Internet to have a positive impact on the IT landscape, as it allows users to pay on bases of use per service. It transfers customer interests from acquisition and maintenance to the use of services made available by cloud service providers [3]. Cloud computing is about moving services, processing or data to an internal or external venue, a centralized facility or a contractor for cost and business advantages. Cloud computing has features that include multi-tenancy and resource pooling. Cloud computing has three basic service types [4]:

1. Software-as-a-Service (SaaS), where Cloud Service Providers (CSPs) make applications available to Cloud users over the Internet.
2. Platform-as-a-Service (PaaS), in which CSPs provide platforms for Cloud users to develop and deploy their own applications.
3. Infrastructure-as-a-Service (IaaS), where CSPs provide Cloud customers with computing, storage, network and other computing services (Fig. 1).

F. P. García Márquez (ed.), *International Conference on Intelligent Emerging Methods of Artificial Intelligence & Cloud Computing*, Smart Innovation, Systems and Technologies 273, https://doi.org/10.1007/978-3-030-92905-3_56

Fig. 1. Service types of cloud computing.

2 Modes of Deployment in Cloud Computing

There are also four deployment models for cloud computing, private cloud, public cloud, community cloud, and hybrid cloud [5].

Private Cloud is owned and managed by an individual organization. The facilities could be on-promise or off-premise. Private Cloud allows for a more secure environment due to internal staff utilization.

Public Cloud is owned and controlled by major CPSs. These providers own large data centers, sometimes spread across different geographic locations, providing various services that free the customer from costly procurement of infrastructure [6].

Community clouds are part of several organizations that come together on the basis of shared interest. Community can manage the community cloud or a third party.

Hybrid Cloud is a combination of public, private or community Cloud. The same technology is used by the hybrid cloud, but the organizations are different.

3 Essential Characteristics of Cloud Computing

The necessary characteristics of Cloud Computing are as follows [7]:

(1) *On-demand self-service*: A user can provision computing resources at the same time as required, such as server time and network storage, without requiring human interaction with each service provider.
(2) *Broad network access*: Capabilities are accessible over the network and accessed via common frameworks that facilitate the use of heterogeneous thin or thick client devices (e.g. mobile phones, laptops, tablets and workstations).
(3) *Resource pooling*: The computing resources of the company are pooled to serve multiple users using a multi-tenant model, with different virtual and physical resources being dynamically delegated and reassigned to meet consumer demands. Resource examples include storage, memory, transmission, and bandwidth of the network [8].

(4) *Rapid elasticity*: Capabilities can be provided and released elastically, in some cases automatically, in order to rapidly scale outward and inward in accordance with demand.
(5) *Measured service*: Cloud systems automatically control and optimize resource use by leveraging metering capabilities at an appropriate level of abstraction for the type of service (e.g., storage, bandwidth, processing and active user accounts).

Use of resources can be monitored, controlled and reported, providing transparency of the service used for both the provider and the consumer.

4 Security Classification in Cloud Computing

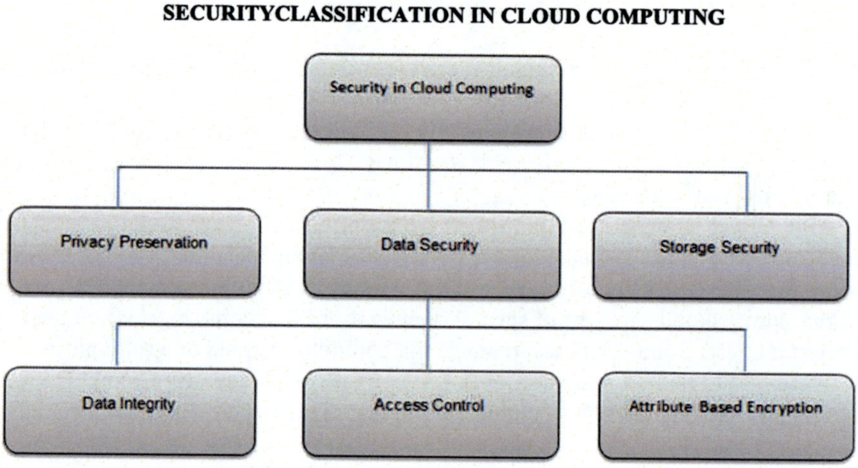

Fig. 2. Security classification in cloud.

Privacy Preservation: Privacy Preservation describes the privacy of data and essential cloud information as it does not trust the cloud servers. Confidentiality and permission are key cloud protection criteria [9] (Fig. 2).

Storage Security: Data Protection describes the function of providing transparency to the shared data stored on dishonest cloud servers through Cloud Storage Security.

Data Security: Data Security describes data or information security as the process of securing data from unauthorized users, preventing changes and limiting access to sensitive information.

Data Integrity: Validations using cryptographic tools such as message digests, hashing and digital signatures, etc. are commonly ensured.

Access Control (AC): The Access Control System includes components and methods for specifying access control policies for legitimate users.

Attribute based Encryption (ABE): Uploaded data is encrypted using ABE, which defines the access policy for data-related attributes. Only authorized users with similar attributes can only decrypt and access the data.

5 Data Storage in Cloud Computing

Cloud computing is a method where software, hardware and/or other resources are provided "as a service." Cloud takes some advantages to its subscribers, such as relief from the burden of storage management, universal access to data, pervasive, lower capital costs, etc. Various problems associated to cloud computing comprise the following [10]:

- Security of data from stealing
- Data Integrity over the Cloud
- Protected transmission of data to and from Cloud sever
- Authenticating files without much overhead/Computation
- Rights management
- Maintain security during sharing.

Data storage consistency, or some time more generally referred to as data integrity assurance, is one of the major security concerns in the cloud. The data may be modified by an unauthorized party without intimating the data owner. How would the data owner gets to know that his or her data has not been changed by other intruders (or, accidentally or intentionally, by the cloud provider itself).Detecting such illegal activity on data is therefore a top priority problem. It is possible to classify data storage correctness schemes into two types, (refer to Table 1).

(1) Without Trusted Third Party (TTP) and
(2) With TTP, based on who is responsible for the verification.

As with traditional network security, they are trying to protect the confidentiality of data in the two phases of the data life cycle and are as follows.

Data at rest: Symmetric key encryption techniques (e.g. AES, TDES, and DES etc.) are recommended for data at rest, which are more time-consuming yet reliable approaches.

Data in transition: For data in transition, it is recommended to use SSL forms which are already available secure protocols. Hash functions such as SHA-1, MD-5 are used to verify integrity.

For their accessibility, confidentiality and fairness of outsourced data in clouds is of paramount importance. The reasons given are as follows:

(1) The CSP, whose primary purpose is to make a benefit and maintain a reputation, has deliberately hidden data loss from an event that consumers never access.
(2) The malicious CSP could delete some data or get all the information easily and sell it to the company's biggest rival.

(3) An intruder who intercepts and steals the communications will find out about the sensitive information of the user as well as some important business secrets.
(4) There are a wide range of internal and external risks to cloud infrastructures.

Table 1. Comparison of present works based on data and storage security in cloud computing

Method	Services	Privacy	Confidentiality	Integrity	Access control	Storage security
Cloud computing using access control architecture	Secure access permission for multiple services	No	No	No	Yes	No
User authentication and access control technique in cloud computing	Cryptographic RBAC	Yes	Yes	No	Yes	No
Fine-grained data access control systems with user accountability	ABE	No	Yes	No	Yes	No
Trust based AC policy	Trust management, role-based access control	No	No	No		No
Temporal access control	Temporal access control encryption, proxy-based re-encryption	No	Yes	No	Yes	No

6 Conclusion

Data security and storage security in cloud computing were explained in detail in this survey work. The proposed strategies explain the different classifications of data and storage protection. Finally, a comparison table with all existing methods is given, tabulated with advantages and disadvantages. From the table it can be assumed that there should be less storage and computational overhead for both data and storage protection. Authentication, confidentiality of authorization and honesty should be maintained in data security.

References

1. A. Lele, Cloud computing, in *Smart Innovation, Systems and Technologies* (2019)
2. X. Xu, From cloud computing to cloud manufacturing. Robot. Comput. Integr. Manuf. (2012). https://doi.org/10.1016/j.rcim.2011.07.002
3. R. Buyya, C.S. Yeo, S. Venugopal, J. Broberg, I. Brandic, Cloud computing and emerging IT platforms: Vision, hype, and reality for delivering computing as the 5th utility. Futur. Gener. Comput. Syst. (2009). https://doi.org/10.1016/j.future.2008.12.001
4. S. Shilpashree, R.R. Patil, C. Parvathi, 'Cloud computing an overview.' Int. J. Eng. Technol. (2018). https://doi.org/10.32628/ijsrset196120
5. G. Eugene, Cloud computing models (2013)
6. P. Pandey, Survey: SOA architecture in cloud computing. Int. J. Data Sci. Technol. **2**(1), 1 (2016). https://doi.org/10.11648/j.ijdst.20160201.11
7. R.G. Kaple, P.S.B. Rathod, Review paper on an efficient encryption scheme in cloud computing using ABE. Int. J. Trend Sci. Res. Dev. (2018). https://doi.org/10.31142/ijtsrd10915
8. D.J. Prince, Introduction to cloud computing. J. Electron. Resour. Med. Libr. (2011). https://doi.org/10.1080/15424065.2011.626360
9. S. Rajeswari, R. Kalaiselvi, Survey of data and storage security in cloud computing, in *IEEE International Conference on Circuits and Systems, ICCS 2017* (2018). https://doi.org/10.1109/ICCS1.2017.8325966
10. K. Hamlen, M. Kantarcioglu, L. Khan, B. Thuraisingham, Security issues for cloud computing. Int. J. Inf. Secur. Priv. (2010). https://doi.org/10.4018/jisp.2010040103

Security Operation Modes for Enhancement of Utility Computer Network Security

Pankaj Saraswat(✉)

SOEIT, Sanskriti University, Mathura, India
pankajsaraswat.cse@sanskriti.edu.in

1 Introduction

Cyber security turns into a developing worry for control system utilities with expanding power systemcomputer organize interconnections, control system mechanization, and open access necessities under government deregulation arrangements. Regular utility system conventions, for example, DNP3 were initially intended for systems that were just open by set number of prepared utility staff. As aresult, DNP3 and other utility system conventions did not execute security, for example, encryption and verification, presently making them be helpless against digital assaults in the increasingly open systems administration condition that has been exasperated with open access utility power systems.

1.1 Cyber-Security for Utilities

With execution of encryption and confirmation security, with just confirmation as of now proposed for the DNP3 detail, utility correspondence is still powerless against assault because of the consistency of information transmission timing and sizes. Traffic-investigation assaults based on scrambled information transmission headers, outline sizes, and timing can be utilized to precisely decide the information transmission type, for example spontaneous insurance information transmissions are probably not going to happen at an ordinary requested information transmission time and control information will utilize less edges than observing information [1–3]. In view of the traffic investigation assaults, digital aggressors might have the option to decide shortcomings to misuse inside the power system, for example, finding gear to harm that will largely affect control systemtasks than what might ordinarily happen or actualizing a forswearing of-administration assault on an organized gadget with the goal that it can't react fittingly to security occasions, for example incapable to transmit a control direction to a gadget to close circuit breakers. Moreover, cryptanalysis assaults can be consolidated with traffic investigation assaults to break the encryption on the information transmissions, permitting a digital aggressor to control information and gadgets straightforwardly, for example affecting the conduct of tasks by adjusting or making information and controlling the tasks of a gadget by actuating transfers. Thusly disappointment to give satisfactory security to control system information transmissions may have extreme ramifications for control system tasks.

F. P. García Márquez (ed.), *International Conference on Intelligent Emerging Methods of Artificial Intelligence & Cloud Computing*, Smart Innovation, Systems and Technologies 273, https://doi.org/10.1007/978-3-030-92905-3_57

1.2 Proposed Security Operation Modes

Another digital security pseudo-layer is proposed to improve encryption security by giving expanded adequacy against traffic examination and cryptanalysis assaults than essentially utilizing encryption security as it were. The proposed security layer is situated beneath the DNP3 information connect layer. The security layer includes three security activity modes that are inferred from techniques used to measurably control the transmission pace of casings. The three security activity modes are: cushioning, outline transmission gathering (FTG), and split-outline transmission gathering (SFTG). The security activity modes give upgraded encryption classification abilities and constrained traffic investigation aversion capacities.

The security activity modes improve commonplace strategies utilized to balance the traffic examination assault potential, which can use cushioning to camouflage the size of the client information and factually control the transmission pace of casings. These techniques guarantee that all information transmissions are all the same size and that information transmissions happen at a particular interim, where information transmissions will be made if no client information is accessible for transmission. With these strategies, joined with encryption security for the whole casing counting the whole casing header, traffic seems, by all accounts, to be totally uniform to an assailant. A digital aggressor can't decide the idea of the information being transmitted, for example, separating among control and insurance situated information transmissions.

The main explanation is because of the particular transmission rate for requested information transmissions, for example control and observing information, which are exceptionally steady and unsurprising in their planning. Thusly, digital assaults would have the option to utilize traffic examination precisely to dispose of the greater part of the casings made by the measurable transmission rate to fill in the information transmission holes when no client information was being transmitted. As a result, DNP3 gadgets would need to process and transmit more information without discernable expanded security.

The subsequent explanation is because of assurance situated information transmissions. The insurance situated information transmission timing is unusual and can't be deferred since they are time-basic to guarantee solid power system tasks. Since the insurance information transmissions are time-basic, DNP3 gadgets utilizing the measurable transmission rate for outlines would need to expand the transmission rate to create a uniform traffic design that won't postpone insurance information transmissions. In any case, this would make a DNP3 gadget use more prominent assets wastefully to make outlines between the run of the mill information transmission rate, for requested control and checking information, which are probably not going to contain any security information. For instance, if the permitted transmission delay for security information was 4 ms and the typical interim between transmission times was 2 s [4–6], the DNP3 outstation gadget would regularly need to deliver 500 pointless casings between client information transmissions inside the 2 s interim. Time-basic assurance information transmission thusly restricts the pertinence of measurably controlling the transmission rate utilized in control system computer systems.

1.3 Application for Enhancement of Power System Controls

The power dissemination system tasks can be altogether upgraded with the utilization of current continuous computercorrespondence and systems administration innovation and cutting edge advanced sign handling innovation. The security activities proposed in this paper have been effectively applied for improvement of research facility model control circulation system solidness control. An outline is given in this paper.

2 Security Operation Modes

2.1 Padding Mode

The cushioning security activity mode is utilized to cushion all DNP3 information connect layer edges to a similar size, which is the greatest 292-octet information interface outline size. Encryption and validation is applied to the information connect layer outline by the security layer after the cushioning has been included into the edge. The security-layer cushioning mode is utilized when an ordinary level of security is required for the force systemcomputer arrange since it doesn't present any critical overhead for the security tasks or transfer speed. This mode would consequently be ordinarily utilized in substation LANs, for generally safe connections such as associating gadgets into the substation, or for insignificant hazard connections, for example, interfaces that have never been assaulted.

The cushioning security activity mode is utilized to cushion all DNP3 information connect layer edges to a similar size, which is the greatest 292-octet information interface outline size. Encryption and validation is applied to the information connect layer outline by the security layer after the cushioning has been included into the edge. The security-layer cushioning mode is utilized when an ordinary level of security is required for the force systemcomputer arrange since it doesn't present any critical overhead for the security tasks or transfer speed. This mode would consequently be ordinarily utilized in substation LANs, for generally safe connections such as associating gadgets into the substation, or for insignificant hazard connections, for example, interfaces that have never been assaulted.

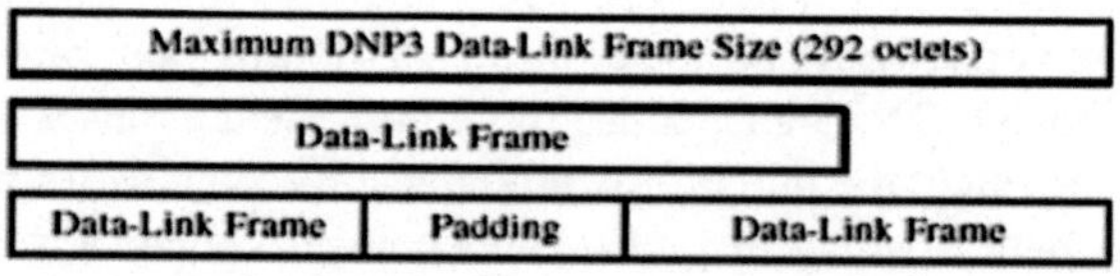

Fig. 1. DNP3 data-link frame padded to maximum 292 octets.

For application layer messages separated into a few information connect layer outlines, the cushioning shrouds the size of the last information interface layer outline in the arrangement. Camouflaging the size of the last piece may give protection from digital assaults if explicit sorts of information consistently have a similar number of octets in the last information connect layer outline utilized for the application layer piece. The cushioning is put arbitrarily inside the information interface layer outline so as to cloud the information limits in the edge, for example the digital aggressor can't expect that the twelfth octet in the casing would be the application header. The cushioning mode activity is appeared in Fig. 1 before encryption and confirmation.

In Fig. 1, cushioning octets are produced for the DNP3 datalink layer outline with the goal that it is the most extreme permitted 292 octets. The cushioning is then arbitrarily set into information interface layer outline anytime.

2.2 Frame Transmission Group Mode

The edge transmission gathering (FTG) security mode is utilized to mask a scrambled information interface layer outline among a gathering of scrambled produced outlines. The FTG security mode incorporates the cushioning mode activity with the goal that the entirety of the outlines in the transmission bunch are a similar size. In the FTG mode, the information connect layer outline containing the client information is set haphazardly inside the gathering of produced outlines, taking after factual control of the information transmission timing. Hence, there are a restricted number of edges being transmitted at the normal transmission time, however with the genuine edge haphazardly set inside those edges [7–9]. A digital assailant is in this manner incapable to correctly figure out which information interface layer outline contains the client information and along these lines must spend assets on inspecting every one of the edges inside the transmission gathering to decide the area of the client information. The FTG security mode is appeared in Fig. 2.

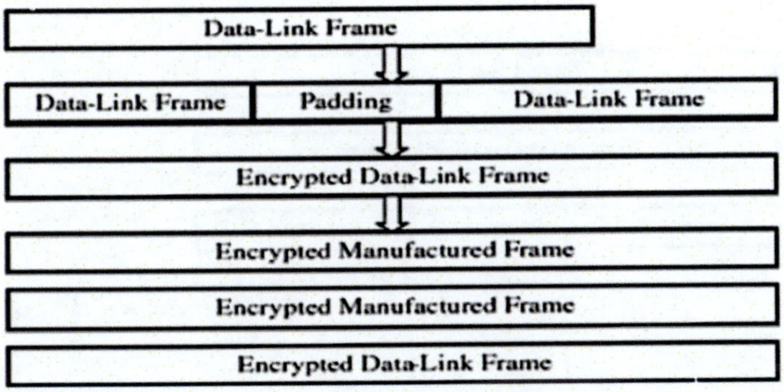

Fig. 2. FTG security mode operations are shown randomly placing the user data within the frame transmission group.

In Fig. 2, the client information connect layer outline is cushioned and afterward encoded and perhaps confirmed. The scrambled client information interface layer outline is then arbitrarily put among the encoded fabricated edges in the transmission gathering. The FTG security mode has a similar confinement as the cushioning mode in which a digital assailant could tally the number of information interface outlines used to pass on the application message piece to decide whether control, checking, or insurance information is being transmitted.

The FTG mode gives more security than the cushioning mode, and is subsequently utilized for joins that are increasingly defencelessto assault, for example, those interfacing with brilliant meters or for joins between significant DNP3 gadgets, for example, between substations. Be that as it may, since the FTG security mode makes different casings to camouflage the client information, this mode requires all the more preparing and data transfer capacity ability which confines its immaterialness to low data transfer capacity gadgets.

2.3 Split Frame Transmission Group Mode

The split casing transmission gathering (SFTG) security activity mode is utilized to camouflage an information interface layer outline among a gathering of produced outlines. The SFTG security mode is gotten from the FTG security mode. Anyway with the SFTG security mode, the encoded and verified client information interface outline is separated into a few produced outlines that are then scrambled as opposed to masking the real edge among a gathering of made edges.

In the SFTG mode, the information interface layer outline containing the client information is scrambled without cushioning. The scrambled datalink layer outline is then separated into various sections, with the quantity of sections being equivalent to the size of the transmission gathering. Each section is irregular in size and each portion is put haphazardly inside a doled out made outline. Generally, the activities for the portions are comparable to the cushioning mode activities. The entirety of the fabricated outlines are then cushioned to a similar size as was acted in the other two security modes before being encoded. The SFTG security mode is appeared in Fig. 3.

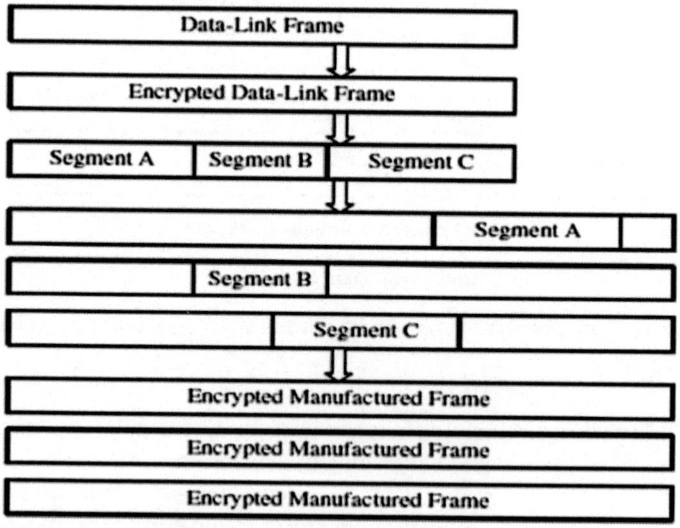

Fig. 3. SFTG security mode operations are shown randomly placing the user data within the frames of the transmission group.

The SFTG security mode transmits a predetermined number of outlines at the normal transmission time for a digital assailant, in any case, with the client information interface layer outline arbitrarily put into those casings. The digital assailant should in this way control the entirety of the casings in the transmission gathering and reassemble the client information interface layer outline before being fit for getting the client information.

The SFTG mode gives more security than the FTG mode since a digital assailant needs to control more information, and is along these lines utilized for joins that are experiencing assault or connections that are basic to control system activities and can subsequently be expected to be assaulted. In any case, since the SFTG security mode produces different casings to camouflage the client information furthermore, utilizes more encryption tasks, this mode requires more preparing and transfer speed ability which restricts its pertinence to low data transmission gadgets.

2.4 Security Header

The security modes necessitate that a security header be annexed onto the information connect layer outline with the goal that the getting DNP3 gadget could figure out which security activities are to be performed on the information connect layer outline. The security layer header is appeared in Fig. 4.

Mode	Type	Sequence Number	Sequence Length	Pad Start	Pad Stop	CRC

Fig. 4. Security header appended onto the DNP3 data-link layer frame before encryption and authentication.

In Fig. 4, there are six fields in the security header, which are: mode, type, succession number, grouping length, cushion start, cushion stop, and CRC.

- *Mode*: Demonstrates the security mode utilized for the information transmission. With this sign inside the casing, source what's more, goal gadgets can autonomously and right away change to another mode without advising the other gadget before the change. This builds the security layer responsiveness to changes in the security prerequisites.
- *Type*: Shows if the casing contains client information or if the edge is produced.
- *Sequence Number*: Gives the edge grouping to the SFTG mode with the goal that the casing containing the client information can be appropriately reassembled from the made casings. This field has no importance to different modes.
- *Sequence Length*: Shows the quantity of casings in the SFTGtransmission gathering. This field gives the adaptability to a security layer to adjust the quantity of casings utilized in SFTG transmission bunch without beforehand arranging this worth. This builds the security layer responsiveness to changes in security prerequisites.
- *Pad Start*: For the cushioning and FTG security modes this field gives the beginning octet area of the cushioning inside the outline. For the SFTG mode, this field shows where the encoded client information interface layer outline portion starts inside the produced edge.

- *Pad Stop*: For the cushioning and FTG security mode this field gives the completion octet area of the cushioning inside the outline. For the SFTG mode, this field shows where the encoded client information connect layer outline portion finishes inside the fabricated edge.
- *CRC*: Gives a cyclic redundancy check to the security header.

2.5 Interoperability and Security Mode Transitions

A trouble in receiving any progressions to a standard, for example, DNP3 is interoperability between gadgets that have not embraced the security, just as gadgets utilizing extraordinary security modes. The interoperability between gadgets utilizing security and those not utilizing security is required since quick selection of the security would be troublesome and costly to perform. Interoperability between gadgets utilizing diverse security modes is important to permit gadgets utilizing lower security abilities to speak with gadgets utilizing higher security abilities. The security layer interoperability issues are taken care of through the security layer activities and the security header affixed onto the DNP3 information connect layer outline before encryption.

For the interoperability between DNP3 gadgets utilizing the security and gadgets that don't, the edge size shows if security has been utilized on the information interface layer outline. All the security modes cushion the information connect layer casing to the most extreme permitted 292 octets. The security header is then included onto the outline. Consequently, if the edge is bigger than 292 octets, security has been utilized generally the security layer modes were not utilized for the information interface layer outline [7, 10, 11].

The adaptability in changing between security modes gives the way to a gadget to work in an ostensible mode and afterward progress to a higher security mode when important. In this way, DNP gadgets with low preparing or transmission capacity ability can at present evoke the FTG and SFTG security modes on an impermanent premise when important, expanding security while diminishing the gadget surveying by an ace to get the essential data transmission to deal with the security mode tasks (either the quantity of surveys or the sort of information being surveyed).

3 Security Layer Analysis

The security layer is intended to give extra quality to encryption and validation tasks in constraining the viability and immaterialness of traffic investigation and cryptanalysis assaults. The security layer expands the sum of information that a digital aggressor must control just as the measure of time for them to get the client information. The security layer isn't intended to secure against assaults, for example, replay assaults. Be that as it may, the DNP3 convention stack gives insurance from replay assaults because of the utilization of two arrangements of succession numbers, the discontinuity arrangement numbers utilized by the vehicle layer and the message part arrangements of the application layer. The vehicle layer succession number space is bigger than the application layer succession number space. Along these lines, a replayed casing is improbable to have arrangement numbers that match both the expected vehicle layer and application layer succession numbers, making these casings be disposed of.

3.1 Padding

The cushioning security mode includes next to no preparing overhead to the security activities and thusly gives about a similar presentation as essentially encoding and verifying the information. The cushioning tasks give fundamental security to restrain the ability of traffic investigation assaults dependent on the edge size. Notwithstanding, as referenced this security must be accommodated every individual information interface layer outline that is transmitted since the security doesn't modify the DNP3 detail. Consequently, a digital aggressor can in any case include the quantity of edges in the information transmission to decide the idea of the information transmission, regardless of whether it is control, observing, or insurance information. This shortcoming could be helped if the application layer cushioned all message pieces to a similar size, be that as it may, this would require a DNP3 determination change which would disregard the limitations made for the security layer.

Since the cushioning activities are about identical in execution and security as to just utilizing encryption on information transmissions, it is utilized as the reason for correlation with the other security modes. The time wherein to break the encryption for a security layer outline utilizing the cushioning activity is surrendered (1) as a general condition for any figure what's more, figure mode for an animal power assault. Cryptanalysis is accepted to abbreviate this time by a scalar factor reliant on the figure, key length and innovation as of now accessible for cryptanalysis assaults. Be that as it may, the overall measure of time to break the encryption is the equivalent for the entirety of the modes, taking out the need to examine confused elements that affectthe quality of the figure, for example, those relating to cryptanalysis.

$$T = ET \tag{1}$$

where T is the all-out time to break the casing encryption, and ET is an ideal opportunity to break the security layer outline encryption for the cushioning mode subject to components, for example, figure, figure mode, key size, and presently accessible cryptanalysis assault methods.

3.2 Frame Transmission Group

The FTG security mode expands the general preparing for the security layer where the security layer creates (N − 1) outlines, where N is the size of the information transmission gathering. Be that as it may, the preparing necessities for the goal gadget are a lot of lower since the security layer can dispose of the produced outlines dependent on the security header.

The quality of the FTG security mode in contrast with the cushioning mode can be grouped into three situations: best case, thinking pessimistically, and normal case. The most ideal situation is yielded (2) and speaks to the digital aggressor assaulting every one of the edges in the transmission bunch before finding the outline containing the client information as decided from the sort field in the security header. For the most ideal situation the security for the information transmission is identical to the number of casing utilized in the transmission gathering.

$$T = N \times ET \tag{2}$$

where T is the complete time to break the casing encryption, N is the quantity of casings in the transmission gathering, and ET is the time to break the security layer outline encryption subordinate on components, for example, figure, figure mode, key size, and right now accessible cryptanalysis assault strategies.

The most dire outcome imaginable is surrendered (3) and speaks to the digital aggressor finding the client information on the main endeavour, where a digital assailant can decide a fruitful assault from the sort field in the security header. The direst outcome imaginable of (3) is equivalent to the cushioning security mode given by (1).

$$T = ET \tag{3}$$

where T is the absolute time to break the casing encryption, and ET is an ideal opportunity to break the security layer outline encryption subject to elements, for example, figure, figure mode, key size, what's more, at present accessible cryptanalysis assault methods.

The normal case situation is surrendered (4) and speaks to the normal number endeavours for an aggressor to find and acquire the client information.

$$T = A \times ET = \frac{\sum_{i=1}^{N} i}{N} \times ET \approx \frac{N}{2} \times ET \tag{4}$$

where T is the complete time to break the casing encryption, An is the normal number of endeavoursto find outline containing the client information in the edge transmission gathering, ET is the opportunity to break the security layer outline encryption subject to factors, for example, figure, figure mode, key size, and as of now accessible cryptanalysis assault procedures, N is the quantity of outlines in the transmission gathering, and I is the edge position inside the gathering.

The FTG security thusly by and large gives N/multiple times more noteworthy security than the cushioning security mode. All together for the FTG mode to give more powerful security than the cushioning security mode, the transmission bunch size must be bigger than 2. In this way FTG in any event requires a data transmission multiple times more noteworthy than what might be required if security was not utilized or if the cushioning security mode was utilized.

The FTG security mode likewise supports littler transmission bunch sizes, which restricts the necessary extra transmission capacity for this security mode. Since (4) demonstrates that the normal measure of time to break the security is half of the transmission bunch size, enormous transmission bunches don't give as a lot of security to the information transmissions in correlation with the measure of time it would take to produce and transmit the edges. For the FTG security mode to be viable, each of the outlines in the transmission bunch must utilize unique encryption, as was proposed for the security layer. On the off chance that the equivalent encryption was utilized for every one of the transmission gathering outlines, the security would be identical to (3), which is no security improvement contrasted with the cushioning security mode. The FTG security mode is likewise still helpless against the traffic examination assault where the absolute number of casings transmitted inside a timeframe is checked to decide the sort of information being transmitted, for example control or checking information.

3.3 Performance

The three security-mode tasks were tried utilizing nonoptimized code consented for Windows 2000 on a Pentium 550 MHz machine utilizing a business visual-based compiler furthermore, business programming encryption segments. Subsequently better would be normal for an upgraded real systemgadget.

In Fig. 5, the cushioning security mode tasks are appeared with different measured information interface outlines, from the base 10 octets to the most extreme 292 octets. The information transmission size hub is set apart for the client information sizes from the vehicle layer, furthermore, thusly go from 0 octets (outline size = 10 octets) to 250 octets (outline size = 292 octets). The encryption utilized with the cushioning mode was the 256-piece Advanced Encryption Standard (AES) utilizing the chain blocking figure (CBC) mode with no confirmation.

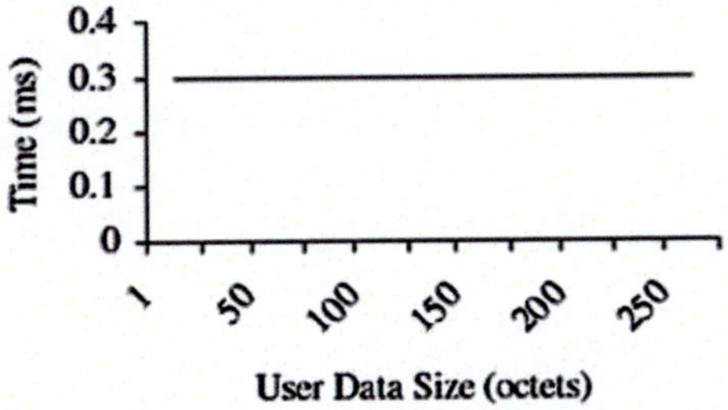

Fig. 5. Padding mode operation performance for various user data sizes

As can been seen from Fig. 5, the cushioning activities (adjusted to the closest tenth of a millisecond) have immaterial consequences for the encryption tasks. Along these lines, just the FTG what's more, SFTG security modes require thought execution savvy for the DNP3. A similar test was performed for the cushioning mode unscrambling tasks, with irrelevant contrasts among encryption and unscrambling tasks.

In Fig. 6, the FTG and SFTG security mode encryption tasks are demonstrated contrasted with one another for different transmission bunch sizes. The base FTG transmission bunch size was N = 3 while the SFTG least transmission bunch size was N = 2. Transmission bunch sizes was constrained to ten since enormous transmission bunch sizes are not pragmatic for DNP3 gadgets, because of the overhead in assembling outlines what's more, the deferrals related with the goal getting the information, particularly with the SFTG mode where the information is separated over the entirety of the casings in the transmission gathering. The equivalent encryption was utilized for the FTG and SFTG modes as the cushioning mode.

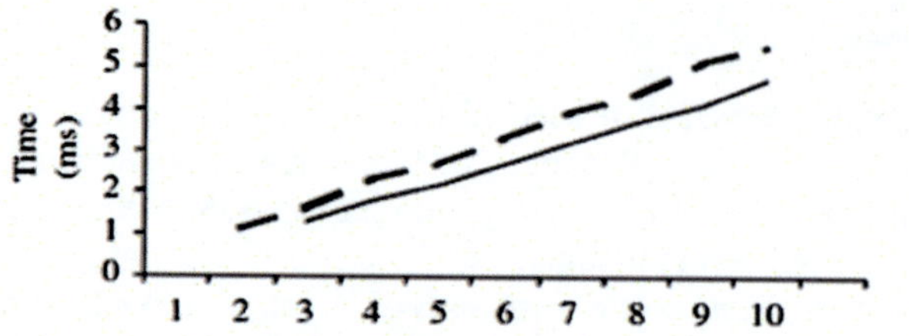

Fig. 6. FTG and SFTG modes encryption operation performance for various frame group sizes. FTG is the solid line and SFTG is the dashed line.

Table 1 gives the numeric outcomes to the FTG and SFTG mode for a progressively itemized examination of qualities, with a transmission bunch size of (N = 3) which is the base gathering size for the FTG mode.

Table 1. FTG and SFTG performance for (N = 3)

Security mode	Encryption (ms)	Decryption (ms)
FTG	1	0.7
SFTG	1.6	0.8

In light of the outcomes from Table 1, and the cushioning mode activity execution, Table 2 demonstrates the by and large execution for every one of the security modes dependent on a solitary client information interface layer outline. In Table 2, the transfer speed transmission time (TT) is given in relationship to the cushioning mode since the other security modes are products of this esteem.

Table 2. Overall mode performance for a link with (N = 3)

Security mode	Overall performance
Padding	0.3 ms + TT
FTG	1.6 ms + $TT \leq$ FTG $\leq$ 2.0 ms + 3TT
SFTG	2.4 ms + 3TT

In Table 2, the FTG has a scope of qualities for the by and large execution, since the client information interface edge will be between the first and last edges in the transmission gathering. On the off chance that the client information interface layer outline is the primary edge, the decoding tasks are comparable to the cushioning unscrambling activity.

4 Conclusion

Three security activity methods of the security layer found underneath the DNP3 information connect layer, which are cushioning security mode, the FTG security mode, and the SFTG security mode, have been proposed to improve the security of intensity systemcomputerorganizes in this paper.

The cushioning mode gives improved security past just utilizing encryption with insignificant impacts on execution. The cushioning mode is along these lines appropriate for lower security requests. The FTG security mode gives a lot higher security than the cushioning mode, however with expanded preparing overhead. The SFTG security mode gave a lot higher security than the FTG mode, however with higher handling request. The FTG and SFTG modes are suggested for joins requiring high security or for joins that briefly require higher security during digital assaults. The change between modes is exceptionally adaptable, permitting joins that require brief more elevated levels of security to effectively progress to a higher security mode and afterward back to the ordinary mode a short time later. The advances between modes can happen quickly without earlier data trades and freely, for example, one mode for each connection or an outstation progressing modes without demand from an ace.

The outcomes from the presentation investigation for the three security modes show that these three security activity modes can give extra solidarity to encryption and confirmation tasks in restricting the adequacy and appropriateness of traffic investigation and cryptanalysis assaults for power system information transmissions. Since the security layer modes are free of explicit encryption figures and validation tasks, the paper doesn't give imperatives on the kinds of encryption figures or validations utilized for the information transmissions.

The security tasks proposed in this paper have been effectively applied to upgrade power system security controls.

References

1. Z. Qin, F. Yu, C. Liu, X. Chen, How convolutional neural networks see the world—A survey of convolutional neural network visualization methods. Math. Found. Comput. (2018). https://doi.org/10.3934/mfc.2018008
2. N. Singhal, J.P.S. Raina, Comparative analysis of AES and RC4 algorithms for better utilization. Int. J. Comput. Trends Technol. 2231–280 (2011).
3. S. Almuhammadi, I. Al-Hejri, A comparative analysis of AES common modes of operation. Can. Conf. Electr. Comput. Eng. (2017). https://doi.org/10.1109/CCECE.2017.7946655
4. M. Budka, K. Deshpande, J. Thottan, *Communication Networks for Smart Grids: Making Smart Grid Real* (2014)
5. S. Duangphasuk, S. Kungpisdan, S. Hankla, Design and implementation of improved security protocols for DHCP using digital certificates, in *ICON 2011—17th IEEE International Conference on Networks* (2011). https://doi.org/10.1109/ICON.2011.6168490
6. M. Su, F. Li, G. Shi, L. Li, An action based access control model for multi-level security. Int. J. Secur. Appl.
7. N. Deng, RFID technology and network construction in the internet of things, in *Proceedings—2012 International Conference on Computer Science and Service System, CSSS 2012*. https://doi.org/10.1109/CSSS.2012.248

8. E. Doynikova, I. Kotenko, Countermeasure selection based on the attack and service dependency graphs for security incident management. Lect. Notes Comput. Sci. (Including subseries Lect. Notes Artif. Intell. Lect. Notes Bioinformat.) (2016). https://doi.org/10.1007/978-3-319-31811-0_7
9. Z. Tsiatsikas, D. Geneiatakis, G. Kambourakis, A.D. Keromytis, An efficient and easily deployable method for dealing with DoS in SIP services. Comput. Commun. (2015). https://doi.org/10.1016/j.comcom.2014.11.002
10. R.J. Campbell, The smart grid and cybersecurity-regulatory policy and issues, in *The Transformation of Electrical Power: Key Issues* (2013)
11. J. Koh, J. Myung, J. Kang, Throughput enhancement of secondary transmitter under target physical layer security rate, in *IEEE International Symposium on Personal, Indoor and Mobile Radio Communications, PIMRC* (2013). https://doi.org/10.1109/PIMRC.2013.6666190

Supervised Machine Learning Algorithm: A Review of Classification Techniques

Pankaj Saraswat(✉)

Department of Computer Science and Engineering, Sanskriti University, Mathura, Uttar Pradesh, India
pankajsaraswat.cse@sanskriti.edu.in

1 Introduction

Machine Learning (ML) algorithm will be considered as a subset of Artificial Intelligence (AI) since those algorithms are often seen as building blocks to make computers learn to behave more intelligently by somehow generalizing instead of just storing and retrieving data items like a database system, and other applications would do [1]. Machine learning possesses its inspiration from a spread of educational disciplines, including engineering, statistics, medical, and psychology. The core function of Machine learning algorithm attempts is to inform computers the way to automatically find a decent predictor supported experiences, and this job is completed by good classifier there are several applications for Machine Learning (ML), the foremost significant of which is data mining. People are often vulnerable to making mistakes during analyses or, possibly, when trying to determine relationships between different features. This makes it difficult for them to search out solutions to certain problems. ML can often be successfully applied to those problems, improving the system efficiency and therefore, the machine designs [2].

Machine Learning (ML) has become one of the mainstays of information technology. With the ever-increasing amounts of information becoming available, there's a decent reason to believe that smart data analysis will become even more pervasive as a necessary ingredient for technological progress. There are different applications for Machine Learning (ML), the foremost significant of which is data mining (DM). People are often at the risk of making mistakes during data analyses or, possibly, when trying to determine relationships between different features. Data mining (DM) and Machine Learning (ML) are Siamese twins from which several insights are often derived through proper learning algorithms [3]. There has been tremendous progress in data mining and machine learning as a result of the evolution of smart and Nanotechnology which led to curiosity to find hidden patterns in data to derive value. The fusion of statistics, machine learning, computer science, and computing has created a solid science, with a firm mathematical base, and with most powerful tools. Machine learning algorithms are organized into a text analysis that supported the required outcome of the algorithm. Supervised machine learning produce a function that maps inputs to desired outputs.

Machine Learning (ML) from experiences is an attribute of humans while computers. Don't have this ability. In supervised machine learning, our main objectives to find out

F. P. García Márquez (ed.), *International Conference on Intelligent Emerging Methods of Artificial Intelligence & Cloud Computing*, Smart Innovation, Systems and Technologies 273, https://doi.org/10.1007/978-3-030-92905-3_58

a target function that will be wont to predict the values of a category [4]. The method of applying a supervised machine learning to a real-world problem is described within the below figure (Fig. 1).

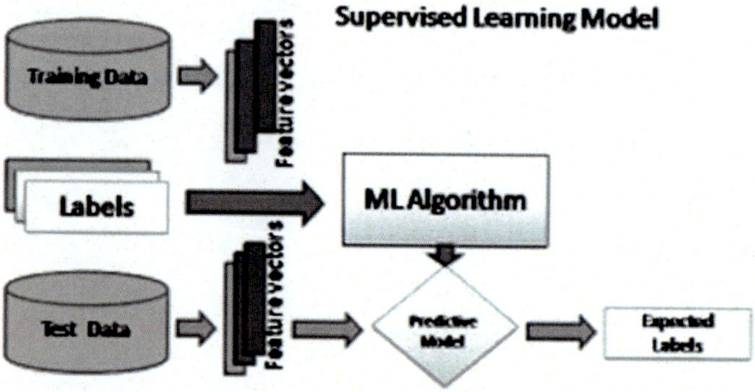

Fig. 1. Supervised machine learning model

Machine learning is finely suitable for the intricacy of handling through the different data sets, and therefore, the wide range of variables in addition because of the amount of information concerned where machines learning prospers on increasing data sets. The additional data supply into a machine learning structure, the more it's ready to be trained and concern the results to the superior value of insights [5]. At the liberty from the confines of individual-level thought and study, Machine Learning (ML) is clever to search out, and show the hidden patterns within the data set.

2 Supervised Machine Learning

In supervised machine learning, the primary step is managing the data set, to perform better training on data set an appropriate expert could suggest an improved feature selection. If the concerned expert isn't in reach, then the opposite approach is, "brute-force", which suggests measuring everything available within the hope that the proper (informative, relevant) features are often isolated. However, a data set collected by the "brute-force" method isn't directly suitable for induction. Ultimately, in most cases, it contains noise and missing feature values, and thus requires significant data pre-processing [6]. Within the next step, data pre-processing may be a key function of the researcher in Supervised Machine Learning (SML). Multiple techniques are proposed by different researchers to influence the missing data problems.

One amongst standard techniques of the supervised machine learning task is the classification problem: The learner is required to find out a function which maps a vector into one amongst different classes by observing different input-output samples of the function. Supervised machine learning is the process of learning a set of rules from instances, or more, generally creating a classifier that will be wont to generalize from new instances [7]. The method of applying a supervised machine learning to a real-world problem is described in Fig. 2. Inductive machine learning is the process of learning a set of rules from instances (examples during a training data set), or more, generally speaking, creating a classifier that may be wont to generalize from new instances. The second step is data preparation and data preprocessing.

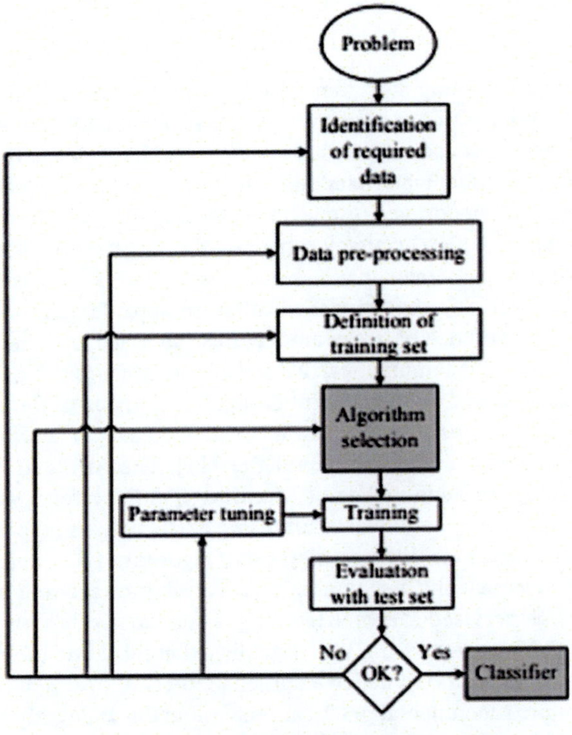

Fig. 2. Process of supervised machine learning

Depending on the circumstances, researchers have multiple methods to decide on from to handle missing the data that has recently introduced a survey of the latest techniques for outlier (noise) detection. These researchers have identified the techniques' advantages and drawbacks. Instance selection isn't only wont to handle noise but to deal with the in the feasibility of learning from very large data sets [8]. Instance selection in these data sets is an optimization problem that attempts to take care of the data mining quality while minimizing the sample size. It reduces data and enables a data mining algorithm to function and work effectively with very large data sets. There's a spread of procedures for data sampling instances from a large data set.

Feature subset selection is the process of identifying and removing as many irrelevant and redundant features as possible. This dimensionality reduction of the information and enables data processing algorithms to work faster and more effectively [9]. The fact that several features depend upon each other often unduly influences the accuracy of supervised machine learning classification models. This problem will be addressed by constructing new features from the fundamental feature set. This method is termed feature construction/transformation. These newly generated features may cause the creation of more concise and accurate classifiers. Additionally, the invention of meaningful features contributes to higher comprehensibility of the produced classifier and an improved understanding of the learned concept.

2.1 Feature of Machine Learning Algorithm

A supervised machine learning algorithm is applicable in various domains. Generally, Support Vector Machine (SVM) and neural networks tend to perform much better when handling multi-dimensions and continuous features. On the opposite hand, logic-based systems tend to perform better when handling categorical features [10]. For artificial neural network models and support vector machines, the massive sample size is required to realize its maximum prediction accuracy, whereas, NB may have a comparatively small data set. There's general agreement that K- Nearest Neighbor (k-NN) is extremely sensitive to irrelevant features, this characteristic is often explained by the way the algorithm works. Moreover, the presence of irrelevant features can make artificial neural network training very inefficient, even impractical. Most decision tree algorithms cannot perform well with problems that need diagonal partitioning. The division of the instance space is orthogonal to the axis of 1 variable and parallel to all other axes. Therefore, the resulting regions after partitioning are all hyper rectangles [11]. The Artificial Neural Network (ANN) and therefore, the Support Vector Machine (SVM) perform well when multicollinearity is present, and a nonlinear relationship exists between the input and output features. The most recent Machine Learning (ML) algorithm fall within the supervised machine learning category and hence aimed at learning an existing pattern between inputs and outputs. Supervised Machine learning is comparatively well-understood at a theoretical level, and it benefits from well-established algorithmic tools. Unsupervised machine learning has to this point defied a unified theoretical treatment. Nevertheless, it arguably poses a more fundamental practical problem in this it directly tackles the challenge of supervised learning by direct observation with none type of explicit feedback. Reinforcement learning has found extensive applications in problems that are characterized by clear feedback signals, like a win/lose outcome in games, which entail searches over large trees of possible action-observation histories [12].

A supervised learning algorithm initially performs the analysis task from the practice data and constructs a contingent function, to map new examples. A maximum setting probably facilitates the algorithm to exact courage the category labels for covered instances, and therefore, the same needs the supervised learning algorithm to reduce from training data to cover situations during a "rational" manner. The supervised learning algorithm is possibly utilized in multiple application areas that include online marketing, financing, manufacturing, testing, stock market prediction, and so on. An Artificial Neural Network (ANN), usually called a Neural Network (NN), maybe a mathematical model that's inspired by the structure, and/or functional aspects of biological neural networks [13]. A neural network consists of an interconnected group of artificial neurons, and it processes information employing a Connectionist approach to computation. In most cases, an ANN is an adaptive system that changes its structure supported external or internal information that flows through the network during the training phase. Modern neural networks are non-linear statistical data modeling tools. They're usually wont to model complex relationships between inputs and outputs or to search out patterns in data.

Decision tree learning, employed in data mining and machine learning, uses a decision tree as a predictive model that maps observations about an item to conclusions about the item's target value. More descriptive names for such tree models are classification trees or regression trees. In these tree structures, leaves represent classifications and branches represent conjunctions of features that result in those classifications. Decision tree algorithm, a decision tree is often used to visually and explicitly represent decision-making in data mining, a decision tree describes data, but not decisions, rather the resulting classification tree may be an input for decision-making. Support Vector Machine (SVM) constructs a set of hyper plane during a high or infinite-dimensional space, which may be used for classification, regression or other tasks. Intuitively, a decent separation is achieved by the hyper plane that has the most important distance to the closest training data points of any class, since generally the larger the margin the lower the generalization error of the classifier. Whereas the initial problem could also be stated during a finite-dimensional space, It often happens that therein space the sets to be discriminated aren't linearly separable.

The supervised machine learning algorithms such as decision trees and support vector machines are capable enough to deal with big data mining tasks. Although the efficiency of the algorithm considerably improving there's a necessity for the adaptive boosting process required to extend the predictive accuracy.

3 Conclusion

This paper describes the best-known supervised machine learning techniques in relative detail. We should always remark that our list of references isn't a comprehensive list of papers discussing supervised methods, our main objective was to provide a review article of the key ideas, instead of a straightforward list of all publications which had discussed or made use of these ideas. Despite this, we hope that the references cited cover the most important theoretical problems, and provide access to the main branches of the literature handling such methods, guiding the researcher in interesting research directions. Supervised machine learning methods are being applied in several domains. Because of the scope of this paper, it's very difficult to debate the strength and weaknesses of every algorithm of machine learning. The selection of the algorithm in machine learning mainly depends on task nature. The performance of the support vector machine and Neural Networks is best when handling multi-dimensions and continuous features. While logic-based systems tend to perform better when handling discrete/categorical features. For artificial neural network (ANN) models a support-vector machine, the outsized sample size is required to realize its maximum prediction accuracy, whereas, NB may have a comparatively small data set.

References

1. A.I. Kadhim, Survey on supervised machine learning techniques for automatic text classification. Artif. Intell. Rev. (2019)
2. J. Martínez Torres, C. Iglesias Comesaña, P.J. García-Nieto, Review: Machine learning techniques applied to cybersecurity. Int. J. Mach. Learn. Cybern. (2019)
3. A. Singh, N. Thakur, A. Sharma, A review of supervised machine learning algorithms, in *Proceedings of the 10th INDIACom; 2016 3rd International Conference on Computing for Sustainable Global Development, INDIACom 2016* (2016)
4. M. Crawford, T.M. Khoshgoftaar, J.D. Prusa, A.N. Richter, H. Al Najada, Survey of review spam detection using machine learning techniques. J. Big Data (2015)
5. M. Usama et al., Unsupervised machine learning for networking: Techniques, applications and research challenges. IEEE Access (2019)
6. P.P. Angelov, X. Gu, Brief introduction to statistical machine learning, in *Studies in Computational Intelligence* (2019)
7. D. Tuia, C. Persello, L. Bruzzone, Domain adaptation for the classification of remote sensing data: An overview of recent advances. IEEE Geosci. Rem. Sens. Mag. (2016)
8. U. Narayanan, A. Unnikrishnan, V. Paul, S. Joseph, A survey on various supervised classification algorithms, in *2017 International Conference on Energy, Communication, Data Analytics and Soft Computing, ICECDS 2017* (2018)
9. J. Richiardi, B. Ng, Recent advances in supervised learning for brain graph classification, in *2013 IEEE Global Conference on Signal and Information Processing, GlobalSIP 2013—Proceedings* (2013)
10. A. Criminisi, J. Shotton, E. Konukoglu, Decision forests: A unified framework for classification, regression, density estimation, manifold learning and semi-supervised learning. Found. Trends Comput. Graph. Vis. (2011)
11. Z. Madhoushi, A.R. Hamdan, S. Zainudin, Sentiment analysis techniques in recent works, in *Proceedings of the 2015 Science and Information Conference, SAI 2015* (2015)
12. H. Hormozi, E. Hormozi, H.R. Nohooji, The classification of the applicable machine learning methods in robot manipulators. Int. J. Mach. Learn. Comput. (2012)
13. M. Schrynemackers, R. Küffner, P. Geurts, On protocols and measures for the validation of supervised methods for the inference of biological networks. Front. Genet. (2013)

Survey on Fog Computing and Its Function in IoT

Pankaj Saraswat(✉)

Department of Computer Science and Engineering, Sanskriti University, Mathura, Uttar Pradesh, India
sanpubip@gmail.com

1 Introduction

The Internet of Things (IoT) is one of the main technologies that have the ability to render the society with limitless benefits. IoT allows many objects to communicate with each other without human interference [1]. Due to low computing power and storage computations IOT encounters several problems like security, reliability, privacy and speed. Cloud of things (COT) is the result of combining IOT with cloud [2]. COT provides the best solution to address most of the issues. The CoT optimizes the analysis and collection of IoT data and offers fast minimal-cost installation and implementation for large data analysis and delivery. IoT's convergence with cloud services offers several benefits to various IoT applications. Nonetheless, as there are many IoT systems with diverse frameworks, it is a challenging task to create new Internet of things technologies [3]. This is due to IoT systems produce vast quantities of data from detectors and other devices. These enormous data are examined to make decisions. It takes excessively high network bandwidth to transfer all of these data to the cloud. Fog computing comes into play to solve these problems. In order to transfer the data to web a high network capacity is needed. Fog computing plays a important role to rectify these issues. Cisco gives the term "fog computing". It's a new technique that brings a lot of advantages to various domains, particularly the IoT. Fog computing, like the cloud, enables services for IoT clients like information handling and storage. Fog computing is focused on supplying local fog devices with information handling and storage capabilities rather than sending them to the cloud [4]. Fog computing in the IoT is intended to improve productivity, performance and decrease the amount of data going to the internet for storing, analyzing and maintaining. The data gathered by detectors will therefore be sent to service interface devices for analysis and transient storage, rather than to the internet, minimizing internet traffic and delay. Combining fog computing with the IoT introduces a new business platform, called fog as a service (FaaS), in which a provider constructs a network of fog nodes throughout its membership base and functions as a tenant for many renters from several different markets. Faas will allow the introduction of new services to customers. Faas allow organization (both big and small) to perform public and private computation in order to fulfil the requirements of large number of different customers (Fig. 1).

F. P. García Márquez (ed.), *International Conference on Intelligent Emerging Methods of Artificial Intelligence & Cloud Computing*, Smart Innovation, Systems and Technologies 273, https://doi.org/10.1007/978-3-030-92905-3_59

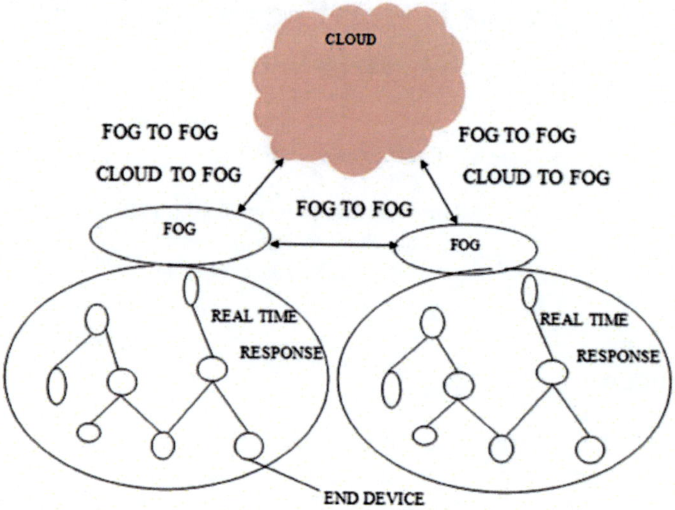

Fig. 1. IoT enabled fog computing

- Fog Computing

Fog computing is a framework with limited capacity like decentralized computing, processing and network services among various end devices and traditional cloud computing. This offers a good approach for bandwidth-sensitive IoT applications [5]. Fog Computing is a globally distributed computing system with a pool of resources composed of one or more omnipresently linked diverse devices at the top of the network and not solely securely supported by cloud computing, jointly delivering distributed computing, processing and connectivity to a large number of customers in disconnected environments [6]. Fog Computing is a fully digital platform that offers services for computing, processing and communication between end devices and conventional cloud infrastructure data centers, usually but not necessarily placed at the boundary of the network. Fog computing is a situation in which a large number of diverse prevalent and distributed devices interact and ultimately collaborate between them and the network to execute processing and computing tasks without third party interference.

- Characteristics

Fog computing is a web extension but nearer to end users. Fog computing acts an interface between cloud to end user taking networking resources, processing to end devices by themselves. These machines are referred to as Fog nodes. With network connection they can be installed anywhere. Any computer with processing, memory and network access, such as industrial controllers, switches, servers, integrated routers and CCTV cameras, can be a fog node (Fig. 2).

Some of the characteristics of cloud computing are-

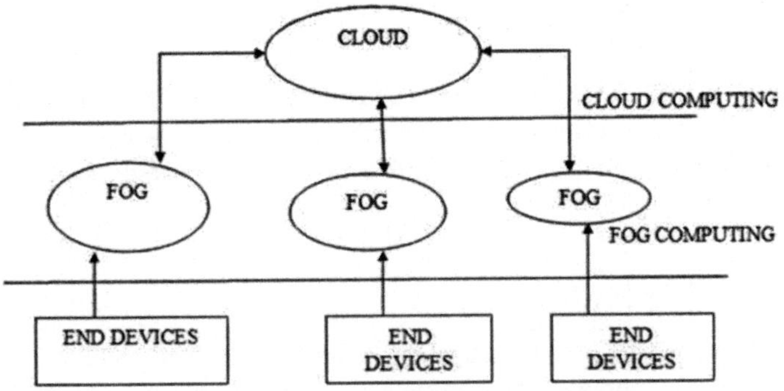

Fig. 2. Fog computing-extension of cloud

- Location awareness and low latency: Fog computing promotes the understanding of location where fog nodes could be installed at various locations. It also has lower latency when storing end system data as the fog is nearer to end devices.
- Geographic distribution: unlike the central cloud, the resources and software offered by the fog are decentralized and released anywhere.
- Scalability: The external environment is monitored by vast detector networks. The fog provides centralized data storage facilities that can function with such a massive computer.
- Connectivity support: One of the important elements of fog technologies is the ability to communicate remotely to portable devices, allowing connectivity methods.
- Interactions in real time: Fog computing systems offer real-time interactions between fog networks instead of cloud-based batch processing.
- Heterogeneity: Fog routers or end devices are built by different producers and must be configured according to their frameworks. The fog has the capacity to perform on various platforms.
- Interoperability: elements of Fog can modularize and function across different fields and providers.

Benefits

- Greater business agility: using the right equipment will easily develop and execute fog computation applications. Such programs can also configure the system to function as per the requirements of the user.
- Low latency: fog has the ability to sustain applications in real time (e.g. games, streaming media).
- Geographical and wide-scale allocation: Fog technology provides broad and widely dispersed applications with decentralised computing and processing resources.
- Lower operating costs: preserve system capacity by storing locally collected data instead of submitting it to the cloud for review.
- Stability and Heterogeneity: Stability and heterogeneity: Fog technology enables various services to communicate between distinct physical settings and systems.

2 Working

Fog computing is yet another computer worldview that stretches to the edge of the network the conventional decentralized computing and management. The working of fog computing involves a layered architecture. The 3 main layers present in the architecture are.

2.1 Terminal Layer

It is the closest layer and physical state to the end client. This includes for example cameras, smart phones, shrewd cars, clever cards, per user, etc. from different IoT tools. Such devices are all widely shared topographically [7]. This layer is responsible for identifying the specific articles or information and distributing this detected information for managing and storing to the top layer.

2.2 Fog Layer

On the bottom of the device is the fog layer. This layer consists of innumerable nodes, many of which include switches, targets, antennas, servers, etc. Such fog nodes are commonly distributed between the endpoint gadgets and the cloud. The end user will effectively connect administration with fog nodes. The administration has the ability to point, relay and, by the way, retain the information gathered.

2.3 Cloud Computing Layer

Various high-performance databases and devices are accessible to the cloud computing layer which provides different software services, such as home automation, smart transport, electric grid, etc.

3 Technologies for Fog Computing

3.1 Computing Technologies

Fog nodes can support both separately and independently local computing and information processing. These computing technologies need to support the ability to manage minimal-latency applications. The most popular computational techniques are: offloading computing, latency management.

3.2 Communication Technologies

In order to connect cloud, end users, and fog nodes the fog node acts an intermediary networking element. There are three categories of communication—Firstly wired or wireless communication takes place between cloud's data center and fog nodes. Secondly connection can be connected or connectionless between fog nodes. Thirdly normally wireless communication takes place between end devices and fog nodes.

3.3 Storage Technologies

For less delay applications fog computing is best. In order to minimize the time required to upload the data from distant databases, fog nodes the decentralized node with useful content. Through effective usage of storage technologies fog computing provides the most valuable services.

4 Applications

4.1 Connected Car

For all Internet-Connected cars, fog computing will be the most efficient solution, as it offers a high level of real-time connectivity[8]. It will also allow vehicles, traffic lights to communicate with each other to provide customers with a great service. Various accidents and crashes can be avoided using fog computing.

4.2 Smart Traffic Lights

Fog computing enables roads to be opened based on the perception of flashing lights [9]. This detects pedestrian's presence and tests neighbouring automobiles range and pace. It may be known that smart traffic lights are fog nodes that are coordinated to give alarm signals to neighbouring automobiles. With Bluetooth, 3 G, smart traffic signals and roadside systems, the fog connections between the car and the routers are enhanced.

4.3 Smart Home

Fog computing has many advantages for applications for home security. This offers a single framework for all separate devices to be incorporated [10]. It also offers versatile tools for processing, production in real time and low delay.

4.4 Wireless Sensor and Actuator Networks

Actuators are used to control the operation of the calculation and modify the nature of a locked-loop device. Actuators can be viewed as fog nodes which provide various activities to monitor sensors on end devices. Such WSNs need less capacity, less storage and also very low latency to operate.

4.5 Healthcare

Fog computing plays an important role in healthcare [11]. Combining large number of services analysis, medical record, distant storage requires secure, inaccessible internet connectivity. Traffic issues and network connectivity are dealt by fog computing.

5 Conclusion

In IoT domains, fog computing is an evolving issue. Experts as well as business practitioners seemed to indicate that fog-enabled IoT is a crucial facilitator for IoT implementations across various industries. IoT devices are complex in nature and have limited capacity for storage and distribution. However, there are several problems with the conventional distributed cloud, such as higher bandwidth and service failure. Fog computing was established as an expansion of the cloud to solve these problems. The paper presented Fog computing concept, Fog computing is significant. And Fog's computing architecture is provided with three layers. Essential technologies have been explored for Fog computing. Here it is known that if fog computing offers most promising performances then the application delay problem can be prevented.

References

1. A. Soia, O. Konnikova, E. Konnikov, The internet of things, in *Proceedings of the 33rd International Business Information Management Association Conference, IBIMA 2019: Education Excellence and Innovation Management through Vision 2020* (2019)
2. B. Negash, A.M. Rahmani, P. Liljeberg, A. Jantsch, Fog computing fundamentals in the Internet-of-Things, in *Fog Computing in the Internet of Things: Intelligence at the Edge*, pp. 3–13 (2017)
3. C.V. Nisha Angeline, R. Lavanya, *Fog computing and its role in the internet of things*, pp. 63–71 (2018)
4. S. Yi, C. Li, Q. Li, *A survey of fog computing* (2015)
5. P. Pagel, S. Schulte, Fog computing. Informatik-Spektrum (2019)
6. D.L. Beaty, Fog computing. ASHRAE J. **60**(1), 68–74 (2018)
7. P. Hu, S. Dhelim, H. Ning, T. Qiu, Survey on fog computing: architecture, key technologies, applications and open issues. J. Netw. Comput. Appl. (2017)
8. M. Sookhak et al., Fog vehicular computing: augmentation of fog computing using vehicular cloud computing. IEEE Veh. Technol. Mag. **12**(3), 55–64 (2017)
9. F. Haouari, R. Faraj, J.M. Alja'Am, Fog computing potentials, applications, and challenges, in *2018 International Conference on Computer and Applications, ICCA 2018* (2018)
10. H. Ghayvat, S. Mukhopadhyay, J. Liu, A. Babu, E. Elahi, X. Gui, Internet of Things for smart homes and buildings: Opportunities and challenges. Aust. J. Telecommun. Digit. Econ. **3**(4), 33 (2015)
11. T.N. Gia, M. Jiang, A.M. Rahmani, T. Westerlund, P. Liljeberg, H. Tenhunen, Fog computing in healthcare Internet of Things: A case study on ECG feature extraction, in *Proceedings—15th IEEE International Conference on Computer and Information Technology, CIT 2015, 14th IEEE International Conference on Ubiquitous Computing and Communications, IUCC 2015, 13th IEEE International Conference on Dependable, Autonomic and Se* (2015)

Understanding Human Emotional Intelligence

Rishi Sikka(✉)

Department of Electronics and Communication Engineering, Sanskriti University, Mathura, Uttar Pradesh, India
rishisikka.ec@sanskriti.edu.in

1 Introduction

People are equipped for creating a large number of facial activities during communication that change in multifaceted nature, power, and importance. Sentiment or aim is frequently conveyed by natural changes in one or a few discrete highlights. The expansion or absence of at least one facial activities may modify its explanation [1]. Furthermore, some outward appearances may have a comparative gross morphology yet show changed significance for various pronunciation powers. So as to catch the tinge of outward appearance in non-verbal correspondence, a current testing tool is utilized which will have the option to capture human feelings by perusing or looking at outward appearances. This calculation consequently separates highlights and their movement data, segregate unobtrusively extraordinary outward appearances, and devicespeech intensity. Emotional perspectives have tremendous effect on Social insight like correspondence understanding, basic leadership and furthermore helps in understanding behaviour of human [2]. Emotionreaction is completed in a miscellaneous manner, it might be verbal or non-verbal. Voice (Audible) is verbal type of correspondence and facial expressions, activity, body stances and motion is non-verbal type of correspondence. While imparting just 7% impact of message is contributes by verbal part all in all, 38% in vocal part and 55% in impact of the speaker's message is contributed by outward appearance. Hence robotized and real-time appearance would assume significant job in human and machine communication. Facial appearance reaction would be valuable from human offices to clinical practices. Examination of facial appearance assumes major jobs for applications which depend on "emotion recognition" like "Human Computer Interaction" (HCI) [3], Robot, Alert System, Animation and Pain checking for patients (Fig. 1).

Categories of Facial Expressions

Human emotions	Face motion	Definition
fear	Outer eyebrow inverted U, U-shaped inner eyebrow, opened mouth, dropped jaw	Danger, fear, horror, panic, nervous, worried
Sadness	Eyebrow downward, inner eyebrow raised, eyes closed, mouth shape pulled downwards	Suffering, misery, hurt, hopelessness, disappointment

(*continued*)

F. P. García Márquez (ed.), *International Conference on Intelligent Emerging Methods of Artificial Intelligence & Cloud Computing*, Smart Innovation, Systems and Technologies 273, https://doi.org/10.1007/978-3-030-92905-3_60

(*continued*)

Human emotions	Face motion	Definition
Happiness	Eyes wide opened, upward shape of mouth, lip corner pulled up, raised cheeks, eye wrinkles	Pride, cherish, thrill, pleasure, hope, relief
Surprise	Eyes opened, raised eyebrows, jaw dropped, open mouth	unexpectation, shock, astonishment, amazed
Anger	Eyebrows downwards, eyes opened, tight teeth, shut lips, lids raised	Harmful, irritation, frustration, dislike, hate, annoyed
Disgust	Corner depressed, wrinkled nose, lip depressed, eyebrows lowered	Taste, sound, smell, tough

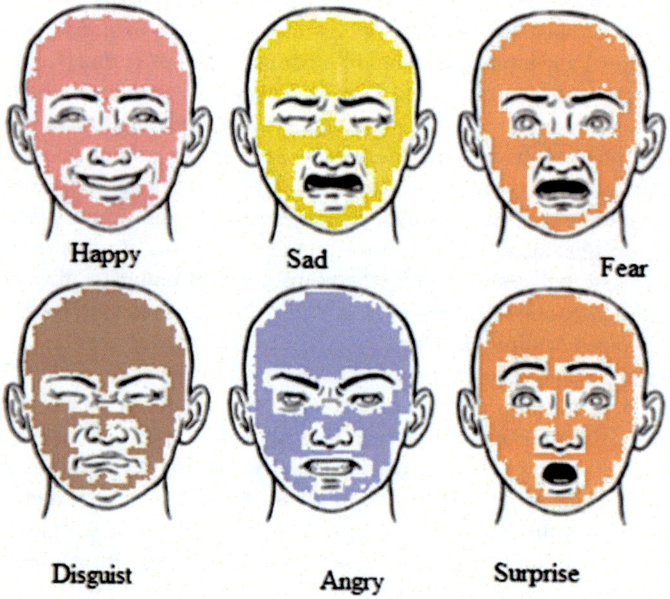

Fig. 1. Human emotions

2 System Methodology

The model which performs acknowledgment of outward appearance is called facial acknowledgment model. "Image Processing" is utilized for recognizing facial expression. With the assistance of "image processing" valuable data from picture can get separated. "Image Processing" changes over picture into advanced frame and play out certain tasks on it to extricate valuable data from picture.

"Facial Recognition model" comprises of following advances.

Image Acquisition

Still picture or picture successions are utilized for appearance recognition. 2-Dimension grey-scale facial picture is generally mainstream for facial picture acknowledgment in spite of the fact that coloured pictures can provide on more data about feeling, for example, reddening. In future coloured pictures will incline toward for the equivalent due to ease accessibility of shading picture supplies [4]. For picture procurement Camera, Cell Phone or other computerized gadgets are utilized.

Pre-processing

Pre-Processing assumes a key role in by and large procedure. Pre-processing stage improves the nature of info picture and finds information regarding emotions by removing noise or distortion and smoothing the picture [5]. It expels excess from picture without the picture detail. Pre-Processing likewise incorporates filtration and standardization of picture which produces uniform size and pivoted picture.

Segmentation

Segmentation segregates thepicture into significant sections. Division of a picture is a technique for separating the picture into homogenous, self-reliable parts comparing to various items in the picture on the basis of surface, edge and power [6].

Feature Extraction

Feature extraction can be considered as "manoeuvring" part in picture. It incorporates data of shape, movement, shading, surface of facial picture. It removes the significant data structure picture. When contrasted with unique picture include extraction altogether decreases the data of picture, which gives storage advantage [7].

Classification

Classification organize pursues the output of feature extraction. It recognizes the facial picture and assembled them as per certain classes and help in their capable acknowledgment. It is an intricate procedure since it might get influenced by numerous elements [8]. Classification stage can likewise be called as highlight choice stage, manages removed data and gathering them as per certain parameters (Fig. 2).

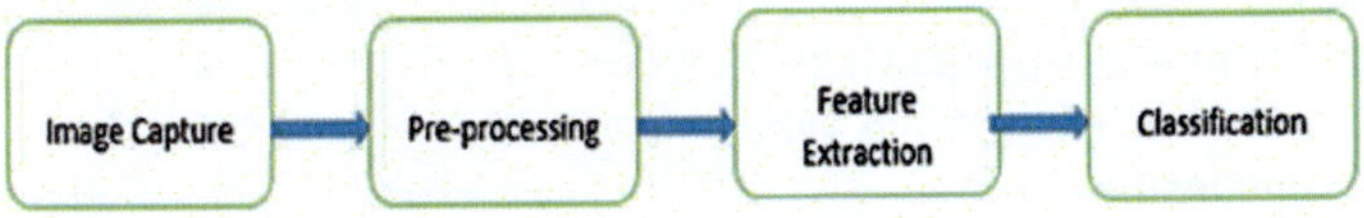

Fig. 2. Basic block diagram

3 Considering Eigen Values for Facial Expression Recognition

An algorithmhas been recommended for facial expression recognition which can characterize the given picture into the categories of one of the seven fundamental expressions for classifications (satisfaction, bitterness, dread, shock, outrage, sicken and nonpartisan). PCA is utilized for decreasing the dimensions in input information while holding those qualities of the informational collection that contribute most to its change, by maintaining lower-control head segments and disregarding higher-request ones. Such low-request parts contain the "most significant" parts of the information. The extricated feature vectors in the decreased space are utilized to prepare the managed Neural Network classifier. This methodology results amazingly ground-breaking since it doesn't require the discovery of any reference point or hub. The proposed strategy is quick and can be utilized for situation based applications. This code has been tried using the "JAFFE Database" [9]. Utilizing 150 pictures arbitrarily chose for preparing and 63 pictures for testing, with no covering, we get an astounding acknowledgment rate more prominent than 83%.

Principal Component Analysis

"Principal component analysis" or "karhunen-loeve"change is a standard system utilized in the measurable design acknowledgment and sign handling for information decrease. As the example regularly contains excess data, mapping it to a component vector can dispose of this excess but protect the greater part of the inherent data substance of the design. These features that have been analysed have extraordinary job in recognizing input designs. A face picture in 2-measurement with size m x m can likewise be considered as one dimensional vector of measurement m 2. Every one of these vectors are of length m 2, this depicts m x m picture and is a direct mix of the first face pictures. As these vectors are the eigenvectors of the covariance matrix comparing to the first face pictures, furthermore, on the grounds that they are face-like in appearance, they are alluded as "Eigen faces" [10]. After estimation of the covariance matrix, noteworthy eigenvectors of the covariance network are registered. The quantity of Eigen-vector relies upon application and precision that the framework needs and it is clear that if number of Eigen-vectors [11] is huge the exactness of the strategy improved however computational intricacy expanded.

- Let n (i) be initialized where i = 1 to x
- Computing the mean "M" and then subtracting it to obtain n (i) = M.
- Computing the matrix $A_{ij} = (n\ (i) - M)\ (N\ (i) - M)^t$
- Determining the Eigenvalues and the Eigen vectors
- Sorting the obtained values.

Significant Highlights

1. Outward appearance acknowledgment dependent on Neural Networks.
2. PCA-based dimensionality decrease.
3. High acknowledgment rate.

4. Instinctive GUI (Graphical User Interface).
5. Simple C/C++ usage.

4 Conclusion

In this paper the confinements have been investigated of existing model. Emotion recognition utilizing mind movement. In emotional recognition utilizing thoughts cerebrum exercises utilizing EEG signals has been utilized which is hardest assignment to do as it become costly, complex and furthermore tedious when this paper attempt to gauge human cerebrum with Electroencephalography (EEG). In any event, when they have utilized existing information their consequence of examination were 31–81 rates right and from which even by utilizing Fuzzy rationale 72–81 rates just for two classes of feelings were accomplished. This paper additionally gives us thought that we can detect human feelings likewise by perusing and contrasting the appearances and pictures or information which is put away in information base. In this paper by utilizing a framework which is prepared by neural systems this paper have accomplished up to 97% precise outcomes.

References

1. P. M. Niedenthal, M. Brauer, Social functionality of human emotion. Annu. Rev. Psychol. (2012)
2. S. Côté, G.A. Van Kleef, T. Sy, The social effects of emotion regulation in organizations, in *Emotional Labor in the 21st Century: Diverse Perspectives on Emotion Regulation at Work*, 2013, pp. 79–100
3. R. Mortier, H. Haddadi, T. Henderson, D. Mcauley, J. Crowcroft, A. Crabtree, Human-data interaction: The encyclopedia of human-computer interaction. Encycl. Hum. Comput. Interact. 1–48 (2016)
4. R. Koprowski, Image acquisition, in *Studies in Computational Intelligence* (2018)
5. C. Kamusoko, Pre-processing, in *Springer Geography* (2019)
6. A. Bellili, S. Larabi, Image segmentation by image analogies, in *Lecture Notes in Computer Science (including subseries Lecture Notes in Artificial Intelligence and Lecture Notes in Bioinformatics)*, vol. 8641 LNCS, pp. 143–151 (2014)
7. S. Sen, A. Dutta, N. Dey, Feature extraction, in *SpringerBriefs in Applied Sciences and Technology* (2019)
8. T.M. Helfer, E. Eixarch, F. Figueras, E. Gratacós, Classification, in *Fetal Growth Restriction: Current Evidence and Clinical Practice* (2018)
9. M. Abdulrahman, A. Eleyan, Facial expression recognition using Support Vector Machines, in *2015 23rd Signal Processing and Communications Applications Conference, SIU 2015—Proceedings* (2015)
10. A.K. Ramaniharan, S.C. Manoharan, R. Swaminathan, Laplace Beltrami eigen value based classification of normal and Alzheimer MR images using parametric and non-parametric classifiers, Expert Syst. Appl. (2016)
11. P. David Polly, A. Michelle Lawing, A.C. Fabre, A. Goswami, Phylogenetic principal components analysis and geometric morphometrics. Hystrix (2013)

Review on Continuous Variable Transmission (CVT)

Ajay Agrawal(✉)

Department of Mechanical Engineering, Sanskriti University, Mathura, Uttar Pradesh, India
ajayagrawal.me@sanskriti.edu.in

1 Introduction

Any transmission is intended to translate engine performance into performance of the vehicle. By providing a variety of gear ratios between the engine crankshaft and the vehicle's output axle, a transmission performs this task. While the engine runs at the same speed or rpm range, increasing gear ratio can result in a different vehicle speed and torque profile. The aim of the transmission is to allow the engine to function in an ideal state of power generation and to efficiently transfer this power to the track by using the correct gear ratio.

A traditional transmission increases the speed of the car while retaining the working range of the engine by using gear sets. By using a clutching system that activates different gears depending on engine and vehicle speed, the transmission will transfer power from the engine crankshaft into usable power at the output shaft. As the gear ratio increases, the output axel's rotational speed or angular velocity increases with respect to the engine crankshaft's angular velocity.

Engine performance in different ranges of the engine creates power band where certain engine speed ranges produce a specific horsepower output level and rate. To gain from this effect, a transmission can be modified. The tuner can choose a power band that results in lower power output and higher fuel consumption for normal driving, or maybe a power band that generates higher power output for aggressive driving at higher engine speeds. The traditional transmission usually works within a single engine power band which causes the engine speed to fluctuate within this power band as each gear shift is done.

The transmission adapts to the drive wheels the performance of the internal combustion engine. These engines will run at a relatively high speed of motion, which is inadequate for beginning the vehicle, stopping and slower movement. The transmission lowers the higher speed of the engine to the slower speed of the axle, increasing the torque. The transmission is usually connected to the engine's crankshaft. The transmission output is transmitted to one or more differentials via driveshaft, which in turn drives the wheels. The need for transmission in an automobile is a result of the internal combustion engine's characteristics.

Engines are typically operate over a period of 600 to around 7000 revolutions per minute (although this differs and generally is lower for diesel engines), while the wheels

F. P. García Márquez (ed.), *International Conference on Intelligent Emerging Methods of Artificial Intelligence & Cloud Computing*, Smart Innovation, Systems and Technologies 273, https://doi.org/10.1007/978-3-030-92905-3_61

of the vehicle spin between 0 rpm and around 1800 rpm. Sometimes, when the car moves from rest or drives gradually, the greater torque is needed, while the most force is required at high speed. It is therefore necessary to have a system which transforms the output of the engine such that it can supply high torque at low speeds but also operate at highway speeds with the engine still operating within its limits.

The process is carried out by transmissions. The transmission's structural elements, primarily casting, also contribute significantly to the power train system and the automobile as a whole. This is important for the lowest noise and vibration engineering. The rigidity of the power train assembly itself is important in determining the magnitude and frequency of the source (engine) vibrations [1–4].

1.1 Optimum RPM Range

An automobile engine runs at its peak at a certain rate of revolutions per minute (RPM) and it is the role of the transmission to insure that the power is transmitted to the wheels while preserving the engine speed within this optimum range. This is achieved by the transmission through different combinations of gears. In addition to the different forward gears, a transmission contains a neutral position that disconnects the engine from the drive wheels, and the reverse position that causes the drive wheels to turn in opposite direction, thus reversing the car's direction.

For more than 100 years, a Continuously Variable Transmission (CVT) has been around but has only recently found its way into automotive applications. In road vehicles, the overwhelming majority of transmissions are either manual or conventional in design of automatic. Such drives utilize meshing gears that provide separate measures between the engine and the speed of the car. However, there are alternative designs that can transmit power and give less ratio change at the same time; in other words, a continuously variable transmission.

Continuously Variable Transmission (CVT) is such a type of transmission that provides an smooth and uninterrupted speed ratio range, unlike a normal transmission that only provides a few discrete ratios. It was used for industrial applications such as operating lathes or light duty drill presses, though, until recently. But the stage was set for CVT's rise in the automotive arena with the introduction of improved materials such as high density belts, advanced hydraulics and more recently, high speed sensors and microprocessors. Most small tractors with basic rubber belt CVTs for home and garden use.

Small utility vehicle John Deere Gator lines use a conical pulley system belt. They can deliver an abundance of power and reach speeds of 10–15 mph (16–24 km/h), all without a clutch or gear shift. CVTs, typically the variable pulley variety/rubber belt, are used by almost all snowmobiles, old and new, and motor scooters.

A Continuously Variable Transmission (CVT) can change through an infinite number of efficient gear ratios between maximum and minimum values. The modern mechanical transmissions on the other side give only a fixed set of gear ratios. A CVT's unique feature allows the driving shaft to maintain a constant angular velocity over a variety of output speeds. This property increases fuel efficiency relative to other transmissions by enabling the engine to run for a variety of vehicle speeds at its most effective revolutions per minute (RPM).

Another possibility is that, by allowing the engine to rotate at the RPM at which it produces the peak power, a CVT can be used to achieve the best automotive performance. This is normally higher than the maximum efficiency of the RPM. Using CVT makes the engine move immediately from idle to pre-programmed rpm, so the engine input is constant and then varies the output speed to accelerate smoothly and seamlessly. CVTs require power to be applied without any jerk. The continuously variable transmission (CVT) was around as long as the engine. Engineers' technical superiority over the multi-ratio gearbox has always been acknowledged.

A CVT allows the engine to run at any desired speed at its most fuel-efficient or power-efficient speed. A sequence of independent ratios no longer bind engine speed and vehicle speed with a CVT. Alternatively, based on engine characteristics and performance specifications, they may function independently across a large and step-less unit. The benefits of this selectivity infinite ratio are enormous. More noticeable in the operation of the IC engine is that the engine can be mounted at cruising speeds into its more fuel-efficient area, and then accelerated to its maximum performance region once peak power is needed, irrespective of vehicle speed [5–7].

1.2 Components of a CVT

- A high power/density belt.
- Cone pulleys set.
- Hydraulic Actuator.
- Mechanical torque sensor.
- Microprocessor.
- Multi-layered clutch or Torque Converter (replacing conventional clutches).

1.3 Working Principle of a CVT

A Continuously Variable Transmission works by varying the working diameter of the transmission's two main pulleys. The pulleys have V-shaped grooves that are mounted on the connecting belt. One side of the pulley is fixed; a hydraulic actuator controls the other side of the pulley. The hydraulic actuator can increase or decrease the space between the pulley's two sides. This makes the belt ride lower or higher, depending on the driving conditions, along the inner walls of the pulley, thus changing the gear ratio.

With no "measures" in between, this action is infinitely variable. Thus a CVT can keep the engine in its optimum rpm range, which in turn increases the efficiency of the engine. As explained above, the two pulley widths in the hydraulic actuator are adjusted by oil pressure that responds to the position of the speed, throttle and other conditions that are sensed by devices such as microprocessors and other sensors.

1.4 Hydraulic Actuators

A hydraulic actuator (also known as a linear hydraulic motor) is a mechanical device used by a unidirectional stroke to provide unidirectional power. Hydraulic actuators derive their strength from the usual oil pressurized hydraulic fluid. The hydraulic cylinder consists of a cylinder barrel in which moves back and forth a piston connected to a piston rod. At each point, the pipe is capped by the bottom of the bottle (also called the end of the cap) and by the top of the cylinder where the piston rod falls comes from the cylinder.

The piston has rings and seals that are sliding. The piston divides the cylinder's inside into two chambers-the lower chamber (cap end) and the side chamber of the piston rod (rod end). The hydraulic pressure for linear work and motion acts on the piston. The piston rod also has mounting attachments for connecting the cylinder to the pushing/pulling object or machine component. The actuator or "motor" side of this system is a hydraulic cylinder. The hydraulic system's "generator" side is the hydraulic pump that brings a regulated or fixed flow of oil to the hydraulic cylinders bottom side (i.e. the bottom chamber) to move the piston rod up.

The piston pushes the oil back into the reservoir in the other chamber. If we assume that the piston rod chamber oil pressure is approximately zero, the piston rod force F is equal to the cylinder pressure P times the piston area A.

The piston moves down when the oil is pumped into the side chamber of the piston rod and the oil flows back to the reservoir without pressure from the bottom chamber area. In the piston rod area chamber the fluid pressure is:

$$Pull\ Force/(Piston\,Area - Piston\,Rod\,Area)$$

Thus the Pull Force,

$$Fp = P(Ap - Ar)$$

Where Fp is the pulling force, P is the fluid pressure, Ap is the piston face area and Ar is the rod cross-section area [7].

1.5 Torque Converter

An engine is connected through a clutch to a transmission. Without this connection, without killing the engine, a car would not be able to come to a complete stop. But CVT cars do not have a clutch that disconnects the engine transmission. They're using an amazing device called a torque converter instead.

Inside a strong and rigid torque converter housing there are 4 main components:

- Pump
- Turbine
- Stator
- Transmission fluid.

The torque converter housing is bound to the engine's flywheel, so it turns at any speed at which the engine is running. Attached to the housing are the fins that make up the torque converter pump, so they also turn at the same speed as the engine. The cutaway illustrates how the torque converter binds it. A type of centrifugal pump is the pump inside a torque converter. Like a washing machine's spin cycle flushes water and clothing to the outside of the wash tank as it turns, fluid is thrown to the outside.

As fluid flows outwards, a vacuum is created at the center that draws more fluid in. Then the fluid reaches the turbine blades that are attached to the transmission. The turbine is spinning the transmission, which essentially moves the car. The turbine blades are curved. It ensures that the fluid approaching the turbine from the outside has to change direction before it leaves the turbine's core. It is this change of direction that allows the engine to turn. User must apply a force to that object in order to change the direction of a moving object.

And whatever the force that causes the object to turn, that force must also feel, but in the opposite direction. The fluid causes the turbine to spin as the turbine causes the fluid to change direction. The fluid exits in the center of the turbine, moving in a direction different from when it entered. Fluid exits the turbine moving opposite the direction the pump and engine is turning. If the pump was allowed to hit the fluid, the engine would be slowed down and power would have been wasted. That's why there is a stator for a torque converter. The stator is located in the torque converter's very center.

Its task is to steer the returned fluid from the generator before it again reaches the pipe. It dramatically increases the torque converter's output. The stator has a very aggressive blade design that reverses the fluid's direction almost completely. A one-way clutch (within the stator) connects the stator to a fixed shaft in the transmission (the direction the clutch allows the stator to spin. Due to this arrangement, the stator can not spin with the fluid it can only spin in the opposite direction, forcing the fluid to change direction as it strikes the stator blades. As belt experiences high stress values it should be very strongly made and gripped properly on the set of pulleys.

1.6 Belt Materials

- Rubber.
- Steel.
- Thin & high strength Metals.

The pulleys of the variable diameter are the center of a CVT. Each pulley consists of two cones of 20–22° facing each other. A belt passes between the two cones in the groove. If the belt is made of rubber, V-belts are preferred. V-belts are named because the belts have a V-shaped cross section, which increases the belt's frictional grip. The transmissions relied on high-density rubber belts in earlier CVT applications, which can slip and stretch, thus reducing their efficiency. The introduction of new materials increases the reliability and efficiency of CVTs.

The creation and production of metal belt for linking the pulleys was one of the most significant developments. Such lightweight belts consist of several thin steel bands (typically 9 or 12) that hold high-strength, bow-tie-shaped metal pieces together. Metal belts do not slip and are highly durable, allowing more engine torque to be handled by CVTs. They're even quieter than CVTs powered by rubber cord.

1.7 Types of CVT

1.7.1 Frictional Type CVTS

Frictional Type CVTS The most common type of CVT is the frictional type, in which two bodies are brought into contact at points of varying distance from their axes of rotation, and allowing friction to transfer motion from one body to the other. Sometimes there is a third intermediary body, usually a wheel or belt. The simplest CVT seems to be the "disk and wheel" design, in which a wheel rides upon the surface of a rotating disk; the wheel may be slid along its splined axle to contact the disk at different distances from its center.

Such a design's speed ratio is simply the wheel radius divided by the distance from the point of contact to the center of the disk. Friction has an important role in frictional CVT designs-the maximum torque transmitted by such design is: Tmax = Cf somewhere FN Ro where To is the torque output, Cf is the friction coefficient between the wheel and the disk, FN is the force pushing the wheel into the disk (normal force), and Ro is the radius of the wheel or disk output. The friction coefficient depends on the materials used; rubber on steel typically ranges from 0.8 to 0.9.

The power is lost in two ways.

- Deformation of the components:
- Differential slip.

Deformation of the components, the two's larger factor, is caused by high normal forces, and can be minimized by using very hard materials that do not deform much, and materials with a very high friction coefficient. A large contact area between the rotating components causes the differential slip; in this example, the wheel riding "footprint" on the disk. The tip of the footprint nearest to the disk's axis of rotation can move along a wider distance than the point furthest from the axis of rotation, allowing the circle and the sides of the footprint to fall away.

Using a hard wheel that produces a small area of contact, differential slip is minimized. The "cone and wheel" design, where the disk is replaced by a cone, is very similar to the "disk and wheel." There is little downside of 4 using a cone rather than a flat disk, except to which the differential slip of the ground surface by reducing the gap in the distance travelled by the interface area's inner and outer sides. Some prototypes utilized different shapes, but it is the same concept.

1.7.2 Toroidal CVT

Two coaxial disks containing annular groves of a semi-circular cross section on their opposing surfaces are the simplest toroidal CVT. Disks arrangement is such that cross-section centers overlap. Two or more idler wheels, of a radius equal to the radius of the cross sections of the grooves, are placed between the disks so that their axes are perpendicular to and cross the axes of the disks. The speed ratio in the picture is varies by moving the wheels in opposite directions around the vertical axis. When the wheels are in contact with the drive disk near the center, they must perforce contact the driven disk near the rim, resulting in a reduction in speed and an increase in torque.

When the wheels are in touch with the drive disk, they will perforce the moved disk near the rim, resulting in speed reduction and torque changes. The opposite happens when they touch the drive disk near the rim. The benefit of this method of transmission is that the wheels do not have to slip on a splined tube, resulting in a smoother, stronger construction. Just as the CVT disk developed into the CVT disc, so too has the toroidal CVT evolved into a cone shape. The result is a transmission that is much more compact. This type is unusual that it is possible to control the speed ratio by directly rotating the wheels or slightly moving them up or down, causing them to rotate and alter the speed ratio on their own.

Pulleys of variable diameter type CVT Variable pulleys of diameter are a variation of the theme. Two twenty-degree cones face each other, a v-belt between them. The angle from the middle that the v belt touches the cones is calculated by the gap between them; the further they are, the lower the drives of the belt and the lower the radius of the pitch. The wider the belt is, the wider is the range of available radius, so this is not often the way to use the usual 4L/A series type belt. Special belts are often used, or even chains with special contact pads on the links. In order to keep the belt tight, variable diameter pulleys always have to come in pairs, one increasing in radius as the other decreases.

Variable diameter pulleys type CVT Variable diameter pulleys are a variation in the theme. Two 20° cones face each other, with a v-belt riding between them. The distance from the center that the vbelt contacts the cones is determined by the distance between them; the further apart they are, the lower the belt rides and the smaller the pitch radius. The wider the belt is, the larger the range of available radii, so the usual 4L/A series belt is not often used in this way. Often special belts, or even chains with special contact pads on the links, are used. Variable diameter pulleys must always come in pairs, with one increasing in radius as the other decreases, to keep the belt tight.

Typically a cam or lever powers one, while a spring simply keeps other one close. A variety of systems have used different diameter pulleys, such as

- power tools
- Snowmobiles,
- Automobiles.

The pulleys of the variable diameter are the core of a CVT. Each pulley consists of two cones of 20° facing each other. A belt runs between the two cones in the groove. If the belt is made of rubber, V-belts are preferred. V-belts are named because the belts 8 have a V-shaped cross section, which improves the belt's frictional strength. When the pulley's two cones become distant (when the distance increases), the belt travels lower in the groove, the belt loop size around the pulley is narrower.

The belt rides higher in the groove when the cones are close together (when the diameter decreases), and the radius of belt loop around the pulley increases.

Hydraulic pressure, spring tension or centrifugal force may be used by CVTs to create the force needed to adjust the pulley halves. Pulleys with variable diameter must always be in pairs. One of the pulleys is attached to the engine's crankshaft, identified as the drive pulley (or running pulley). The moving pulley is also referred to as the input pulley because this is where the motor energy reaches the transmission. The second pulley is called the pulley operated because it is powered by the first pulley. The powered pulley passes control to the driveshaft as an output pulley [8–10].

2 Discussions

CVTs have limitless gear ratios and performance enhancement. Within the CVT, pulleys and a belt change the gear ratios seamlessly without any "shift shock" or delay. The infinite ratios help to maintain a steady speed of driving, reduce fuel consumption and thus increase fuel economy. A CVT helps to keep the engine in its optimum rpm range due to its ability to constantly change the ratios, thus increasing the fuel efficiency. In reality, the Honda Jazz 2012 model sold in the UK promised marginally better fuel consumption for the CVT variant than the manual version. CVTs accelerate faster than a traditional system. For a supplier, a major benefit of a CVT is that its output costs less than an integrated standard multi-speed because it requires fewer parts. CVT prevents manual transmission gear changes and associated engine speed up and down.

CVTs use steel metal belts that have less capacity to transmit torque and therefore cannot be used in heavy vehicles. Its use is limited to light vehicles. Friction between the pulley and the belt causes greater wear. The transmission fluid is a little expensive.

3 Conclusion

A Continuously Variable Transmission or CVT blends the simplicity of an automatic transmission with the reliability of a manual transmission. Honda Motors description outlines the definition of CVT in its entirety. With its higher fuel performance, infinite gear ratios, lower production prices, steady running speeds and improved acceleration capabilities, CVT is certainly a future technology. This technology has only recently found such wide applications. So for a better appreciation, most of us need to get used to the dynamics of a CVT-equipped vehicle. CVT use in the 800cV Formula One Canon-Williams-Renault in 1993 was the most amazing functional proof of the CVT results. The experimental CVT Formula One was one second faster per lap, without so much development as the MT version.

References

1. T. Hofman, C.H. Dai, Energy efficiency analysis and comparison of transmission technologies for an electric vehicle, in *2010 IEEE Vehicle Power and Propulsion Conference, VPPC 2010* (2010)
2. C.T. Chung, Y.H. Hung, Energy improvement and performance evaluation of a novel full hybrid electric motorcycle with power split e-CVT. Energy Convers. Manag. (2014)
3. E. Morimoto, N. Niguchi, K. Hirata, Control method for a magnetic-CVT motor, in *Proceedings - 2016 22nd International Conference on Electrical Machines, ICEM 2016* (2016)
4. K.S. Ivanov, Synthesis of toothed continuously variable transmission (CVT), in *Mechanisms and Machine Science* (2012).

5. D. Rockwood, N. Parks, D. Garmire, A continuously variable transmission for efficient urban transportation. Sustain. Mater. Technol. (2014)
6. N. Cholis, S. Ariyono, G. Priyandoko, Design of single acting pulley actuator (SAPA) continuously variable transmission (CVT), in *Energy Procedia* (2015)
7. V. Seelan, Analysis, design and application of continuously variable transmission (CVT). Int. J. Eng. Res. Appl. (2015)
8. J.G.L.M. van Spijk, A. Englisch, The variable pulley CVT, in *Encyclopedia of Automotive Engineering* (2014)
9. W. Zeng, Y. Wang, H. Xia, Y. Qing, Design and optimum of novel mechanical CVT. Zhongguo Jixie Gongcheng/China Mech. Eng. (2013)
10. K. Okubo, T. Fujii, Fundamentals of power transmitting belts and belt type CVTs (continuously variable transmissions). Toraibarojisuto/J. Jpn. Soc. Tribol. (2018)

An Automated Solution for Test Optimization Using Soft Computing Techniques

Baswaraju Swathi[1(✉)] and Harshvardhan Tiwari[2]

[1] Jain University, Bengaluru, Karnataka, India
[2] Centre for Incubation, Innovation, Research and Consultancy (CIIRC), Jyothy Institute of Technology, Bengaluru, Karnataka, India

1 Introduction

1.1 Software Testing

The influence of the software has extremely growing due to the digitalization, as it increases standards, improves the business perspective thereby impacting the smart world. Nonetheless software needs to be fail proof. Reports across prove that software failures lead to irrecoverable loss which includes major enterprises. Software testing deals with thorough analysis to discover and evaluate the software faults and bugs such that the application needs to meet the requirements. The current trend of Software Testing ensures the application quality by test-centric and test-driven development. Testing is a continuous process and ensures at each phase the performance and effectiveness of the application. Software testing includes primarily three phases: Test Planning, which deals with choosing test strategies and testing models; Test Designing, which consists of test case development; Test Implementation which majorly concerns preparing and executing the test suit and analysing the test results. This article primarily focussed on optimization of test design. In the mentioned testing activities test case generation [1] and test case execution stands to be the most censorious activity. Test case generation primarily focuses on creating test cases and test data. Test suits needs to be derived to a certain extent that they need to cover more test scenarios or product use cases. Deriving a minimal set of test cases to cover maximum test scenarios [2] and executing the derived test cases within the specified time bounds remain as a challenging tasks to the software testers (Fig. 1).

Sprint 1 of SDLC cycle starts with the planned product requirements and specifications to create and execute the test cases. In Sprint 2 besides the planned requirements of current sprint, effort grows considering the previous sprint regression and bugs, which need to be planned in testing activities.

1.2 Soft Computing

Soft computing is an approach based on biological inspired algorithms [3] to solve uncertainty problems. These biological algorithms majorly use genetics, behaviour, and swarm of particles for reasoning and random search. Artificial Neural Networks describes

F. P. García Márquez (ed.), *International Conference on Intelligent Emerging Methods of Artificial Intelligence & Cloud Computing*, Smart Innovation, Systems and Technologies 273, https://doi.org/10.1007/978-3-030-92905-3_62

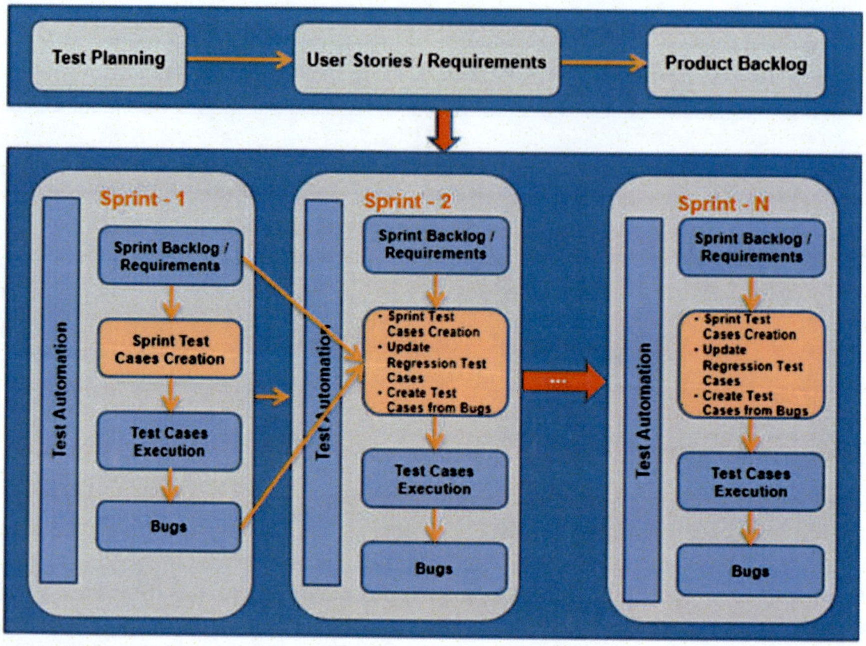

Fig. 1. Sprint methodology and the cumulative effort of test design and execution activities, which is the challenge for overcoming the risk of compromising on the product quality delivery

the real world conditions and the way to solve these problems by a system similar to a human neural system. Fuzzy logic is applicable to the systems with the values specified by true or false, uses fuzzy rules and reasoning. Artificial Intelligence uses symbolic processing of the data by symbolic languages such as LISP. Classification problems and Regression problems are solved by Support vector machine. Intelligent algorithms are used for progressively enhancing the efficiency of the specific tasks. Genetic algorithms and Particle swarm optimization are enhancing techniques which imitate biological evolution. These algorithms are appropriate in machine intelligence [4] and to optimize the solution space.

1.3 Soft Computing Techniques in Software Testing

As software testing is a continuous and integral part of continues delivery of software, there is always a continuous need to improve and optimize the process without compromising on the product quality. Optimization can be achieved by implementing an intelligent model/algorithm in place where the tests can be automatically created, generated and executed. Also the tests created aim to cover the various test scenarios and execute the tests optimally. Optimization techniques considers the various factors such as cost, productivity, time and effort. As per the study [5] the primary factors which can be considered for achieving optimization are the test case creation, reducing the duplicates test cases, designing automation frameworks, identifying effective Regression test suits.

1.4 Data Mining Techniques in Software Testing

Data mining technology is considered to discover knowledge from testing data, hence increase software quality. Classification is a data mining method which examines the numerous features connected with diverse forms of data. Clustering is an investigative procedure that depend on graphic methods to understand data. Clustering methods use graphics to demonstrate the dissemination of data relative to diverse categories of indicators. Data mining representations can be applied for improving system requirements, classifying equivalence classes in product inputs, scheming a nominal set of regression examinations, and assessing the accuracy of software outputs [6]. Study proves that Data mining techniques can inevitably yield a set of non-duplicate test cases, hence covers the maximum communal functional associations prevailing in software.

2 Background

Rayadurgam [7] defines a framework, focuses on the ways to achieve structural coverage criteria with model checkers. The method proposed is an automated solution for test suit generation by means of model checkers on prescribed specification of functionalities. Robert Nilsson [8] proposes a heuristic driven method to derive test cases. The technique targets testing the functionality of the user interaction screens for web applications, applies Genetic algorithm. The technique [9] considers parsing the teat that classify constraints and individual satisfactory standards by means of text examination on the artifacts designed, therefore passing dependencies upon individuals. The method [10] describes a hybrid particle swarm optimization and genetic algorithm for automated generation of test cases. Coverage criteria considered here is branch coverage Pandey [11] compares the various soft computing techniques applicable to software testing and the evaluation criteria in terms of coverage and the effectiveness. A novel method [12] to combine data mining techniques for test case reduction. Further K-means algorithm was proposed for the test reduction. Kumar and Bhatia [13], incorporated fuzzy techniques to optimize the test cases. Fuzzy clustering techniques are used to test reduction. Genetic algorithm in combination with mutation testing [14] is applied to reduce the test cases. In [15] focuses on the various data mining techniques applicable on test case generation. Classification and clustering data mining techniques are used to optimally generate test cases.

2.1 Optimization Techniques Used in Test Case Generation

2.1.1 Pair Wise Testing

The conventional test case derivation mechanisms like boundary value testing, equivalence testing guarantee the test cases but cannot be applied on large set of data. In addition to this, the test cases designed may not depend on a single parameter. Nevertheless the test cases with possible combinations can lead to a large test data hence would be tedious task. Pair wise testing [16] is considered to be one of the best approaches where the amount of test cases can be drastically reduced with considerable test coverage. Pair wise testing is applicable with the systems with an outsized range of combination of values. Considering prominent Triangle problem as a program under test with three sides of a triangle say a, b, c, and the constraints on the program would be the length of sides and the type of triangle formed. For an instance the total possible combination of tests with range of values from 1 to 10 would be 10*10*10 which results 1000 test cases. For the mentioned values in the Table 1 (randomly chosen), the possible combinations resulted into 64, whereas pairwise reduced the above test cases to 16 combinations. (An online pairwise tool is considered.)

Table 1. Values of sides of a triangle (randomly chosen)

a	b	c
1	10	5
9	2	3
7	5	6
4	8	9

Table 2. Combination of values from pairwise tool

a	b	c
1	10	5
1	2	3
1	5	6
1	8	9
9	2	6
9	5	9
9	8	5
9	10	3
7	5	5
7	8	3
7	10	6
7	2	9
4	8	6
4	10	9
4	2	5
4	5	3

Pair wise testing operates and tries to minimize the combination of values for a set of input elements, the relation over the elements is a challenge. Hence, there should be a way of optimising the tests which can relate the data elements. The same holds true with the user interface elements in a web page (Table 2).

2.1.2 Clustering Techniques

The duplicate test cases and the missed test cases derived during testing process increases the considerable amount of time and cost. Clustering techniques [17] are one form of data mining approach which can be applied in test optimization. The strategy behind this clustering technique is to categorize the test cases which belong to similar set and consider a particular test case to execute. Suitably for white box testing the test cases which cover the same independent path can all be grouped into a cluster. Hierarchical clustering can be applied for test set clustering. Agglomerative approach here starts with a group of test sets followed by similar clusters and merging them to form a new test set. The source code of a program is represented as a program graph, the graph algorithms can be applied as well.

2.1.3 Genetic Algorithm in Test Case Optimization

Genetic algorithm works on principle of biological evolution where the operations selecting the preliminary population, performing cross over, mutation operations to further evolve the generations has been applicable to test case optimization. Study proves that genetic algorithm [14] has enhanced the optimization process suitable for both white box and black box testing.

Algorithm for GA:

Input: test suits (input data for a test) and

Output: Optimal test suit

Algorithm:

Initial Population: Set of test cases chosen randomly, represented in terms of bits or strings.

Fitness Function: Proportional to coverage criteria

Selection: Chromosomes with maximum coverage criteria is considered for next generation

Cross over and mutation operations are used to generate diversified range of population.

Termination: For a specific number of generations

2.1.4 Particle Swarm Optimization

PSO solves the optimization problem by enhancing the population, is a metaheuristic algorithm and is applicable in a larger space. As per [18], PSODGT a technique which generates the test data based on particle swarm approach and the test coverage as condition decision coverage. The general form of Particle Swarm optimization suitable for test case generation is given below.

Algorithm for PSO:

Input: Initial group of test cases.

Output: Optimal test suit

Calculate the fitness function of the test case using the parameters which includes coverage criteria.

Update the individual finest value

Update best global value set, for the next iterations

Consider the new population

Revise the position and velocity of the swarm

3 Proposed Methodology

Proposed method at the sprint level integrates Genetic algorithm in to the test generation process as a part of a particular sprint and validated with the baseline test cases. Further the automated test suits are evaluated with test coverage and code coverage (Fig. 2).

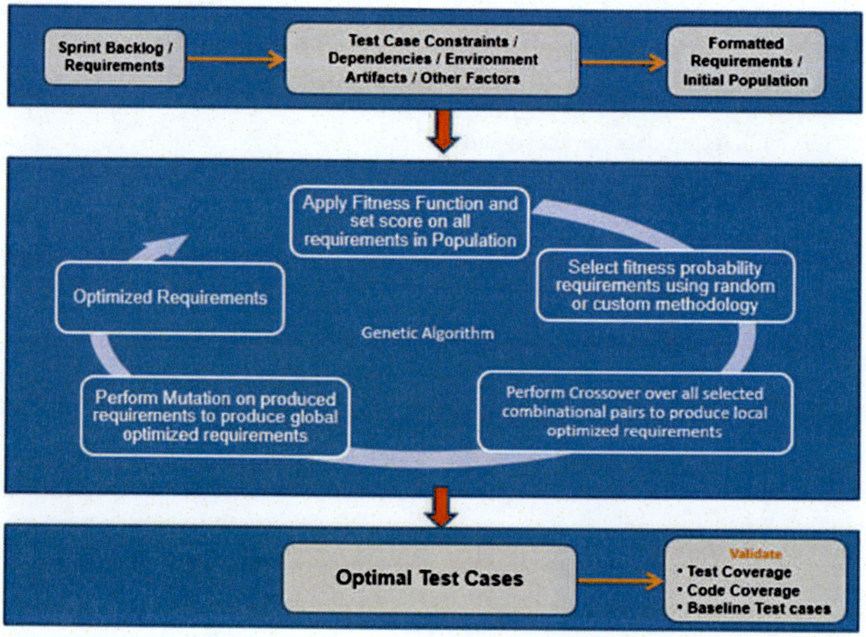

Fig. 2. Integration of GA soft-computing algorithm in the test design activity in a sprint model

Further, the proposed Cluster Genetic Algorithm considers the test cases as initial population. Later clustering techniques are used to group the tests and the GA operations are applied such that new population is generated. The paper [13] proposes a fuzzy clustering technique to lessen the number of test suits. As per [19] a K means algorithm assuming the test cases test suit to be {T_1, T_2, T_3,...Tn}, and describes a mapping test suit > {1,2,..K}. The idea is to group the test cases having similar test coverage and to separate dissimilar test cases which will continue with the next iteration in GA (Fig. 3).

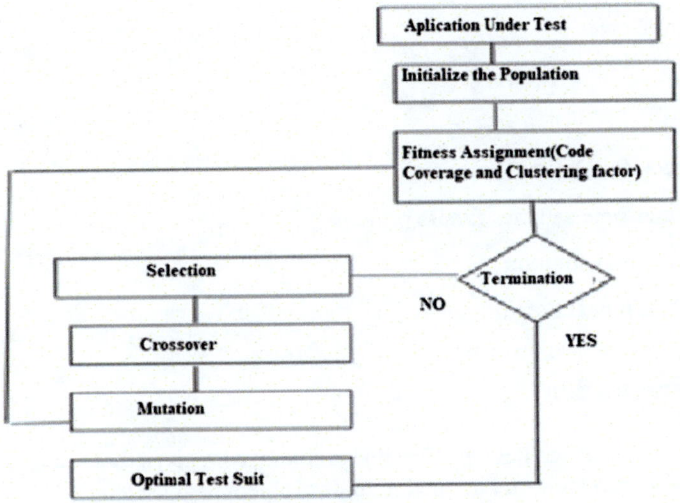

Fig. 3. CGA based test case generation

Algorithm

STEP 1. Generate initial population

Generate a "gene pool" or population of genes consisting of digits used to generate potential lengths of triangle sides

STEP 2. Create Fitness function measure

Set fitness function (aggregate of code coverage and clustering factor)

Clustering factor is calculated based on average code coverage of the test suit.

STEP 3. Initialize chromosomes

Create initial candidate Test Suites (chromosomes) by randomly choosing from initial population

STEP 4. Select chromosome pairs for Genetic Operation based on fitness function.

STEP 5. Apply Genetic Operators

Do for certain Chromosome pairs

While (conditions are covered)

Do crossover

Do mutation

Eliminate Duplicates

End While

End For

STEP 6. Check for Optimal solution

If (new chromosome satisfies fitness principle)

STOP with optimal solution

Else go back to repeat Step 5

4 Implementation

The algorithm considered here is a Triangle problem, with given three sides of a triangle. Pseudo code for the above mentioned triangle problem is given as:

If all three sides of a triangle are equal

Triangle is "Equilateral"

If all sides are different

Triangle is "Scalene"

If any of the two sides are equal

Triangle is "Isosceles"

From the above, program has three independent execution paths corresponding to one decision point. Initially the random test cases for a, b, c with the values 1, 2, and 3 are considered to form the chromosomes. Fitness value is assigned based on the coverage and clustering value of each path. Selection strategy used here is rank based selection. Cross over and mutation is performed on generation of a random value between specified ranges as 0–1,number generated is compared with a threshold value which is a mean value such that a gene from parent 1 is added as offspring, otherwise a gene from parent 2 is added. A random gene addition is considered to be a mutation operation.

The experiment results in the further section describe the values of fitness, code coverage and branch coverage over a certain specified range of generations. NUnit tool is used to record the code coverage of the test suits for each generation. At a basic level the method was applied to two console programs triangle problem and commission problem and achieved the below specified result (Figs. 4, 5 and 6).

5 Results and Graphs

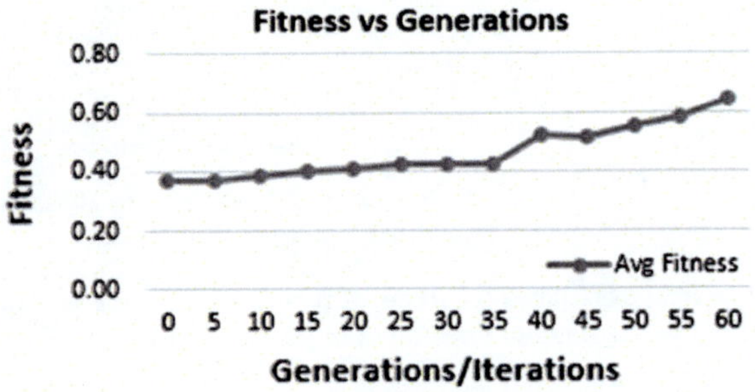

Fig. 4. The average fitness of the test suits considered for each generation starting with a random population in generation-0

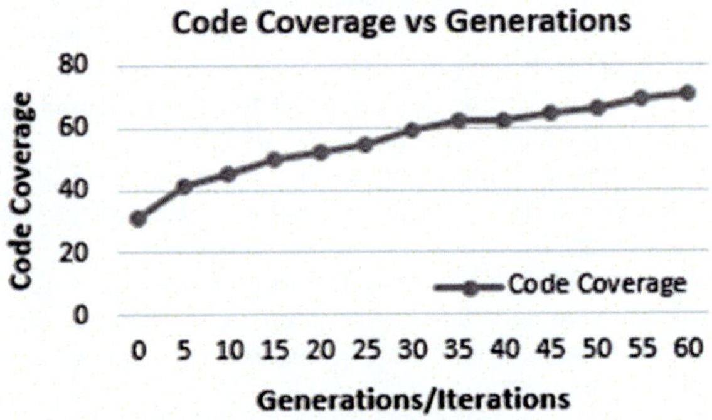

Fig. 5. The code coverage achieved (in terms of percentage) for the test suits considered for each generation starting with a random population in generation-0

6 Conclusion

The paper proposed an automated solution for test suit generation with soft computing practices primarily genetic algorithm. The initial section of the paper conveys the various techniques used in software testing with respect to test case generation. The proposed method customized genetic algorithm and the well-known Triangle problem is considered to evaluate the method. Future examination of the problem stated in the article is to apply the proposed algorithm to the various console applications and a small scale web application.

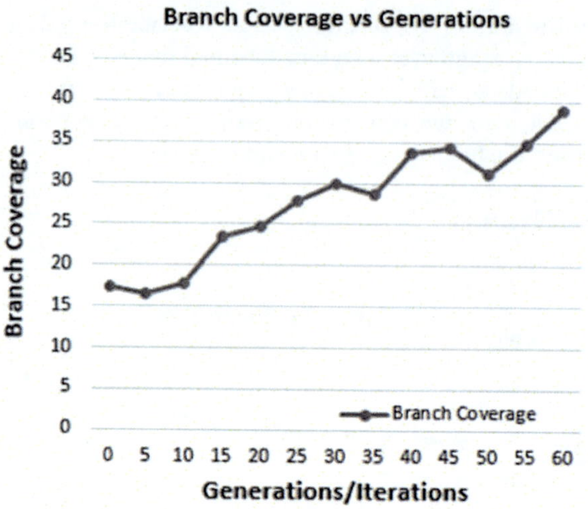

Fig. 6. The branch coverage achieved (in terms of percentage) for the test suits considered for each generation starting with a random population in generation-0

References

1. G. Gongge, Z. Hao, Y. Jing, Research on automatic generation of test cases, in *2012 IEEE/ACIS 11th International Conference on Computer and Information Science*, Shanghai (2012), pp. 428–431. https://doi.org/10.1109/ICIS.2012.91
2. A. Isabella, E. Retna, Study paper on test case generation for GUI based testing. Int. J. Softw. Eng. Appl. (IJSEA) **3**(1), 139 (2012). https://doi.org/10.5121/ijsea.2012.3110
3. A. Darwish, Bio-inspired computing: algorithms review, deep analysis, and the scope of applications. Futur. Comput. Inform. J. **3**(2), 231–246 (2018). ISSN 2314-7288. https://doi.org/10.1016/j.fcij.2018.06.001
4. H. Hourani, A. Hammad, M. Lafi, The impact of artificial intelligence on software testing, in *2019 IEEE Jordan International Joint Conference on Electrical Engineering and Information Technology (JEEIT)*, Amman, Jordan (2019), pp. 565–570. https://doi.org/10.1109/JEEIT.2019.8717439
5. B. Swathi, H. Tiwari, Test case generation process using soft computing techniques. Int. J. Innov. Technol. Explor. Eng. (IJITEE) **9**(1) (2019). ISSN: 2278-3075
6. H. Wang, L. Bai, M. Jiezhang, J. Zhang, Q. Li, Software testing data analysis based on data mining, in *2017 4th International Conference on Information Science and Control Engineering (ICISCE)*, Changsha (2017), pp. 682–687. https://doi.org/10.1109/ICISCE.2017.148
7. S. Rayadurgam, M.P.E. Heimdahl, Coverage based test-case generation using model checkers, in *Proceedings. 8th Annual IEEE International Conference and Workshop On the Engineering of Computer-Based Systems-ECBS 2001*, Washington, DC, USA (2001), pp. 83–91. https://doi.org/10.1109/ECBS.2001.922409
8. R. Nilsson, J. Offutt, J. Mellin, Test case generation for mutation-based testing of timeliness. Electron. Notes Theor. Comput. Sci. **164**(4), 97–114 (2006). ISSN 1571-0661. https://doi.org/10.1016/j.entcs.2006.10.010
9. S. Masuda, T. Matsuodani, K. Tsuda, Automatic generation of test cases using document analysis techniques. Int. J. New Technol. Res. (IJNTR) **2**(7), 59–64 (2016). ISSN: 2454-4116

10. P. Chawla, I. Chana, A. Rana, A novel strategy for automatic test data generation using soft computing technique. Front. Comput. Sci. **9**, 346–363 (2015). https://doi.org/10.1007/s11704-014-3496-9
11. B. Pandey, R. Jain, Soft computing based approaches for software testing: a survey. Int. J. Soft Comput. Eng. (IJSCE) **4**(2) (2014). ISSN: 2231-2307
12. K. Muthyala, R. Naidu, A novel approach to test suite reduction using data mining. Indian J. Comput. Sci. Eng. **2**(3), 500–505 (2011)
13. G. Kumar, P.K. Bhatia, Software testing optimization through test suite reduction using fuzzy clustering. CSIT **1**, 253–260 (2013). https://doi.org/10.1007/s40012-013-0023-3
14. B. Swathi, H. Tiwari, Genetic algorithm approach to optimize test cases. Int. J. Eng. Trends Technol. **68**(10), 112–116 (2020)
15. Y. Dubey, D. Singh, A. Singh, Amalgamation of automated test case generation techniques with data mining techniques: a survey. Int. J. Comput. Appl. **134**, 18–22. https://doi.org/10.5120/ijca2016907950
16. C.B.A.L. Monteiro, L.A.V. Dias, A.M.D. Cunha, A case study on pairwise testing application, in *2014 11th International Conference on Information Technology: New Generations*, Las Vegas, NV (2014), pp. 639–640. https://doi.org/10.1109/ITNG.2014.42
17. S. Chaudhary, A. Jatain, Performance evaluation of clustering techniques in test case prioritization, in *2020 International Conference on Computational Performance Evaluation (ComPE)*, Shillong, India (2020), pp. 699–703. https://doi.org/10.1109/ComPE49325.2020.9200083
18. Y.H. Jia, W.N. Chen, J. Zhang, J.J. Li, Generating software test data by particle swarm optimization, in *Simulated Evolution and Learning. SEAL 2014*, ed. by G. Dick et al. Lecture Notes in Computer Science, vol. 8886 (Springer, Cham, 2014). https://doi.org/10.1007/978-3-319-13563-2_4.
19. K. Muthyala et al., A novel approach to test suite reduction using data mining. Indian J. Comput. Sci. Eng. (IJCSE) **2**(3) (2011). ISSN: 0976-5166

Emergence of Internet of Things (IoT) and Its Smart Application

Mrinal Paliwal(✉)

Department of Computer Science and Engineering, Sanskriti University, Mathura, Uttar Pradesh, India
mrinalpaliwal.cse@sanskriti.edu.in

1 Introduction

Internet of Things (IoT) techniques proposed a general concept for the ability of network devices to sense and collect information from around the real world, then share that information across the web server wherever it may be processed and utilized for various interesting purposes. The IoT is composed of smart machines interacting and communication with different machines, objects, environments, and infrastructures [1]. Currently, a day every person is connected with one another using a lot of communication approaches. Wherever the most popular communication approach is the internet, therefore, in different words, this paper is able to say the web that connects peoples. This paper tends to reside in a new era of computing, data, and communication technology that, as several are an expression, pushes forward seamless interaction between humans, nature, and physical objects and is captured inside the ecosystems of the Internet of Things (IoT). The anticipated rich connectivity of environment-to-machine, machine-to-machine, and machine-to-backbone web infrastructure starts to be seen and developed almost all over within the world. As a result, large amounts of information are being propagated, [2] that is being afterward processed into some helpful decision-making and intelligent actions to form our lives easier and safer. Concurrently, however, the large amount of information being changed additionally because the method it's being processed within the cloud and edge devices of IoT platforms might not continuously leverage upon secure and reliable protocols and mechanisms.

The essential concept of the Internet of Things (IoT) has been around for nearly two decades and has attracted many researchers and industries due to its great estimated. Impact on improving our daily lives, and society. Once things like the house appliances are connected to a network, they will work along in cooperation to produce the perfect service as an entire, not as a group of independently operating devices [3]. This can be helpful for several of the real-world applications and services, and one would, for instance, apply it to make a sensible residence; windows will be closed automatically once the air conditioning is turned on or maybe opened for oxygen once the gas kitchen appliance is turned on. The concept of IoT is particularly valuable or persons with disabilities, as IoT technologies will support human activities at an on large scales like

F. P. García Márquez (ed.), *International Conference on Intelligent Emerging Methods of Artificial Intelligence & Cloud Computing*, Smart Innovation, Systems and Technologies 273, https://doi.org/10.1007/978-3-030-92905-3_63

building or society because of the devices will mutually collaborate to act sort of total system. Among the submissions received, all of that went through a rigorous peer-review method [4].

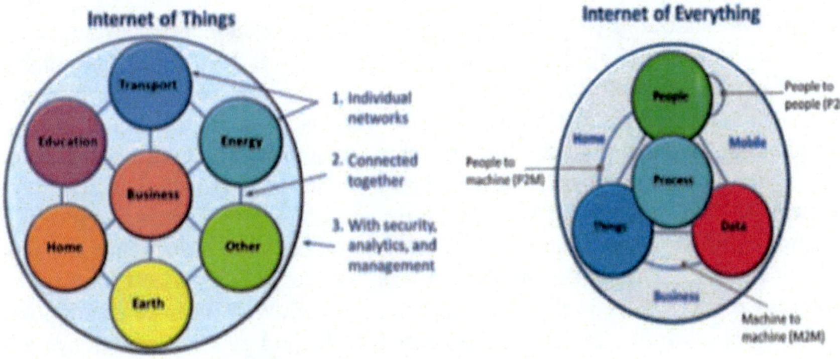

Fig. 1. IoT global networking and IoT

The Internet of Things (IoT) is developing quickly and turning into a progressively grown topic that creates excitement and anxiety around the globe. There are lots of indications showing that the Internet of Things can modification several sectors, as well as higher education institutions, particularly universities. Now, universities have an opportunity to guide the technical development, and therefore, the innovation models for the Internet of Things (IoT), and to make the leaders of the Internet of Things into the future, furthermore, on address, the TIPPSS risks that stand for Trust, Identity, Privacy, Protection, Safety, and Security associated with the IoT. The Internet of Things (IoT) could be a world physical network that connects devices, objects, and things to the web infrastructure to speak or interact with the internal and therefore, the external environment as illustrated in Fig. 1, and for the aim of exchanging data through the data sensing devices consistent with specific protocols [5]. Thus, the Internet of Things (IoT) is enabling connectivity for all the world, and for anyone to be networked around the world anytime and anywhere using any network or any service to realize the goal of Intelligence distinguishing, tracking, and managing things.

2 Related Work

The Internet of things (IoT) produces such a lot of services for sensing and information sharing purposes. During this proposed method, a resource model is intended for procuring sensors similarly to cloud stores toward the client's end. This model has taken under consideration an emerging IoT based mostly multi-sensing function and internet of things at the cloud for supporting a business model [6]. Supported the heuristic algorithm, it will integrate the allocation of best-suited devices consistent with the requirements of the application at an equivalent time. A sensing element device will produce data and information which will simplify the period decision-making and actuate independent action and policies. During this paper, the authors theorized two options, specifically frames of references, and frames of performing. Their approach inherently implies frames-aided communications between modelers and analysts, additionally as relevant parties, and more differentiates the objective and use of every model that may confide the purpose of the system needs [7].

The Internet of things (IoT) is emerging technology researchers to develop smaller and more cost-effective wireless technology that consumes less power and might be integrated into virtually any kind of device. There are three IoT components that enable seamless connections that are: Hardware: created from sensors, actuators, and embedded communication hardware, Middleware: on-demand storage and computing tools for data analytics, and Presentation: data visualization and interpretation tools which may be wide accessed on completely different platforms and which may be designed for various applications [8]. There are a number of potential methods for introducing low-power communications to an IoT node, ranging from purpose-designed protocols like ZigBee to low-power variants of Bluetooth, Wi-Fi, and NFC. Though Wi-Fi is that the most popular type of integrated wireless technology and therefore the best power-per-bit transmission efficiency, Internet of things IoT enhances different formats as well as radiofrequency Identification (RFID) technology that is employed throughout the business, industry and private technology systems and enables the design of microchips for wireless data communication [9].

2.1 Overview and Components of IoT

In the current time, everywhere like at an airport, train depot, shopping malls, in colleges an information desk is necessary that has information regarding the train schedule, promotional offers, and necessary notice immediately. From an academic organization perspective, the problem is that it requires some employees that are dedicated to that purpose which should have up to this point information regarding the institute and therefore, the recent happenings within the institute. The second problem is that someone has to move to the institute at a desk to induce information from them. The solution to the present is to use technology and build technology responsible to answer all the queries asked by individuals [10] (Fig. 2).

The most effective tool is mobile phones that are accessible to almost everybody which is connectable to the internet to download the most recent information [11]. If the data isn't updated over the internet, in those cases wherever the data isn't being updated over the internet, this paper need to call the customer service center for support. Some authors designed a device that has all the data held on in its database, whenever somebody desires information they need to use that device and find the related data from through that device. For this to work, the device should be accessible to a user who needs any facilitate or support [12].

There is a unit combination of problems concerned once considering accomplice domestic structure. It has to offer a patron inviting interface at the host include, all collectively that the signs, and symptoms are reliably essentially set up, watched, and controlled. What's additional, the general structure should be comfortable vibrant to know the important energy of far away headway. Eventually, [13] the system should see possible which is able to allow its application in domestic implementation. To scale back

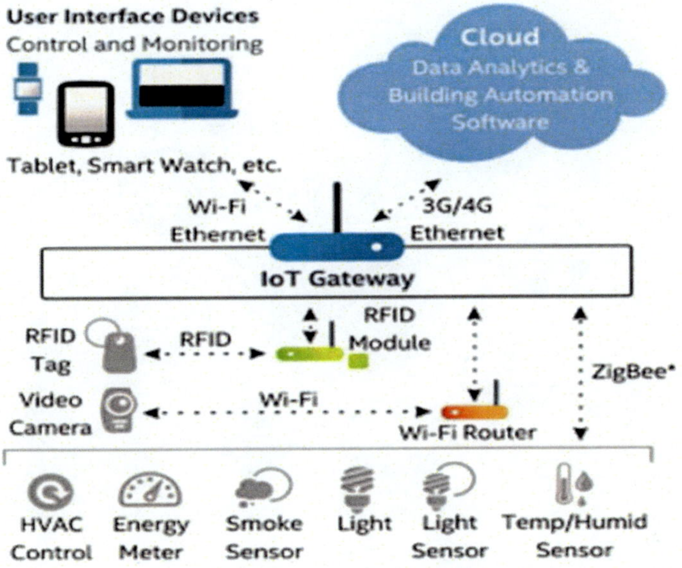

Fig. 2. Overview of conceptual architecture

the inadequacies of each appliance and to surmount the appearance troubles harbinger documented, this mission organizes regionally and remotely managed systems with the usage of Cloud information make. This grips the device to paintings at the same time as not the dependence of a versatile provider, asserts the shape to be used with sundry preservationist sorts out, and underpins the structure to work domestically as presently as a smartphone or computer get to is not improved [14].

This paper use the most completely unique wireless sensors to indicate the information to fathom the characteristic conditions and dismissing introduce any defect in enhancements. It's the principal to behave signs and symptoms per the private goal. The emergence of Iot is described in Fig. 3. The home laptop firmly displays the sensor's regards and therefore, the headways in like way. If the scenario determined record lower back to a cloud server. Right here client will change multiple settings, and optically watch the unit's judgment and control. This paper are ready to whilst all is alleged in performed packaging one Registration software anyplace Technicians, food market traders, and optional comfort provider can pick on it. They deliver their nuances just like the fashion of agreement and their comfort transient approach, [15] etc. as possible Cloud Server can observe records planning on gaining knowledge of units. It, however, mail or SMS Technician and send information to the owner (mail or SMS). We'll interface any methods of clients on cloud servers alongside these strains it helps multi-purchaser system houses. Here this paper will patron just a lone cloud server at any charge we'll relate a few of proportions of consumers there to through laptop or any robot appearances. Figure 4 shows Real-Time Analytics and the Internet of Things.

Fig. 3. Emergence of IoT

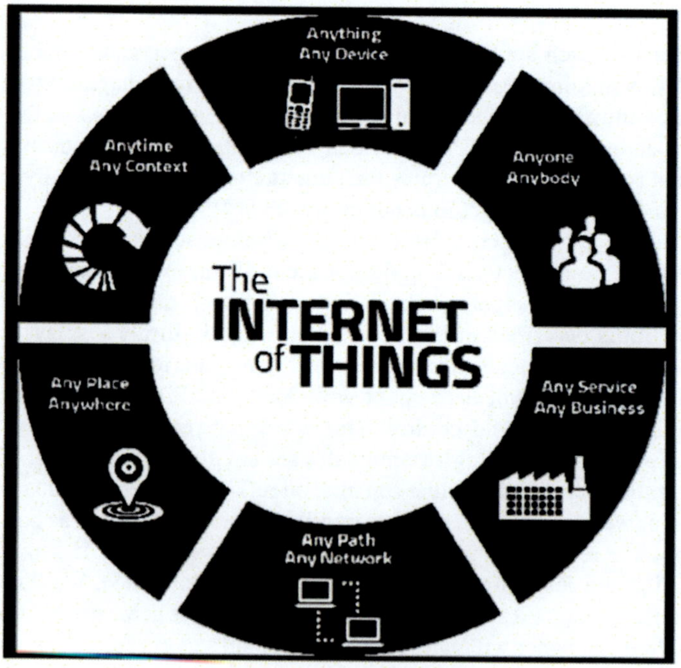

Fig. 4. Real time analytics and the IoT

3 Conclusion

The internet of things (IoT) assurance to deliver a step-change in an individual's "quality of life and smart enterprises" productivity. Through a widely distributed, locally intelligent network of smart devices, the Internet of things has the potential to enable extensions and enhancements to fundamental services in a smart home, transportation, logistics, security, education, healthcare, and other areas, while providing a replacement ecosystem for application development. A concerted effort is required to move the industry beyond the first stages of market development towards maturity, driven by a standard understanding of the distinct nature of the opportunity. This global market has distinct characteristics within the areas of service distribution, business and charging models, capabilities required to deliver IoT services, and therefore, the differing demands these services will place on cell phone networks. The internet of things has been gradually bringing a sea of technological changes in our daily lives, which successively helps to form our life simpler and easier, though various technologies and applications. Though the Internet of things (IoT) has abundant benefits, there are some flaws within the IoT governance and implementation level.

References

1. M. Noura, M. Atiquzzaman, M. Gaedke, Interoperability in internet of things: taxonomies and open challenges. Mob. Netw. Appl. (2019)
2. M. Jia, A. Komeily, Y. Wang, R.S. Srinivasan, Adopting internet of things for the development of smart buildings: a review of enabling technologies and applications. Autom. Constr. (2019)
3. M.S. Mahdavinejad, M. Rezvan, M. Barekatain, P. Adibi, P. Barnaghi, A.P. Sheth, Machine learning for internet of things data analysis: a survey. Digit. Commun. Netw. (2018)
4. A.H. Alavi, P. Jiao, W.G. Buttlar, N. Lajnef, Internet of things-enabled smart cities: state-of-the-art and future trends. Meas.: J. Int. Meas. Confed. (2018)
5. P. Porambage, J. Okwuibe, M. Liyanage, M. Ylianttila, T. Taleb, Survey on multi-access edge computing for internet of things realization. IEEE Commun. Surv. Tutor. (2018)
6. H.N. Saha, A. Mandal, A. Sinha, Recent trends in the internet of things, in *2017 IEEE 7th Annual Computing and Communication Workshop and Conference, CCWC 2017* (2017)
7. K. Chen et al., Internet-of-things security and vulnerabilities: taxonomy, challenges, and practice. J. Hardw. Syst. Secur. (2018)
8. N. Kaur, S.K. Sood, An energy-efficient architecture for the internet of things (IoT). IEEE Syst. J. (2017)
9. C. Qu, M. Tao, R. Yuan, A hypergraph-based blockchain model and application in internet of things-enabled smart homes. Sensors (Switzerland) (2018)
10. K.K. Patel, S.M. Patel, P.G. Scholar, Internet of things-IOT: definition, characteristics, architecture, enabling technologies, application & future challenges. Int. J. Eng. Sci. Comput. (2016)
11. A. Bader, H. Ghazzai, A. Kadri, M.S. Alouini, Front-end intelligence for large-scale application-oriented internet-of-things. IEEE Access (2016)
12. F. Dalipi, S.Y. Yayilgan, Security and privacy considerations for IoT application on smart grids: survey and research challenges, in *Proceedings - 2016 4th International Conference on Future Internet of Things and Cloud Workshops, W-FiCloud 2016* (2016)
13. M.A. Rahman, A.T. Asyhari, The emergence of internet of things (IoT): connecting anything, anywhere. Computers (2019)

14. K. Pallavi, J.D. Mallapur, K.Y. Bendigeri, Remote sensing and controlling of greenhouse agriculture parameters based on IoT, in *2017 International Conference on Big Data, IoT and Data Science, BID 2017* (2018)
15. A. Gyrard, P. Patel, S.K. Datta, M.I. Ali, Semantic web meets internet of things and web of things, in *26th International World Wide Web Conference 2017, WWW 2017 Companion* (2019)

Analysis System of City Price Based on Big Data

Madhav Solanki(✉)

SOEIT, Sanskriti University, Mathura, India
madhavsolanki.cse@sanskriti.edu.in

1 Introduction

As indicated by the advanced universe inquiring about report given by International Data Corporation (IDC), by 2020, 35 ZB (1 ZB = 1 billion TB) information will be accessible around the world. Numerous nations and ventures have appended incredible significance to the estimation of the information. For instance, the US discharged "Huge Data innovative work activity", which expects to improve the capacity of acquiring information and experiences by utilizing a lot of complex information, and in excess of 200 million dollars will be contributed for it. China has additionally sent the advancement technique of huge information from the administration level. Apparatuses like the conventional database have been not able adapt to such an enormous size of information, in certain territories of the individuals' occupation, more consideration ought to be paid to the ideas, procedures and systems of large information. As of late, with China's quick financial improvement, the cost level is additionally progressively rising, regardless of whether the improvement of big business or the endurance of individuals are profoundly grieved by significant expenses [1–3]. While the contributing elements of expansion brought about by the rising cost are unpredictable, for example, request pull, cost push, financial arrangement, etc. Due to the mind boggling data, the national full scale control is a long way from fulfilling need of the open. In any case, Big information can assemble data and knowledge in all perspectives with its one of a kind points of interest, for example, mass (Volume), speed (Velocity), assorted (Variety), and so forth, what's more, by the system development, data assortment, information examination, it could help chiefs accomplish exact guideline in the part of cost.

2 Systematic Level of City Price Analysis

At present, the condition the full scale control confronting has been changed significantly. The multifaceted nature of the monetary system, the assorted variety of the assortment of wares, and the assorted variety of open interest, all of these components ought to be considered during the cost guideline in order to improve the specialized structure and information preparing capacity of the administration. The examination arrangement of city cost structured in this paper focuses on directing city value precisely, and it applies the large information investigation as its centre, and utilizes data innovation, to build a Human-Computer Helpful system for value choice and counsel. Its principle work is

F. P. García Márquez (ed.), *International Conference on Intelligent Emerging Methods of Artificial Intelligence & Cloud Computing*, Smart Innovation, Systems and Technologies 273, https://doi.org/10.1007/978-3-030-92905-3_64

helping government unbiasedly assess a wide range of products and the advancement pattern of cost in various regions, track the cost variances whenever, handle the law of value changes furthermore, break down the pattern of cost later on, lastly utilize strategy preferences of government's full scale control.

This paper structures three layers for the investigation system of city cost. The first is the choice layer, and it's in general objective is serving the basic leadership. Among all of the business information, local information and product information, helpful data ought to be removed from the mass of information for the administration so as to do proficient basic leadership, along these lines, and it ought to be the centre of the examination arrangement of city value, it has a place with choice reaction. The capacity of the basic leadership system for the most part centre around improving the relevance and viability of large scale control, improving the government's capacity of dissecting the value pattern and decreasing the danger of misconstruing the circumstance [4–6].

The subsequent one is the supporting layer, its capacity for the most part centre around gathering and examining the significant information immediately, at that point choosing out the key components and data, and changing them into viable data quickly, starting here of view, it has a place with the sound examination. To be progressively explicit, the fundamental capacity of the supporting layer is gathering information rapidly from better places and areas, finding the unusual information quickly, and afterward surveying and examining the potential danger to the general cost, through the organize examination, information mining and time arrangement strategies, the viable data could be found.

The last one is the indicating layer, which for the most part incorporates observing the anomalous information, cautioning about the value circumstance, nonstop data following, unique sorts of basic leadership, and the data security etc. The primary capacity of the demonstrating layer is checking on the dynamic condition, for example, the recorded information, outside information and current information, and afterward advising leaders the unusual information, the pattern of the appropriation and dispersion by representation innovation, lastly making admonitions about the pattern of cost changes (expansion and collapse). Simultaneously, the indicating layer ought to likewise ensure the data security, via doing information data assurance through the specialized way, in order to counteract the obstruction of outside data and the interruption of destructive data.

3 Big Data Transmission Path of the Analysis System of City Price

As per PC engineering, proposed by Von Neumann, the PC contains five essential areas, that is input, memory, math and rationale, control, and yield, this five essential areas accomplish elements of info, capacity, figuring, control and yield of data individually. As indicated by the qualities of huge information, this paper separates the investigation arrangement of city cost in light of large information into five subsystems, that is the information procurement subsystem, the information the executives subsystem, the information examination subsystem, the information administration subsystem what's more, the data security subsystem, and this five subsystems accomplish elements of information obtaining, information the board, information investigation, information administration and information insurance separately, in this way, it would appear that a comparable PC design (as appeared in Fig. 1). Meanwhile, the investigation arrangement of city cost dependent on huge information could likewise be isolated into two sections, the inside some portion of the assortment (that is the piece of the inner systems administration of the value division system) and the outer transmission part (the social network's observations and decisions of the circumstance of cost). By gathering information from various areas, enterprises, wares, and through an assortment of terminal hardware, stockpiling stage, investigation system and outer interface, a division of work system comprising of incalculable assortment terminals, stockpiling stages, application examination and administration segments could be framed [7–9]. Through the assortment and examination of interior data, the transmission and acquisition of outside data, and so on, the administration office needs to get together the inner assets and outside assets, and by choosing, utilizing and assessing, the administration could build up a major database of value examination. The government ought to think about the macroeconomic circumstance what's more, open requests guarantee the powerful association of each piece of information data and successful information transmission investigation.

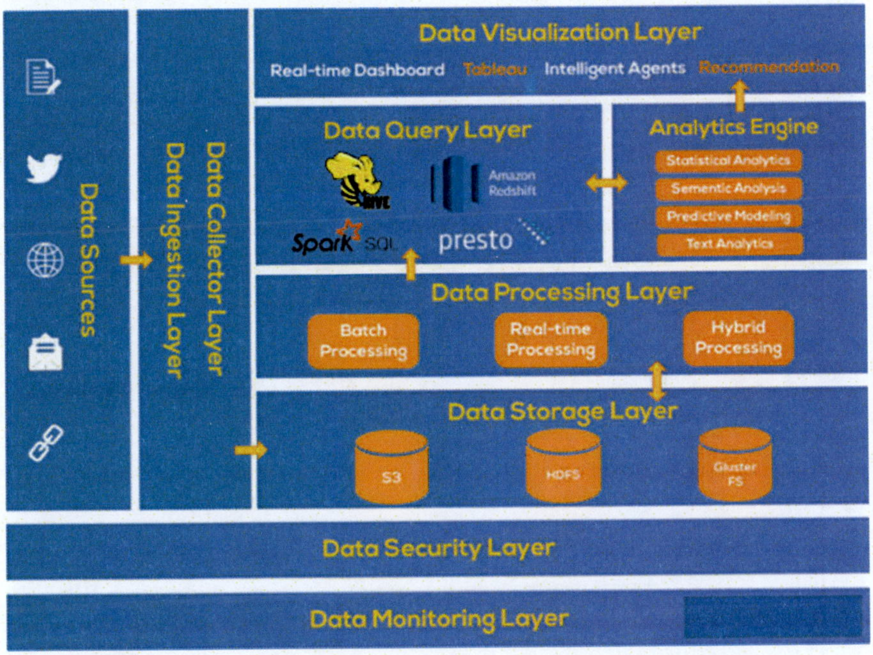

Fig. 1. The technical path of the analysis system of city price diagram

4 The Technical Path of the Analysis Subsystem of City Price

4.1 The Data Acquisition Subsystem

The information procurement subsystem of city cost dependent on enormous information comprises of two significant information sources and two primary information assortments (as appeared in Fig. 2). Two major information sources imply that the information assortment focuses (official) set up by the value division in urban communities of the entirety nation and the transmission stage (informal) for the data of open recognition about cost. The authority information assortment focuses should get through the limits of the Department, and should cover ventures, customs, markets, banks, etc. to the extent conceivable. The outer information stage essentially originates from the remarks on the Internet, the open's response to the value circumstance, the free research organizations' investigation about the pattern of cost.

This paper set forward a sort of procedure for enormous log information preparing, including log assortment, pre-process, capacity, and investigation. The gathering procedure is liable for unite the log information from disseminated web servers, at that point anyone can the board and use the log information advantageously. Pre-processing methodology is answerable for cleaning and changing the unformatted log [7–9]. Away methodology, everyone use Hadoop also, hive to store log information. In light of the past three ventures, in the examination technique, all use hive SQL to examination log information.

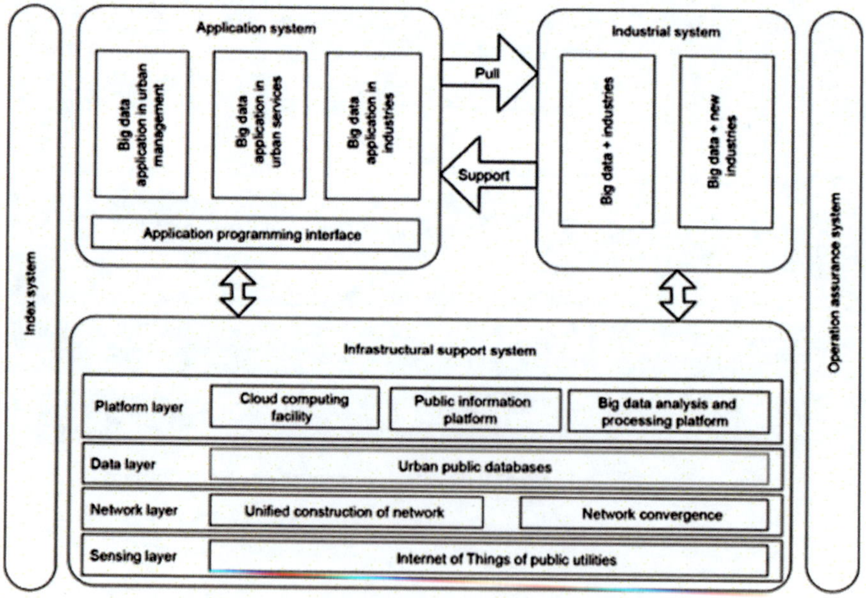

Fig. 2. The technical path for the data of city price acquisition subsystem

4.2 The Data Management Subsystem

The information the board subsystem is certainly not a straightforward put away information the board, yet a progression of the board exercises of information determination, protection, support and documenting, and attempting to utilize the information. With an increasingly exhaustive depiction on qualities and tasks, Study on the programming model for the enormous information processing stage Current heterogeneous mufti-centre reconfigurable stages are confronting issues, for example, challenges being developed, absence of adaptability and versatility.

With deference of the basic circumstance, a major information application arranged heterogeneous stage programming model; which could give a simple and elite structure stream to the cross breed system. The board procedure of value information as appeared in Fig. 3: the gathered unique information must be prepared by the channel, counting organized information cleaning, semantic investigation and include extraction of unstructured information, and so on, and the sifted information will be stacked into the information distribution centre. Information with much mistake ought to be screened out during the board procedure of information, and the subjective information ought to be handled quantitatively; in the meantime, it needs to keep up judicious administration.

Examination of Web log structure, joined with the real circumstance of Internet Co log examination request, the plan of the procedure of information examination of huge information log, the system engineering and module division are conveyed out dependent on the stage improvement mode [10–12]. Stage based each procedure of log investigation to give the client with the type of configurable undertakings. At the point when the client employments the stage to log investigation, just need to give the stage to finish the undertaking of the arrangement of the interface. Stage assists clients with finishing the log stockpiling furthermore, figuring.

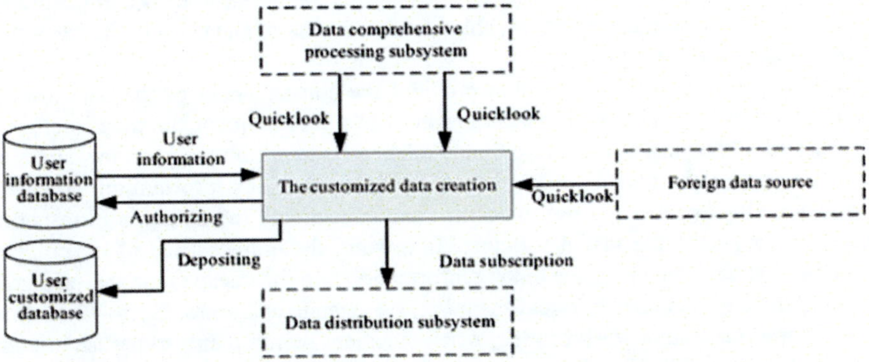

Fig. 3. The technical path for the data of city price management subsystem

4.3 The Data Analysis Subsystem

Hive is an information distribution centre, based over Hadoop. Hive was conceived in Facebook, utilized for the executives and examination of gigantic system information.

At present, Hive has become the top apache venture. The plan objective of Hive is to make the investigator who is capable in SQL better question what's more, examination of enormous informational collections put away in HDFS. At present, the fruitful use of large information is the Hadoop strategy. This paper attempts to build up a model of investigation arrangement of city cost dependent on the structure of Hadoop. At the base of Hadoop is HDFS, and it stores records on all capacity hubs in the bunch of Hadoop, in the interim, as a database for segments, Base runs on HDFS, with the point of finding the necessary information rapidly from billions of columns of information inside the host and access it, and Base employments Map Reduce to manage gigantic measures of information. Catalogue serves for the information table and capacity the executives and gives a common model to them; Hive what's more, Pig gives a significant level language support for Base, also, it makes the procedure of information insights basic; The capacity of Scoop is principally bringing in information from the social database to Hadoop; the capacity of Flume is bringing in the stream information of log information into HDFS legitimately as well as HFDS or Hive; while Zookeeper and Oozie are answerable for organizing the work process.

4.4 The Data Decision Transmission Subsystem

Because of the all-encompassing Big Data-as-a-Service (BDaaS) administration model with the qualities of huge information has not however showed up, Extended system based Big Information as-a-Service model has been proposed. By include properties of the information sources, information types, administration activity to construct numerous sorts, for example, search, examination, representation, and to make administration powerfully. Case study appears, contrasted and the current information administrations, BDaaS with an increasingly far reaching portrayal on properties and tasks. Additionally, it has abilities such as solid semantic perception and programmed administration structure, and incorporated the remarkable blend activities of BDaaS into the execution of information benefits flawlessly.

The information choice subsystem provides the consequences of the data investigation for the decision creator in a clear way, and offers help for the data intelligent preparing simultaneously, and it principally takes the perception program for data handling to complete the above work. In the period of enormous information, countless information about cost ought not to be firm and conceptual; the dynamic information should be mapped to the visual structure. In a word, the quintessence of information choice subsystem is changing the unique information into the visual structure, primarily including three stages, that is visual mapping, see change and show. Each information square is appropriated on various rack, when the client requirements to peruse or compose the document, HDFS will compute the least burden and the ongoing utilization of the server to give access to the client [5, 6, 13]. HDFS can be parallel to peruse and compose from the server group. The system disappointment is unavoidable, so how to do the information recuperation after disappointment and deficiency resilience is significant. HDFS enables information to be upheld up, and different reinforcements are put away in physical areas. The unwavering quality of the information is ensured by diverse different servers.

5 Conclusion

The huge information period has come; the ideas and innovation of huge information ought to not exclusively be utilized in business markets, yet in addition in people in general help field, its quintessence is giving references to chiefs after the enormous information mining by embracing data innovation. Large information applications in government and open administrations can successfully advance government work, government divisions to improve administration effectiveness, basic leadership and social administration level has Great social worth. To make a long story short, large information will give an amazing new instrument to individuals, with the goal that individuals can all the more effectively handle the law of things, increasingly precise comprehension of the world, to foresee the future and change the world. It is especially significant for the government to control the cost with far reaching and auspicious data, while the investigation arrangement of city cost can ace the full scope of value data after an arrangement of activities like information assortment, information the executives, information investigation, information transmission, information security, and afterward it can give valuable reference to the administration's macroeconomic guideline and control.

References

1. M.M. Rathore, A. Ahmad, A. Paul, S. Rho, Urban planning and building smart cities based on the internet of things using big data analytics. Comput. Netw. (2016). https://doi.org/10.1016/j.comnet.2015.12.023
2. Y. Zhang, H. Luo, Y. He, A system for tender price evaluation of construction project based on big data. Procedia Eng. (2015). https://doi.org/10.1016/j.proeng.2015.10.114
3. R. Mehmood, R. Meriton, G. Graham, P. Hennelly, M. Kumar, Exploring the influence of big data on city transport operations: a Markovian approach. Int. J. Oper. Prod. Manag. (2017). https://doi.org/10.1108/IJOPM-03-2015-0179
4. Z. Lv et al., Managing big city information based on WebVRGIS. IEEE Access (2016). https://doi.org/10.1109/ACCESS.2016.2517076
5. B. Tang, Z. Chen, G. Hefferman, T. Wei, H. He, Q. Yang, A hierarchical distributed fog computing architecture for big data analysis in smart cities, in *ACM International Conference Proceeding Series* (2015). https://doi.org/10.1145/2818869.2818898
6. P. Bi, M. Li, Research on analysis system of city price based on big data, in *Proceedings of 2016 IEEE International Conference on Big Data Analysis, ICBDA 2016* (2016). https://doi.org/10.1109/ICBDA.2016.7509819
7. B. Tang et al., Incorporating intelligence in fog computing for big data analysis in smart cities. IEEE Trans. Ind. Inform. (2017). https://doi.org/10.1109/TII.2017.2679740
8. A.P. Plageras, K.E. Psannis, C. Stergiou, H. Wang, B.B. Gupta, Efficient IoT-based sensor BIG data collection–processing and analysis in smart buildings. Futur. Gener. Comput. Syst. (2018). https://doi.org/10.1016/j.future.2017.09.082
9. A. Kumar, R. Shankar, L.S. Thakur, A big data driven sustainable manufacturing framework for condition-based maintenance prediction. J. Comput. Sci. (2018). https://doi.org/10.1016/j.jocs.2017.06.006
10. T. Rabl, M. Sadoghi, H.A. Jacobsen, S. Gómez-Villamor, V. Muntés-Mulero, S. Mankovskii, Solving big data challenges for enterprise application performance management. Proc. VLDB Endow. (2012). https://doi.org/10.14778/2367502.2367512

11. N. Ferreira, J. Poco, H.T. Vo, J. Freire, C.T. Silva, Visual exploration of big spatio-temporal urban data: a study of New York city taxi trips. IEEE Trans. Vis. Comput. Graph. (2013). https://doi.org/10.1109/TVCG.2013.226
12. Z. Lv, T. Yin, X. Zhang, H. Song, G. Chen, Virtual reality smart city based on WebVRGIS. IEEE Internet Things J. (2016). https://doi.org/10.1109/JIOT.2016.2546307
13. Z. Li, S. Zhu, H. Hong, Y. Li, A. El Saddik, City digital pulse: a cloud based heterogeneous data analysis platform. Multimed. Tools Appl. (2017). https://doi.org/10.1007/s11042-016-4038-2

Deep Learning: Potato, Sweet Potato Protection and Leafs Diseases Detections

Hany S. Elnashar(✉)

Beni-Suef University, Beni Suef, Egypt

1 Introduction

Global food demands are expected to be increased day by day, based on the fast world's population increasing. Cassava is considered as the second-largest provider of carbohydrates in Africa [1, 2]. Also, Cassava is a most important food security crop, which is grown by smallholders because it can withstand the weather harsh conditions [3]. North Africa has similar crops such as potatoes and sweet potatoes that are suffering from the same viral diseases. These diseases have a destructive rule on the crops and have a major reason behind poor yields and productions as well. Data Science Research areas can help identifying these common diseases and provide required support in treating and controlling. Since there are known cassava diseases classified in five types, disease detection methods require farmers to seek help from their governments. Meanwhile, the governments are suffering from many challenges such as labor-intensive, low supply, and high costs national problems. Hence, effective solutions for farmers must cover significant limitations. It is known that African farmers can only have access to limited capabilities from mobile cameras or low- bandwidth. Although, farmers across Africa are interested in the cassava from large scale to the minimum small areas. Even, a wide range of environmental climatic conditions are affecting food crops, industrial crops, and food security.

In Uganda, where data are collected, CMD is the most common, which shows symptoms of yellowing and wrinkled leaves, while CBB, which leads to root rot. These two diseases are among the serious threats to Sub- Saharan Africa's food security [1]. This means looking for better automatic approaches to support farmers to have an early protection and more prevention of these diseases and solve leakage of detecting cassava diseases and locations [2]. Using machine learning technologies to solve this kind of problem nowadays is very useful; In Machine learning techniques, deep learning is widespread in agriculture and applied research. Disease prediction is attracting more attention with high accuracy [3]. In this paper, the performance has shown promising results. There are still limited trials and experiments in this type of classification for cassava diseases and the relative crops as potatoes and sweet potatoes. The paper also uses deep learning method based on RsNet50 network technique to achieve high accuracy score that reaches 95% with a very limited image imbalanced dataset of leaf disease categories and labeled images.

F. P. García Márquez (ed.), *International Conference on Intelligent Emerging Methods of Artificial Intelligence & Cloud Computing*, Smart Innovation, Systems and Technologies 273, https://doi.org/10.1007/978-3-030-92905-3_65

1.1 Literature Review

Plant Disease Detection and Classification by Deep Learning [4]. Detection and Classification of Plant Leaf Diseases Using Image processing Techniques [5] as survey. A methodology for model-based greenhouse design: Part 3, sensitivity analysis of a combined greenhouse climate- crop yield model [6]. Automatic plant disease diagnosis using mobile capture devices, applied on a wheat use case [7]. Symptom based automated detection of citrus diseases using color histogram and textural descriptors [8]. All of them are plant disease classification and for applied research using Computer and data scientists there are few researches, starts for examples by, Plant disease diagnoses using deep learning now are underestimation and analysis [9, 10]. Used leaf images as a dataset didn't specify the data were balanced or imbalanced [11], is a practical and applicable solution for detecting the class of diseases and their locations in tomato plants. It's a suitable deep learning architecture unit used rather than the collecting items of physical samples (part of plants, leaves) and read and diagnosis them in the laboratory manually works. In [12], used transfer learning on a dataset of images collected from Tanzania fields to train a deep convolutional neural network and identify three types of diseases with two types of pests damages, and prove that transfer learning is so good a tool in automated disease detection.

2 Methods

2.1 The Cassava Image Datasets Descriptions:

Datasets used in the experiment were adopted from Kaggle. This dataset consisted of 5 fine-grained cassava leaf Dataset of 21,367 labeled images collected during a regular survey in Uganda. Most images were crowdsourced from farmers taking photos of their gardens and annotated by experts at the National Crops Resources Research Institute (NaCRRI) in collaboration with the AI lab at Makerere University, Kampala [13]. This is in a format that most realistically represents what farmers would need to diagnose in real life [2]. It has train files that contain (image_id, label) where image _id refers to an image file name, and label refers to the predicated ID code of the disease type. As shown in Table 1, Lable_num _to disease_map.json is the mapping between disease code and real disease name, as shown in Figs. 1 and 2).

Table 1. Image per disease labeling ID

	Image_id	Label
0	1000015157.jpg	0
1	1000201771.jpg	3
2	100042118.jpg	1
3	1000723321.jpg	1
4	1000812911.jpg	3

```
{0: 'Cassava Bacterial Blight (CBB)',
 1: 'Cassava Brown Streak Disease (CBSD)',
 2: 'Cassava Green Mottle (CGM)',
 3: 'Cassava Mosaic Disease (CMD)',
 4: 'Healthy'}
```

Fig. 1. Disease codding labels And its shape in forms on the table ns 21,397 with two columns, so it is said as the shape of (21,397, 2)

As shown, the most common class tends to one of the diseases and not to healthy plants. And by looking at the image that reflects different classes as per labeling. It is shown in Fig. 3. Based on this representation, shape data is limited in some classes a very limited image dataset of 5 fine-grained cassava leaf disease categories having 10,000 labeled images. Additionally, these images were highly imbalanced, being heavily biased towards classes—CMD and CBSD.

One of the common symptoms of the disease is a color change [14], usually yellow color with different patterns [15]. So will need to be careful with augmentations as a type of data increases with each pattern [16]. For example, how different augmentation changes the images. In the first row, where are original images, in the second row, there are augmented images. Frist in Fig. 4, using Flip Augmentation by Horizontal Flip process [17], is show that.

Second, using Shift Augmentation by Shift Scale Rotate [18] as shown in Fig. 5.

Thread type of Augmentation is the Cutout augmentation process [18] with max_h_size and max_w_size. It will be as shown in Fig. 6.

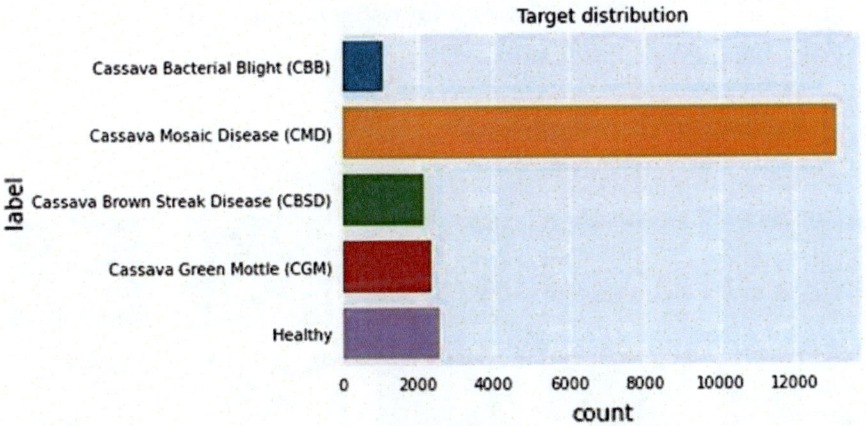

Fig. 2. Target distribution as per labelling

Fig. 3. Cassava image disease labeling

Fig. 4. Flip augmentation by horizontal flip process

Fig. 5. Shift scale rotate of image

Fig. 6. Cutout scaled with a maximum of image

2.2 Approach

2.2.1 Data Loading Methods

Loading training, validation, and test datasets, and with a print out the number of images in each one, it could be sizing as that training images contains 13,380 images, validation data are 8017 images, and test data is one image. for data shapes, these Images formatted as the shape of (128,512,512,3) (128,) with some sort of labeling from zero to four as its type of diseases, and for validation, it same as training but deferent size as shown in Fig. 7 with the same shape and labeling formats but for the testing set it will different it in the shape of (1,512,512,3)(1,) without labeling but with image ID and it is shown in Fig. 8.

2.2.2 Building Models

ResNet50 neural-net has batch-normalization (BN) layers and using the pre-trained model this BN layer using statistics of training data, instead of one used for inference batch-normalization layers of a pre-trained model as trainable, meaning allowing weight updates to happen during training [19, 20]. So using a pre-trained ResNet50 model as a base model, and cross-entropy as loss functions with accuracy as a metrics value as based on doing multi- classification dataset [21]. Resnet50 model has built as 48 layers of Convolution along with one MaxPool and one Average Pool layer [22, 23]. It has more than 3 × 10^9 operations using Floating points [23], with structure developed to the science of computer vision and deep learning [24]. It consist of a 1 + 9 + 12 + 18 + 9 + 1 = 50 layers of Deep Convolutional network, this architecture is

Fig. 7. Test image

used on computer vision tasks [25]. In this paper used for task of image classification based on sparse_categorical_crossentropy as a loss function with the same as categorical_crossentropy. It just has a different interface there did not one-hot encode labels. Figure 9 show model.

Model is training with BATCH_SIZE = 16 * strategy.num_replicas_in_sync, IMAGE_SIZE = [512, 512], CLASSES = ['0', '1', '2', '3', '4'], EPOCHS = 25, HEIGHT = 512, and WIDTH = 51, Fig. 10 show when epoch 1 to epoch 25.

And the model summary is shown in Fig. 10.

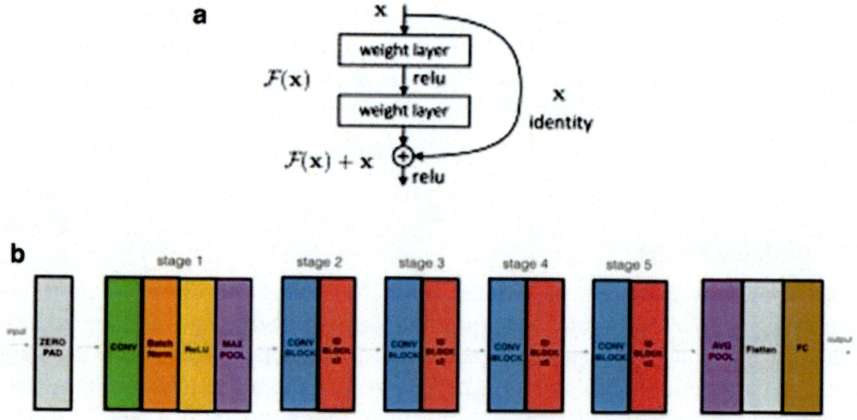

Fig. 8. **a** Model architecture. **b** Model architecture with layers stages

```
Epoch 1/25
184/184 [==============================] - 68s 652ms/step - sparse_categorical_accuracy: 0.4
733 - loss: 1.5701 - val_sparse_categorical_accuracy: 0.6082 - val_loss: 1.2182
Epoch 25/25
184/184 [==============================] - 52s 495ms/step - sparse_categorical_accuracy: 0.6
519 - loss: 0.9486 - val_sparse_categorical_accuracy: 0.6542 - val_loss: 0.9334
```

Fig. 9. Printout for some epochs

2.2.3 Performance Measurements and Results: Performance

Network starts to learn at start learning at around initial_learning_rate = 1e-5, with decay_steps = 10,000, and decay_rate = 0.9 Fig. 12 show Cyclic Learning Rate (Fig. 11).

And the variable available as per classification task are "sparse_categorical_accuracy', 'loss', 'val_loss', and 'val_sparse_categorical_accuracy', So draw learning curves to evaluate model performance as shown in Fig. 13.

Analyzing the Predictions

```
Model: "sequential"
_________________________________________________________________
Layer (type)                 Output Shape              Param #
=================================================================
batch_normalization (BatchNo multiple                  21
_________________________________________________________________
lambda (Lambda)              multiple                  0
_________________________________________________________________
resnet50 (Model)             (None, None, None, 2048)  23587712
_________________________________________________________________
global_average_pooling2d (Gl multiple                  0
_________________________________________________________________
dense (Dense)                multiple                  2098176
_________________________________________________________________
dense_1 (Dense)              multiple                  5125
=================================================================
Total params: 25,691,034
Trainable params: 2,103,307
Non-trainable params: 23,587,727
_________________________________________________________________
```

Fig. 10. Model summary

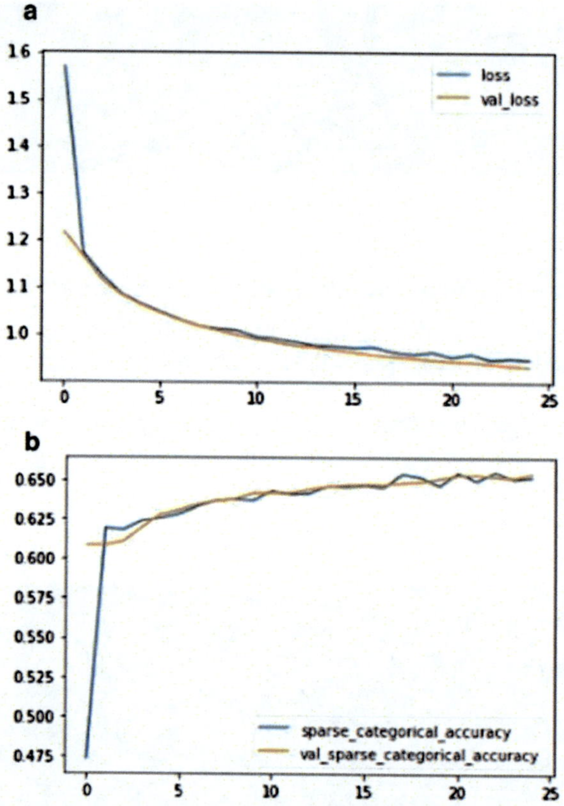

Fig. 11. Learning curves of model performance. **a** loss, val_loss. **b** sparse_categorical_accuracy/val_sparse_categorical_accuracy

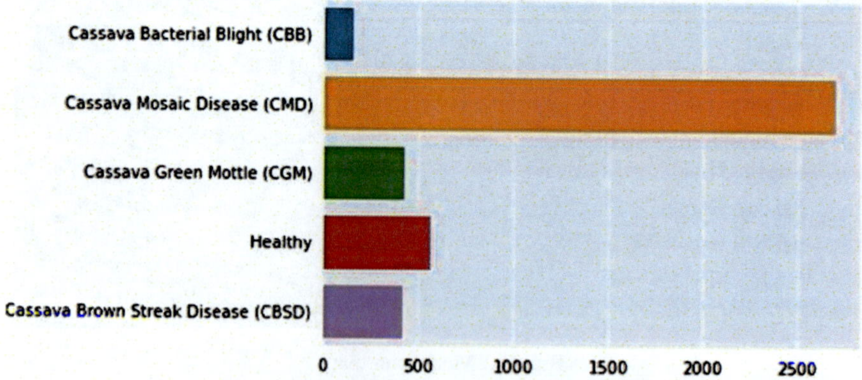

Fig. 12. Predictions on validation distributions

```
id,label
2216849948.jpg,3
```

Fig. 13. Prediction using test data results

Let start make predictions on validation data and collect them. Which be shown in Table 2 and Fig. 12 that the distribution of predictions is similar to the distribution original classes.

Table 2. Predictions, targets lists

	Target	Prediction	Logits	Image_paths
0	0	0	1.496523	../input/cassava-leaf-disease-classification/t...
1	4	0	1.710990	../input/cassava-leaf-disease-classification/t...
2	3	3	5.492183	../input/cassava-leaf-disease-classification/t...
3	3	3	7.024399	../input/cassava-leaf-disease-classification/t...
4	3	3	6.161258	../input/cassava-leaf-disease-classification/t...

And now to make predictions on test data using this model to look for healthy images in dataset, and give image name and label ID as shown in Fig. 13.

3 Conclusions

In this paper, the research has used resnet50 methodology in classification and predication of the plant diseases, mainly Cassava which already has a dataset. Although, there has been imbalanced data, the study has reached to a prediction and also identified the infected and uninfected leaves. Since, potatoes and sweet potatoes are the same family plant that have been included in this study, the study accuracy was %100 in identifying healthy plants and has reached to %98 in identifying infected plants due to changes affected leaves. This study can be applied in North Africa and the countries which produce these crops and the similar crops which make plenty of these plant products and the relative industry development; it also protects food security from the challenges of increasing population and in the meantime has an important role in infection control in the current time. Therefore, this study may be applied in exploring and searching in infection rates in infected agricultural areas, by using the available tools with them or via satellite imagery, so there can be instant and preventive protection. In this work, the ResNet50 model is redeveloped and trained with a very limited dataset having high-class imbalance plants. Three For similar plants with same category of leafs pants. The data was heavily biased towards CMD and CBSD classes, hence, necessitated various techniques to counter the class imbalance. ResNet50 model performance showed promising results and to the best. Techniques applied were class weight, focal loss, SMOTE, and different image dimensions which we found the input shape of a vector (128, 512, 512, 3) giving the best performance. Therefore, this model can be integrated into mobile devices such as smartphones or others to be used by farmers or agricultural extension workers for real-time monitoring and early warning of cassava infections using the mobile application. It is very interesting how the model performance for the imbalanced dataset can be highly improved by class-weight, SMOTE, focal loss techniques, and large input shape dimensions of images. From this experiment, So it got an increment of over 10% in accuracy and a log loss that dramatically reduced to 0.08% from over 20% when class-imbalanced rectification techniques coupled with data augmentation and large input image dimensions were used. The more Effectiveness of this model that it could be used for other plants that have leaves and this leaves effects on its density productions, it could be used by small farmers and for governmental farms for massing productions and wide range protection even IOT applications in farms or satellites scanning.

References

1. H. Makurira, Rainfed Agriculture in Sub-Saharan Africa. Water Product. Rainfed Agric. **181** November 1947, pp. 9–21, 2010, https://doi.org/10.1201/b10823-3
2. A. Kalyebi, S. MacFadyen, H. Parry, W.T. Tay, P. De Barro, J. Colvin, African cassava whitefly, Bemisia tabaci, cassava colonization preferences and control implications. PLoS ONE **13**(10), 1–14 (2018). https://doi.org/10.1371/journal.pone.0204862
3. D. C. Cires¸an, U. Meier, J. Schmidhuber, *Multi-column Deep Neural Networks for Image Classification* (2012)
4. M. H. Saleem, J. Potgieter, K. M. Arif, Plants Plant Disease Detection and Classification by Deep Learning. https://doi.org/10.3390/plants8110468.
5. S. N. Ghaiwat, P. Arora, Detection and classification of plant leaf diseases using image processing techniques: a review. Int. J. Recent Adv. Eng. Technol. ISSN (Online, no. 2, pp. 2347–2812, 2014)
6. J. Cañadas, J.A. Sánchez-Molina, F. Rodríguez, I.M. del Águila, Improving automatic climate control with decision support techniques to minimize disease effects in greenhouse tomatoes. Inf. Process. Agric. **4**(1), 50–63 (2017). https://doi.org/10.1016/j.inpa.2016.12.002
7. A. Johannes et al., Automatic plant disease diagnosis using mobile capture devices, applied on a wheat use case. Comput. Electron. Agric. **138**, 200–209 (2017). https://doi.org/10.1016/j.compag.2017.04.013
8. H. Ali, M.I. Lali, M.Z. Nawaz, M. Sharif, B.A. Saleem, Symptom based automated detection of citrus diseases using color histogram and textural descriptors. Comput. Electron. Agric. **138**, 92–104 (2017). https://doi.org/10.1016/j.compag.2017.04.008
9. A. Fuentes, S. Yoon, S. C. Kim, D. S. Park, A robust deep-learning-based detector for real-time tomato plant diseases and pests recognition. Sensors (Switzerland), **17**(9) (2017). https://doi.org/10.3390/s17092022
10. V. Maeda-Gutiérrez et al. Comparison of convolutional neural network architectures for classification of tomato plant diseases. Appl. Sci. **10**(4) (2020). https://doi.org/10.3390/app10041245
11. A. Fuentes, S. Yoon, S. Cheol Kim, D. Sun Park, *A Robust Deep-Learning-Based Detector for Real-Time Tomato Plant Diseases and Pests Recognition*. https://doi.org/10.3390/s17092022
12. K. P. Ferentinos, Deep learning models for plant disease detection and diagnosis (2018). https://doi.org/10.1016/j.compag.2018.01.009

13. V. Nelson, J. Morton, T. Chancellor, P. Burt, B. Pound, Climate Change, Agricultural Adaptation and Fairtrade Identifying the Challenges and Opportunities," no. 1, p. 45 (2010). www.nri.org
14. S. Golwalla, M. Nadkar, A. Golwalla, S. Golwalla, Infectious Diseases and Infections, Golwalla's Med. Students, 693–693 (2017) https://doi.org/10.5005/jp/books/13059_11
15. P. A. Miceli, W. D. Blair, M. M. Brown, *Isolating Random and Bias Covariances in Tracks* (2018)
16. C. Shorten, T. M. Khoshgoftaar, A survey on image data augmentation for deep learning. J. Big Data **6**(1) (2019). https://doi.org/10.1186/s40537-019-0197-0
17. J.S.M. Moragas, R.R. Reddy, F.H. Alfaro, M.Y. Mommaerts, Systematic review of 'filling' procedures for lip augmentation regarding types of material, outcomes and complications. J. Cranio-Maxillofacial Surg. **43**(6), 883–906 (2015). https://doi.org/10.1016/j.jcms.2015.03.032
18. O. To, D Ata a Ugmentation in T Raining Cnn S" no. 2017, pp. 1–9 (2020)
19. J. Brownlee, Better deep learning. train faster, reduce overfitting, and make better predictions. Mach. Learn. Mastery With Python, **1**(2), 539, (2018)
20. A. Jeddi, "Tackling the Problem of Limited Data agmennd Annotations in Semantic Setation," arXiv, 2020.
21. C. Lea, R. Vidal, A. Reiter, and G. D. Hager, "Temporal convolutional networks: A unified approach to action segmentation," Lect. Notes Comput. Sci. (including Subser. Lect. Notes Artif. Intell. Lect. Notes Bioinformatics), vol. 9915 LNCS, pp. 47–54, 2016
22. V. Sangeetha and K. J. R. Prasad, "Syntheses of novel derivatives of 2-acetylfuro[2,3-a]carbazoles, benzo[1,2-b]-1,4-thiazepino[2,3- a]carbazoles and 1-acetyloxycarbazole-2-carbaldehydes," Indian J. Chem. - Sect. B Org. Med. Chem., vol. 45, no. 8, pp. 1951–1954, 2006. https://doi.org/10.1002/chin.200650130.
23. K. He, X. Zhang, S. Ren, and J. Sun, "Deep residual learning for image recognition," Proc. IEEE Comput. Soc. Conf. Comput. Vis. Pattern Recognit., vol. 2016-Decem, pp. 770–778, 2016. https://doi.org/10.1109/CVPR.2016.90.
24. K. Simonyan and A. Zisserman, "Very deep convolutional networks for large-scale image recognition," 3rd Int. Conf. Learn. Represent. ICLR 2015 - Conf. Track Proc., pp. 1–14, 2015.
25. A. Romero, N. Ballas, S. E. Kahou, A. Chassang, C. Gatta, and Y. Bengio, "FitNets: Hints for thin deep nets," 3rd Int. Conf. Learn. Represent. ICLR 2015 - Conf. Track Proc., pp. 1–13, 2015.

Social Distance Measurement and Face Mask Detection Using Deep Learning Models

Md Mahabub Alam(✉), Md Naimul Islam Suvon, and Riasat Khan

North South University, Dhaka, Bangladesh
{mahabub.alam01,naimul.suvon,riasat.khan}@northsouth.edu

1 Introduction

The novel coronavirus outbreak started in Wuhan, China, from December 2019, and then it consequently influenced numerous nations worldwide. The World Health Organization (WHO) declared it a pandemic sickness as the infection spread through 114 nations, caused more than 111 million dynamic cases till February 2021. The infection fundamentally spreads in those individuals, who are in close contact with one another (inside 6 feet), which is known as person-to-person transmission. Someone can transmit it without having any symptoms, and they are referred to as the asymptomatic carriers. Hence, it is important to wear a mask over the nose and mouth and maintain 6 feet distance from others, specially outside the home, regardless of whether individuals do not have any indications or not.

This paper has implemented a robust system that will automatically monitor the people maintaining adequate physical and social distance and wearing facemask outside their home. We will detect humans from video sequences with SSD-MobileNet and Faster R-CNN ResNet50 deep learning algorithms in the first proposed system. Next, the video frame will be converted into an overhead or bird view to estimate the distance between people accurately. It offers comprehensive coverage of the total view of the image and eliminates the effect of different barriers. Finally, an approximation of adequate physical separation to the image pixels is established to check the possible violation of the social distance. We will train a custom dataset model containing mask and no mask data for the facemask detection system. Then we will detect the region of interest (ROI) of an input image and detect facemask in every ROI using the custom model.

We have arranged the paper in the following sections: first, the related works are discussed in Sect. 2. Section 3 contains the system methodology of our work in detail. Section 4 includes assessing the individual and the total system's performance. Finally, Sect. 5 concludes the research work with a short description of future works.

F. P. García Márquez (ed.), *International Conference on Intelligent Emerging Methods of Artificial Intelligence & Cloud Computing*, Smart Innovation, Systems and Technologies 273, https://doi.org/10.1007/978-3-030-92905-3_66

2 Related Work

In the last few years, there are many works on image processing, computer vision, machine and deep learning, which have been implemented on the object detection and distance measurement tasks very successfully. In [1], Deore et al. used the face detection approach for security purposes. The authors used the Viola-Jones algorithm for facial part detection and the pinhole camera model to measure the distance from the camera. A projection histogram is used in their project for eye line detection. They achieved higher accuracy in measuring the distance from the camera and eye line detection. On the other hand, they were not able to increase facial part detection and eye detection accuracy. The accuracy of these two tasks was less than 50%.

Tathe and Narote detected humans from moving cameras using several algorithms [2]. They utilized projection function and pixel count methods to detect the eye region, a shift algorithm, and the Kalman filter technique for tracking humans. This work utilizes a hierarchical Gaussian process to detect humans, which helps them detect people more accurately than it would be possible from one frame. Finally, the authors used the Markov model to extend their work into people tracking in video sequences. This approach covered a small number of frames at once, and it also helped to possibly longer real-time tracking by reducing the detection time.

In [3], Kajabad and Ivanov proposed a system to detect distance between people using the video frames from a static camera. This paper uses the RCNN algorithm [11] to detect moving objects (people), which applies a sequential learning data approach from the video frame. The image is represented by pixels, and a pixel consists of numbers, mostly between 0 to 255. The work used a background subtraction rule (known as Gaussian Mixture algorithms), optical flow methodology (heat map color technique to analyze every frame), and median filter to eliminate noise throughout the tracking and detect people in their system.

Hou et al. proposed a profound CNN model on object detection in [4]. They moderated the computational unpredictability issues by figuring the location with a solitary regression issue. The authors used the You only look once (YOLO) model [5] to obtain the high-speed performance. They trained the YOLO model with the COCO dataset. To report the YOLO model's result, they used box coordinates, object confidence, and people object class for people detection. For camera view, the ROI of a picture centers around the person on foot strolling road was changed, which is discussed briefly in this paper [6]. Overall, they have shown a well-structured way of measuring distance between people in an image as the distance between pixels and real life is not the same.

3 System Methodology

In this paper, a social distance detection system has been implemented first to measure the distance between people in various outdoor and indoor environments. The proposed social distance detection is divided into three sections.

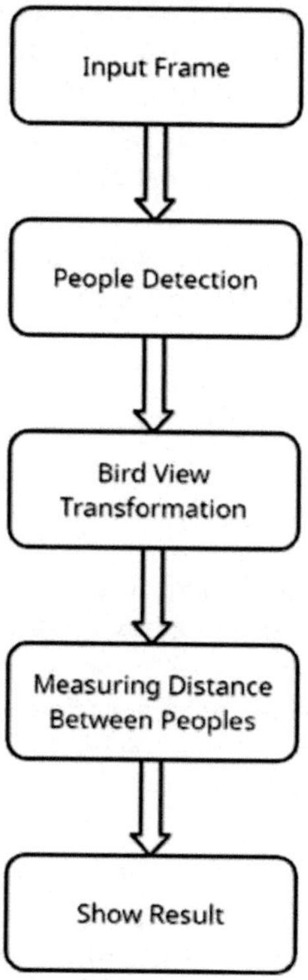

Fig. 1. Proposed social distancing detection process

As shown in Fig. 1, the proposed system will take first the video images as an input frame from the camera; then, it will detect all the people in that frame. Next, it will convert the image into a bird view image, which will give us the top view or overhead view of the image. We will then use this image to measure the distances between the people and demonstrate the images' results by showing an alert if the threshold distance is violated. The top or overhead view approach of the image leads to a better approximation of the distance by overcoming the occlusion effect.

In Fig. 2, we have partitioned our face mask detection system into two parts, where in the first phase, we will focus on loading the dataset and train our dataset to make a model and then serialize the face mask detector. Finally, during the second stage, we can move into stacking our indicator model, perform face discovery and ordering each face as either Mask and No Mask. In the subsequent sections, the social distance measurement and face mask detection processes have been discussed in detail.

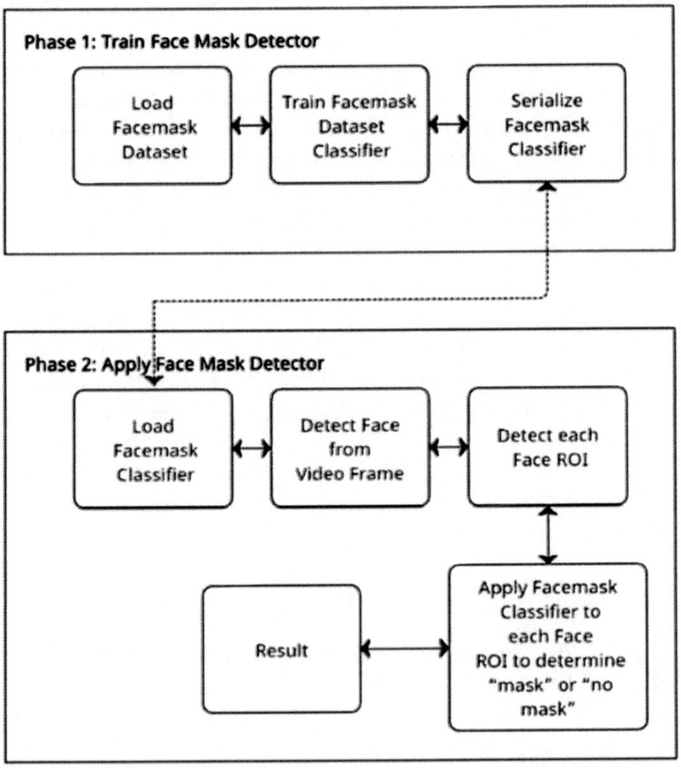

Fig. 2. Proposed face mask detection system

3.1 Social Distancing Detection

Model Selection. For people detection, first of all, we have to select an effective dataset and then a perfect object detection model has to be applied. There are already many pre-trained models that can detect people with high accuracy. We have used the model trained by the COCO (Common objects in context) dataset [7]. This dataset has 120,000 images with a total of 88,000 labeled objects in these images. All the models of COCO are trained to detect 90 different kinds of objects. From these, one of the objects is people's class, which is needed for our system. Many trained models have already been used this dataset using different CNN architectures, e.g., SSD-MobileNet [8], Inception_v3

[9], Resnet50, Fasten RCNN, and more. These models exhibit different performances depending on the speed of the model. We have done some tests on some of these models to evaluate our model's quality, especially its frames per second. As we will do real-time video analysis, finally, we have chosen the SSD-MobileNet_v1_COCO model for the people detection, which has a faster execution speed [10].

People Detection. To use our SSD-MobileNet model to detect people, we load it into a TensorFlow graph and define the output in our case. After that, we have passed all our input frame images to obtain the desired outcome. We specified some of the parameters during this process, including the type of input required for the model and the specific output we want from our model. In our case, we need the outputs of bounding box coordinates, the confidence of each prediction and prediction object class. In the COCO model, the required people class is one from the 90 object classes. After detecting, we have filtered out the weak prediction results by selecting a threshold of 0.75.

Bird View Transformation and Measuring Distancing. For transforming an image into a bird view, we have first taken the 4 points on the original image that we are going to convert. The topics have to form a rectangle with at least two opposite sides being parallel. It has been done to maintain the same proportion when the transformation takes place. For the people detection, we have found a location of the bounding box for each person. Next, from that bounding box, we have to calculate the lowest point, which is the bottom-center point of that box, and it will help us find the ground point of a person. Next, the transformation matrix is used to get each person's real GPS coordinates to each of these points.

After finding each person's ground point, the distance between each pair of points is computed. We have used Eq. (1), which is the Euclidian distance, to measure the distance between the points.

$$d(p, q) = ((p_1 - q_1)^2 + (p_2 - q_2)^2 + (p_3 - q_3)^2)^{\frac{1}{2}} \tag{1}$$

We have chosen 120 pixels as our social distance threshold, which is approximately equal to 2 feet.

3.2 Face Mask Detection

Dataset and Model Selection. We have used the open-source dataset of Bhandary [12] for face mask detection. There are in total 1,376 images, and among them, 690 and 686 images are with and without the mask, respectively. We have taken almost 1:1 ratio for both of the classes. Our aim is to make a custom model to detect whether a person is wearing a mask or not. We have chosen MobileNetV2 convolutional neural network architecture for the face mask detection model, which is a highly efficient architecture, especially for implementing embedded devices (Raspberry Pi, Google Coral, NVIDIA Jetson Nano) with minimum computational capacity.

Data Preparation and Training. Before training, we have preprocessed the images by resizing them into 224 × 224 pixels, converting them to array format, and scaling the pixel intensities in the input image to the range [−1, 1] by using the convenience function. Then we have appended the preprocessed data and associated label to labels and data lists, respectively. Next, we have done a one-hot encoding to our class labels, where each element of our array consists of an array in which only one index is "hot". We have divided our face mask detection data into 80:20 for training and testing, respectively. We have applied an on-the-fly mutation to our data at the time of our training to improve our generalization, known as data augmentation, where random zoom, rotation shear, shift, and flip parameters are initialized. After preparing the dataset, next, we must fine-tune the MobileNetV2 architecture. To fine-tune the MobileNetV2, we have followed three steps. First, we have loaded the MobileNetV2 model with pre-trained ImageNet weights by leaving off the head of network. Next, we have constructed a new fully connected head and append it to the base in place of the old head. Finally, we freeze the base layer of the network where the weights of this base layer will not be updated during the process of backpropagation.

Implementation of Face mask detector with OpenCV. Finally, after training our face mask detector model, we will start the second phase of our system, which loads the model and detects faces and the region of interest (ROI). We have first taken the input image and preprocess it using OpenCV blobFromImage function of the deep neural network module. We have resized the image into 300 × 300 pixels and performed a subtraction. After this, we detect the faces from the images to find their precise location. Next, we have extracted all the face ROIs. We have then computed the bounding box values for each face and compare them to see if the boxes fall within the boundaries of the images. Finally, we obtained faces. We also attain a percentage of detection for each face to know the accuracy of each face mask detection.

4 Performance Evaluation

In this section, we have evaluated the performance of the people detection and social distance measurement and face mask recognition systems. The social distance measurement performance has been assessed in two steps. First, we have evaluated the performance of human detection and then, we have shown the performance of the overall system.

4.1 Performance of People Detection

For the people detection, we have used a pre-trained COCO model. As we know, for object detection, we cannot measure the accuracy metrics like the classification problem. There exist 12 different accuracy metrics for object detection; the mAP (mean average precision) is one of them. mAP is the average measurement area under the precision-recall (PR) curve. It indicates the detection model's accuracy and compares the detected bounding box with the ground-truth box. The higher value of the mAP score indicates the better accuracy of the detection system. For our implemented SSD-MobileNet_v1_COCO model for people detection, we obtained an average mAP of 21% with a 30 ms execution time, which is fast compared to the other detection models. We choose this model because we have to analyze it in real-time, and we have a plan to implement our proposed system in an embedded system, so our frames should execute very fast. We have also tried the Faster R-CNN ResNet50 (COCO) model, which has better accuracy of 30% mAP, but the execution time is 89 ms, which is slow for the real-time analysis. In Table 1, we have shown the performance metrics of the pre-trained COCO models with their execution time and mAP.

Table 1. Model accuracy metrics for various human detection algorithms

Accuracy metrics	SSD-MobileNet	Faster R-CNN ResNet50
mAP	21%	30%
Execution time	30 ms	89 ms

4.2 Performance of Overall Social Distance Measurement System

After detecting the people, the image has been transformed into a top or bird view. Finally, the distance between the people will be measured using that bird view image, and we have found the overall result.

In Fig. 3, we have shown an image by implementing our proposed SSD-MobileNet algorithm. We can see the bird view on a small screen (lower left corner of the image), which is the four corner points that we take from the original image and transform it into a bird view with a ground point for each person. From Fig. 3, we can observe that when the person is too close to each other (below the predefined threshold of 120 pixels), the bounding box around the individual will be indicated with a red box, and the circle of the bird view image turns to red.

Fig. 3. Social distancing detection evaluation on video with the bird view transformation by the SSD-MobileNet algorithm

4.3 Performance of Face Mask Detection

We have used 34 epochs during the training and validate our model with 276 (20%) sample images for the face mask detection model. Figure 4 demonstrates the training and testing accuracy and loss curves of the face mask detection model. It illustrates the high accuracy of the detection system with the validation loss lower than the training loss. We have achieved an accuracy of approximately 99% in our test dataset.

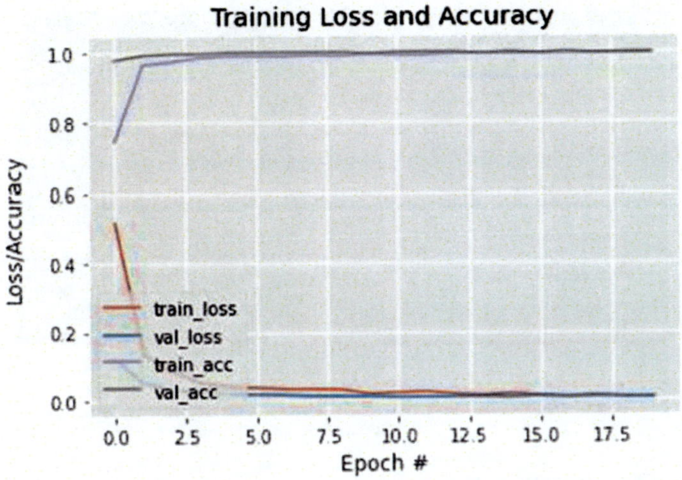

Fig. 4. Training loss versus accuracy of the face mask detection model

Table 2 demonstrates different performance metrics (precision, recall, f1-score, and support) of the face mask detection system. Finally, we have implemented our face mask recognition model in a real environment by using a mobile IP camera, which is connected to our detection program, to see if it is working well.

Table 2. Model accuracy metrics for face mask detection

	Without masks	With masks	Macro avg	Weighted avg	Accuracy
Precision	0.99	1.00	0.99	0.99	–
Recall	1.00	0.99	0.99	0.99	–
F1-score	0.99	0.99	0.99	0.99	0.99
Support	138	138	276	276	276

Figure 5a shows that our model can detect a face with the mask accurately. The images are outside our training and testing dataset in real-time video frames, and it is detecting the face mask continuously. In Fig. 5b, we have also demonstrated that our model can also detect a face without mask in real-time with high accuracy. The frame was a little bit blurry, but still, it could detect the face without mask.

5 Conclusion

In this paper, we have implemented two safety systems for the current and future COVID-19 situation, i.e., physical distance measurement and face mask detection. We have detected people from video frames with the SSD-MobileNet and Faster R-CNN

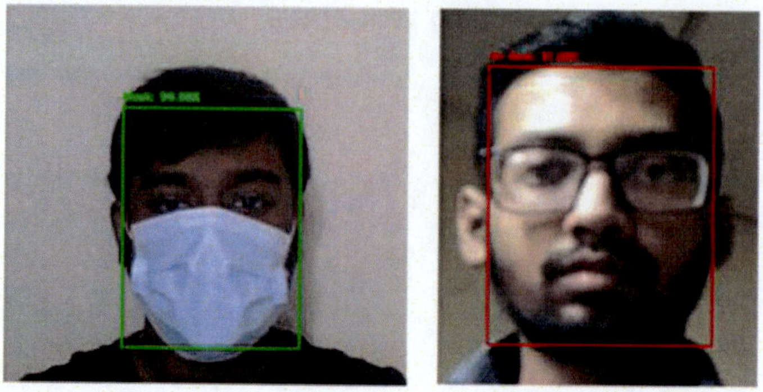

(a) (b)

Fig. 5. Real-time face mask detection with **a** mask and **b** without mask

ResNet50 deep learning algorithms in the first system. Next, we measured the distance between them by transforming the image into an overhead or bird view approach. An approximation of social distance into pixel is considered, and a threshold value is established to detect its violation. This system's accuracy greatly depends on people detection, where we have used a pre-trained people detection model that has above 95% accuracy, with a fast execution time. The second preventive system is implemented to detect face masks by implementing MobileNetV2 convolutional neural network architecture. Our model of mask detection has 99% accuracy, and our system can also recognize faces in real-time.

In the future, we can efficiently run our proposed deep learning-based models in an embedded system, e.g., NVIDIA Jetson Nano, and make a wireless face mask detector and social distance detector device. Moreover, we can also make a parallel implementation of these two systems to work into a single device that will reduce the cost of the total system. The human detection system's accuracy can be further increased by adding an overhead dataset with the pre-trained model.

References

1. G.Deore, R. Bodhula, V. Udpikar, V. More, Study of masked face detection approach in video analytics, in *Conference on Advances in Signal Processing (CASP)*, Pune, India (2016), pp. 196–200
2. S.V. Tathe, S.P. Narote, Realtime human detection and tracking, in *Annual IEEE India Conference (INDICON)*, Mumbai, India (2013), pp. 1–5
3. E.N. Kajabad, S.V. Ivanov, People detection and finding attractive areas by the use of movementdetection analysis and deep learning approach. Procedia Comput. Sci. **156**, 327–337 (2019)
4. Y.C. Hou, M.Z. Baharuddin, S. Yussof, S. Dzulkifly, Social distancing detection with deep learning model, in *8th International Conference on Information Technology and Multimedia (ICIMU)*, Selangor, Malaysia (2020), pp. 334–338
5. J. Redmon, S. Divvala, R. Girshick, A. Farhadi, You only look once: Unified, real-time object detection, in *Proceedings of the IEEE Conference on Computer Vision and Pattern Recognition*, Las Vegas, United States (2016), pp. 779–788

6. A.H. Ahamad, N. Zaini, M.F.A. Latip, Person detection for social distancing and safety violation alert based on segmented ROI, in *10th IEEE International Conference on Control System, Computing and Engineering (ICCSCE)*, Penang, Malaysia (2020), pp. 113–118
7. T.Y. Lin, M. Maire, S. Belongie, J. Hays, P. Perona, D. Ramanan, P. Dollár, C.L. Zitnick, Microsoft COCO: common objects in context, in *European Conference on Computer Vision* (Springer, Cham, 2014), pp. 740–755
8. Y.C. Chiu, C.Y. Tsai, M.D. Ruan, G.Y. Shen, T.T. Lee, Mobilenet-SSDv2: an improved object detection model for embedded systems, in *2020 International Conference on System Science and Engineering (ICSSE)*, Kagawa, Japan (2020), pp. 1–5
9. Documentation of Inception_v3, https://www.tensorflow.org/api_docs/python/tf/keras/applications/inception_v3. Accessed 22 Feb 2021
10. TensorFlow Model Garden, https://github.com/tensorflow/models. Accessed 22 Feb 2021
11. R. Girshick, J. Donahue, T. Darrell, J. Malik, Region-based convolutional networks for accurate object detection and segmentation. IEEE Trans. Pattern Anal. Mach. Intell. **38**(1), 142–158 (2016)
12. P. Bhandary, Face mask dataset, GitHub repository (2020), https://github.com/prajnasb/observations/tree/master/experiements/data

Design and Analysis of Digital Compressed ECG Sensing Encoder for IoT Health Monitoring Devices

Shivangi Srivastava(✉) and M. Sabarimalai Manikandan

Biomedical System Lab, School of Electrical Sciences, Indian Institute of Technology Bhubaneswar, Odisha 752050, India
{ss109,msm}@iitbbs.ac.in

1 Introduction

Modern connected wearable or portable medical devices are capable of acquiring single- or multi-lead electrocardiogram (ECG) signals for analysis and diagnosis of various cardio- vascular diseases (CVDs). In the wearable body area network (WBAN), wireless ECG sensing node generally consists of analog-front end (AFE), analog-to-converter (ADC), processor, on-broad memory and wireless radios for sensing, digitizing, processing, storing and transmitting the processed or acquired data wirelessly to the edge or cloud health monitoring and management servers [1, 2]. Most internet of medical things (IoMT) applications, the wireless ECG sensing nodes are widely explored for continuous long-term monitoring with the major objectives of improving the outcomes of acute cardiovascular disease and reducing unnecessary death or disability. Moreover, the Smartphone can be used as the data aggregator and to transmit the acquired ECG data to a remote remote computing servers through WiFi or cellular network [3, 4]. Most wearable devices have limited battery power, data rate (bandwidth) and on-broad memory, which need to be addressed when high-resolution single-channel or multichannel ECG signals are acquired for diagnosis of various kinds of cardiovascular diseases. Therefore, there is a high demand for exploring energy-efficient on-broad data compression technique which can reduce the amount of acquired data needs to be stored or transmitted wirelessly and also alleviate problems of both bandwidth and memory space constraints.

Most state-of-the-art ECG compression methods were based on discrete wavelet transform (DWT) and efficient encoding of wavelet coefficients [18–20]. Although DWT-based methods provides better compression efficiency with minimal reconstruction errors, digital CS-based data reduction methods recently explored because tremendous energy saving can be achieved with moderate compression ratios that can be easily implemented with use of simple addition and comparison operations on the resource-constrained ECG sensing nodes.

This work was carried out under the Scheme for Promotion of Academic and Research Collaboration (SPARC) project, Ref. SPARC/2018- 2019/P468/SL, Ministry of Human Resource Development (MHRD) (Now Ministry of Education (MoE)), Government of India.

F. P. García Márquez (ed.), *International Conference on Intelligent Emerging Methods of Artificial Intelligence & Cloud Computing*, Smart Innovation, Systems and Technologies 273, https://doi.org/10.1007/978-3-030-92905-3_67

In this study, we investigate the performance of digital compressed sensing (DCS)-based ECG encoder with four binary sensing matrices such as deterministic binary block diagonal (DBBD) matrix, random Bernoulli sensing matrix (RBeSM), random sparse binary sensing matrix (RSBiSM) and random binary sensing matrix (RBiSM) in terms of compression ratio (CR), reconstruction quality (signal-to-noise ratio (SNR) and percentage root-mean-square difference (PRD), and real-time feasibility (computational operations and processing time). In addition, we further demonstrate that the DCS-based ECG encoder can achieve higher CRs with minimal reconstruction errors with selection of a suitable measurement quantization bit and demonstrate the impact of a sampling rate in achieving the better compression performance by using the four sensing matrices that are not addressed jointly in the past studies.

The rest of this paper is organized as follows. Section 2 summarizes the existing digital compressed sensing (DCS) methods. Section 3 presents the DCS-based ECG encoder. Section 4 presents evaluation results obtained for four sensing matrices, different sampling rates and measurement quantization bits by using a wide variety of ECG signals. Finally, Sect. 5 presents conclusions and future directions.

2 Existing Compressed Sensing Methods

The concepts and excellent reviews of compressed sensing methods were presented in Refs. [3] and [8]. In this Section, we summarize few DCS methods reported for ECG signals that are related to our present study. In [3], H. Mamaghanian et al. proposed real-time system-level comparison between a new digital CS-based ECG data compression and the discrete wavelet transform (DWT)-based ECG compression methods by implementing on the embedded Shimmer wireless body sensor networks (WBSN) mote. In [5], Dixon et al. studied several design considerations such as the encoder architecture and measurement matrix design for CS-based ECG telemonitoring via a WBAN. Results showed the high CRs can be achieved by using a 1-bit Bernoulli measurement matrix. In [4], binary permuted block diagonal (BPBD) matrix was used because it requires simple hardware and is incoherent with wavelet and DCT basis functions. Results showed that the proposed CS encoder achieved CRs (N/M)of 6:1 which save 82.9% of the energy consumption of transceivers. In [10], CS based ECG compression was presented based on the deterministic binary block diagonal (DBBD) matrix and a modified Kronecker-based recovery technique with different sparsifying basis. From the results, it was observed that compressed sensing with DBBD matrix and DCT sparsifying basis provides the best recovery for ECG signals.

In [11], the CS encoder includes the Bernoulli random matrix, uniform scalar quantizer and lossless Huffman coding (HC) with an optimal codebook. Results showed that the proposed encoder had rate of 1.41 bps and rate of 3.27 bps for ADSNR = 25 dB at quantization of 10 bit with ossless Huffman coding (HC) and without HC. In [12], the rakeness-based sensing approach had CR (N/M) of 5.33 and 4 with the average PRD of 10.7% and 7.42%, respectively whereas the minimal mutual coherence (MMC) sensing approach had average PRD of 13.3% and 8.85%, respectively for the input signal length of 521 samples. Further, the rakeness-based sensing approach had CR (N/M) of 2 and 2.5 with the average Pearson's Correlation Coefficient (PCC) of 0.936 and 0.858, respectively whereas the random sparse binary (RSB) matrix based approach had the average PCC of 0.876 and 0.793, respectively.

In [13], addressed the problem of high decoding complexity by the using weighted regularized least squares (WRLS) and the with the CS encoding schemes based on the random sparse binary (RSB) matrix and random memodulator (RD) sensing matrix (block-diagonal matrix). For the RSB sensing matrix, the proposed method achieved "Very Good" (VG) grade grade with at least 32% higher CR than other methods. In [14], results showed the signal-adaptive sensing matrix based approach had PRD vales of 2.4%, 5.5%, 8.6% and 10.9% whereas method based on the sparse distribution and the orthonormal Symlet-4 wavelet basis had PRD values of 5%, 9%, 10% and 15% for the CR values of 2, 3. 4, and 5, respectively. An information-enhanced sparse binary matrix (IESBM) was presented based on the area of interest (AI),'QRS complex detection' and the enhanced measurement (EM) model [15]. In this study, for MIT-BIH arrhythmia database with $N = 512$ and $M = 256$, results showed the PRD values of 3.16%, 3.36%, 3.33%, 3.27%, 6.56%, and 9.19% for the sparse binary matrix (SBM), Bernoulli binary matrix (BBM), Toeplitz binary matrix (TBM), circulant binary matrix (CBM) and the 6-bit Gaussian matrix (GM), respectively by using BSBL algorithm with DWT sparsifying basis.

In [16], results showed that the DBBD matrix based CS provides a better performance in terms of reconstruction quality compared with the Gaussian and binary permuted block diagonal (BPBD) matrices with an improvement of SNR from 6 to 20 dB when CR values from 90% to 50% for N = 500 samples with the OMP algorithm for reconstructing the compressed ECG signal with DCT sparsifying basis. Results further showed that the proposed thresholding based recovery algorithm is 23 and 12 times faster than the OMP and stagewise OMP (StOMP) algorithms, respectively. In the CS algorithm with signal-adpative sensing matrix [17], the optimized dictionary matrix based reconstruction approach had PRD values of 1.51%, 2.24%, 5.29%, 6.24%, and 7.79% and the optimized dictionary matrix based approach had the PRD values of 1.70%, 3.09%, 6.25%, 11.58% and 19.47% for the CR values of 2, 4, 6, 8, and 10.

3 Compressed Sensing Based Ecg Encoder

In this study, we present the digital compressed sensing (DCS)-based ECG encoder which consists of preprocessing, sensing matrix generation, measurement generation, and measurement quantization by selecting the suitable sensing matrix and measurement quantization bit and sparsifying matrix.

A. *Preprocessing*

In some of the DCS encoder design methods, the ECG signal was filtered using 4th order Butterworth high-pass and low-pass filters with cutoff frequencies are 0.5 Hz and 40 Hz, respectively [13]. In our study, the preprocessing performs the operations of amplitude normalization, mean subtraction and moving averaging (MA) filtering. In this study, the input ECG signal is processed to smooth out high-frequency noises by using the 10-points MA filter.

B. *Compressed Sensing Matrices*

From the studies, the optimal sensing matrix design must consider the following parameters [15]: the type of random distribution (Gaussian, Bernoulli, and Uniform), the bit precision of sensing coefficients (Non-binary (full precision, Quantized) and Binary), the sparseness of random binary sensing matrix (exact number of ones, $d \ll M$ that randomly located in each of the columns/rows of sensing matrix and unequal number of ones, $d \gg M$ that are randomly located in each of the columns/rows), the structure of the sensing matrix (Block Diagonal, Toeplitz, Circulant or Permuted Block Diagonal), the very low coherence with the sparsifying basis and the computational hardware requirements for CS encoding stage.

(1) *Random Bernoulli Sensing Matrix:* Random Bernoulli sensing matrix includes $\pm$ 1 entries which are drawn from a Bernoulli distribution with entries (1, –1). The Bernouilli matrix sensing approach requires $M \times \{N{-}1\}$ additions, $M \times N$ boolean comparisons, 2's complement operations for '–1' and $M \times N$ storage elements. The need of multiplications are eliminated and but it requires the two's complement operation for '-1' to use only adders.
(2) *Random Dense Binary Sensing Matrix:* Although the random non-binary sensing matrices are difficult to generate with full precision non-binary values and also to store them in a resource-constrained wearable sensor nodes [3]. Therefore, binary sensing matrices with 1-bit were explored to reduce the hardware complexity and save energy. It just requires adders and no multipliers, and less memory footprint for storing 1 bit, which are not possible with random non-binary sensing matrices. For generating the sensing matrix with '1' and '0' entries, the Linear Feedback Shift Register (LFSR) is widely used with initial seed to ensure the same binary sequence with randomly located 1's and 0's.
(3) *Random Sparse Binary Sensing Matrix:* The random sparse binary (RSB) sensing matrix (Φ_{SB}) consists of 'd' entries of 1's which are randomly located in each of the columns of Φ matrix, while other entries of the sensing matrix are all zero. It was found that the mutual coherence of RSB matrix with DWT and DCT basis can be low as comparable with the mutual coherence of the random Gaussian sensing matrix [3]. The limited number of 1's, $d \ll N$ can reduce the memory footprint by storing only the positions of the non-zero entries instead of storing all the values of the matrix. From the preliminary results as shown in Figs. 1 and 2, the compression ratio and reconstruction quality depends on the two optimal parameters such as (*i*) the number of non-zero entries and (*ii*) positions of 1's in each of the columns. In addition, we need to find the minimum number of non-zero entries that can lead to best trade-off between the reconstruction error and processing time although the optimal number of non-zeros can be estimated for specific validation database(s) in offline. Furthermore, the random positions of 1's would also affect the reconstruction quality with the same number of non-zero entries for the compression of same ECG signal.
(4) *Deterministic Binary Block Diagonal Sensing Matrix:* The deterministic binary block diagonal (DBBD) sensing matrix can be constructed with the diagonal blocks

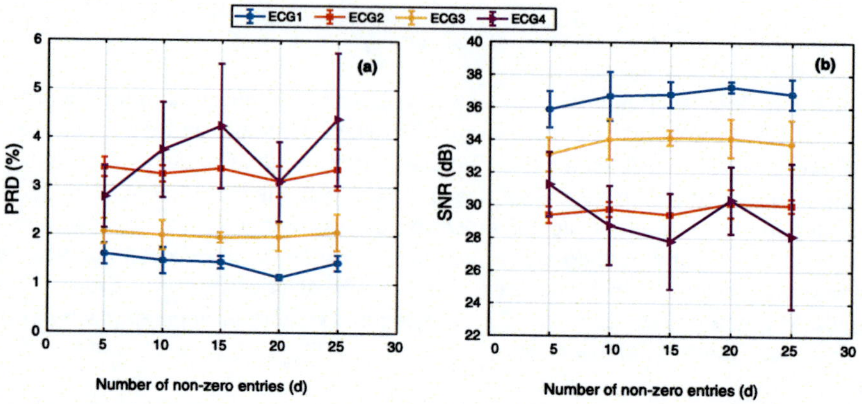

Fig. 1. Performance of random sparse binary sensing matrix based CS enoder for different number of non-zero entries ('*d*').

having the elements of consecutive 1's. The number of 1's is computed as $\left[\frac{N}{M}\right]$. For example, if $N = 16$ and $M = 8$ the number of consecutive 1's is 2 for each row. Since the DBBD sensing matrix is binary and has limited number of consecutive 1's, the CS encoder requires just addition operations unlike multiplications as in the case of random sensing matrices and also only two positions need to be stored for each of row which requires lesser on-broad memory as compared to the RSB sensing and non-binary sensing matrices.

III. *Quantization of CS Measurements*

Most existing methods were not considered the effect of quantization of CS measurements in the performance analysis of the CS based encoder. Although the higher resolution bit (equivalent to resolution bit of sample) was used for encoding the measurements, optimal selection of quantization bit may provide better compression efficiency with minimal reconstruction errors. Therefore, the measurements y are processed using the uniform scalar quantizer to further reduce the rate of the encoder. In this study, effect of the measurement quantization is investigated with measurement precision of 10, 8, and 6 bits. The quantization is implemented as

$$\hat{y} = Q[y] = \left[\frac{y}{\Delta}\right], \tag{1}$$

where Δ is the quantization step size which is computed as

$$\Delta = \frac{y_{max} - y_{min}}{2^{B_y}} \tag{2}$$

where B_y denotes the number of bits used for quantization of measurements.

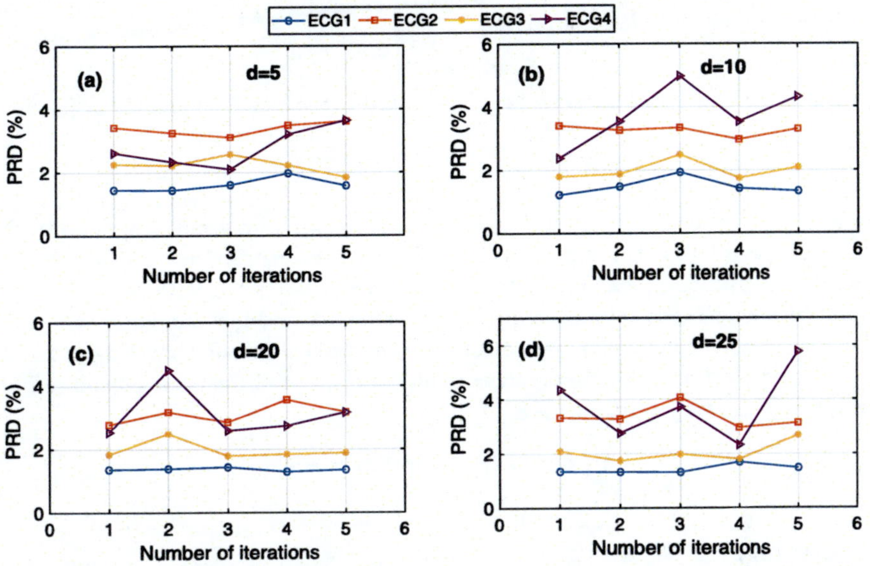

Fig. 2. Performance of random sparse binary sensing matrix based CS enoder for same number of number of non-zero entries (‘d’) but located in different positions of columns of sensing matrix.

IV. *Sparsifying or Dictionary Matrix*

Any arbitrary signal $\mathrm{x} \in \mathbb{R}^n$ can be represented as

$$\mathrm{x} = \sum_{i=1}^{N} \alpha_i \psi_i = \Psi\alpha \tag{3}$$

where α is the transformed coefficients for the sparsifying matrix Ψ and N denotes the length of the signal to be processed. The estimated $\alpha \in \mathbb{R}^n$ is the sparse vector when the vector contains only few non-zero elements, K. Depending on the application scenarios such as the ECG reconstruction from few measurements; simultaneous ECG reconstruction and signal decomposition; and specific signal component extraction in the signal recovery process, the dictionary matrix can be constructed with different kinds of basis functions for better extraction of both time-localized and frequencylocalized waves of the ECG signal. From our past studies reported in [18–20], it is observed that the discrete wavelet transform (DWT) with Biorthogonal wavelet basis reported higher compression ratio with minimal errors. Results showed that the DWT can provide highly sparse representations of the the ECG signals. Some of the methods used DCT as the sparsifying matrix in the reconstruction process. The dictionary matrix was constrcuted by using the the signal properties [16]. In this study, sparsifying matrix is constructed based on the discrete cosine, discrete sine and impulse functions.

E. *Signal Recovery or Reconstruction*

In the digital compressed sensing, the compressed signal can be expressed as [3–8]:

$$y = \Phi x = \Phi \Psi \alpha = D\alpha \tag{4}$$

where $y \in \mathbb{R}^M$ is the compressed signal or measurements with vector length of M, $\Phi \in \mathbb{R}^{M \times N}$ is a sensing matrix with $M \ll N$, and dictionary matrix $D = \Phi\Psi \in \mathbb{R}^{M \times N}$. In order to reconstruct the ECG signal by using the expression (3) with best sparsifying matrix (Ψ), we need to estimate α for the predefined D and few measurements y. The computation of α is not direct approach like the traditional transform based techniques. The ℓ_1-minimization can be approached with the concept of the basis pursuit (BP) by solving the following equation [3–8]:

$$\arg\min \|\alpha\|_1 \quad \text{subject to} \quad y = D\alpha \tag{5}$$

4 Results and Discussion

A. *Test ECG Database*

In the past studies, the MIT-BIH arrhythmia database (https://www.physionet.org/content/mitdb/1.0.0/) was widely used for the performance evaluation of CS-based ECG encoders [13, 15, 17]. It consists of two-channels of 48 ECG recordings obtained from 47 subjects. Each of the records is slightly over 30 min long. It contains different types of cardiac arrhythmias and kinds of noises and artifacts. The recordings were processed using the band-pass filter with a passband from 0.1 to 100 Hz. The ECG signals were digitized with a sampling rate of 360 samples per second and 16-bit sample amplitude resolution over a 10 mV range. The analog-to-digital converter (ADC) was unipolar, with 11-bit resolution over a $\pm$ 5 mV range. The sample values of the ECG signal vary from 0 to 2047 inclusive, with a value of 1024 corresponding to zero volts. The ECG records labeled 100 to 124 and 200 to 234 (some numbers are missing). Many methods were tested using the selected records number 100, 112, 123, 200 and 222 for our experiments. In this study, we used first 5 s ECG signals from 48 recordings for performance evaluation.

B. *Compression Performance Metrics*

In most methods, quality of the reconstructed ECG signal was assessed using the percentage root-mean-square difference (PRD) and the signal-to-noise ratio (SNR) [14, 16, 17]. Existing CS based ECG compression methods considered the PRD value less than 9% for assessing the clinical acceptability of the reconstructed ECG signal. The PRD is defined as

$$\text{PRD}[\%] = \frac{\|\mathbf{x} - \widetilde{\mathbf{x}}\|_2}{\|\mathbf{x}\|_2} \times 100 \tag{6}$$

where $\mathbf{x}$ and $\widetilde{\mathbf{x}}$ are the original and recovered ECG signals, respectively. The SNR is computed as

$$\mathrm{SNR[dB]} = 20\log_{10}\left(\frac{\|\mathbf{x}\|_2}{\|\mathbf{x}-\widetilde{\mathbf{x}}\|_2}\right) \tag{7}$$

where, higher SNR indicates the better quality of ECG reconstruction. In this study, the compression ratio is defined as.

$$\mathrm{CR} = \frac{NB_o}{NB_c} \tag{8}$$

where $N B_o$ denotes the number of bits used for encoding the original signal and $N B_c$ denotes the number of bits used for encoding the compressed measurements.

C. *Performance Evaluation Study Conditions*

In this study, we investigated the performance of the digital compressed sensing (DCS) based ECG encoder by using different kinds of sensing matrices using a wide variety of ECG signals taken from the MIT-BIH arrhythmia database. The performance evaluation of the DCS-based ECG encoder was carried out with the following test conditions and parameters:

- Original ECG signal was digitized using an ADC with a sampling rate of 360 Hz and bit resolution of 11 bit. In addition, the ECG signals were resampled with rate of 250, 200, and 100 Hz for performance evaluation.
- Sensing matrices such as the random binary sensing matrix (RBiSM), random Bernoulli sensing matrix (RBeSM), random sparse binary sensing matrix (RSBiSM) and deterministic binary block diagonal (DBBD).
- Signal duration is 5 s and Signal length of 1800 samples.
- Measurement quantization is implemented with 10, 9, 8, 7, 6 an 5 bits.
- Number of measurements (*M*): 900, 450, 300, and 225.
- Sparsifying matrix: constructed based on the discrete cosine, discrete sine and impulse matrices.
- Total number of ECG segments: 148 (2–3 ECG segments from two-channel 48 recordings).
- Performance metrics: Compression ratio (CR), percentage root- mean-square difference (PRD), signal-to-noise ratio (SNR) and processing time.

D. *Performance of the CS based ECG Encoder*

Since the existing traditional lossy ECG compression methods were evaluated using the ECG signals with a sampling rate of 360 samples/s. In this subsection, we present the performance of the CS based ECG encoder with four sensing matrices for different measurement lengths including M = 900, 450, 300, and 225 and measurement quantization of 10, 8, and 6 bits. Evaluation results of this study were summarized in Table 1 which includes the average and standard deviation of the SNR and PRD values obtained for the 148 ECG segments. From the distortion metric values, it is noticed that the CS-ECG encoder with the deterministic binary block diagonal (DBBD) sensing matrix shows higher compression ratio with acceptable level of PRD value below 9%. Results further showed that the compression performance of the CS- ECG encoder with RSBiSM, RBeSM and RBiSM can be improved the CR with acceptable level of PRD value below 9% by selecting a suitable quantization bit for encoding the measurements. This study showed that the CR of 4 to 5 can be achieved with acceptable PRD value by using three sensing matrices such as the random binary, random sparse binary and random Bernoulli matrices whereas the DBBD sensing matrix based ECG encoder can provide higher compression ratio of 12 to 16 with acceptable level of PRD for most of the records.

E. *Signal Redundancy Impact on the Performance of the CS- based ECG Encoder*

The main purpose of this study to investigate the effect of the sampling rate in achieving the better compression performance by using the four sensing matrices. Results of this study are shown in Figs. 3 and 4 for the sampling rates of 360, 250, 200 and 100 Hz and the measurement quantization of 10, 9, 8, 7, 6 an 5 bits and number of measurements (M) of 900, 450, 300, and 225. From the results, it is observed that the CS-based ECG encoder with DBBD sensing matrix provides better compression ratio and PRD values as compared to the other three sensing matrices such as the RSBiSM, RBeSM and RBiSM for all the sampling rates with the same measurement quantization and number of measurements. Results further showed that ECG signals with higher sampling rate can provide higher compression ratio range for a given PRD value. For example, the

Table 1. Performance of compressed sensing ECG encoder with four sensing matrices, Fs = 360 Hz, Quantizer with 10, 8, and 6 bits and M: 900, 450, 300, and 225

Quantization Bit, B = 10 bits								
Sensing Matrix	M = 900, N = 1800		M = 450, N = 1800		M = 300, N = 1800		M = 225, N = 1800	
	B = 10, CR = 2.4		B = 10, CR = 4.4		B = 10, CR = 6.6		B = 10, CR = 8.8	
	SNR	PRD	SNR	PRD	SNR	PRD	SNR	PRD
DBBD	39.7 ± 2.6	1.0 ± 0.3	39.6 ± 3.2	1.12 ± 0.4	36.8 ± 4.5	1.6 ± 0.4	25.7 ± 4.6	5.8 ± 2.9
RSBiSM	31.4 ± 5.0	3.2 ± 2.2	15.1 ± 3.9	19.12 ± 7.8	8.7 ± 3.1	38.8 ± 7.8	6.4 ± 2.6	49.8 ± 13.2
RBeSM	24.1 ± 3.9	6.4 ± 1.7	13.7 ± 2.6	21.83 ± 7.3	13.7 ± 3.2	21.8 ± 7.6	6.4 ± 2.5	50.5 ± 2.5
RBiSM	24.7 ± 3.7	6.8 ± 3.2	14.4 ± 3.61	20.48 ± 7.7	8.7 ± 2.9	38.2 ± 11.7	6.2 ± 2.6	49.6 ± 12.9
Quantization Bit, B = 8 bits								
	M = 900, N = 1800		M = 450, N = 1800		M = 300, N = 1800		M = 225, N = 1800	
Sensing Matrix	B = 8, CR = 2.75		B = 8, CR = 5.5		B = 8, CR = 8.25		B = 8, CR = 11	
	SNR	PRD	SNR	PRD	SNR	PRD	SNR	PRD
DBBD	35.6 ± 1.6	1.6 ± 0.3	35.1 ± 3.3	1.7 ± 1.5	27.6 ± 3.3	4.4 ± 1.5	25.3 ± 4.3	6.0 ± 2.8
RSBiSM	30.5 ± 4.4	3.4 ± 2.4	15.1 ± 3.9	19.1 ± 7.8	8.7 ± 3.0	38.8 ± 3.6	6.4 ± 2.6	49.7 ± 13.2

(continued)

Table 1. (*continued*)

Quantization Bit, B = 10 bits								
Sensing Matrix	M = 900, N = 1800		M = 450, N = 1800		M = 300, N = 1800		M = 225, N = 1800	
	B = 10, CR = 2.4		B = 10, CR = 4.4		B = 10, CR = 6.6		B = 10, CR = 8.8	
	SNR	PRD	SNR	PRD	SNR	PRD	SNR	PRD
RBeSM	22.3 ± 4.4	8.8 ± 1.7	13.7 ± 3.1	21.8 ± 7.6	13.7 ± 3.2	21.8 ± 7.7	6.2 ± 2.6	50.5 ± 2.5
RBiSM	23.7 ± 3.9	7.2 ± 2.4	14.4 ± 3.5	20.5 ± 7.7	8.7 ± 2.8	38.3 ± 11.7	6.4 ± 2.5	49.6 ± 12.9
Quantization Bit, B = 6 bits								
Sensing Matrix	M = 900, N = 1800		M = 450, N = 1800		M = 300, N = 1800		M = 225, N = 1800	
	B = 6, CR = 3.67		B = 6, CR = 7.33		B = 6, CR = 11		B = 6, CR = 14.67	
	SNR	PRD	SNR	PRD	SNR	PRD	SNR	PRD
DBBD	25.1 ± 1.4	5.6 ± 0.9	24.8 ± 1.3	5.7 ± 0.9	23.3 ± 1.8	5.5 ± 0.9	22.1 ± 2.4	8.0 ± 2.3
RSBiSM	25.8 ± 2.1	5.2 ± 1.6	14.7 ± 3.5	19.8 ± 7.5	8.6 ± 3.0	39.1 ± 7.5	6.3 ± 2.5	49.9 ± 13.1
RBeSM	19.9 ± 2.1	9.8 ± 1.5	13.2 ± 3.2	23.0 ± 7.7	13.2 ± 2.9	23.0 ± 7.3	6.2 ± 2.6	50.7 ± 2.4
RBiSM	21.2 ± 3.5	9.7 ± 1.6	14.0 ± 3.3	21.1 ± 7.5	8.7 ± 2.8	38.4 ± 11.6	6.3 ± 2.4	49.7 ± 12.9

CS-based ECG encoder had CR ranges of 8.8 to 14.6 and 8.8 to 11.0 for the ECG signals with the sampling rates of 360 Hz and 250 Hz, respectively for a PRD value less than 9%. Further, the CS-based ECG encoder had CR ranges of 6.6 to 9.46 and 4.4 to 5.5 for the ECG signals with the sampling rates of 200 Hz and 100 Hz, respectively for a PRD value less than 9%. Although the CS-based ECG encoder had higher CR ranges for higher sampling rates for the DBBD sensing matrix, the energy consumption analysis is significant to conclude the impact of the sampling rate in achieving the better compression ratio with minimal reconstruction error.

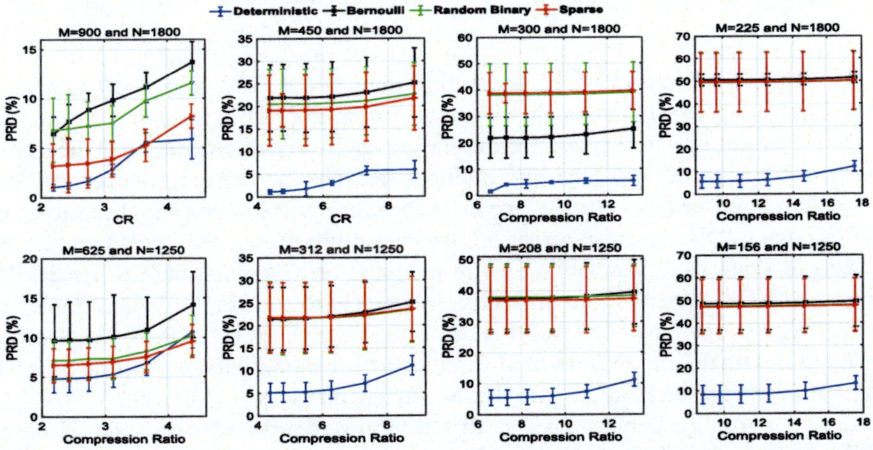

Fig. 3. Performance of DCS-based ECG encoder with four sensing matrices in terms of PRD metric with the quantizer of 10, 9, 8, 7, 6 an 5 bits and number of measurements (M) of 900, 450, 300, and 225. **a–d** Signals with sampling rate of 360 Hz. **e–h** Signals with sampling rate of 250 Hz

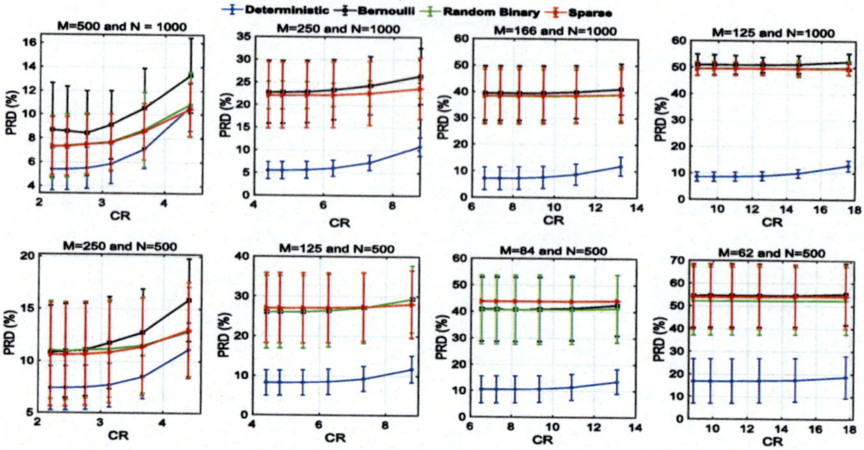

Fig. 4. Performance of DCS-based ECG encoder with four sensing matrices in terms of PRD metric with the quantizer of 10, 9, 8, 7, 6 an 5 bits and number of measurements (M) of 900, 450, 300, and 225. **a–d** Signals with sampling rate of 200 Hz. (e)-(h) Signals with sampling rate of 100 Hz

From this study, we further noticed that the PRD values are significantly different for each of the sensing matrices with M = N/2 measurements and measurement quantization bits from 10 to 6 (Please see plots of the Fig. 3a, e and plots of the Fig. 4a, e) but there is no much variation in the PRD values if the $M \ll N$ (Please see plots of the Fig. 3d, f–h and the plots of the Fig. 4c, d, f–h). Evaluation results summarized in Table 1, and Figs. 3 and 4 demonstrated that the DCS-based ECG encoder with DBBD sensing matrix can provide higher CRs with minimal reconstruction error.

F. *Computational Time Analysis*

In this study, we investigated the processing time of the ECG signal reconstruction algorithm for four sensing matrices with same sparsifying matrix including the discrete cosine waveforms, discrete sine waveforms and impulse waveforms although the reconstruction is done on the high speed computing server. The analysis of reconstruction time is most important for timely display the ECG waveform and also perform the analysis of the ECG signal at the diagnosis server or hospital. Results of this study were summarized in Table 2. From the computation time results, it is observed that the CS-based ECG encoder with DBBD sensing matrix had lesser processing time for the reconstruction of the ECG signal as compared to the other sensing matrices such as the RSBiSM, RBeSM and RBiSM with different values of M. Results further showed that processing time has huge variations for the RBeSM and RBiSM sensing matrices.

In the future directions, we design digital hardware architecture for the CS-based ECG encoder with DBBD sensing matrix to further analysis of processing energy consumption at the encoder stage including the wireless data transmission consumption. It

Table 2. Processing time for the signal reconstruction algorithm

M = 900 N = 1800				M = 450 N = 1800		
Sensing Matrix	PT ($\mu \pm \sigma$) (ms)	Min PT (ms)	Max PT (ms)	PT ($\mu \pm \sigma$) (ms)	Min PT (ms)	Max PT (ms)
DBBD	5.91 ± 1.62	3.870	8.910	3.61 ± 0.78	2.820	5.560
RSBiSM	23.85 ± 3.94	16.530	29.140	17.37 ± 5.15	10.510	25.920
RBiSM	44.89 ± 23.04	13.220	85.570	17.89 ± 3.36	12.130	22.210
RBeSM	24.67 ± 16.76	2.490	49.610	13.09 ± 4.41	6.630	18.130

is also observed that there is no randomness in the key generation and thus it cannot be directly used for simultaneous data compression and encryption. We further study the feasibility in introducing the randomness in sensing measurement without increasing the data rate budget.

5 Conclusion

In this paper, we investigated the performance of the digital compressed sensing (DCS) based ECG encoder for IoT enabled health monitoring integrated with wireless wearable ECG sensing notes which are constrained with limited battery power, bandwidth (data rate), processor speed and memory space. In this study, the DCS-based ECG encoder was developed with four sensing matrices such as DBBD, RSBiSM, RBeSM and RBiSM, and their performances were evaluated by using a wide variety of ECG signals, in terms of most widely used performance metrics such as CR, SNR, PRD and processing time. Evaluation results were obtained for different sampling rates, measurement quantization and number of measurements. Results demonstrated that the DCS-based ECG encoder with DBBD sensing matrix can provide higher compression ratios with minimal reconstruction errors and processing time as compared to other three sensing matrices. In the future direction, we attempt to present an energy-efficient digital hardware for the optimal DCS-based ECG encoders in terms of compression ratio and PRD, encoding time and energy consumption.

References

1. U. Satija, B. Ramkumar and M. Sabarimalai Manikandan, Real-Time Signal Quality-Aware ECG Telemetry System for IoT-Based Health Care Monitoring, *IEEE Internet of Things J.*, vol. 4, no. 3, pp. 815–823, 2017.
2. J. Chen, S. Sun, N. Bao, Z. Zhu and L. -b. Zhang, Improved Reconstruction for CS based ECG Acquisition in Internet of Medical Things, *IEEE Sensors Journal,* doi: https://doi.org/10.1109/JSEN.2021.3055635.
3. H. Mamaghanian et al., Compressed Sensing for Real-Time Energy- Efficient ECG Compression on Wireless Body Sensor Nodes, *IEEE Trans. Biomed. Eng.*, vol. 58, no. 9, pp. 2456-2466, Sept. 2011.

4. A. Ravelomanantsoa et al., Simple and Efficient Compressed Sensing Encoder for Wireless Body Area Network, *IEEE Trans. Instr. and Measur.,* vol. 63, no. 12, pp. 2973-2982, Dec. 2014
5. A. M. Dixon et al., Compressed sensing system considerations for ECG and EMG wireless biosensors, *IEEE Trans. Biomed. Circuits Syst.,* vol. 6, no. 2, pp. 156-166, Apr. 2012
6. H. Mamaghanian, N. Khaled, D. Atienza and P. Vandergheynst, Structured sparsity models for compressively sensed electrocardiogram signals: A comparative study, in *Proc. IEEE Biomed. Circuits Syst. Conf.,* pp. 125–128, Nov. 2011.
7. Z. Zhang et al., Compressed sensing for energy-efficient wireless telemonitoring of noninvasive fetal ECG via block sparse bayesian learning. IEEE Trans. Biomed. Eng. **60**(2), 300–309 (2013)
8. D. Craven et al., Compressed Sensing for Bioelectric Signals: A Review. IEEE J. Biomed. and Health Inform. **19**(2), 529–540 (2015)
9. Kan Luo et al., A Digital Compressed Sensing-Based Energy-Efficient Single-Spot Bluetooth ECG Node, *J. Healthcare Engineering,* vol. 2018, 2018.
10. D. Mitra et al., Investigation of Kronecker-Based Recovery of Compressed ECG Signal, *IEEE Trans. Instr. and Measur.,* vol. 69, no. 6, pp. 3642-3653, June 2020
11. V. Cambareri et al., A Case Study in Low-Complexity ECG Signal Encoding: How Compressing is Compressed Sensing?, *IEEE Signal Processing Letters,* vol. 22, no. 10, pp. 1743-1747, Oct. 2015
12. M. Mangia et al., Rakeness-Based Design of Low-Complexity Compressed Sensing, *IEEE Trans. Circuits and Systems I: Regular Papers,* vol. 64, no. 5, pp. 1201-1213, May 2017
13. Ahmad M. Rateb, A Fast Compressed Sensing Decoding Technique for Remote ECG Monitoring Systems, *Access IEEE,* vol. 8, pp. 197124- 197133, 2020.
14. E. Balestrieri et al., A Novel Method for Compressed Sensing Based Sampling of ECG Signals in Medical-IoT Era, in *Proc. of IEEE Int. Symp. on Medical Measurements and Applications (MeMeA),* pp. 1–6, 2019.
15. K. Luo et al., Information-enhanced sparse binary matrix in compressed sensing for ECG. Electron. Lett. **50**(18), 1271–1273 (2014)
16. A. Ravelomanantsoa et al., Compressed Sensing: A Simple Deterministic Measurement Matrix and a Fast Recovery Algorithm, *IEEE Trans. Instru. and Measur.,* vol. 64, no. 12, pp. 3405-3413, Dec. 2015
17. E. Picariello et al., A New Method for Dictionary Matrix Optimization in ECG Compressed Sensing, *2020 IEEE Int. Symposium on Medical Measurements and Applications (MeMeA),* Bari, Italy, 2020.
18. M. Sabarimalai Manikandan, and S. Dandapat, Wavelet-based electrocardiogram signal compression methods and their performances: A prospective review, *Biomed. Signal Process. Control,* vol. 14, pp. 73- 107, Nov. 2014
19. M. Sabarimalai Manikandan and S. Dandapat, Wavelet threshold based TDL and TDR algorithms for real-time ECG signal compression, *Biomed. Signal Process. Control,* vol. 3, no. 1, pp. 44–66, Jan. 2008.
20. M. Sabarimalai Manikandan and S. Dandapat, Wavelet threshold based ECG compression using USZZQ and Huffman coding of DSM, *Biomed. Signal Process. Control,* vol. 1, no. 4, pp. 261–270, October 2006.

Author Index

F. P. García Márquez (ed.), *International Conference on Intelligent Emerging Methods of Artificial Intelligence & Cloud Computing*, Smart Innovation, Systems and Technologies 273, https://doi.org/10.1007/978-3-030-92905-3